Wildlife Ecology, Conservation, and Management

To our colleagues Graeme Caughley, Peter Yodzis, and James N.M. Smith who have influenced both our approach to wildlife biology and the writing of this book.

Wildlife Ecology, Conservation, and Management

Second Edition

Anthony R.E. Sinclair PhD, FRS
Biodiversity Research Centre, University of British Columbia,
Vancouver, Canada

John M. Fryxell PhD
Department of Integrative Biology, University of Guelph, Guelph,
Canada

Graeme Caughley PhD
CSIRO, Canberra, Australia

Blackwell
Publishing

BLACKWELL PUBLISHING
350 Main Street, Malden, MA 02148-5020, USA
9600 Garsington Road, Oxford OX4 2DQ, UK
550 Swanston Street, Carlton, Victoria 3053, Australia

First edition published 1994
Second edition published 2006 by Blackwell Publishing Ltd

1   2006

Library of Congress Cataloging-in-Publication Data

Sinclair, A.R.E. (Anthony Ronald Entrican)
  Wildlife ecology, conservation, and management / Anthony R.E.
Sinclair, John M. Fryxell, Graeme Caughley. – 2nd ed.
    p. cm.
  Rev. ed. of: Wildlife ecology and management / Graeme Caughley,
Anthony R.E. Sinclair.
  Includes bibliographical references and index.
  ISBN-13: 978-1-4051-0737-2 (pbk. : alk. paper)
  ISBN-10: 1-4051-0737-5 (pbk. : alk. paper)
  1. Wildlife management.   2. Wildlife conservation.   3. Animal ecology.
I. Fryxell, John M., 1954–   II. Caughley, Graeme.   III. Caughley, Graeme.
Wildlife ecology and management.   IV. Title.=20

  SK355.C38 2005
  639.9–dc22

                                                    2005007229

A catalogue record for this title is available from the British Library.

Set in 9.5/12pt Berkeley
by Graphicraft Limited, Hong Kong
Printed and bound in India
by Replika Press

The publisher's policy is to use permanent paper from mills that operate a sustainable forestry
policy, and which has been manufactured from pulp processed using acid-free and elementary
chlorine-free practices. Furthermore, the publisher ensures that the text paper and cover board
used have met acceptable environmental accreditation standards.

For further information on
Blackwell Publishing, visit our website:
www.blackwellpublishing.com

Contents

Preface   xi

1   Introduction: goals and decisions   1
    1.1   How to use this book   1
    1.2   What is wildlife conservation and management?   2
    1.3   Goals of management   3
    1.4   Hierarchies of decision   5
    1.5   Policy goals   7
    1.6   Feasible options   8
    1.7   Summary   8

Part 1   Wildlife ecology   9

2   Biomes   11
    2.1   Introduction   11
    2.2   Forest biomes   12
    2.3   Woodland biomes   14
    2.4   Shrublands   14
    2.5   Grassland biomes   15
    2.6   Semi-desert scrub   17
    2.7   Deserts   17
    2.8   Marine biomes   17
    2.9   Summary   18

3   Animals as individuals   19
    3.1   Introduction   19
    3.2   Adaptation   19
    3.3   The theory of natural selection   19
    3.4   Examples of adaptation   21
    3.5   The effects of history   23
    3.6   The abiotic environment   27
    3.7   Genetic characteristics of individuals   27
    3.8   Applied aspects   33
    3.9   Summary   35

4   Food and nutrition   36
    4.1   Introduction   36
    4.2   Constituents of food   36
    4.3   Variation in food supply   40
    4.4   Measurement of food supply   42
    4.5   Basal metabolic rate and food requirement   46
    4.6   Morphology of herbivore digestion   49
    4.7   Food passage rate and food requirement   51
    4.8   Body size and diet selection   52
    4.9   Indices of body condition   53
    4.10   Summary   59

5   The ecology of behavior   60
    5.1   Introduction   60
    5.2   Diet selection   60
    5.3   Optimal patch or habitat use   66
    5.4   Risk-sensitive habitat use   69
    5.5   Quantifying habitat preference using resource
           selection functions   70
    5.6   Social behavior and foraging   72
    5.7   Summary   77

6   Population growth   78
    6.1   Introduction   78
    6.2   Rate of increase   78
    6.3   Fecundity rate   82
    6.4   Mortality rate   82
    6.5   Direct estimation of life-table parameters   84
    6.6   Indirect estimation of life-table parameters   85
    6.7   Relationship between parameters   87
    6.8   Geometric or exponential population growth   88
    6.9   Summary   89

7   Dispersal, dispersion, and distribution   90
    7.1   Introduction   90
    7.2   Dispersal   90
    7.3   Dispersion   92
    7.4   Distribution   93
    7.5   Distribution, abundance, and range collapse   98
    7.6   Species reintroductions or invasions   99
    7.7   Dispersal and the sustainability of metapopulations   104
    7.8   Summary   108

8   Population regulation, fluctuation, and competition within
    species   109
    8.1   Introduction   109
    8.2   Stability of populations   109
    8.3   The theory of population limitation and regulation   111

8.4   Evidence for regulation   116
8.5   Applications of regulation   120
8.6   Logistic model of population regulation   121
8.7   Stability, cycles, and chaos   125
8.8   Intraspecific competition   131
8.9   Interactions of food, predators, and disease   134
8.10 Summary   134

9   Competition and facilitation between species   135
9.1   Introduction   135
9.2   Theoretical aspects of interspecific competition   136
9.3   Experimental demonstrations of competition   138
9.4   The concept of the niche   143
9.5   The competitive exclusion principle   146
9.6   Resource partitioning and habitat selection   146
9.7   Competition in variable environments   153
9.8   Apparent competition   153
9.9   Facilitation   154
9.10 Applied aspects of competition   159
9.11 Summary   162

10   Predation   163
10.1 Introduction   163
10.2 Predation and management   163
10.3 Definitions   163
10.4 The effect of predators on prey density   164
10.5 The behavior of predators   165
10.6 Numerical response of predators to prey density   169
10.7 The total response   170
10.8 Behavior of the prey   176
10.9 Summary   178

11   Parasites and pathogens   179
11.1 Introduction and definitions   179
11.2 Effects of parasites   179
11.3 The basic parameters of epidemiology   180
11.4 Determinants of spread   183
11.5 Endemic pathogens   184
11.6 Endemic pathogens: synergistic interactions with food
      and predators   184
11.7 Epizootic diseases   186
11.8 Emerging infectious diseases of wildlife   187
11.9 Parasites and the regulation of host populations   188
11.10 Parasites and host communities   190
11.11 Parasites and conservation   191
11.12 Parasites and control of pests   194
11.13 Summary   195

12	Consumer–resource dynamics	196
	12.1 Introduction	196
	12.2 Quality and quantity of a resource	196
	12.3 Kinds of resources	196
	12.4 Consumer–resource dynamics: general theory	197
	12.5 Kangaroos and their food plants in semi-arid Australian savannas	200
	12.6 Wolf–moose–woody plant dynamics in the boreal forest	207
	12.7 Other population cycles	212
	12.8 Summary	215
Part 2	**Wildlife conservation and management**	**217**
13	Counting animals	219
	13.1 Introduction	219
	13.2 Estimates	219
	13.3 Total counts	219
	13.4 Sampled counts: the logic	221
	13.5 Sampled counts: methods and arithmetic	226
	13.6 Indirect estimates of population size	235
	13.7 Indices	241
	13.8 Summary	243
14	Age and stage structure	244
	14.1 Age-specific population models	244
	14.2 Stage-specific models	247
	14.3 Sensitivity and elasticity of matrix models	248
	14.4 Short-term changes in structured populations	251
	14.5 Summary	252
15	Model evaluation and adaptive management	253
	15.1 Introduction	253
	15.2 Fitting models to data and estimation of parameters	254
	15.3 Measuring the likelihood of models in light of the observed data	256
	15.4 Evaluating the likelihood of alternative models using AIC	258
	15.5 Adaptive management	264
	15.6 Summary	267
16	Experimental management	268
	16.1 Introduction	268
	16.2 Differentiating success from failure	268
	16.3 Technical judgments can be tested	269
	16.4 The nature of the evidence	272
	16.5 Experimental and survey design	274
	16.6 Some standard analyses	279
	16.7 Summary	287

17   Conservation in theory   289
     17.1   Introduction   289
     17.2   Demographic problems contributing to risk of extinction   289
     17.3   Genetic problems contributing to risk of extinction   291
     17.4   Effective population size (genetic)   297
     17.5   Effective population size (demographic)   298
     17.6   How small is too small?   299
     17.7   Population viability analysis   300
     17.8   Extinction caused by environmental change   305
     17.9   Summary   310

18   Conservation in practice   312
     18.1   Introduction   312
     18.2   How populations go extinct   312
     18.3   How to prevent extinction   321
     18.4   Rescue and recovery of near extinctions   323
     18.5   Conservation in national parks and reserves   324
     18.6   Community conservation outside national parks and reserves   332
     18.7   International conservation   332
     18.8   Summary   334

19   Wildlife harvesting   335
     19.1   Introduction   335
     19.2   Fixed quota harvesting strategy   335
     19.3   Fixed proportion harvesting strategy   341
     19.4   Fixed escapement harvesting strategy   344
     19.5   Harvesting in practice: recreational   346
     19.6   Harvesting in practice: commercial   346
     19.7   Age- or sex-biased harvesting   347
     19.8   Bioeconomics   347
     19.9   Game cropping and the discount rate   352
     19.10 Summary   353

20   Wildlife control   355
     20.1   Introduction   355
     20.2   Definitions   355
     20.3   Effects of control   356
     20.4   Objectives of control   356
     20.5   Determining whether control is appropriate   357
     20.6   Methods of control   358
     20.7   Summary   364

21   Ecosystem management and conservation   365
     21.1   Introduction   365
     21.2   Definitions   365
     21.3   Gradients of communities   366
     21.4   Niches   366
     21.5   Food webs and intertrophic interactions   366

21.6   Community features and management consequences   368
21.7   Multiple states   370
21.8   Regulation of top-down and bottom-up processes   371
21.9   Ecosystem consequences of bottom-up processes   373
21.10   Ecosystem disturbance and heterogeneity   374
21.11   Ecosystem management at multiple scales   376
21.12   Biodiversity   377
21.13   Island biogeography and dynamic processes of diversity 379
21.14   Ecosystem function   381
21.15   Summary   383

Appendices   385
Glossary   389
References   401
Index   450

Preface

Our objective in writing this book is modest. We seek to provide a text that can be used in both undergraduate and graduate level courses in wildlife management and conservation. Conservation is becoming an increasingly important component in the management of animal populations and their habitats. We have recognized this development by including conservation in the new name for the text.

New quantitative methods, developed over the last 10 years, are now so fundamental to management that we have included them at the most basic levels. In addition, several chapters in the book will be useful to practicing wildlife managers. For example, we have included modern approaches to censusing, the use of age- and stage-structured data in demography studies, and the use of models as efficient methods for making decisions. We emphasize, in the last chapter, that all wildlife problems have to be addressed in the context of the whole ecosystem, and cannot be solved in isolation of other species and environments.

In this edition we have rearranged the sequence of chapters better to reflect the progression from individuals to populations, communities, and ecosystems. We have also included four new chapters. Chapter 5 deals with how animals find their food and the consequences that this will have on their populations. Chapter 14 addresses the increasing use of age- and stage-structured information in populations as a method to identify rates of increase, a valuable tool in conservation situations where censusing is impractical. Chapter 15 explains how to use modern statistical methods to choose between alternative models, for example different models that describe a population that is changing. We have added a final chapter (Chapter 21) that provides an overview of community and ecosystem ecology as background to the way we manage whole systems.

Modern approaches to wildlife ecology, conservation, and management often demand sophisticated quantitative methods of data analysis and modeling. We have therefore provided an accompanying CD that illustrates in close detail how to calculate most of the mathematical concepts discussed in the book, including all of the simulation models. To further the development of problem-solving skills, we also include a series of computer labs, touching on several key concepts. All of the quantitative material has been developed using Mathcad, a powerful computer-aided design package for mathematics. A free evaluation copy of Mathcad 11 is provided with the book to assist the reader in the development of these skills.

Anne Gunn and David Grice were invaluable in bringing together the first edition of this book after Graeme Caughley fell ill. Fleur Sheard prepared the line drawings for the book. David Grice has also helped us in picking up the strands of the first edition 10 years later. We would like to thank the following people for their help with material and constructive comments: Sue Briggs, Andrea Byrom, Steve Cork, Charles Krebs, Graham Nugent, John Parkes, Roger Pech, Laura Prugh, Wendy Ruscoe, Dolph Schluter, Julian Seddon, Grant Singleton, David Spratt, Eric Spurr, Vernon Thomas, and Bruce Warburton.

The CSIRO Sustainable Ecosystems Division, Canberra, Australia, and Landcare Research, Christchurch, New Zealand, were most generous in providing facilities while preparing this second edition. The librarians in both institutions were most accommodating in finding material for us. We also thank the Natural Sciences and Engineering Research Council of Canada for supporting us during the writing. A.R.E.S. was supported by a Senior Killam Research Fellowship from the Canada Council.

Our close friend and colleague, Graeme Caughley, died in 1994. In this edition we have retained the substance and spirit of his scholarship, expanding in the fields where advances have occurred since the first edition. For this new edition we are indebted to Anne Sinclair who obtained most of the new reference material, typed the manuscript, checked it and put it together. Without her it would have taken much longer.

1 Introduction: goals and decisions

This book is structured as two interlocking parts. The first part provides an overview of wildlife ecology, as distinct from that portion of applied ecology that is called wildlife management and conservation. We have observed that many courses offered in wildlife management do not stipulate a solid grounding in ecology as a prerequisite. The chapters on wildlife ecology (Chapters 2–12) are there to remedy that defect. These chapters cover such topics as growth and regulation of wildlife populations, spatial patterns of population distribution, interactions among plants, herbivores, carnivores, and disease pathogens. While these topics are often covered in introductory biology or ecology courses, they rarely focus on the issues of most concern to a wildlife specialist. We view wildlife management and conservation as applied ecology. You will have trouble applying it unless you know some. In particular, you will need an understanding of the theory of population dynamics and of the relationship between populations, their predators, and their resources if you are to make sensible judgments on the likely consequences of one management action versus another.

The second part deals with wildlife conservation and management (Chapter 13 onwards). These chapters cover census techniques, how to test hypotheses experimentally, how to evaluate alternative models as tools for conservation and management, and the three aspects of wildlife management: conservation, sustained yield, and control. We close with a chapter that places the problems of wildlife management into the context of the ecosystem. Species populations cannot be managed in isolation because they are influenced by, and themselves influence, many other components of the ecosystem. In the long run, wildlife management becomes ecosystem management.

Many of the key issues in wildlife ecology are of a quantitative nature: processes of population growth, spatial distribution, or interactions with the physical environment or other organisms. Coping with these topics demands conceptual understanding of quantitative ecology. Mathematical models are also an essential component to decision-making in both wildlife conservation and management, for the simple reason that we rarely can rely on previous experience to identify the most appropriate choices. Every problem is unique: new species, new sets of challenges and constraints, all taking place in a continually changing physical environment. Mathematical models provide a useful tool to deal appropriately with these uncertainties. Moreover, mathematical models help to clarify the logic that guides our thinking.

To assist in developing the requisite skills, all the quantitative material in the book is elaborated and demonstrated through a set of "interactive" computer programs. These are written using MATHCAD modelling software. MATHCAD is a computer-aided design package which provides a powerful set of integrated tools for numerical computation, graphic depiction of results, and word processing. The CD in the back of this book includes an Evaluation Version of MATHCAD® 11 Single User Edition, which is reproduced by permission. This software is a fully functional trial of MATHCAD that will expire 120 days from installation. For technical support and more information about purchasing MATHCAD or upgrading from previous editions see http://www.mathcad.com. Choose "install demo" to install MATHCAD from the CD. Once installed, the specific files relating to material covered in this book should be readily usable.

A set of tutorials is also provided on the accompanying CD. These tutorials are meant to help you to learn how to use MATHCAD and how to develop simple ecological models. They complement the quantitative material covered in the book. The tutorials are designed to allow you to hone your problem solving skills. The text is fully self-explanatory if you do not use the accompanying CD. By working through the MATHCAD files, however, you will greatly expand your familiarity with the mathematical principles involved, which can prove invaluable in future professional endeavors.

1.2 What is wildlife conservation and management?

The remainder of this chapter explains what wildlife management is, how it relates to conservation, and how it should operate. You are faced with the difference between value judgments and technical judgments and how these relate to goals and policies compared with options and actions. We take you through the various steps involved in deciding what to do and how it should be done. We describe decision analysis and matrices and how they help to evaluate feasible management options.

"Wildlife" is a word whose meaning expands and contracts with the viewpoint of the user. Sometimes it is used to include all wild animals and plants. More often it is restricted to terrestrial vertebrates. In the discipline of wildlife management it designates free-ranging birds and mammals and that is the way it is used here. Until about 25 years ago wildlife was synonymous with "game," those birds and mammals that were hunted for sport. The management of such species is still an integral part of wildlife management but increasingly it embraces other aspects such as conservation of endangered species.

"Wildlife management" may be defined for present purposes as "the management of wildlife populations in the context of the ecosystem." That may be too restrictive for some who would argue that many of the problems of management deal with people and, therefore, that education, extension, park management, law enforcement, economics, and land evaluation are legitimate aspects of wildlife management, and ought to be included within its definition. They have a point, but the expansion of the definition to take in all these aspects diverts attention from the core around which management activities are organized: the manipulation or protection of a population to achieve a goal. Obviously people must be informed as to what is being done, they must be educated to an understanding of why it is necessary, their opinions must be canvassed and their behavior may have to be regulated with respect to that goal. However, the most important task is to choose the right goal and to know enough about the animals and their habitat to assure its attainment. Hence wildlife management is restricted here to its literal meaning, thereby emphasizing the core at the expense of the periphery of the field. The broader extension and outreach aspects of wildlife management are dealt with thoroughly in other texts devoted to those subjects (Lyster 1985; Geist and McTaggert-Cowan 1995; Moulton and Sanderson 1999; Vasarhelyi and Thomas 2003).

1.2.1 *Kinds of management*

Wildlife management implies stewardship, that is the looking after of a population. A population is a group of coexisting individuals of the same species. When stewardship fails, conservation becomes imperative. Under these circumstances, wildlife management shifts to remedial or restoration activities.

Wildlife management may be either **manipulative** or **custodial**. Manipulative management does something to a population, either changing its numbers by direct means or influencing numbers by the indirect means of altering food supply, habitat, density of predators, or prevalence of disease. Manipulative management is appropriate when a population is to be harvested, or when it slides to an unacceptably low density, or when it increases to an unacceptably high level.

Custodial management on the other hand is preventative or protective. It is aimed at minimizing external influences on the population and its habitat. It is not aimed necessarily at stabilizing the system but at allowing free rein to the ecological processes that determine the dynamics of the system. Such management may be appropriate in a national park where one of the stated goals is to protect ecological processes and it may be appropriate for conservation of a threatened species where the threat is of external origin rather than being intrinsic to the system.

Regardless of whether manipulative or custodial management is called for, it is vital that (i) the management problem is identified correctly; (ii) the goals of management explicitly address the solution to the problem; and (iii) criteria for assessing the success of the management are clearly identified.

1.3 Goals of management

A wildlife population may be managed in one of four ways:
1 make it increase;
2 make it decrease;
3 harvest it for a continuing yield;
4 leave it alone but keep an eye on it.
These are the only options available to the manager.

Three decisions are needed: (i) what is the desired goal; (ii) which management option is therefore appropriate; and (iii) by what action is the management option best achieved? The first decision requires a judgment of value, the others technical judgments.

1.3.1 *Who makes the decisions?*

It is not the function of the wildlife manager to make the necessary value judgments in determining the goal any more than it is within the competence of a general to declare war. Managers may have strong personal feelings as to what they would like, but so might many others in the community at large. Managers are not necessarily provided with heightened aesthetic judgment just because they work on wildlife. They should have no more influence on the decision than does any other interested person.

However, when it comes to deciding which management options are feasible (once the goal is set), and how goals can best be attained, wildlife managers have the advantage of their professional knowledge. Now they are dealing with testable facts. They should know whether current knowledge is sufficient to allow an immediate technical decision or whether research is needed first. They can advise that a stated goal is unattainable, or that it will cost too much, or that it will cause unintended side effects. They can consider alternative routes to a goal and advise on the time, money, and effort each would require. These are all technical judgments, not value judgments. It is the task of the wildlife manager to make them and then to carry them through.

Since value judgments and technical judgments tend to get confused with each other it is important to distinguish between them. By its essence a value judgment is neither right nor wrong. Let us take a hypothetical example. The black rat (*Rattus rattus*) is generally unloved. It destroys stored food, it is implicated in the spread of bubonic plague and several other diseases, it contributes to the demise of endangered species, and it has been known to bite babies. Suppose a potent poison specific to this species were discovered, thereby opening up the option of removing this species from the face of the earth. Many would argue for doing just that, and swiftly. Others would argue that there are strong ethical objections to exterminating a species, however repugnant or inconvenient that species might be. Most of us would have a strong opinion one way or the other but there is no way of characterizing either competing opinion as right or wrong. That dichotomy is meaningless. A value judgment can be characterized as hardheaded or sentimental (these are also value judgments), or it may be demonstrated as inconsistent with other values a person holds, but it cannot be declared right or wrong. In contrast, technical judgments can be classified as right or wrong according to whether they succeed in achieving the stated goal.

1.3.2 *Decision analysis*

In deciding what objective (goal) is appropriate we consider a range of influences, some dealing with the benefits of getting it right and others with the penalties of getting it wrong. Social, political, biological, and economic considerations are each examined and given due weight. Some people are good at this and others less so. In all cases, however, there is a real advantage, both to those making the final decision and to those tendering advice, to have the steps of reasoning laid out before them as a decision is approached.

At its simplest, this need mean no more than the people helping to make the decision spelling out the reasons underpinning their advice. However, with more complex problems it helps to be more formal and organized, mapping out on paper the path to the decision through the facts, influences, and values that shape it. That process should be explicit and systematic. Different people will assign different values (weights) to various possible outcomes and, particularly if mediation by a third party is required, an explicit statement of those weights allows a more informed decision. It helps also to determine which disagreements are arguments about facts and which are arguments about judgments of value.

Table 1.1 is an objective/action matrix in which possible objectives are ranged against feasible actions. The objectives are not mutually exclusive. It comes from the response of the Department of Agriculture of Malaysia to the attack of an insect pest on rice (Norton 1988). It allows the departmental entomologists and administrators to view the full context within which a decision must be made. Each of the listed objectives is of some importance to the department. The next step would be to rank those objectives and then to score the management actions most appropriate to each. The final outcome is the choice of one or more management actions that best meet the most important objective or objectives. Such very simple aids to organizing our thoughts are often the difference between success and failure.

Table 1.1 Possible objectives and management actions for public pest management. The initial problem is to assess how each action is likely to meet each objective.

Actions	Objectives					
	Improve farmers' ability to control pest	Improve farmers' incentives	Strengthen political support	Keep department's cost low	Reduce damage	Reduce future pest outbreaks
Short term						
1 Warn and advise farmers						
2 Advise and provide credit						
3 Advise and subsidize pesticides						
4 Advise, subsidize and supervise spraying						
5 Mass treat and charge farmers						
6 Mass treat at department's cost						
Medium term						
7 Intensive pest surveillance						
8 Implement area-wide biological control						
9 Training courses for farmers						

After Norton (1988).

Another such aid is the feasibility/action matrix. Table 1.2 is Bomford's (1988) analysis of management actions to reduce the damage wrought by ducks on the rice crops of the Riverina region of Australia. The feasibility criteria are here ranked so that if a management action fails according to one criterion there is no point in considering it against further criteria. Note how this example effortlessly identifies areas of ignorance that would have to be attended to before a rational decision is possible.

Our third example of decision aids is the payoff matrix (Table 1.3). It expresses the state of nature (level of pest damage in this example) as rows and the options for management action as columns (Norton 1988). The problem is to assess the probable outcome of each combination of level of damage and the action mounted to alleviate it. Note that the column associated with doing nothing gives the level of damage that will be sustained in the absence of action. It is the control against which the net benefit of management must be assessed. The cells of this matrix are best filled in with net revenue values (benefit minus cost) rather than with benefit/cost ratios because it is the absolute rather than relative gain that shapes the decision.

1.4 Hierarchies of decision

Before we begin manipulating a wildlife population and its environment we must ask ourselves why we are doing it and what it is supposed to achieve. In management theory that decision is usually divided into hierarchical components.

At the bottom, but here addressed first, is the management action. It might be to eliminate feral pigs (*Sus scrofa*) on Lord Howe Island off the coast of Australia. The management action must be legitimized by a technical objective, for example to halt the decline of the Lord Howe Island woodhen (*Tricholimnas sylvestris*) on Lord Howe Island. Above that is the policy goal, a statement of the desired endpoint of the

Table 1.2 A matrix to examine possible management actions against criteria of feasibility.

Control options	Feasibility criteria					
	Technically possible	Practically feasible	Economically desirable	Environmentally acceptable	Politically advantageous	Socially acceptable
1 Grow another crop	1	0				
2 Grow decoy crop	1	1	?	1	1	1
3 Predators and diseases	0					
4 Sowing date	1	1	?	1	1	1
5 Sowing technique	1	1	?	1	1	1
6 Field modifications	1	1	?	1	1	1
7 Drain or clear daytime refuges	?	0				
8 Shoot	1	1	?	1	?	1
9 Prevent access, netting	1	1	0	1	1	1
10 Decoy birds or free feeding	?	1	?	1	1	
11 Repellents	1	0				
12 Deterrents	1	1	?	1	1	1
13 Poisons	1	1	?	0		
14 Resowing or transplanting seedlings	1	?	1	?	1	1

1, Yes; 0, no; ?, no information.
After Bomford (1988).

Table 1.3 A payoff matrix for pest control.

State of nature	Actions			
	Do nothing (0)	Pest control strategies		
		(1)	(2)	(3)
Level of pest attack				
Low (L)	Outcome L, 0	Outcome L, 1	Outcome L, 2	Outcome L, 3
Medium (M)	Outcome M, 0	Outcome M, 1	Outcome M, 2	Outcome M, 3
High (H)	Outcome H, 0	Outcome H, 1	Outcome H, 2	Outcome H, 3

From Norton (1988).

exercise, which in this example might be to secure the continued viability of all indigenous species within the nation's national park system.

In theory the decisions flow from the general (the policy goal) to the special (the management action), but in practice that does not work because each is dependent on the others, in both directions. Nothing is achieved by specifying "halt a species' decline" as a technical objective unless there is available a set of management actions that will secure that objective. Obviously a management action cannot be specified to cure a problem of unknown cause. All three levels of decision must be considered together such that the end product is a feasible option.

A feasible option is identified by answering the following questions:

1 Where do we want to go?
2 Can we get there?

3 Will we know when we have arrived?

4 How do we get there?

5 What disadvantages or penalties accrue?

6 What benefits are gained?

7 Will the benefits exceed the penalties?

The process is iterative. There is no point in persevering with the policy goal thrown up by the first question if the answer to the second question is negative. The first choice of destination is, therefore, replaced by another, and the process is then repeated.

Question (3) is particularly important. It requires formulating stopping rules. That does not mean necessarily that management action ceases on attainment of the objective, rather that management action is altered at that point. The initial action is designed to move the system towards the state specified by the technical objective; the subsequent action is designed to hold the system in that state. If we cannot determine when the objective has been attained, either for reasons of logic (ambiguous or abstract statement of the objective) or for technical reasons (inability to measure the state of the system), the option is not feasible.

1.5 Policy goals

Policies are usually couched in broad terms that provide no more than a general guide for the manager. The specific decisions are made when the technical objectives are formulated. However, there are two types of policy goals that the manager must know about in case they clash with the choosing of those objectives.

1.5.1 The non-policy

Non-policies stipulate goals that are not clearly defined. They are usually formulated in that way on purpose so that the administering agency is not tied down to a rigidly dictated course of action. Policies are usually formulated by the administering agency whether or not they are given legislative sanction. If the agency has not developed a policy it may fill the gap with a non-policy that commits it to no specified action. Take, for example, the goal of "protecting intrinsic natural values." It reads well but is entirely devoid of objective meaning.

1.5.2 The non-feasible policy

In contrast to the relatively benign non-policy, the non-feasible policy can be damaging. Although it may give each interest group at least something of what they desire, sometimes the logical consequence is that two or more technical objectives are mutually incompatible.

An example is provided by the International Convention for the Regulation of Whaling of 1946 which was "to provide for the proper conservation of whale stocks" and "thus make possible the orderly development of the whaling industry." This pleased both those people concerned about conservation of whales and those people wishing to harvest whales. Unfortunately the goal is a nonsense because, for reasons that are elaborated in Chapter 19, species with a low intrinsic rate of increase are not suitable for sustainable harvesting. The two halves of the policy goal contradict each other. The history of whaling since 1948, in which the blue (*Balaenoptera musculus*), the fin (*B. physalus*), the sei (*B. borealis*), the Brydes (*B. edeni*), the humpback (*Megaptera novaeangliae*), and the sperm (*Physeter macrocephalus*) were reduced to the level of economic extinction, is a direct consequence of choosing a policy goal that was not feasible.

Another form of the non-feasible policy is that in contrast to the non-policy, the policy is so specific that it actually determines technical objectives and sometimes even management actions. If these are unattainable in practice, the policy goal itself is also unattainable. An example is provided by the now defunct policy to exterminate deer in New Zealand. It was always an impossibility.

1.6 Feasible options

Objectives must be attainable. It is the wildlife manager's task to produce the attainable technical objectives by which the policy goal is defined. In contrast to the goal, which may be described in somewhat abstract terms, a technical objective must be stated in concrete terms and rooted in geographic and ecological fact. It must be attainable in fact and it should be attainable within a specified time. A technical objective should, therefore, be accompanied by a schedule.

1.6.1 *Criteria of failure*

It follows as a corollary that there must be an easy way of recognizing the failure to attain an objective. The most common is to measure the outcome against that specified by the technical objective. Another is to compare the outcome with a set of **criteria of failure**, set before the management action is begun. These two are not the same. Comparison of outcome with objective can produce assessments like "not quite" or "not yet." Not so with criteria of failure. They take the form: "the operation will be judged unsuccessful, and will therefore be terminated, if outcome x has not been attained by time t."

1.7 Summary

We view wildlife management as simply the management of wildlife populations. Three important points underlie any management: (i) the management problem is identified correctly; (ii) the goals of management explicitly address the solution to the problem; and (iii) criteria for assessing the success of the management are clearly identified.

Four management options are available: (i) to make the population increase; (ii) to make it decrease; (iii) to take from it a sustained yield; or (iv) to do nothing but keep an eye on it. We have first to decide our goal for the population, and that will be largely a value judgment. To help us steer through social, political, and economic influences we use a decision analysis to reveal those influences and their effect on goals and policies. A series of questions about the selected option must be posed and answered to ensure that it is feasible and that its success or failure can be determined.

Part 1
Wildlife ecology

2 Biomes

2.1 **Introduction**

This chapter provides a brief overview of the main ecological divisions in the world and will supply a background of natural history for the chapters that follow.

The earth, or biosphere, can for convenience be divided up into major regions. On land these regions are characterized by a similarity of geography, landform, and major floral and faunal groupings. Thus, we can talk about the tundra – high latitude, cold, usually flat or rolling relief, and with low-growing shrubs like willows and mat-forming herbs. Tropical lowland forest is very different – moist, warm regions near the equator dominated by dense forest. Regions with similar characteristics are called **biomes**. They are divided further into units of greater similarity, called **ecosystems**, based on environment and groupings of plants and animals. Ecosystems are the main functional units of the biosphere, largely self-contained apart from inputs of energy and nutrients from outside. Some organisms, such as migrants and dispersers, can move between ecosystems. They vary in size from parts of oceans to small water-sheds on land.

Ecosystems comprise the abiotic environment and the biotic groupings of plant and animal species called **communities**. Each of the species in a community has a characteristic density (or range of densities) and it is the interaction of these various populations that gives a particular community its special features. Populations have their own features, for example age and sex ratios, and these are affected by both the environment in which the animals live and the particular adaptations of the individuals, their morphology, physiology, and behavior. Thus, in the study of wildlife ecology and management we need to understand both the large-scale spatial and temporal events occurring in biomes and ecosystems, and the smaller-scale characteristics of individuals and populations.

Habitat is the suite of resources (food, shelter) and environmental conditions (abiotic variables such as temperature and biotic variables such as competitors and predators) that determine the presence, survival, and reproduction of a population. In Chapters 5, 8, and 12 we shall examine the relationships between populations and their resources. In Chapters 8, 9, and 10 we examine how some components of their habitat, such as competitors and predators, impinge on the populations and their role in wildlife management.

We will now review the main features of the various biomes and some of the wildlife forms that inhabit them. Although biomes are characterized by many different properties, they can be summarized conveniently according to mean annual temperature and rainfall (Fig. 2.1). Biomes are groupings of ecosystems with similar environment and vegetation structure (physiognomy). There are six major terrestrial biomes distinguished by their physiognomic characteristics: forests, woodlands, shrublands,

11

Fig. 2.1 World biome types in relation to rainfall and temperature. (After Whittaker 1975.)

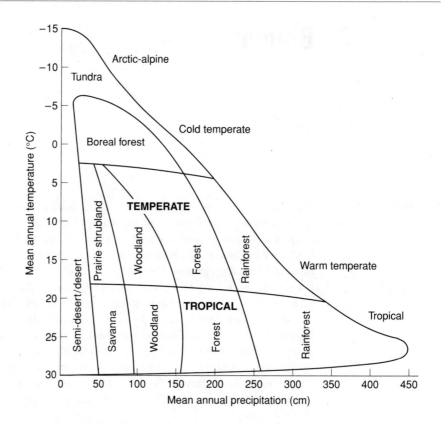

grasslands, semi-desert scrub, and deserts. We include one group of marine biomes. Walter (1973) provides more detailed descriptions.

2.2 Forest biomes

2.2.1 Boreal forest

Taiga in Eurasia or **boreal forest** in North America starts where 10°C mean daily temperature is exceeded for more than 30 days per year. Tundra takes over where this temperature is exceeded for less than 30 days. The boreal forest is dominated by several species of conifer of the genera *Pinus* (pine), *Picea* (spruce), *Abies* (fir), and *Larix* (larch), although only white spruce (*Picea glauca*) spans the whole of North America. Eastern Asia also has many species of conifer, but, in contrast, only two species, Norway spruce (*Picea abies*) and Scots pine (*Pinus sylvestris*), predominate in Europe.

In dense boreal forest the shrub layer is almost absent and mosses dominate the herb layer. In openings of the forest, and in wetter areas where trees are absent, there is a sparse shrub layer of willows (*Salix*), birches (*Betula*), and alders (*Alnus*). Soils are acid, low in nutrients, and have a thick humus layer that takes a long time to decompose.

Boreal forest is the main habitat of the snowshoe hares (*Lepus americanus*) and their main predators the lynx (*Lynx canadensis*), great horned owls (*Bubo virginianus*), and goshawks (*Accipiter gentilis*). Among other birds, ravens, swallows, chickadees, woodpeckers, and forest grouse are common. During the Pleistocene this was also the habitat for browsing mastodons (*Mammut americanum*), woolly rhino (*Coelodonta antiguilatis*), and giant ground sloths (*Megalonyx*).

2.2.2 Temperate forest

Temperate forests can be divided into deciduous forests, rainforests, and evergreen forests. Deciduous trees drop their leaves as an adaptation to winter. Leaves being delicate structures are likely to be damaged by freezing. Thus, nutrients are withdrawn from the leaf and stored in the roots. The dead leaf is then shed. Because trees need to regrow their leaves in spring they require a growing season of 4–6 months with moderate summer rainfall and mild winters. They avoid the extremes of the wet maritime and cold continental climates. These forests are found mostly in the mid-latitudes of the northern hemisphere, particularly in western Europe, eastern North America, and eastern Asia. There is a large variety of tree species with oak (*Quercus*), beech (*Fagus*), maple (*Acer*), and elm (*Ulmus*) being common. Forests of western Europe are not as rich in tree species because of extinctions during the last ice age.

Small mammals such as voles (*Microtus*), mice (*Clethrionomys*), and shrews (*Sorex*) are numerous although with relatively few species. Large mammals are represented by deer (*Odocoileus, Cervus*) and bison (*Bison*). The majority of the insectivorous bird species, such as thrushes (*Turdus*) and old and new world warblers (*Phyloscopus, Dendroica*), migrate to the tropics or southern hemisphere.

Temperate rainforests occur along the western coasts of North America, Chile, New Zealand, and southern Australia, in maritime climates with high year-round rainfall. They are known for their large trees (60–90 m high) such as redwoods (*Sequoia sempervirens*) in California, Douglas fir (*Pseudotsuga douglasii*) in British Columbia, eucalypts (*Eucalyptus regnans*) in Australia, and podocarps (*Podocarpus*) in New Zealand. Diversity is often low for both plant and animal species. Large vines, lianes, palms, and epiphytes are rare, but tree ferns in the southern hemisphere are common.

Temperate evergreen forests vary around the world. In this biome are included the dry sclerophyll forests of eastern Australia dominated by *Eucalyptus* species with their tough, elongated leaves; the dry pine forests of western North America including Monterey (*Pinus radiata*) and ponderosa pine (*P. ponderosa*); and the dry forests of southeast Asia. Canopies are open and the understory vegetation is sparse and often adapted to dry conditions. Evergreen forests in New Zealand are quite different for they occur in wet regions. Although close to Australia, these forests must have been separated by continental drift before the Australian flora developed for there are no eucalypts or acacias. The forests are dominated by evergreen conifers, notably the kauri (*Agathis*), in the warmer north, and several species of *Podocarpus* and *Dacrydium* in the south. There are five evergreen species of southern beech (*Nothofagus*). New Zealand forests are noted for their endemic birds and the absence of indigenous terrestrial mammals.

2.2.3 Tropical forest

Daily temperatures in tropical forests remain similar year round (24–25°C), and day length varies less than an hour. Seasons are determined by rainfall, there being some months with less rain than others – 100 mm of rain is a dry month. In Malaysia, Indonesia, and some parts of the Amazon basin (Rio Negro) all months have more than 200 mm of rain, some receiving over 450 mm. In Africa and India there is a short dry season. High temperatures in these forests cause high transpiration rates, and plants have adaptations to overcome water loss through thickening of the cuticle, producing leathery leaves. Examples are the rubber tree (*Ficus elastica*) and *Philodendron*. Leaves in the shade are large; those in the light are smaller.

In contrast to the relative paucity of species in temperate forests there is a high diversity of plants and animals in tropical rainforests. The most extensive rainforest is found in the Amazon basin of South America, but other forests are found in central and west Africa, southeast Asia, Indonesia, and northern Australia. One can find more than 200 species of trees in small areas. Leaf shapes are similar between species of tree. The canopy is high and closed, and 70% of plant species are trees. Most of the other plant species are also concentrated in the canopy: associated climbing lianes and epiphytes such as orchids form part of the canopy. The lack of light results in a relatively sparse understory. The roots of the large trees do not reach far into the soil because it is permanently wet. These giant trees therefore develop buttress roots reaching 9 m up the trunks to support them. Individual trees have a periodicity of growth and flowering, but two individuals of the same species can be out of phase. Periodicities of growth differ between species; they are not related to the annual cycle and vary between 2 and 32 months.

Most of the animal species are adapted to the canopy. The greatest diversity of primates occurs in these forests, and in South America other mammals such as sloths (*Brachydura*) are also adapted to feeding in the canopy. The diversity of bird species is high, the highest being in the Amazon forests. Feeding and breeding of many bird and bat species are adapted to the flowering periodicities of their preferred feeding trees.

2.3 Woodland biomes

Tropical broadleaf woodlands are an extension of the tropical forests in drier seasonal climates and low-nutrient soils. As an adaptation to this climate trees have large leaves which they shed during the dry season. A few species, such as *Balanites* in Africa and *Eucalyptus* in Australia, have small xeromorphic leaves which are retained throughout the year. Trees often flower at the end of the dry season before leaf formation. The dense herb layer leads to frequent fires in the dry season so that shrubs and trees have evolved fire resistance.

Typical of this biome are the extensive *Colophospermum* and *Brachystegia* woodlands of southern Africa and *Isoberlinia* woodlands in west Africa. The canopy varies from 3 to 10 m and is relatively open. Soils and grasses are low in nutrients; ungulate species are also at low density, although some, for example roan (*Hippotragus equinus*) and sable antelope (*H. niger*), are adapted to this habitat. Similar vegetation occurs in Brazil, India, and southeast Asia. The Indian and Asian woodlands are the centers of radiation for the cattle group (*Bos*) – gaur, banteng, kouprey, and yak.

As in the tropics, temperate woodlands occur in drier environments than the forests. This biome covers a heterogeneous collection of small conifer and deciduous tree habitats in the Mediterranean and Mexico, but none of them are very extensive.

2.4 Shrublands

The best known of these types is what is called the Mediterranean vegetation – a scrub adapted to the dry conditions of a Mediterranean climate, which consists of dry hot summers and cool wet winters. Similar types are found in South Africa, southern Australia, central Chile, and southern California. Shrubs are low, with sclerophyllous leaves. Many are adapted to annual fires, regrowing from the root stock.

Typical trees and shrubs of the Mediterranean are various oaks (*Quercus*), holly (*Ilex*), the evergreen pines and junipers (*Juniperus*), and olive (*Olea*); in California *Quercus*, *Cupressus* and chaparral shrubs (*Ceanothus*); in Chile various cacti (*Trichocereus*); in South Africa *Elytropappas* and the major radiation of *Protea*; and in Australia the mallee scrub made up of *Eucalyptus* shrubs as well as "grass trees" (*Xanthorrhoea*, *Kingia*), cycads (*Macrozamia*), the evergreen *Casuarina*, and several members of the Proteaceae. All are adapted to a period of slow growth and the prevention of water loss by closing stomata during the summer drought. The leaves are hard and leathery, characteristic of sclerophyllous vegetation. In isolated areas, such as southwest Australia or South Africa, plants show a high degree of speciation and many of the species are endemic. There are several small mammals and passerine birds adapted to the regime of summer drought, but their diversity is usually low. For example, in California chaparral there is the wrentit (*Chamaea fasciata*) and kangaroo rat (*Dipodomys venustus*). In the Mediterranean there is the Sardinian warbler (*Sylvia melanocephala*), in South Africa the Cape sugarbird (*Promerops cafer*) on proteas, and in Australia the western spinebill (*Acanthorhynchus superciliosus*) feeding on banksias. The southwestern USA and Mexico are the centers of radiation for oak (*Quercus*) and juniper (*Juniperus*) woodlands. There is an associated fauna of birds and mammals that form a hotspot of biodiversity in North America (Dobson *et al.* 1997).

2.5 Grassland biomes

It is no accident that all the large herds of ungulates occur in the grassland biomes – caribou in the Canadian tundra, saiga on the Asian steppes, bison on the American prairies, and various antelopes on the African savannas. They have in common the ability to migrate in response to the seasonal climate and changing vegetation and so place themselves in the areas of highest food production at the time. They avoid thereby many of their predators who cannot migrate to the same extent. These two abilities – to find temporary food patches and to avoid predators – allow a higher density of animals than if the population did not migrate in a similar area; the large herds are not just a consequence of the extensive area of these biomes.

2.5.1 Tropical savanna

Tropical savanna comprises grassland with scattered trees. Often the trees are sparse, as on the open plains of East Africa, or quite dense with up to 30% canopy cover, as in some of the *Acacia* savannas of Africa and Australia. Although temperature is fairly constant, rainfall is highly seasonal and falls in the range of 500–1000 mm. Grasses are mostly perennial, 20–200 mm in height, and are usually burned each dry season. Most savannas in Africa are maintained by fire rather than by soil moisture; examples of the latter (edaphic grassland) are seen in the flood plains of the larger rivers such as the Zambesi and Nile, or shallow lake beds of Africa, and the llanos of the Orinoco in Venezuela.

The African savannas support a wide range of large mammal species, some communities having as many as 25 ungulate and seven large carnivore species as well as many rodent and lagomorph herbivores, mongooses, civets, and other small carnivores. The Australian savannas support an array of macropod herbivores (kangaroos) but no large carnivores, although the three that used to occur have become extinct on the mainland in the last 30,000 years. Small carnivorous marsupials are represented by dasyurids. In the birds, finches, parrots, and emus (*Dromaeus novaehollandiae*) are common. South American wet savannas, in Venezuela and the Pantanal of Brazil for example, have a range of large rodents such as capybaras (*Hydrochoeris hydrochoeris*) and coypus (*Myocastor coypus*) that partly take the place of ungulates in Africa, but the drier pampas has very few large herbivores. There may be historical reasons for their absence: in the Tertiary there were many endemic herbivores belonging to the Notoungulate group which have since died out. Of the birds, pipits, buntings, and tinamous are characteristic.

2.5.2 *Temperate grasslands*

Temperate grasslands are similar to the tropical savannas in that they support perennial grasses and are often maintained by fire. They are seasonal in both precipitation (rain or snow) and temperature. They occur in dry climates in the centers of the North American and Asian continents. In South America we see this vegetation as the pampas of Argentina. Temperate grasslands experience cold winters with low snowfall, spring rains, and a summer drought. Like tropical savannas they support large herds of ungulates – bison and pronghorn (*Antilocapra*) on the American prairies, saiga (*Saiga*) and horses on the Asian steppes – and carnivores such as wolves (*Canis lupus*). Nonetheless the number of species is low. Birds are represented by larks, pipits, buntings, grouse, buzzards (*Buteo*), and falcons.

2.5.3 *Tundra*

Arctic tundras occur north of the tree line in both North America and Eurasia. There is a maximum of 188 days with mean temperature above 0°C but sometimes as few as 55 days. The growing season spans the four summer months and is determined locally by when the snow melts. Exposed areas have longer growth whereas those under snow drifts have shorter seasons, and so a mosaic of vegetation is maintained. Plant communities consist of a complex mixture of sedges, grasses, lichens, mosses, and dwarf shrubs.

In the Arctic, soils are frozen in permafrost except for a shallow layer at the surface which thaws in summer. Lemmings (*Lemmus*) feed on the vegetation year round, being protected under the snow in winter. Geese nest in large numbers and impose a heavy grazing impact in summer. Ptarmigan (*Lagopus*) are another abundant bird group. Because of the permafrost the ground snow does not drain easily in the summer and much of the tundra is swampy; these swamps provide ideal breeding grounds for mosquitos, which form dense swarms in late summer. This abundance of insects, combined with the almost constant daylight, provides good breeding conditions for insectivorous birds – many shorebirds (plovers, sandpipers) and passerines (e.g. snow bunting, *Plectrophenax nivalis*) migrate to this biome to breed. Large mammals include muskoxen (*Ovibos moschatus*) and caribou (*Rangifer tarandus*); small mammals such as arctic hare (*Lepus arcticus*) are sometimes numerous, and wolves, arctic foxes (*Alopex lagopus*), and snowy owls (*Nyctea scandiaca*) are common predators.

2.5.4 *Alpine*

In contrast to the tundra where precipitation is low and drainage poor, many alpine areas have high precipitation, good drainage, and a high degree of fragmentation. In temperate regions this leads to relatively high growth. In tropical regions temperature varies considerably during the day and forces special adaptations by plants.

Alpine meadows have a similar vegetation structure to that of the tundra but because they are confined to mountain tops they are often found in small scattered patches. Fewer bird and mammal species use these areas for breeding in comparison to the tundra. In North America the characteristic mammals are marmots (*Marmota*), pikas (*Ochotona*, a small lagomorph), and voles (*Microtus*) instead of lemmings. Elk (*Cervus elaphus*), moose (*Alces alces*), caribou (*Rangifer tarandus*), and bears (*Ursus*) use the meadows in summer. In Asia the Himalayan alpine zone is the center of evolutionary radiation for the goats and sheep. These species form the prey of snow leopard (*Panthera uncia*). Pikas have also diversified here.

Alpine meadows on the tropical mountains of Africa produce some extraordinary adaptations in the vegetation. The weather is extreme: it freezes every night and becomes relatively hot every day. Several plant types (*Senecio*, *Lobelia*) show gigantism – plant genera which are small herbs in temperate regions become large trees in this environment. The leaves are fleshy and store water. Few animal species are adapted to these conditions, but one is the hill chat (*Cercomela sordida*).

2.6 Semi-desert scrub

Warm semi-desert scrub is most extensive in a band surrounding the Sahara and extending through Arabia, Iran, and to India. The Somali horn of Africa and the Namibian zone of southwest Africa have, in prehistory, been joined to the Sahara. The vegetation is scattered thorn bush (*Acacia*) and succulents, with a sparse herb layer. Several of the antelopes in the Somali–Sahara area are browsers with convergent adaptations of long necks and the ability to stand up on their back legs (the dama gazelle (*Gazella dama*), dibatag (*Ammodorcas clarkei*), and gerenuk (*Litocranius walleri*)). In both Asia and Africa, the main arid-adapted small mammals are the gerbils (*Gerbillus*, *Tatera*) and jerboas (*Jaculus*, *Allactaga*).

North American semi-desert scrub surrounds the Sonoran and Mojave deserts. Creosote bush (*Larrea divaricata*) is common and there is a wide variety of other spiny and succulent plants such as prickly pear (*Opuntia*). A number of arid-adapted small mammals such as pocket mice (*Perognathus*) and kangaroo rats (*Dipodomys*) live on seeds. Ground-feeding birds such as doves, new world sparrows, and juncos are characteristic. The equivalent Australian vegetation is dominated by shrubs of the family Chenopodiaceae. Small mammals include hopping mice (*Notomys*) and the marsupial jerboa pouched mouse (*Antechinomys*). However, most of the mammals and birds are derived from the temperate woodlands and are recent invaders. These areas are known for the large flocks and nomadic movements of Australian finches (Ploceidae) and budgerigars (*Melopsittacus undulatus*) following the unpredictable pattern of rainfall.

At higher latitudes in the rain shadow of the Rocky Mountains and the Himalayas, a cool semi-desert vegetation is characterized by low, aromatic shrubs such as sagebrush (*Artemisia*) and perennial tussock grasses. Small mammals and birds are similar to those in the warm semi-deserts. Ground squirrels (*Spermophilus*) are common in this type of vegetation in North America.

2.7 Deserts

Deserts tend to occupy the mid-latitudes and extend from the west towards the middle of continents – the Sahara in Africa, the Gobi in Asia, and the deserts of Australia, southern California, and Arizona are examples. They receive on average less than 250 mm of rain per year. Smaller ones include the Namib desert of southern Africa, the Sonoran and others in southwest USA, and the Atacama of Chile. Below 20 mm annual rainfall there is no vegetation, and from 20 to 100 mm it is very sparse: plants have typically xeric adaptations – many species lie dormant as seeds for periods of several years, but germinate, flower, and set seed again in quick succession after a rain storm. At this time the desert comes to life as insects breed and nomadic birds move in to take advantage of the high seedset. Few large mammals are adapted to this environment but the addax (*Addax nasomaculatus*) in the Sahara, the camel (*Camelus*) in Asia, and the red kangaroo (*Macropus rufus*) of Australia are examples.

2.8 Marine biomes

Marine biomes can be divided into open ocean (pelagic), sea floor (benthos), and continental shelf.

2.8.1 *Pelagic*   The surface layers of the pelagic biome receive light and so support phytoplankton, small single-cell algae, and diatoms. These support zooplankton, a mixture of small crustaceans, molluscs, worms, and many other forms which are fed upon by fish. Small fish are transparent as a way of avoiding predation. Larger species such as tuna (*Thunnus*) are fast swimming and move in large shoals. The essential chemicals for growth (nutrients) in these waters are not high and so the amount of plant and animal material is also low.

In the deep pelagic zone there is no light and the animals have to survive on the dead material that sinks from the surface layers. These are called heterotrophic systems because they depend on food from outside sources rather than on plants, which trap their own light and make carbohydrates (these are called autotrophic systems). One still finds crustacea, colonial protozoans (foraminifera, radiolaria), and fish, many of which cannot see or have extraordinary adaptations to lure other fish within catching distance. This biome also contains the giant squids.

2.8.2 *Benthos*   The deep ocean benthos is one of the most extreme of all environments: cold, dark, and pressured. Nevertheless, a diversity of animals live in the bottom mud. Some are attached to the mud (sea anemones, sponges, and brachiopods), others are burrowers, and yet others crawl over the surface.

2.8.3 *Continental shelf*   The continental shelf and the surface waters above it are the richest in nutrients, plankton, and animal life. Dense algal forests can grow because light reaches the sea bottom and these in turn support communities of inshore fish. The higher density of marine invertebrates and fish in these environments supports larger mammal predators such as seals, sealions, and some whales, but mostly in temperate regions. Tropical continental shelves are less productive and support fewer mammals: the dugong (*Dugong dugon*) and manatees (*Trichechus*) which graze on submarine vegetation.

Cold-water currents high in nutrients well up at the edge of the continental shelf. Upwellings occur particularly in arctic and antarctic waters, but there are some in the tropics such as the Humboldt current off Peru. The upwellings are rich in plankton, and a wealth of fish, seabirds, and whales feed on them.

Coral reefs are a special biome forming a rim around oceanic islands. Although not usually associated with a continental shelf, they have similar ecological characteristics.

2.9 **Summary**   The world can be divided into broad ecological divisions, each of which has a characteristic vegetation and wildlife. The forest biomes are diverse, being subdivided into boreal, temperate, tropical, woodland, and shrubland. Grassland biomes include tropical savanna, temperate grassland, alpine grassland, and tundra. The deserts constitute a further biome. Each of these can be divided further into ecosystems and communities based on groupings of plants and animals. Within these larger groupings each animal species selects its habitat.

3 Animals as individuals

3.1 Introduction

In order to manage a population we need to know something about the character-istics of its members. We seek knowledge of their morphological and physiological adaptations to environment, their behavior, particularly with respect to dispersal, repro-duction, and use of habitat, and the genetic variability among them.

In this chapter we begin broadly by outlining the mechanisms by which these adaptations come about: the evolutionary process of speciation, convergence, and radiation. We then focus in on the methods by which the genetic constitution of individuals, or groups of individuals, can be determined and the importance of such information in wildlife management.

3.2 Adaptation

To understand why a population of a species lives where it does, that is to explain its distribution in nature, we should know how an individual is adapted to its environ-ment, what types of environment it encounters, and what resources are available. An adaptation is defined as "a trait that increases fitness relative to an alternative trait" (Schluter 2000). When we talk about the adaptations of individuals we mean the way in which an animal fits into its environment and uses its resources. The adaptive characters that describe an individual – its physical attributes (morphology), physiology, and behavior – are determined first by the processes of natural selection and secondly by its history over evolutionary time, its phylogeny.

The physical environment – temperature, humidity, and other features that we call the **abiotic** environment – together with the effects of other species which form the food, competitors, and predators (the **biotic** environment), acts through natural selec-tion to produce a suite of adaptations which are called life-history traits.

3.3 The theory of natural selection

The term "evolution" refers simply to change in a population over time. It does not necessarily mean speciation (although this may be an outcome) and it does not imply a mechanism of change. The idea of evolution was already being talked about in Europe in the early 1800s, albeit as a radical concept. Charles Darwin described a mechanism for this change in his book *On the Origin of Species* in 1859. It was called **natural selection** and proposed jointly by Darwin and A.R. Wallace in 1858. Darwin based his theory on three observations:

1 Populations increase geometrically through reproduction.
2 All individuals are different – the genetic mechanism for this was demonstrated later by Gregor Mendel (Mendel 1959).
3 Populations remain constant (at least within broad limits) due to a lack of resources. The relative stability of populations was first noted by Malthus (1798) in his essay on populations. From these observations there follow two postulates.

(a) There is competition for resources between individuals.

(b) Those individuals that are most capable of obtaining the resources, and can survive and reproduce, are those that will leave the most progeny. The next generation will contain a greater proportion of those types.

The **selection** comes from the relative success of the different types in leaving progeny. The process of natural selection is the replacement of types (or morphs) that produce fewer successful offspring with those that are more successful. The more successful types are described as **fitter** than less successful types. **Fitness** is defined as "the relative reproductive success of an individual in the long term" where "reproductive success" includes births, survival, and reproduction of the offspring, "long term" means over several generations, and "relative" means in comparison with other members of the population. Fitness is measured by the comparison of reproductive rates and survival rates between types. Indirect indicators of fitness can be morphological, physiological, or behavioral traits that are correlated with these rates.

This is the theory of natural selection at its simplest. It carries the following corollaries:

1 Natural selection results in **adaptation** to the environment because the types leaving more progeny are by definition better at surviving and reproducing in that environment. The most successful types are the **fittest** individuals.

2 Since no population has all possible varieties, natural selection cannot produce perfect adaptation – only the best among those available, and these best may be quite imperfectly adapted.

3 Natural selection results in adaptation to past and present conditions, not to future conditions. It cannot anticipate future conditions or select for individuals preadapted to them. If the changing conditions suit a currently rare individual type, it is through chance alone and does not indicate predetermined design.

4 Natural selection acts only on the inherited components of an individual, namely the **genes**. For these purposes genes are those elements of the chromosome that segregate independently, and therefore may include several DNA groups if they are linked. Natural selection cannot maintain either whole **phenotypes** or whole **genotypes**. The genotype is the total complement of genes in the individual. The phenotype is the individual organism, which is a product of the genotype interacting with the environment during development. Phenotypic variation is reflective of genotypic and environmental variation.

5 A favorable gene can have both advantageous and disadvantageous effects within the same individual due to **pleiotropy** and **polygenic effects**. Pleiotropy describes a gene affecting more than one character in the individual, and some effects may be beneficial while others are disadvantageous. Polygenic effects implies that a character is affected by several genes, some good some bad. All that is required is that the beneficial effects outweigh the detrimental ones.

6 Natural selection does not guard against the extinction of species. Many adaptations do indeed promote the continued existence of a species but there are also many that result in extreme specialization to unusual environments, restricted habitats, or isolated areas. These species are vulnerable to environmental change. On the island of Hawaii in the Pacific Ocean the extinction of many species of the Hawaiian honeycreepers has resulted in the extinction or near extinction of all species in the plant genus *Hibiscadelphus*. The honeycreepers, with their long, curved bills (see Fig. 3.1), were the pollinators of the curved, tubular flowers of the *Hibiscadelphus* (Diamond and Case 1986).

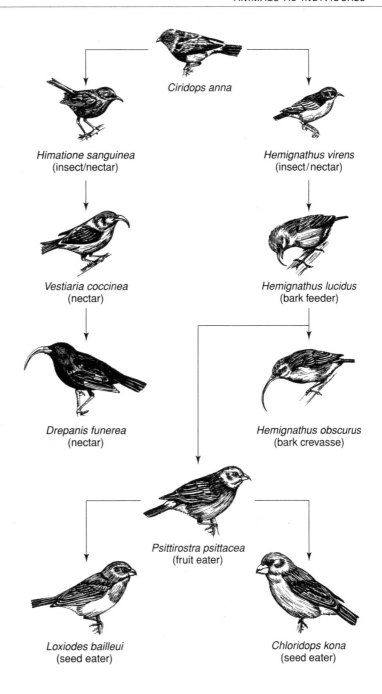

Fig. 3.1 Adaptive radiation of the Hawaiian honeycreepers (Drepanididae). Only a few of the species are illustrated here. There are four main functional groups: insect feeders in foliage, nectar feeders, bark feeders, and seed eaters. (After Futuyma 1986 with nomenclature from Pratt *et al.* 1987.)

Ciridops anna

Himatione sanguinea
(insect/nectar)

Hemignathus virens
(insect/nectar)

Vestiaria coccinea
(nectar)

Hemignathus lucidus
(bark feeder)

Drepanis funerea
(nectar)

Hemignathus obscurus
(bark crevasse)

Psittirostra psittacea
(fruit eater)

Loxiodes bailleui
(seed eater)

Chloridops kona
(seed eater)

3.4 Examples of adaptation

3.4.1 Convergence

Convergence occurs when organisms of different ancestry (i.e. from different phyletic groups) adapt to similar environments and thus develop similar characteristics. One of the classic examples is the placental mammals and the marsupials that have evolved similar morphology and behavior even though they are quite unrelated (Fig. 3.2).

The rock ringtail possum (*Pseudocheirus dahli*), a marsupial of northern Australia, lives in the crevices of large rock piles. Bruce's hyrax (*Heterohyrax brucei*), from the very different placental order Hyracoidea confined to Africa and Arabia, has precisely

Fig. 3.2 Examples of convergent evolution between placental and marsupial mammals.

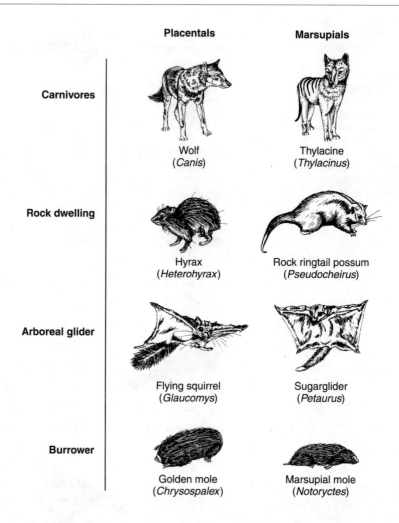

Placentals   **Marsupials**

Carnivores

Wolf
(*Canis*)

Thylacine
(*Thylacinus*)

Rock dwelling

Hyrax
(*Heterohyrax*)

Rock ringtail possum
(*Pseudocheirus*)

Arboreal glider

Flying squirrel
(*Glaucomys*)

Sugarglider
(*Petaurus*)

Burrower

Golden mole
(*Chrysospalex*)

Marsupial mole
(*Notoryctes*)

the same homesite. In North America the hoary marmot (*Marmota caligata*), a rodent of similar size, lives in rock piles on the mountains and feeds on surrounding vegetation. The three species have converged in form and ecology.

There are many examples of convergence in birds. The yellow-throated longclaw (*Macronyx croceus*), a member of the pipit family Motacillidae, lives in the dry open grasslands of eastern Africa. It is brown, yellow below, with a black chest band. It sits on bushes and sings constantly. In North America the western meadowlark (*Sturnella neglecta*) is similar in appearance, behavior, and habitat but it belongs to the entirely different new world family Icteridae. Penguins (Spheniscidae) of the southern hemisphere are the ecological equivalents of the unrelated Alcidae (auks, murres, puffins, guillemots) of the northern hemisphere.

3.4.2 *Adaptive radiation*

Adaptive radiation is the name given to the divergence of a single lineage to provide a variety of forms. Adaptive radiation is the evolution of ecological and phenotypic diversity within a rapidly multiplying lineage. It is the differentiation of a single ancestor into an array of species that inhabit a variety of environments and that differ in the morphological, physiological, and behavioral traits used to exploit those environments (Schluter 2000). The best-known example is that of Darwin's finches on the Galapagos Islands (Grant 1986; Schluter 2000). They were probably founded by a finch species from South America. Subsequently they diverged into types that feed on insects and others that feed on seeds; some species live on the ground, others in trees, and still others among cacti. Adaptation follows the vagaries of unpredictable environmental change, which can be detected in studies of 30 years or more as Grant and Grant (2002) have done with the Galapagos finches.

Another example of adaptive radiation is seen in the endemic honeycreepers (Drepanididae) of Hawaii (Fig. 3.1). Many of these species have become extinct in the past 150 years. They appear to have evolved from a thin-billed insect eater. From this type one group of species developed into long-billed nectar feeders while another group evolved into long-billed bark-crevasse feeders. Yet another group developed thick bills and feeds on fruit and seeds. This one family has filled the niches normally filled on continents by many families of birds. Schluter (2000) gives several other good examples of adaptive radiation.

3.5 The effects of history

3.5.1 Phylogenetic constraints

In discussing how animals fit into their environment we have considered the process of adaptation through natural selection. We have noted that adaptation is not perfect because conditions change and animals are constantly trying to catch up. Further, the organisms are limited by the evolutionary pathways that their ancestors have followed: both birds and mammals have evolved from reptiles but selection on mammals today cannot produce feathers. The potential for growing feathers was lost a long time ago.

Natural selection is constrained in what it can produce by what is currently available. The giant panda (*Ailuropoda melanoleuca*) of China is a large herbivore that eats bamboo shoots almost exclusively. These bamboos provide low-quality food. Most large mammal herbivores, such as horses, deer, and kangaroos, have long intestines and special fermentation mechanisms in the gut to allow maximum digestive efficiency (see Section 4.6); in contrast, carnivores such as cats and omnivores such as bears have relatively short guts. The giant panda probably evolved from bears in the Miocene about 20 million years ago (O'Brien *et al.* 1985a) and changed to a herbivorous diet. Because of its carnivore ancestry it has a short gut and cannot now make the evolutionary jump to the longer, more complex digestive system of herbivores. Hence the giant panda has one of the least efficient digestive systems known for a terrestrial vertebrate: only 18% of the food is digested compared with 50–70% for horses, antelopes, and deer. The giant panda compensates with other adaptations, in particular by eating a very large amount of food and spending most of the day doing so. This prolonged feeding in turn leads to behavioral adaptations: pandas are solitary and spend little time in social and mating activities.

3.5.2 Geographical constraints

Movement of the continents

Towards the beginning of last century, Alfred Wegener, a meteorologist, proposed that the continents were at one time joined together and subsequently drifted apart (Wegener 1924). Wegener's idea was generally rejected. The discovery in the 1960s that the earth's surface is made up of plates, and that these move, proved that Wegener was essentially correct. Volcanic activity and earthquakes along mid-oceanic ridges produce prodigious amounts of submarine basalt and this spreads the sea floor. The continents which float on these basaltic plates are thereby forced apart.

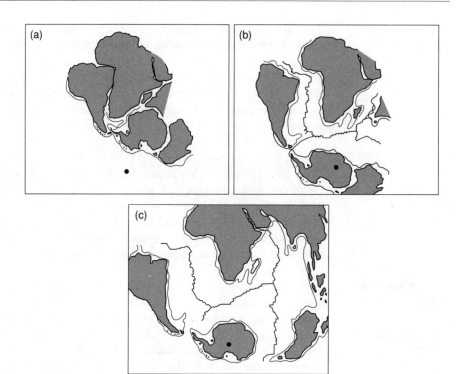

Fig. 3.3 Gondwanaland at different time periods before the present. The thin line around each continent is the limit of water less than 1000 m deep. The dot indicates the South Pole. (a) At 150 million years BP the southern continents were joined. (b) At 65 million years BP Africa had separated from the other continents, which were still joined. (c) The present-day distribution of continents. (After Norton and Sclater 1979.)

Some 150 million years ago there were two great landmasses, Laurasia in the north and Gondwana in the south. Figure 3.3 shows how Gondwanaland split apart. The process began about 115 million years ago, with Africa and India breaking away first. These and Madagascar separated 65 million years ago, while South America, Antarctica, and Australia were still joined. Australia finally separated from Antarctica much later, about 40 million years ago.

These historical movements explain some of the more peculiar distributions of animal groups, for example why marsupials are found today only in Australia, New Guinea, and the Americas. A fossil land mammal from an extinct marsupial family occurred in Antarctica 40 million years ago (Woodburne and Zinsmeister 1982). This supports the idea that Australian marsupials originated from South America via Antarctica before 56 million years ago.

Figure 3.4 shows the distribution of the large flightless ratites (ostriches, rheas, emus, and their extinct relatives). Similar distributions of tree-ducks, penguins, and parrots attest to the breakup of the southern continent.

The joining of North and South America in the Pliocene provides another example of historical events determining the nature of faunas. South America originally had a remarkably diverse mammalian fauna resembling the radiation of the ungulate fauna of Africa today. It included a wide range of marsupial carnivores such as big sabertooth types (*Thylacosmilus*) and hyena types of the family Borhyaenidae, and smaller mongoose types represented by the Didelphids (the group which includes the opossum *Didelphis marsupialis*). These carnivores fed on the herbivorous notoungulates, a huge placental group now entirely extinct.

Fig. 3.4 Present-day evidence of Gondwanaland can be seen in the distribution of ratite birds on the southern continents. The moa in New Zealand and elephant bird of Madagascar became extinct only in the last few centuries. (After Diamond 1983.)

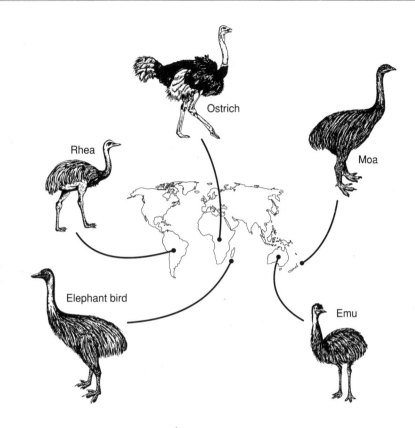

After the two continents joined, a few South American forms moved north, for example the armadillos and sloths, but most died out as a result of competition and predation from North American invaders. The present-day deer, camels, bears, cats, and wolves of South America are all derived from northern forms.

The ice ages: historical effects of climate
During the Pleistocene (from 2 million to 10,000 years ago) the earth went through a series of cold and warm periods. Ice-caps developed over Canada, the northern palearctic (Europe and Asia), and on the main mountain chains such as the Alps, Rockies, Andes, and Southern Alps of New Zealand. Sea levels dropped as much as 100 m and "land bridges" were formed across the Bering Strait between Asia and North America, and across the English Channel between Britain and France. The cold and warm periods in temperate regions were paralleled by dry and wet periods in the tropics.

The ice ages had a significant influence on the present-day distribution of animals. The Beringian land bridge across the present-day Bering Strait allowed an earlier invasion of North America by mammoths (*Mammuthus primigenius*, *M. columbi*), mastodons (*Mammut americanum*) and sabertooth cats (*Smilodon fatalis*), and later invasions by more modern forms such as beaver (*Castor*), sheep (*Ovis*), muskoxen (*Ovibos*), caribou (*Rangifer*), elk (*Cervus*), moose (*Alces*), bison (*Bison*), brown bear (*Ursus arctos*), and wolf (*Canis lupus*). There was a smaller reverse migration from North America into Asia of horses and camels, both of which subsequently became

Fig. 3.5 Patches of high species endemism in birds of the Amazon rainforest (hatched) coincide with the areas that were forest refuges in the past during periods of maximum aridity (stippled). (After Brown 1987.)

1000 km

⧅⧅⧅ High bird species endemism

▒ Forest refuges

extinct in North America. Typical American mammals are deer (*Odocoileus*), mountain goat (*Oreamnos*), and pronghorn (*Antilocapra*). Most of the others are Eurasian forms.

During the last glaciation (12,000 years ago) a few areas within the northern ice-sheets were free of ice and some animals survived and evolved in these "refuge" areas. The northern end of Vancouver Island in Canada was a refuge for elk and marmots which differentiated into new races.

The climatic fluctuations causing the ice ages also caused the expansion and retreat of the tropical forests of South America and Africa. The South American forests contain the highest diversity of bird species anywhere. The centers of endemism within these forests match fairly closely the forest refuge patches left by ice ages (Fig. 3.5). In general, the ice ages have accounted for many present-day distributions of mammals and birds.

The invasion of people

There is one other historical influence that determined the distribution of the larger mammals and birds: the spread of people over the world. They spread into Eurasia from Africa some 200,000 years before present (BP), reaching Australia some 35,000 years BP, North America during the last ice age at 12,000 years BP, and New Zealand,

Madagascar, Hawaii, and Easter Island only about 1000 years ago (Martin and Steadman 1999).

Although there is considerable debate on the effects of these human migrations, one school of thought, discussed in Martin and Klein (1984), MacPhee (1999), and Worthy and Holdaway (2002), holds that the arrival of people resulted in the extinction of large mammals either directly through hunting or indirectly through habitat change. Thus, in North and South America mammoths and giant ground sloths disappeared, in New Zealand the large ratites (moas) were hunted to extinction, in Madagascar both giant ratites (elephant birds, *Aepyornis*) and giant lemurs (*Megaladapsis*) vanished, and in Polynesia a variety of birds such as the giant flight-less galliform (*Sylviornis neocaledoniae*), twice the size of a turkey, became extinct with the arrival of people (Martin and Steadman 1999). Another school of thought holds that rapid climate change caused their extinction (Guthrie 1990). For example, the giant Irish elk (*Megaloceros giganteus*) is thought to have died out at the end of the ice age coincident with change in climate.

Knowledge of past events allows us to answer such questions as why Africa has a wide diversity of large mammals whereas North America and Europe do not (Owen-Smith 1999). When we ask questions concerning the distribution of a species, for example why the white-tailed deer (*Odocoileus virginianus*) is found in South America or why the nine-banded armadillo (*Dasypus novemcinctus*) is found in Texas, we need to know not only about their individual adaptations of habitat selection, diet, and behavior, but also about their historical distributions due to the movement of continents and the effects of the ice ages.

A new evolutionary force now affects animal communities: intensive agriculture and industrialization. This is a post-Pleistocene development which has altered many habitats through pollution and large-scale clearing for agriculture and industry (Morrison *et al.* 1992).

3.6 The abiotic environment

The abiotic environment includes the sets of conditions that determine where an animal can live and reproduce. **Conditions** are those factors such as temperature and rainfall that affect an animal but which are not themselves influenced by the population. Because environments are not constant, animals are adapted to a range of conditions, and usually the less constant the conditions the wider the range (Stevens 1989). The limits of adaptation are called the **tolerance limits** for the animal, and we need to specify whether we mean the limits for reproduction or for occupation. The latter are usually broader. Section 7.4 discusses the conditions, and the adaptations to these by individuals, that determine the distribution of a population and the position of the range boundary.

3.7 Genetic characteristics of individuals

Individuals differ from each other genetically as well as physically and behaviorally in sexually reproducing species. As an example of genetic differences leading to differing behavior, females of 13-lined ground squirrels (*Spermophilus tridecemlineatus*) mate with several males at the start of the breeding season (Schwagmeyer and Wootner 1985). What advantage is this to the males? Do they all stand a chance of producing some offspring, does the first male to mate contribute to all or most of the conceptions, or does the last male to do so score most of the conceptions? It turns out that the first male to mate contributes to 75% of the conceptions (Foltz and Schwagmeyer 1989) and so being first is clearly an advantage. There is some advantage in being second because those males contribute to the other 25%, but subsequent males receive no benefit. What advantage is this mating with several males to the female? Since males are intolerant of juveniles that are not their own offspring, is this a tactic of females aimed at ensuring the cooperation of all surrounding males?

Waterbuck (*Kobus defassa*) in Africa defend territories through which female herds pass while grazing. The male mates with any estrous females when the female herd is in his territory. He also has to defend his territory against other territorial males, and bachelor males that have no territory. In some areas a territorial male allows one other male into his territory (Wirtz 1982). What advantage is there to the territory holder in allowing the second male in? One suggestion is that the second male helps to defend the territory and so allows more opportunities for the primary male to mate. In return the second male may be able to "steal" some matings when the primary male is occupied elsewhere. A similar situation is described for long-tailed manakins (*Chiroxiphia linearis*), a small neotropical bird. Two males defend a territory on a lek, one being dominant and obtaining almost all matings. The unrelated subordinate may benefit by inheriting the territory and obtaining a few matings in the meantime (McDonald and Potts 1994).

So far we do not have the answers to most of these questions. To obtain the answers we must identify the individual parents of offspring. Recent genetic techniques have allowed us to do this.

| 3.7.1 *Methods for studying genetic variability* | **Allozyme gel electrophoresis** |

Allozyme gel electrophoresis

Until recently the standard technique for detecting genetic variability within and between individuals was to measure differences in amino acid composition of allozymes or proteins encoded by different alleles at a locus. Blood or tissue homogenate from individuals is placed on a gel matrix, such as cellulose acetate, and an electric charge applied. The proteins migrate along the gel at rates dependent on their total electric charge. Changes in amino acid composition, the result of mutation, are often reflected as changes in electric charge. The electric current is switched off and the gel is stained for a particular protein after a set time. Differences between individuals are evident as different configurations of the protein bands on the gel.

The method has been used to measure differences between races and between species by assessing the variability in many proteins from several individuals in each population. It is useful because inheritance patterns are generally known. Phylogenetic trees have been constructed by this method.

The technique has several limitations. First, some proteins with different mutations can move at the same rate, thereby appearing to be the same. This problem becomes greater the more distant the relationship between individuals or species. Second, much of the genetic variability is not evident at the protein level because of the redundant nature of the genetic code. Other techniques assess the genetic diversity present in the individual's DNA itself. We examine these next.

Polymerase chain reaction

Taxonomy, population genetics, and molecular ecology have advanced rapidly as a result of a technique called the **polymerase chain reaction** (PCR). This allows millions of copies of a particular target sequence of DNA to be produced so that DNA amplification can now be used easily to identify individuals or groups of organisms.

PCR is the first step in most genetic analyses, from genetic fingerprinting to creating genetic phylogenies.

For studies of phylogeny, highly conserved mitochondrial DNA is used (Parker *et al.* 1998). For identification of individuals it is the highly variable DNA found in the **major histocompatibility complex** (MHC) that is used. Once a target region has been chosen, a short piece of DNA is synthesized to act as a primer (a piece onto which new DNA is attached). We start with a mixture of the original double-stranded DNA, the primers, free nucleotides, and a heat-stable DNA polymerase. The mixture is heated and the strands of the double DNA are separated. Upon cooling, the primers attach themselves at one end of the target DNA and serve as starting points from which the polymerase builds the copy. A new cycle of heating starts the process again and is repeated for 25–49 rounds, depending on the protocol, to make over a million copies of the selected DNA region.

PCR has been used to amplify DNA from extinct animals so as to elucidate phylogenetic relationships. Samples of ancient DNA from extinct species show how the quagga (*Equus quagga*) is placed in the zebras (Higuchi *et al.* 1984) and the sabertooth cat (*Smilodon*) is within the Felidae (Janczewski *et al.* 1992). Samples from the moas (extinct ratites of New Zealand) show that they are an ancient lineage not closely related to modern-day kiwis (*Apteryx*). This indicates that there were two invasions of New Zealand by ratites (Landweber 1999). PCR was also used to show that genetic variation in humpback whales (*Megaptera novaeangliae*) was not reduced when the population went through low numbers from commercial harvesting (Baker *et al.* 1993).

PCR has replaced several older techniques that are now going out of use. For historical interest some of these are **restriction fragment length polymorphisms** (RFLPs), **random amplified polymorphic DNA** (RAPDs), and **variable number tandem repeats** (VNTRs). RFLPs take advantage of mutations in the DNA that can be detected by the presence or absence of a cleavage site revealed by an enzyme called a restriction endonuclease. Restriction enzymes cleave or cut the DNA at particular recognition sites located at random along the DNA molecule. Total DNA is isolated from a tissue sample, challenged with restriction enzymes, and then electrophoresed on an agarose gel matrix. Differences between individuals are detected from the distribution of fragment lengths. This technique was used to detect genetic disorders in humans such as thalassemia (Weatherall 1985), and in studies of the reproductive behavior of lesser snow geese (*Cheu caerulescens*) in northern Canada (Quinn *et al.* 1987).

The use of RAPDs for population genetic inference is being questioned because of problems with reproducibility, dominance, and homology. They can be used for genetic mapping studies or for species diagnostic markers.

Mitochondrial DNA techniques

Mitochondria in cells have their own DNA (mtDNA) whose strands are relatively short (1.6×10^4 base pairs compared with 10^9 base pairs for the nuclear DNA). Parts of the mtDNA mutate at a fast rate and are highly variable, such as the MHC or the d-loop (though tandem repeat areas of nuclear DNA are even more variable). Regions of mtDNA can be monitored for mutations with radioactive probes in the same way as nuclear DNA. Genetic variability accumulates rapidly and large differences between populations are thereby often evident. mtDNA is inherited by matrilineal descent only, thus permitting an assessment of novel sources of variability. For example, there are areas in Texas, California, Montana, and Alberta where the ranges of white-tailed and mule deer (*Odocoileus hemionus*) overlap. Examination of male deer in Texas shows that although they look like mule deer their mtDNA resembles that of white-tailed deer. This suggests that at some point in the past these populations were derived from matings between male mule deer and female white-tailed deer: the female mtDNA was retained although the other characters are those of mule deer (Carr *et al.* 1986; Derr 1991). The process of taking some feature of one species into the genome of another is called **introgression**. In other areas of overlap, such as in Montana, there is little introgression (Cronin *et al.* 1988). Similarly, mtDNA demonstrated introgression in a hybrid zone of indigenous red deer (*Cervus elaphus*) and exotic Japanese sika deer (*C. nippon*) in Scotland (Goodman *et al.* 1999).

Mitochondrial DNA analysis of brown bears (*Ursus arctos*) in North America revealed four phylogeographic clades that do not correlate with present taxonomic classifications based on morphology (Waits *et al.* 1998). The four clades probably evolved before migration of this species into North America. They provide managers with more reliable locations for the conservation of evolutionary units.

3.7.2 *Genetic divergence between geographic regions*

Population genetics studies are interested in the genetic structure of populations, the genetic divergence between populations in different areas, and the gene flow between populations. A simple approach assumes that individuals live in separated clumps or **demes** (the whole being a **metapopulation**). Differences between these local subpopulations can be detected using statistical tests of divergence in allele frequencies, such as heterogeneity tests or estimates of F_{st}, the standardized variance in allele frequency. The F_{st} statistic is based on the "island model" (see below): it is the variance in allele frequencies between populations, σ_p^2, standardized by the mean allele frequency (p) at that locus. Thus,

$$F_{st} = \sigma_p^2/[p(1-p)]$$

and this can be measured relatively easily by sampling allele frequencies in the field. F_{st} is a good measure of genetic differentiation among populations, which is essential to understanding evolutionary change.

However, F_{st} has also been used to measure the number of migrants between populations because of the relationship

$$F_{st} = 1/(1 + 4N_e m)$$

where N_e is the effective population size (see Chapter 17 for an explanation of this) and m is the migration rate. There are a number of assumptions, including (i) the mutation rate is low; (ii) the populations are at equilibrium between migration and genetic drift; (iii) there is random mating; and (iv) all individuals have the same potential to migrate. Given these then one can obtain an estimate of the average number of migrants among populations ($N_e m$). We should be aware that the estimate of $N_e m$ is based on a large number of assumptions that are unlikely to be true. Hence estimates of $N_e m$ tend to be unrealistic. Therefore, it is wise to restrict the use of F_{st} to measures of genetic differentiation and avoid its use for measures of genetic migration. Instead, direct observations of migrants should be employed (Whitlock and McCauley 1999).

F_{st} is unity if there is no migration and very small if there is much migration. Thus, F_{st} is a measure of isolation. For example, statistics from DNA fingerprinting and mtDNA indicated little genetic differentiation between populations of mule deer on the north and south rims of the Grand Canyon, Arizona (Travis and Keim 1995). Genetic differences between populations can be of three forms:

1 There are no differences with distance between populations, suggesting high gene flow and random mating (**panmixia**).

2 There is a positive linear relationship between genetic divergence and distance. This suggests that the species distribution is sufficiently wide that individuals do not move across the whole range, gene flow is restricted, and local adaptation occurs. This is called **isolation-by-distance**. For example, the genetic distance of coconut crabs (*Birgus latro*) on the Pacific islands shows such a relationship (Moritz and Lavery 1996). The greenish warbler (*Phyloscopus trochiloides*) in Asia also shows a clear trend in genetic divergence with distance (see below) (Irwin *et al.* 2005).

3 A population may have a genetic divergence greater than that predicted by distance, suggesting that it has been cut off from the rest of the population for a long period. It is a genetic island, it reflects the effects of time (i.e. history), and so this is called the **island model**. Thus, coconut crabs on Christmas Island in the Indian Ocean show a greater genetic divergence than expected by distance relative to crabs on the Pacific islands (Moritz and Lavery 1996).

Populations whose individuals interbreed freely (a **panmictic** population) will have similar allele frequencies throughout their range. Individuals should be adapted to the average conditions of the range, ideally those of the center. At the edge of the range where conditions are extreme, the individuals should be less well adapted; they cannot develop local adaptations to these edge conditions because any tendency towards genetic divergence is swamped by gene flow from the center. However, if gene flow through a population is slow relative to the rate of local genetic adaptation, the individuals at one end of the range can differ from those at the other end, with intermediate forms between (the isolation-by-distance model). This gradual trend in appearance or behavior is called a **cline**. A good example of this is provided by *Cervus elaphus* (Fig. 3.6). In Europe it is called red deer and is a relatively small dark animal. Males produce a deep-throated roar during the rut. At the other end of its range in North America the same species is called elk. Here it is larger, lighter colored, and the mating call of males is a high-pitched whistle or "bugle." Forms intermediate in both morphology and behavior occur across Asia.

The range of some species extends around the world. Where the two ends meet the animals have diverged sufficiently such that individuals no longer interbreed and they behave as separate species. These are called **ring species**. The classic example is the black-backed gull/herring gull pair. In Europe the herring gull (*Larus argentatus*) is light gray on the upper surface of its wings and back, but these parts gradually become darker through Asia and North America so that they are entirely black on the eastern seaboard of the USA. This form has crossed the Atlantic and lives in Europe as the black-backed gull (*L. fuscus*) without interbreeding with the herring gull: it behaves as a separate species. An elegantly documented example of a ring species by Irwin and co-workers is that of the greenish warbler, a small insectivorous bird of forests in central Asia (Irwin 2000; Irwin *et al.* 2001a). The parent population exists near the Himalayan mountains. Two branches of this population have spread around the Tibetan plateau, one northwest in western Russia and northern Europe,

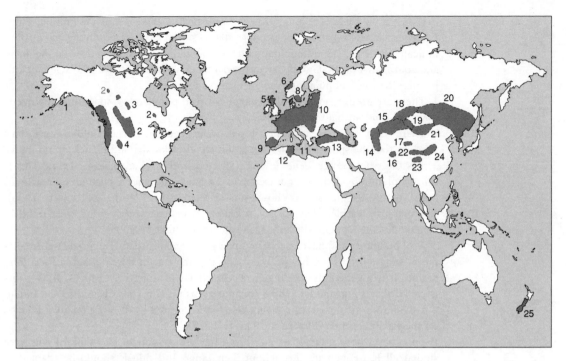

Fig. 3.6 The distribution of *Cervus elaphus* shows a cline from the small red deer of Scotland to the large elk of North America. The numbers and their associated shaded areas indicate different races in the cline. (After Whitehead 1972.)

the other northeast into China and Siberia. The two arms differ progressively in morphology, ecology, and song characteristics. The arms meet again in central Siberia; in the overlap zone the two no longer recognize each other's song and do not interbreed. These physical differences are due to genetic differences, an example of genetic differentiation-by-distance that we discussed above (Irwin *et al.* 2005). Irwin *et al.* (2001b) review other cases of ring species.

Populations isolated by geographical barriers are called **allopatric**. In isolation they can become genetically different through adaptation to their different areas. If the populations then meet and overlap in range (become **sympatric**) they may not interbreed if they have diverged too far, in which case they have become separate species; or they may interbreed and form a zone of hybrids, an area of higher genetic variability with many intermediate forms. The parent types would then be called **races** or **subspecies**. An example is seen in waterbuck in Africa which has a northern form with a white rump patch and a southern form with a thin white ring on the rump. They overlap in a narrow zone in Kenya where they interbreed, producing various rump patterns.

3.7.3 *Individual variability within geographic regions*

So far we have considered genetic variability within and between populations. There is another form of variability which we call a **polymorphism**. The formal definition of this by E.B. Ford (1940) is "the occurrence together in the same habitat of two or more discontinuous forms of a species in such proportions that the rarest of them cannot merely be maintained by recurrent mutation or immigration." This means that the different morphs, often quite visibly different, live together in the same habitat instead of living geographically apart as described above for subspecies. For example, the lesser snow goose has two color morphs, blue and white, which occur together in the same population. This species is polymorphic for color. The common guillemot or murre (*Uria aalge*) has a "bridled" morph with a ring of white feathers around the eye. The normal morph has no white marks. Along the coast of Europe the frequency of the bridled morph increases with latitude and humidity from 0.5% to over 50% (Southern 1951).

Where the frequencies of the morphs in these populations have remained relatively constant, as in the bridled guillemot, the state is called a **stable polymorphism**. There are also cases where one morph displaces another, perhaps because of changing conditions or because two races, originally separate, have recently become sympatric. Such may have happened to the lesser snow geese where the two color morphs used to spend the winter in separate areas but now share their wintering grounds in the central USA (Cooke 1988). The temporary state in which one morph is replacing another is called a **transient polymorphism**.

The selective advantages accruing to the various morphs are usually unknown. Mechanisms that could maintain the polymorphism include:

1 *Heterozygote advantage.* The heterozygote has a selective advantage over both homozygotes. Often the rare allele is a genetic dominant, lethal or disadvantageous in homozygous form.

2 *Frequency-dependent selection.* The rarest morph has a selective advantage over the others. This could occur, for example, where predators have a **search image** for the common morph and so overlook the rare one.

3 *Alternating selection.* Different morphs are advantageous under different environmental conditions. Some morphs may be adapted to wet and cool conditions, others to hot and dry conditions, such that they are advantaged seasonally.

3.8 Applied aspects

3.8.1 *Adaptation*

Why do we need to know about adaptation from the point of view of wildlife management? Many species are becoming rare through loss of their habitat. To rectify this by preserving habitats we need to know their physiological and behavioral adaptations and constraints. For example, to improve the breeding ponds for ducks in the Canadian prairies we need to know the tolerances the ducks have for levels of alkalinity and salinity. Recently hatched ducklings of mallard (*Anas platyrhynchos*) and blue-winged teal (*A. discors*) require fresh water for survival because their salt glands are not completely functional until the ducklings are 6 days old. Growth of mallards in the first month of life is slowed when they live in moderately saline water. Although dabbling ducks often nest on islands in lakes of high salt content, female mallards lead their ducklings to freshwater lakes, and gadwall (*A. strepera*) ducklings use freshwater seepage zones (Swanson *et al.* 1984).

The social organization of animals is an adaptation to habitat, food supply, avoidance of predators, and courtship. Jarman (1974) compares the social organization of the African antelopes (see Section 4.8). At one extreme is the tiny dikdik (*Rhynchotragus kirkii*) that lives in pairs jointly defending a territory in thick scrub. They avoid predators by staying very still when predators are around, and they run only at the last moment. They rely on concealment. Their food is high protein shoots, buds, and flowers on bushes. Being small (5 kg), dikdik do not eat large amounts of food. However, the food is sparsely distributed and so dikdik are also dispersed – they cannot live in large groups. Equally, this sparse food supply should be defended to prevent others from eating it and so they are territorial. Since females are widely dispersed (because of the food they eat), males can obtain mates only by keeping a female in his territory, and so a monogamous pair bond develops.

At the other end of the scale is the African buffalo (*Syncerus caffer*) (500 kg), which lives in herds of several hundred animals, often in open savanna, and which eats abundant but low-quality grass. By living in herds they obtain protection from predators – only lions (*Panthera leo*) are big enough to attack them. Because grass tends to be closely and uniformly distributed, buffalo are able to live in herds and still find enough to eat. It is not worth their while defending a patch of grass because there is plenty more beyond. Thus the mating system has evolved into a dominance hierarchy among males in which the dominant males obtain most of the matings – this is a polygynous system. These comparisons suggest that an adaptation to one thing, say food type, leads to complementary adaptations to habitat utilization, antipredator responses, and social behavior.

3.8.2 *Uses of genetic techniques*

Mitochondrial DNA can be used to determine the geographical areas from which specimens have been taken. Thus, black rhinoceros (*Diceros bicornis*) have become highly endangered because of illegal hunting for their horns. Conservation and enforcement strategies are helped because mtDNA samples from individuals can identify their geographic origins (O'Ryan *et al.* 1994). This approach is used in detecting illegal hunting of other species (Manel *et al.* 2002). Similarly, differences in the frequency of mtDNA genotypes among conspecific nesting populations of green turtles (*Chelonia mydas*), loggerhead turtles (*Caretta caretta*), and hawksbill turtles (*Eretmocheles imbricata*) show little interrookery exchange of maternal lineages. Thus, evidence is consistent with natal homing by females (Bowen and Avise 1996).

Mitochondrial DNA has been used to examine genetic differences among wolf populations. Wayne *et al.* (1992) and Wayne (1992) used mtDNA to conclude that the red wolf (*Canis rufus*) is a modern hybrid of the gray wolf (*Canis lupus*) and the coyote (*Canis latrans*), with a predominance of the latter. Nowak (1992) has suggested that the red wolf is a true species and not just a hybrid, but that it has recently hybridized in some areas with the other canids.

Allozyme and mtDNA data explained the peculiar distribution of clouded salamanders (Jackman 1998). One species, *Ameides vagrans*, is found in California and Vancouver Island, British Columbia, while a second species, *A. ferreus*, occurs in Oregon between the disjunct distribution of the former. The DNA evidence shows that the Vancouver Island population was introduced from California in the 1800s.

Analysis of mtDNA in gray seal (*Halichoerus grypus*) populations in the western North Atlantic (Canadian coast) and eastern North Atlantic (Norway, Baltic Sea) showed no shared haplotypes and an estimated divergence of 1.0–1.2 million years ago. In contrast, Norwegian and Baltic stocks diverged 0.35 million years ago, while populations along the Canadian coast show no divergence (Boscovik *et al.* 1996). Degrees of divergence are important factors when considering issues such as the conservation of genotypes and the reintroduction of lost populations.

Endangered populations are often at the edge of a species range and are subject to unusually high abiotic and biotic pressures. These may contribute to a population decline and range contraction. Mitochondrial DNA is not suitable for identifying causes of these sorts. In contrast, quantitative traits such as morphology, reproductive capacity (clutch size), and behavior may provide useful information because they expose variation in key limiting traits (Storfer 1996). In painted turtles (*Chrysemys picta*) the sex ratio is skewed towards females under warmer conditions. Populations are now managed so as to raise males and balance the sex ratio as global warming proceeds (Janzen 1994). Conservation, therefore, should focus on preserving quantitative traits associated with stress resistance (Storfer 1996).

Genetic information on individuals and populations can be obtained from small quantities of DNA obtained non-invasively. For example, hairs from mammals can be obtained from hair traps. Individuals can then be identified for censusing, as in the very rare hairy-nosed wombat (*Lasiorhinus krefftii*) of Australia (Sloane *et al.* 2000). DNA from fecal samples was used to differentiate feces of Mexican gray wolves (*Canis lupus baileyi*) from those of coyotes (*C. latrans*) (Reed *et al.* 2004). DNA from fecal samples can be used to identify individuals, and from this estimate population size. This technique was used to estimate a population of European badgers (*Meles meles*), by estimating individuals and applying rarefaction analysis (Frantz *et al.* 2004). DNA from fecal samples has the added advantage that dietary information can also be obtained. Coyote populations in Alaska declined when their staple prey, the snowshoe hare, declined. Individuals (identified by DNA analysis of feces; Prugh *et al.* 2005) which ate other, less risky food species had greater chances of surviving the period of food scarcity than those eating risky species such as porcupines (*Erethizon dorsatum*) (Prugh 2004). We should, however, be careful with interpretations, because there are several possible biases in sampling populations (Mills *et al.* 2000).

3.9 Summary

The characteristics of individual animals are shaped by the process of evolution through the associated process of speciation. Geographic barriers, earth movements, and the migration of climatic zones split up the distribution of species, the separated components then adapting to their own disparate environments. Evolution of higher order taxa leads to convergence on one hand and radiation on the other.

These large processes determine the detailed characteristics of the individuals of a population, their morphological and behavioral traits differing within populations and among populations according to genetic programming. Molecular methods allow us to determine more accurately the genetic constitution of individuals and the genetic differences among races, species, and higher taxa. These techniques can be used for assessing parentage and genetic relatedness, censusing, identification of species in law enforcement, and determining genotypes and phylogenies for conservation.

4 Food and nutrition

4.1 Introduction

The three main areas of wildlife management (conservation, sustained yield, and control) require knowledge of the food and nutrition of animal populations. Some of the important questions are:

1 Is there enough food to support and conserve a particular rare or endangered species?
2 What is the food supply needed to support a particular sustained yield?
3 Can we alter the food supply so as to provide more effective control of pest populations?

The field of animal nutrition covers subjects such as anatomy, physiology, and ecology, and there are several good reviews of these areas – for example, Hofmann (1973) deals with the anatomy of ruminants, Robbins (1983) addresses the physiology of wildlife nutrition, and Chivers and Langer (1994) review the form, function, and evolution of the digestive system in mammals. From the point of view of wildlife management, however, we are interested in two main types of information to answer the above questions: we need to know the availability of the food and the requirements of the animals. By matching the two sets of information we can answer the questions. Sections 4.2–4.4 deal with availability, and Sections 4.5–4.9 address animal requirements.

4.2 Constituents of food

4.2.1 Energy

Energy is measured in units of calories or joules (1 cal = 4.184 J). Energy content of foods can be found by oxidizing a sample in a bomb calorimeter. Differences in the energy content of different plant and animal materials are due to the differences in their constituents. The energy content of some of the common components of food is given in Table 4.1. We can see that fats and oils have the highest content (over 9 kcal/g), with proteins coming next (around 5 kcal/g), and sugars and starches

Table 4.1 Approximate energy content of food components.

Food component	Energy (kcal/g)
Fat	9.45
Protein	5.65
Starch	4.23
Cellulose	4.18
Sucrose	3.96
Urea	2.53
Leaves	4.23
Stems	4.27
Seeds	5.07

From Robbins (1983).

(carbohydrates) close to 4 kcal/g. The gross energy of tissues depends on the combination of these basic constituents, particularly in animals. In plant tissues, energy content remains relatively uniform and in the region of 4.0–4.2 kcal/g. Plant parts with a high oil content such as seeds (over 5 kcal/g), or evergreen plants with waxes and resins such as conifers and alpine plants (4.7 kcal/g), are the exceptions (Golley 1961; Robbins 1983).

Energy flow through animals can be measured with isotopes of hydrogen (^{3}H) and oxygen (^{18}O) by the **doubly labeled water method** (Nagy 1983; Bryant 1989). First, water labeled with ^{3}H and ^{18}O is injected and allowed to equilibrate in the animal, this taking 2–8 hours depending on body size. A blood sample is then collected to establish the starting concentrations of the two isotopes. Analysis of ^{3}H is carried out by liquid scintillation spectrophotometry and ^{18}O by proton activation of ^{18}O to ^{18}F (the isotope of fluoride) with subsequent counting of γ-emitting F in a γ-counter. A second blood sample is collected several days later. The timing of the second collection does not need to be exact but should occur when approximately half of the isotope has been flushed from the body. Thus, timing depends on body size and the flow rates of the isotopes. Oxygen leaves the body via carbon dioxide and water, and this rate is measured by dilution of the ^{18}O. Rate of water loss is measured from the dilution of ^{3}H. Thus the difference between the total oxygen loss and the oxygen loss in water gives the rate of carbon dioxide production, which is a measure of energy expenditure. The method and its validation are described by Nagy (1980, 1989).

4.2.2 *Protein*

Protein is a term covering a varied group of high molecular weight compounds: these are major components in cell walls, enzymes, hormones, and lipoproteins. They are made up of about 25 amino acids which are linked together through nitrogen–carbon peptide bonds. Most animal species have a relatively similar gross composition of amino acids. For carnivores, the nutrient composition of their prey is usually well balanced to a consumer's specific needs, whereas in herbivores the foods eaten may be deficient in key nutrients (Wright and Mulkey 1997).

Animals with simple stomachs require 10 **essential amino acids**, these being the forms that cannot be synthesized by the animal and must be obtained in the diet: arginine, histidine, isoleucine, leucine, threonine, lysine, methionine, phenylalanine, tryptophan, and valine. **Non-essential amino acids**, therefore, are ones which can be synthesized in the body. Ruminants, and other species that rely on fermentation through the use of microorganisms, synthesize many of the amino acids themselves and so have a shorter list of essential amino acids.

Although there is some variability in the nitrogen content of amino acids (ranging from 8% to 19%), the average is 16%. Thus, in analyzing tissues for **crude protein**, the proportion composed of nitrogen is multiplied by the constant 4.25 (i.e. 100/16). Crude protein content of plant material tends to vary inversely with the proportion of fiber. Since one of the major constituents of fiber is the indigestible compound lignin, fiber content can be used as an index of the nutritive value of the plant food. In many plant tissues such as leaves and stems, protein and digestible energy content (i.e. the non-fiber component) tend to vary together. However, some plant parts such as seeds are high in energy but quite low in protein.

4.2.3 *Water*

The water content of birds and mammals is a function of body weight (W) to the power of 0.98 when comparing across species, but more restricted groups vary in the exponent. Robbins (1983) found that the water content of white-tailed deer and several rodents varied as a function of $W^{0.9}$.

Water is obtained from three sources:

1 **free water** from external sources such as streams and ponds;

2 **preformed water** found in the food;

3 **metabolic water** produced in the body from the oxidation of organic compounds. Preformed water is high in animal tissues such as muscle (72%) and succulent plants, roots, and tubers. Because of this carnivores may not have to drink often; and herbivores such as the desert-adapted antelope, the oryx, which eat fleshy leaves and dig up roots can also live without free water (Taylor 1969; Root 1972).

The highest rate of production of metabolic water in animals is from the oxidation (catabolism) of proteins because of the initially high water content of these tissues. Catabolism of fats produces 107% of the original fat weight as water, but the low preformed water content (3–7%) means that the absolute amount produced is less than that from protein (Robbins 1983).

Measures of free water intake from drinking underestimate total water turnover and more accurate methods use the ^3H or deuterium oxide isotopes of water. A known sample of isotopic water is injected into an animal, and after a period of 2–8 hours (depending on size of animal) for equilibration, a blood sample is collected. The concentration of isotope in the blood is then measured using a liquid scintillation spectrometer. A second blood sample is collected a few days to a few weeks later, again depending on body size, to obtain a new value of isotope concentration. Because water is lost through feces, urine, and evaporation the isotope is diluted by incoming water. Therefore, the rate of dilution is a measure of water turnover. These techniques are described by Nagy and Peterson (1988) and have been used on a wide range of animals including eutherian mammals, marsupials, birds, reptiles, and fish.

4.2.4 Minerals

Minerals make up only 5% of body composition but are essential to body function. Some minerals (roughly in order of abundance: calcium, phosphorus, potassium, sodium, magnesium, chlorine, sulfur) are present or required in relatively large amounts (mg/g) and are called **macroelements**. Those that are required in small amounts (μg/g) are called **trace elements** (iron, zinc, manganese, copper, molybdenum, iodine, selenium, cobalt, fluoride, chromium). So far very little is known about the mineral requirements for wildlife species, but Robbins (1983) has provided a summary of available information. It is assumed that most native species are adapted to their environment and so can tolerate the levels of minerals found there (Fielder 1986). However, some mineral deficiencies have been observed. Selenium deficiency increases the mortality of juvenile, preweaned mammals (Keen and Graham 1989). Flueck (1994) supplemented wild black-tailed deer in California and increased preweaning fawn survival threefold.

Calcium and phosphorus are essential for bones and eggshells. Cervids have a very high demand for these minerals during antler growth. Calcium is also needed during lactation, for blood clotting, and for muscle contraction. Phosphorus is present in most organic compounds. Deficiencies of calcium result in osteoporosis, rickets, hemorrhaging, thin eggshells, and reduced feather growth. Carnivores that normally eat flesh of large mammals need to chew bone to obtain their calcium. Mundy and Ledger (1976) found that the chicks of Cape vultures (*Gyps coprotheres*) in South

Africa developed rickets when they were unable to eat small bone fragments. This has an important management consequence: bone fragments from large carcasses are made available to vultures by large carnivores, in this case lions and hyenas. Where carnivores were exterminated on ranch land, carcasses were not dismembered and bones were too large for the chicks to swallow. This is a good example of how the interaction of species should be considered in the management and conservation of habitats.

Sodium is required for the regulation of body fluids, muscle contraction, and nerve impulse transmission. Sodium is usually in low concentrations in plants, so herbivores face a potential sodium deficiency. In areas of low sodium availability, herbivores consume soil or water from mineral licks (Weir 1972; Fraser and Reardon 1980). Carnivores can easily obtain sodium from their food, and so are unlikely to experience sodium deficiency. Isotopic sodium has been used as a measure of food intake rates of carnivores such as lions (Green et al. 1984), seals (Tedman and Green 1987), crocodiles (Grigg et al. 1986), and birds (Green and Brothers 1989). This approach is possible because sodium remains at a relatively constant concentration in the food supply. The technique is similar to that for isotopic water described in Section 4.2.3.

Both potassium and magnesium are abundant in plants, and deficiencies in free-living wildlife are therefore unlikely. The same is true for chloride ions and for sulfur. Trace element deficiencies are unusual under normal free-ranging conditions, but they occur locally from low concentrations in the soil: there are some reports of iodine and copper deficiencies and of toxicity from too much copper and selenium (Robbins 1983).

4.2.5 Vitamins

Vitamins are essential organic compounds which occur in food in minute amounts and cannot normally be synthesized by animals. There are two types of vitamins, fat soluble (vitamins A, D, E, K) and water soluble (vitamin B complex, C, and several others). Fat-soluble vitamins can be stored in the body. Water-soluble vitamins cannot be stored and hence must be constantly available. Overdose toxicities can arise only from the fat-soluble vitamins.

Vitamin A, a major constituent of visual pigments, can be obtained from β-carotene in plants. Vitamin D is needed for calcium transport and the prevention of rickets. Vitamin E is an antioxidant needed in many metabolic pathways. It is high in green plants and seeds, but decreases as the plants mature. Vitamin K is needed to make proteins for blood clotting. Deficiencies are unlikely to occur because it is common in all foods. The vitamin K antagonist, warfarin, causes hemorrhaging. It is used as a rodenticide.

Little is known about the B-complex vitamins and whether deficiencies occur in free-living wildlife species, although cases of thiamin (B1) deficiency have been reported for captive animals (Robbins 1983). Vitamin C differs from the others in that most species can synthesize it in either the kidneys or the liver. Exceptions include primates, bats, guinea pigs, and possibly whales. Vitamin C is not as commonly available as the B vitamins but is found in green plants and fruit. It is absent in seeds, bacteria, and protozoa.

Other physiological constraints that may not be called vitamins nevertheless provide limits to animal nutrition. For example, old world starlings and flycatchers cannot digest sucrose (Martinez del Rio 1990).

4.3 **Variation in food supply**

4.3.1 *Seasonality*

Food supply varies with season. To some degree all environments are seasonal, including those of the tropics. Food supply is greatest for herbivores when plants are growing, during the summer at higher latitudes (temperate and polar regions) and during the rainy season in lower latitudes (tropics and subtropics). Protein in grass and leaves declines from high levels of 15–20% in young growth to as little as 3% in mature flowering grass, or even 2% in dry, senescent grass. Leaves from mature dicots maintain a higher protein content of about 10%. Thus herbivores such as elk in North America and eland and elephant in Africa will switch from grazing in the growing season to browsing in the non-growing season. Many forest-dwelling Australian marsupials are mycophagous, that is they prefer to feed on the sporocarps of hypogeous fungi. They feed on dicot fruits and leaves when fungi are not about. Growth rates of pouch young in the Tasmanian bettong (*Bettongia gaimardi*) are directly related to periods of sporocarp production (Johnson 1994).

Winter is the main period of stress for animals in higher latitudes. Low temperatures create higher energy demands just when energy is less available. For example, energy intake of moose in Norway declines by 15–30% during winter, and results in a deficiency of 20–30% relative to their requirements. Energy intake is less (573 kJ/kg$^{0.75}$ per day) in poor habitats compared with good habitats (803 kJ/kg$^{0.75}$ per day) (Hjeljord *et al.* 1994). Even greater reductions in food intake rates during winter have been recorded for black-tailed deer (Gillingham *et al.* 1997).

Animals adjust their breeding patterns so that their highest physiological demands for energy and protein occur during the growing season. Thus northern ungulates give birth in spring so that lactation can occur during the growing period of plants, whereas tropical ungulates produce their young during or following the rains, allowing the mother to build up fat supplies to support lactation. Although most birds complete their entire breeding cycle during one season, the timing of breeding is closely associated with food supply (Perrins 1970). Very large birds such as ostrich behave like ungulates and start their reproductive cycle in the previous wet season so that the precocial chicks hatch at the start of the next wet season (Sinclair 1978).

Carnivores also adapt their breeding to coincide with maximum food supply. Thus, wolves which follow the caribou on the tundra of northern Canada have their young at the time caribou calves are born. Schaller (1972) records that lions have their cubs on the Serengeti plains of Tanzania when the migrant wildebeest are giving birth (Estes 1976). In the same area birds of prey have their young coinciding with the appearance of other juvenile birds and small mammals which form their prey (Sinclair 1978).

4.3.2 *Year-to-year variation in food supply*

A particular kind of variability in food supply occurs with the production of prolific seed crops by some tree species. This seed is termed **mast**. It occurs when the majority of trees in a region synchronize their seed production. Beech trees (*Fagus, Nothofagus*) and many northern hemisphere conifers (e.g. white spruce, *Picea glauca*) produce their seeds at the same time, these mast years occurring every 5–10 years. Birds that depend on these conifer seeds, for example the crossbill (*Loxia curvirostra*), breed throughout the winter when a mast cone crop occurs. In the following year, when few cones are produced, the crossbills disperse to find regions with a new mast crop, sometimes traveling many hundreds of kilometers (Newton 1972).

Red squirrels (*Tamiasciurus hudsonicus*) also respond to cone masts in white spruce. This species caches unopened cones in food tunnels in the ground and uses them throughout the next winter. Survival of squirrels is high during these mast winters.

An unusual form of variability in food supply occurs in the bamboo species which form the main food of the giant panda (*Ailuropoda melanoleuca*). The bamboo synchronized flowering in much of southern China during the early 1980s (Schaller *et al.* 1985). The plants died after flowering and there was little food available for a few years. With the giant panda now confined to a few protected areas, the population suffered from this sudden drop in food supply. Knowledge of such events is important for conservation. It tells us that reserves must be sufficiently diverse in environment, habitat, and food species to avoid the type of restriction in food supply produced by the synchronous flowering of bamboo. Presumably in prehistoric times giant pandas were able to range over a much wider area and so take refuge in regions where bamboo was not flowering. They cannot now move in this way and most of their former range in the lowlands is no longer available.

In the Canadian boreal forest, lynx and great horned owls breed prolifically during the peak of the 10-year snowshoe hare cycle, and cease breeding during the low phase (Rohner *et al.* 2001).

4.3.3 *Plant secondary compounds*

Many plants produce chemicals which deter herbivores from feeding on them. These chemicals are called **secondary compounds**. Their production is associated with growth stage, but this association differs between plant species. Although secondary compounds are found in some grasses (monocots), most are found in dicots. Tannins are low in young oak leaves but are abundant in mature leaves (Feeny and Bostock 1968). Conversely, various secondary compounds are abundant in juvenile twigs of willows, birches, and white spruce in Alaska and Canada, but sparse in mature twigs 3 years and older (Bryant and Kuropat 1980). Thus the palatability and availability of food for herbivores differs between seasons and between years because of changes in the concentration of secondary compounds.

There are three major classes of secondary compounds: terpenes; soluble phenol compounds; and alkaloids, cardenolides, and other compounds.

Terpenes

These are cyclic compounds of low molecular weight and usually with one to three rings. They inhibit activity of rumen bacteria (Schwartz *et al.* 1980) and are bitter tasting or volatile. Examples are essential oils from citrus fruits, carotene, eucalyptol from eucalyptus, papyriferic acid in paper birch, and camphor from white spruce. Camphor and papyriferic acid act as antifeedants to snowshoe hares (Bryant 1981; Sinclair *et al.* 1988), and α-pinene from ponderosa pine deters tassel-eared squirrels (*Sciurus alberti*) (Farentinos *et al.* 1981).

Soluble phenol compounds

The main groups of chemicals are the hydrolyzable and condensed tannins (McLeod 1974). They act by binding to proteins and thus making them indigestible. The name "tannin" comes from the action of polyphenols on animal skins, turning them into leather that is not subject to attack by other organisms, a process called tanning.

Tannins are widespread amongst plant species, occurring in 87% of evergreen woody plants, 79% of deciduous woody species, 17% of annual herbs, and 14% of perennial herbs. Tannins have negative physiological effects on elk (Mould and Robbins 1982), and may determine food selection by browsing ungulates in southern Africa (Owen-Smith and Cooper 1987; Cooper *et al.* 1988) and by snowshoe hares in North America

(Sinclair and Smith 1984). Domestic goats (*Capra hircus*) learn to avoid young twigs of blackbrush (*Coleogyne ramosissima*) because of condensed tannins (Provenza *et al.* 1990).

Alkaloids, cardenolides, and other compounds

These are cyclic compounds with nitrogen atoms in the ring. They occur in 7% of flowering plants, and some 4000 compounds are known (Robbins 1983). Some alkaloids are nicotine, morphine, and atropine. They have several physiological effects, but act more as toxicants or poisons than as digestion inhibitors. Some alkaloids, such as cardenolides in milkweed (Asclepiadaceae), are sequestered by insects like the monarch butterfly (*Danaus plexippus*) whose larvae feed on milkweed. These noxious cardenolides act as emetics to birds. Young, inexperienced blue jays (*Cyanocitta cristata*) at first eat these insects then regurgitate them. From then onwards they avoid these insects (Brower 1984). Cyanogenic glycosides, which release hydrocyanic acid on hydrolysis in the stomach, are sequestered by *Heliconius* butterflies from their passionflower (*Passiflora* species) food plants. These insects are avoided by lizards, tanagers, and flycatchers (Brower 1984).

4.4 Measurement of food supply

4.4.1 *Direct measures*

The amount of food available to animals may be measured directly. For carnivores, some form of sampling of their food may be used: insect traps for insectivores; counts of ungulates available to large carnivores. For grazing ungulates, McNaughton (1976) clipped grass in exclosure plots to measure the available production for Thomson's gazelle on the Serengeti plains. Winter food supply for snowshoe hares was estimated from the abundance of twigs with a diameter of 5 mm on the two most common food plants, gray willow (*Salix glauca*) and bog birch (*Betula glandulosa*) (Smith *et al.* 1988; Krebs *et al.* 2001a). Pease *et al.* (1979) used a different approach by feeding a known quantity of large branches to hares in pens and measuring the amount eaten from these branches. Using this measure as the edible fraction from large branches, they then estimated the total available biomass of edible twigs from the density of large branches in the habitat of hares.

The most serious problem with direct measures is that they all depend on the assumption that we can measure food in the same way that the animal comes across it. It is rare that this assumption is valid: insects that enter pitfall traps or are collected by sweepnets are not the same fraction as that seen by a shrew or bird; ungulate censuses do not indicate which animals are actually available to carnivores, for we can be sure that not all are catchable.

If the food supply is relatively uncomplicated, such as the short green sward which is grazed uniformly by African plains antelopes, then we can clip grass in a way resembling the feeding of animals. With woody plants we cannot measure food in the same way as an animal feeds. Thus in most cases our estimates are simply crude indices of food abundance. Our errors can both over- and underestimate the true availability of food: we may include material that an animal would not eat, so producing an overestimate; or we may overlook food items because animals are better at searching for their own food than we are, so producing an underestimate. We can never be sure on what side of the true value our index lies unless we calibrate it with another method.

4.4.2 *Fecal protein and diet protein*

A second method, which has been applied so far only to herbivores, allows the animal to choose its own food, and so avoids the problems discussed above. Diet

Fig. 4.1 Seasonal changes in energy and body weight of the jerboa (*Allactaga elater*) in the cis-Caspian, Russia, during 1985. (a) Percentage energy of forage in the stomach. (b) Daily energy intake, and daily energy requirement. (c) Body weight. (After Abaturov and Magomedov 1988.)

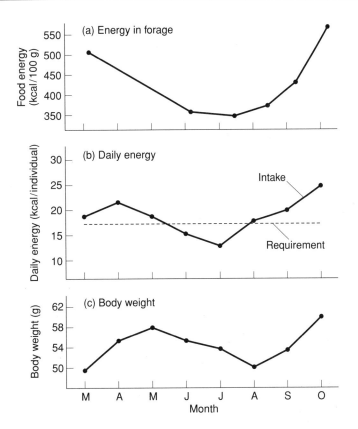

protein, energy, or other nutrients can be estimated by observing what animals eat and then determining the chemical composition of that diet. These indirect estimates of intake are compared with an estimate of requirements either from direct physiological experiments or inferred from the literature. Examples are from reindeer on South Georgia Island (Leader-Williams 1988) and greater kudu (*Tragelaphus strepsiceros*) in South Africa (Owen-Smith and Cooper 1989). Energy intake for the jerboa (*Allactaga elater*) in north cis-Caspian, Russia (Fig. 4.1) dropped below requirements in mid-summer and so body weight declined (Abaturov and Magomedov 1988). Similarly, energy measured from fecal collections showed that energy intake of moose during winter in Norway dropped below requirements by 25–30% (Hjeljord *et al.* 1994). For greater kudu (Fig. 4.2), energy intake during winter was below requirements, but protein intake was sufficient. In contrast, protein intake of African buffalo in tropical dry seasons was below requirements (Fig. 4.3).

These indirect measures of food intake can often be inaccurate because they are an amalgam of several different measurements. One way around this is to use a physiological index from the animal to indicate the quality of the food it has eaten. Nitrogen in the feces predicts nitrogen in the diet down to the minimum level of nitrogen balance. If nitrogen intake falls below this level it is not reflected in the feces because metabolic nitrogen (from microorganisms and gut cells) continues to be passed out irrespective of intake.

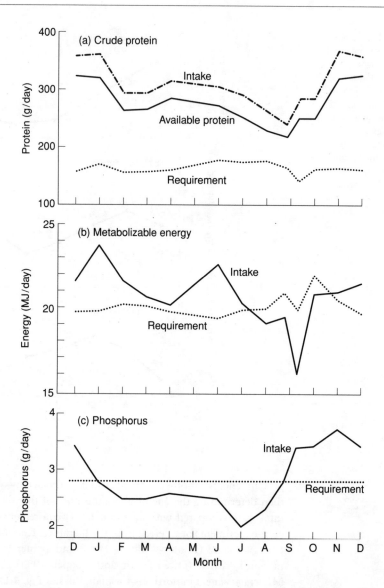

Fig. 4.2 Monthly changes in the estimated daily nutrient intakes of greater kudu relative to estimated maintenance requirements. (a) Crude protein intake (dashed line); available protein (solid line); protein requirement for metabolic turnover, fecal loss, and growth (dotted line). (b) Metabolizable energy intake (solid line); metabolizable energy requirement for resting, activity, and growth (dotted line). (c) Phosphorus intake (solid line); phosphorus requirement (dotted line). (After Owen-Smith and Cooper 1989.)

In tropical regions this relationship has been found for cattle (Bredon *et al.* 1963), buffalo, and wildebeest (Sinclair 1977), and in North America for cattle, bighorn sheep (*Ovis canadensis*), elk, and deer (Fig. 4.4) (Leslie and Starkey 1985; Howery and Pfister 1990). These relationships apply to ruminants eating natural food. Similar relationships have been found for experimental diets in Australian rabbits (Myers and Bults 1977), snowshoe hares (Sinclair *et al.* 1982), elk, and sheep (Mould and Robbins 1981; Leslie and Starkey 1985), although the slopes of the regression lines differ from the natural diets.

A potential problem with this approach is that plant secondary compounds such as tannin may obscure the relationship by causing higher amounts of metabolic nitrogen to be passed out (Robbins *et al.* 1987; Wehausen 1995). This has been observed in experimental diets with high amounts of these compounds (Mould and Robbins

Fig. 4.3 The proportion of crude protein in the diet of African buffalo declines below the estimated 5% minimum requirement in the dry season. Estimates from diet selection with 95% confidence limits (solid line); estimates from fecal protein (broken line). (After Sinclair 1977.)

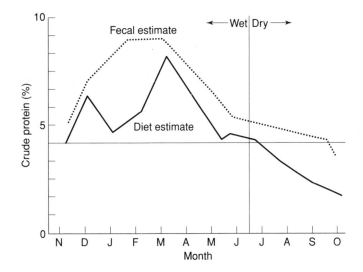

Fig. 4.4 Correlation of dietary nitrogen with fecal nitrogen in (a) elk and (b) black-tailed deer. Nitrogen increases with season. Spring (▲); summer (●); fall (△); winter (○). (After Leslie and Starkey 1985.)

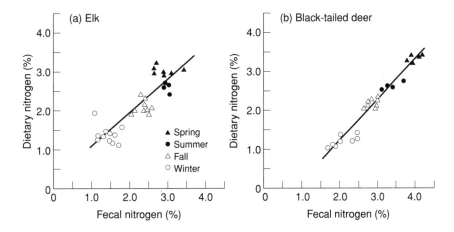

1981; Sinclair *et al.* 1982). However, these are abnormal situations and when animals are allowed to choose their own diet the relationship holds up. The regression has been determined for only a few species on natural diets, so more work is needed in this area. A second potential problem could arise if fecal samples are exposed to the weather and the nitrogen leached out. For white-tailed deer feces in autumn the bias is minimal if samples are collected more than 24 days after defecation (Jenks *et al.* 1990).

The relationship between fecal nitrogen and dietary nitrogen can be used to estimate whether animals are obtaining enough food for maintenance. In African buffalo the estimate of dietary nitrogen was compared with estimates of dietary nitrogen from rumen contents (Fig. 4.3). The two are similar.

A similar approach has related fecal nitrogen directly to weight loss. Thus Gates and Hudson (1981) found that elk lost weight below about 1.6% fecal nitrogen (Fig. 4.5a) during late winter when there was deep snow (Fig. 4.5b).

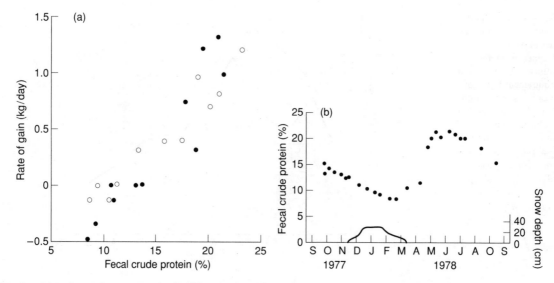

Fig. 4.5 (a) Body weight gain of male elk (●) and calves (○) in Alberta can be predicted from the percentage of fecal crude protein. (b) Seasonal changes in the percentage of fecal crude protein are related to snow depth. (After Gates and Hudson 1981.)

4.5 **Basal metabolic rate and food requirement**

4.5.1 *Energy flow*

The flow of energy through the body is illustrated in Fig. 4.6. Energy starts as consumption energy or intake energy. Part of this is digested in the gut and passes through the gut wall as digestible energy, the rest being passed out in the feces as fecal energy. Part of the digestible energy is lost in the urine, and the remainder, called metabolic or assimilated energy, can then be used for work. The work energy can be divided into two: respiration energy which is used for the basic maintenance of the body (resting energy) and for activity, and production energy for growth and reproduction.

The flow chart for protein is similar except that protein is normally used only for production. Protein is not used in respiration except under special conditions of food shortage when protein is broken down (catabolized) to provide energy.

Metabolic energy (M) can be measured in two ways:

1 in the laboratory by measuring resting energy and activity to obtain the respiration component (R), and from growth and population studies to obtain production (P), so that:

$$M = R + P$$

2 in the field by measuring consumption (C), fecal (F), and urinary (U) outputs, so that:

$$M = C - F - U$$

4.5.2 *Basal metabolic rate*

Basal metabolism is the energy needed for basic body functions. The energy comes from oxidation of fats, proteins, and carbohydrates to produce water and carbon dioxide. Thus maintenance energy can be measured from expired air volume and composition because intake air has a stable composition of 20.94% oxygen, 0.03% carbon dioxide, and 79.03% nitrogen. Since 6 moles of carbon dioxide and water are

Fig. 4.6 Flow chart of energy through the body.

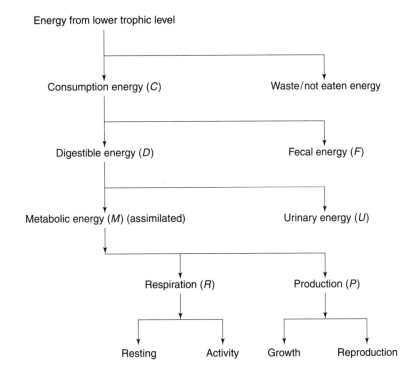

Energy from lower trophic level

Consumption energy (C)   Waste/not eaten energy

Digestible energy (D)   Fecal energy (F)

Metabolic energy (M) (assimilated)   Urinary energy (U)

Respiration (R)   Production (P)

Resting   Activity   Growth   Reproduction

Fig. 4.7 Relationship of basal metabolic rate and body weight in different groups of small mammals. (After Clutton-Brock and Harvey 1983, which is after Mace 1979.)

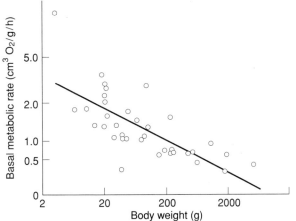

produced with 673 kcal of heat, the carbon dioxide in expired air can be used to calculate the rate of energy used for maintenance. Measurements can be obtained either in chambers or from gas masks, and the animal must be in its thermoneutral zone (not shivering, panting, or sweating), resting, and not digesting food. Such conditions give the basal metabolic rate.

Basal metabolic rates (BMR) of different eutherian mammalian groups, such as those in Fig. 4.7, when plotted against log of body weight, fall on a line whose slope is approximately 0.75. Thus Kleiber (1947) produced the general equation:

$$BMR = 70W^{0.75}$$

where BMR is in kilocalories per day and W is body weight in kilograms. This is an average over all mammals. Specific groups may differ – desert-adapted mammals have lower rates, marine mammals higher rates. Large non-passerine birds are similar to eutherians but the smaller passerines are 30–70% higher. The constant 70 also differs; in marsupials it is 48.6 and in the echidna (a monotreme) it is 19.3 (Robbins 1983). McNab (1988) predicted that animals feeding on lower energy foods should lower their BMR. Experiments with the burrowing rodent (*Octodon degus*) in Chile fed on low or high fiber diets have confirmed this prediction (Veloso and Bozinovic 1993).

Hibernating mammals, such as ground squirrels, can lower their body temperatures to a few degrees above ambient temperature, but no lower than about 0°C. Hummingbirds can lower their body temperature to about 15°C, a process called torpor. Both hibernation and torpor save energy (Kenagy 1989; Kenagy *et al.* 1989).

To this point we have discussed resting or maintenance requirements. Activity adds a further energy cost to maintenance. Standing is on average 9% more costly than lying for mammals, and 13.6% more costly for birds (Robbins 1983). The cost of locomotion is similar for bipedal and quadrupedal animals (Fedak and Seeherman 1979). Cost of locomotion (LC) expressed as kilocalories per kilogram per kilometer, declines linearly with increasing log body size. Thus:

$$LC = 31.10W^{-0.34} \ (W \text{ in grams})$$

Hence, the cost of moving is higher per unit body mass for smaller species and for juveniles.

Average daily metabolic rate (ADMR, the sum of resting and activity rates) is approximately $2 \times BMR$ in captive mammals, but it is difficult to measure for free-living animals. For captive passerines ADMR is $1.31 \times BMR$ and for captive non-passerines it is $1.26 \times BMR$. As a rough approximation, free-living birds and small mammals have a metabolic rate two to four times the BMR.

4.5.3 *Variation in food requirements*

The ADMR or other average measures of metabolic rate hides seasonal fluctuations in food and energy demands. The costs of reproduction add considerably to those for normal daily activity. In the red deer or the wildebeest the rut imposes a considerable energetic cost upon males, which spend several weeks fighting, defending territories, and herding females while eating very little (Sinclair 1977; Clutton-Brock *et al.* 1982). Males put on large amounts of body fat before the rut and use it to cover the extra energy requirements of the rut. Mule deer males (Fig. 4.8a) deposit kidney fat in fall and use it in November during mating (Anderson *et al.* 1972).

Female mammals use additional energy for lactation and for growing a fetus. Like males they accumulate body fat, especially in the mesentery and around the kidneys, before birth and lactation. During the last third of gestation metabolic costs are twice the ADMR and during lactation they are three times the ADMR. In female mule deer (Fig. 4.8b) fat is built up in fall and early winter, and used between late winter and summer during gestation, birth, and lactation. Thus the timing of reproduction in ungulates is influenced in part by the need to obtain good food supplies and to build up fat reserves.

Fig. 4.8 Seasonal changes in the kidney fat index of mule deer are closely associated with reproduction and season. (a) Males; (b) females. (After Anderson *et al.* 1972.)

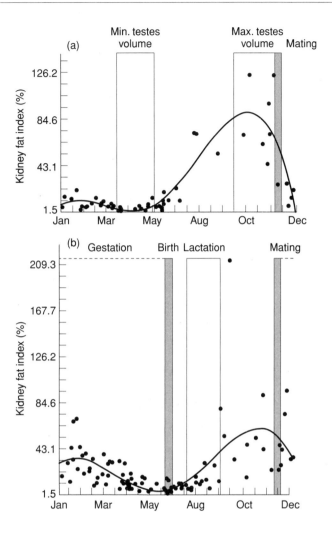

4.6 Morphology of herbivore digestion

4.6.1 *Strategies of digestion*

Carnivores and omnivores digest their food in the stomach and small intestine. The small intestine is relatively short in these species. Herbivores, which make up most (about 90%) of the mammals (Björnhag 1987), need to digest large amounts of fairly indigestible cellulose and hemicellulose, and to do so they have adapted the gut to increase retention time. One strategy is to evolve a much longer small intestine. An exception is the giant panda which evolved from bears and has retained the short intestine. In this species organic matter digestibility is only 18%, one of the lowest recorded (Schaller *et al.* 1985). Another adaptation is to use microorganisms (bacteria, fungi, protozoa) which digest cellulose through fermentation. Plant material must be retained in a fermentation chamber long enough for the microorganisms to cause fermentation. Squirrels eat high energy foods such as seeds, fruits, and insects and so do not need such mechanisms. Some species have unusually low metabolic rates and hence longer retention times. Most are arboreal folivores: koalas (*Phascolarctos cinereus*) (Dawson and Hulbert 1970), sloths, and hyraxes (Rubsamen *et al.* 1979; Björnhag 1987). Reviews of digestive adaptations can be found in Hume and Warner (1980), Hornicke and Björnhag (1980), Robbins (1983), Björnhag (1987), and Chivers and Langer (1994).

4.6.2 *Ruminants*

True ruminants, which include the bovids (cattle, sheep, antelopes), cervids (deer), tylopodids (camels), and giraffes, have an extension of the stomach divided into three chambers. One of these is the rumen, which acts as the fermentation chamber. Plant food is gathered without chewing and stored in the chamber during a feeding period. This is followed by a rumination period during which portions of compacted food (bolus) are returned to the mouth for intensive chewing. In this way coarse plant material is broken down mechanically and made available to the microorganisms for fermentation. The amount of fiber in the food determines how coarse it is, and the coarser the food the longer the process of grinding and fermentation. There is a limit to how coarse the food can be before fermentation takes so long that the animal uses more energy than it gains. On average a ruminant retains food in the gut for about 100 hours.

Microorganisms break down cellulose into short-chain fatty acids, and proteins into amino acids and ammonia, using these to produce more microorganisms. The host animal obtains its nutrients by digesting the dead microorganisms in the stomach and short intestine. The system is efficient, and digestibilities of organic matter and protein of around 65–75% are achieved for medium- to good-quality food (i.e. relatively low in fiber). Another advantage is that nitrogen can be recycled as urea. A disadvantage is that microorganisms digest nutrients that could be used directly by the host, and this leads to a loss of energy through production of methane. Another is that ruminants cannot digest very high fiber diets.

4.6.3 *Hindgut fermenters*

In contrast to the foregut fermenters, or ruminants, a number of animal groups have developed an enlarged colon or cecum or both to allow fermentation. Large animals (> 50 kg) are in general colon fermenters, while small ones (< 5 kg) which feed on fibrous food are cecum fermenters.

Colon fermenters

In most cases both the colon and the cecum are enlarged to hold fiber for microbial digestion. There is little separation of material into small particles and microbes on the one hand and fiber on the other, and there is little evidence that microbial proteins are digested and absorbed, but fatty acids can be absorbed.

Animals in this group are perissodactyls (horses, rhinos, tapirs), macropods (kangaroos), and perhaps elephants, wombats (*Vombatus ursinus*), and dugongs (*Dugong dugon*). These are all large animals and so do not need to ingest high energy and protein per unit of body weight (see Section 4.5.2). Since food material can be retained in the gut for longer periods in large animals, the rate of passage may be slow enough for fermentation and absorption of fatty acids to take place. None of these animals eat their feces, a practice called **coprophagy**.

Cecum fermenters

Small animals (< 5 kg) have a relatively high metabolic rate. Those species which feed on high fiber diets such as grass and leaves need to use the microbial protein produced by hindgut fermentation. They do this by coprophagy. In conjunction with this process there is a sorting mechanism in the colon that separates fluids, small particles of food, and microbes from the fiber. The fluids and microbes are returned by antiperistaltic movements to an enlarged cecum for further fermentation and diges- tion. This mechanism, therefore, retains the nutrients long enough for fermentation. It is necessary because small animals cannot hold food material long enough for fermentation under normal passage rates.

Dead microbial material is passed out in the form of special soft pellets, **ceco- trophs**, and these high nutrient feces are eaten directly from the anus, a behavior called **cecotrophy**. The sorted high fiber is passed out as hard pellets which are not reingested.

Animals that both ferment food in the cecum and practice cecotrophy include myomorph rodents (voles, lemmings, brown rats), lagomorphs (hares, rabbits), some South American rodents (coypu, guinea pig, chinchilla), and some Australian marsupials such as the ringtail possum (*Pseudocheirus peregrinus*) (Chilcott and Hume 1985).

Two marsupials, the koala and greater glider (*Petauroides volans*), feed on arboreal leaves and have cecal fermentation and a colonic sorting mechanism (Cork and Warner 1983; Foley and Hume 1987). Neither practice cecotrophy. At least in the koala, both the metabolic rate and the passage time are slow enough that cecotrophy is not necessary.

Björnhag (1987) identifies four strategies employed by small mammals that feed on plants:

1 They eat only highly nutritious plant parts such as seeds, berries, birds, and young leaves. Squirrels fall in this group.

2 They have a low metabolic rate for their size so that fermentation is prolonged. Koala and tree sloths are examples of this group.

3 Digesta are separated in the colon and easily digestible food particles plus microorganisms are retained to allow fermentation and fibrous material to be sorted and passed out.

4 Only the microorganisms are separated and retained to allow rapid fermentation. Both (3) and (4) recirculate protein-rich fecal material by reingestion through cecotrophy. Examples are voles, lemmings, and lagomorphs. Foley and Cork (1992) review these strategies of digestion and their limits.

4.7 Food passage rate and food requirement

The passage rate of food through an animal depends on the **retention time**, which is the mean time an indigestible marker takes to pass through. Various markers can be used, for example dyes, glass beads, radioisotopes, and polyethylene glycol. Certain rare earth elements (samarium, cerium, lanthanum) bind to plant fiber and provide useful markers to measure passage times of fiber (Robbins 1983).

The rate of food intake by herbivores depends on the nutritive quality of the food. For example, in domestic sheep (Fig. 4.9) and white-tailed and mule deer, intake rate first increases then decreases as the energy quality of the food declines (Sibly 1981; Robbins 1983). This relationship occurs because both energy and protein are inversely related to fiber content.

Estimation of fecal protein can be used as a means of determining whether a population is obtaining enough food because protein intake is related to the amount of protein in the feces (see Fig. 4.4). This method has been used to predict the change in body weight of elk (see Fig. 4.5b) (Gates and Hudson 1981) and to monitor food requirements in snowshoe hares (Sinclair *et al.* 1988).

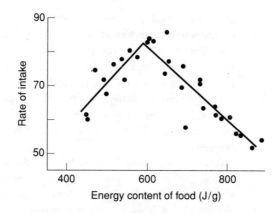

Fig. 4.9 Relationship of intake rate to energy content of food in domestic sheep. Below an energy content of 590 J/g, intake rate falls because of a finite gut capacity and declining fermentation rates. Rate of intake is dry matter/day/body weight$^{0.75}$. (After Sibly 1981, which is after Dinius and Baumgardt 1970.)

4.8 Body size and diet selection

The gut, for example the rumen, large intestine, and cecum, crops of hummingbirds, and cheek pouches of heteromyid rodents, has a capacity which is a linear function of body weight ($W^{1.0}$) (Clutton-Brock and Harvey 1983; Robbins 1983). Energy requirements, however, are a function of metabolic body weight ($W^{0.75}$). Thus, the difference between the exponents ($W^{1.0}/W^{0.75} = W^{0.25}$) means that a larger animal can eat more food relative to requirement than a smaller animal. This can be expressed in two ways: (i) on the same quality of diet a larger animal needs to eat less food per unit of body weight than the smaller; and (ii) a larger animal on a lower nutrient diet can extract the same amount of nutrient per unit of body weight as a smaller animal on a higher nutrient diet. Thus larger animals can eat higher fiber diets, a feature that allows resource partitioning in African ungulates (Bell 1971; Jarman and Sinclair 1979).

Jarman (1974) extended the relationship between body size and diet of African ungulates to explain interspecific patterns of social and antipredator behavior. We can identify five categories, from selective browsers to unselective grazers:

1 Small species (3–20 kg), solitary or in pairs, which are highly selective feeders on flowers, birds, fruits, seed pods, and young shoots. Their habitats are thickets and forest which provide cover from predators. There is little sexual dimorphism and both species help in defending a territory. This group includes duikers (*Cephalophus* species), suni antelope (*Nesotragus moschatus*), steinbuck (*Raphicerus campestris*), dikdik (*Madoqua* species), and klipspringer (*Oreotragus oreotragus*).

2 Small to medium species (20–100 kg) that can be both grazers and browsers, but are very selective of plant parts as in (1). Habitat is riverine forest, thicket, or dense woodland. Group size is larger, from two to six, one male and several females. They are usually territorial and include lesser kudu (*Tragelaphus imberbis*), bushbuck (*T. scriptus*), gerenuk (*Litocranius walleri*), reedbuck (*Redunca* species), and oribi (*Ourebia ourebia*). There is some sexual dimorphism. Predators are avoided by hiding and freezing.

3 Medium size species (50–150 kg) that are mixed feeders, changing from grazing in the rains to browsing in the dry season. Habitats are varied and range from dense woodland and savanna to open flood plains. There is one male per territory. Female group size is variable (6–200) and these groups do not defend a territory but wander through the male territories. Non-territorial male groups are excluded from territories and behave like female groups. Females have a large home range that is smaller in the dry season than in the wet season. Species typical of this group include

impala, greater kudu, sable, kob (*Kobus* species), lechwe, and gazelles (*Gazella* species). They are sexually dimorphic in the extreme. Predators are avoided by group vigilance and by running.

4 Medium to large species (100–250 kg) that are grazers selecting high-quality grass leaves. Males are single and territorial or form large bachelor groups. Female groups range from six to many hundred. They have a large home range often divided into wet and dry season ranges separated by a considerable distance. Habitats are generally open savanna and treeless plains. Predators are avoided by group vigilance and by running. Sexual dimorphism is present but less extreme than in (3). Wildebeest, hartebeest, topi (*Damaliscus korrigum*), and Grevy's zebra are in this group.

5 Large species (> 200 kg) are unselective grazers and browsers of low-quality food. Habitats are closed woodland and open savanna. Movements are seasonal. Males are non-territorial and form a dominance hierarchy. Females form groups of 10 to several hundred and have a large home range. Active group defense against predators is shown by African buffalo and African elephant, while other species use group vigilance and running to avoid predation. Burchell's zebra (*Equus burchelli*), giraffe (*Giraffa camelopardalis*), eland (*Taurotragus oryx*), gemsbok (*Oryx gazella*), and roan antelope (*Hippotragus equinus*) are included in this group.

Jarman's (1974) categories relate body size inversely to food supply because low-quality food is more abundant. In turn this allows species to form larger groups to avoid predators, and the size of group then determines how a male obtains his mate. In small species males keep females in their territories year round and this may be the only way of finding females in estrus. When female groups are larger (group 3), females cannot remain within one territory. Hence males compete for territories within the females' home range to provide an opportunity for mating when females move through the territory. These territories are for mating and not to provide year-round food.

Finally, interspecific competition for male mating territories may have led to larger males with elaborate weapons. Since these selection pressures have not operated on the females, which have remained at a smaller size, sexual dimorphism develops. Thus we see a connection between body size, food quality, group size, predator defense, and mating system.

In other groups of animals gut size can be phenotypically plastic, varying with food availability and season, particularly in birds (Piersma and Lindstrom 1997). Thus, garden warblers (*Sylvia borin*) migrating over the Sahara reduce their gut size and hence food intake (McWilliams *et al.* 1997).

4.9 Indices of body condition

4.9.1 *Body weight and total body fat*

Body weight and fat reserves affect survival and reproduction in mammals (Hanks 1981; Dark *et al.* 1986; Gerhart *et al.* 1996) and birds (Johnson *et al.* 1985). Body weight can be measured directly for small mammals such as the jerboa (Fig. 4.1) and for birds. Weight changes seasonally in response to changes in food supply and hence intake.

Body weight is a function of genetic determinants, age, and the amount of fat and protein stored in the body. To monitor fat and protein changes it is better to take out the effects of body size (the genetic and age components). This can be done by using a ratio of weight to some body measure that is a function of size. Thus, for cottontail rabbits (*Sylvilagus floridanus*), Bailey (1968) found a relationship between predicted body weight (PBW in g) and total length (L in cm) such that:

$$PBW = 16 + 5.48(L^3)$$

The condition index for the rabbits would then be the ratio of observed weight to predicted weight. A similar relationship has been found between foot length and weight for snowshoe hares (O'Donoghue 1991). Murray (2002) found that bone marrow fat of snowshoe hares was predicted by the ratio of body weight to foot length. Kruuk *et al.* (1987) used a general equation for European otters (*Lutra lutra*) where the index of body condition (K) was related to mean body weight (*W* in kg) and body length (*L* in m) by:

$$K = W/(aL^n)$$

with $a = 5.02$ and $n = 2.33$.

At the other extreme of size, blubber volumes of fin whales (*Balaenoptera physalus*) and sei whales (*B. borealis*) have been calculated from body length, girth, and blubber thickness measured at six points along the carcass (Lockyer 1987).

Although body weight alone is a satisfactory measure of condition for such birds as sandhill cranes (*Grus canadensis*) and white-fronted geese (*Anser albifrons*) (Johnson *et al.* 1985), it is usually better to account for body size using some measure such as wing length, tarsus length, bill length, or keel length.

In female mallards (*Anas platyrhynchos*) fat weight (F), an index of condition, is related to body weight (BW) and wing length (WL) by:

$$F = (0.571BW) - (1.695WL) + 59.0$$

and a similar relationship holds for males (Ringelman and Szymczak 1985).

In maned ducks (*Chenonetta jubata*) of Australia, body weight and total fat of females increase before laying. Some 70% of the fat is used during laying and incubation. Protein levels, however, do not change (Briggs 1991). Among northern hemisphere ducks there are four general strategies for storing nutrients before laying:

1 Fat is deposited before migration and is supplemented by local foods on the breeding grounds, as demonstrated by mallard.

2 Reserves are formed entirely before migrating to the breeding area, as in lesser snow geese (*Chen caerulescens*).

3 Reserves are built up entirely on the breeding grounds with no further supplementation, as illustrated by the common eider (*Somateria mollissima*).

4 Body reserves are both formed in the breeding area and supplemented by local food, as seen in the wood duck (*Aix sponsa*) (Owen and Reinecke 1979; Thomas 1988).

Both ducks and game birds can alter the length of their digestive system in response to changes in food supply. Under conditions of more fibrous diets during winter, gut lengths increased in three species of ptarmigan (Moss 1974), gadwall (*Anas strepera*) (Paulus 1982), and mallard (Whyte and Bolen 1985).

In passerine birds energy is stored in various subcutaneous and mesenteric fat deposits, and protein is stored in flight muscles. The latter varies with total body fat, as in the yellow-vented bulbul (*Pycnonotus goiavier*) in Singapore (Ward 1969), and the house sparrow (*Passer domesticus*) in England (Jones 1980). In the gray-backed camaroptera (*Camaroptera brevicaudata*), a tropical African warbler, both total body fat and flight muscle protein vary in relation to laying date (Fig. 4.10) (Fogden and Fogden 1979).

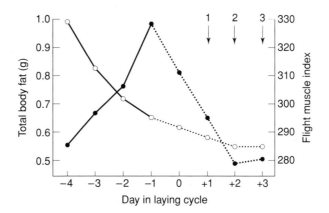

Fig. 4.10 Total body fat (●) and flight muscle index (○) are related to laying day in the female gray-backed camaroptera (a tropical African warbler). Eggs are laid on days 1–3. Flight muscle index is the ratio of (lean dry flight muscle weight)/(flight muscle cord^3). The broken line indicates estimates. (After Fogden and Fogden 1979.)

As in body weight measures, flight muscle weights are corrected for body size by dividing by a standard muscle volume (SMV). Davidson and Evans (1988) used the formula:

$$SMV = H(L \times W + 0.433C^2)$$

for shorebirds of the genus *Calidris*, where H is the height of the keel of the sternum, L is the length of the keel of the sternum, W is the width of the raft of the sternum (one side only), and C is the distance from the keel to the end of the coracoid.

Direct measures of body weight are feasible with birds and small mammals, but impractical for large mammals where some other index of body condition and food reserves must be used. These have been reviewed by Hanks (1981) and Torbit *et al.* (1988). Large mammals store fat subcutaneously, in the gut mesentery, around the kidneys and heart, and in the marrow of long bones. The fat stores are used up in that order (Mech and DelGiudice 1985). Because of this sequential use no single fat deposit is a perfect indicator of total body fat. In caribou, for example, a combination of body mass and a visual index of condition provided the best predictor of fatness (Gerhart *et al.* 1996). Particular fat stores are of interest for specific purposes, such as reproduction (kidney fat) or starvation (bone marrow fat) (Sinclair and Duncan 1972). For these purposes they provide a reasonable guide for managers, total body fat being less useful.

4.9.2 Kidney fat index

Ungulates accumulate fat around the kidney and in other places in the body cavity in anticipation of the demands of reproduction. We saw in Section 4.5.3 how the fat reserves of mule deer change according to the stage of the reproductive cycle (Fig. 4.8), the timing of these changes differing between the sexes.

Although there is little relationship between kidney fat and total body fat in some species (Robbins 1983), others, such as most African ruminants, show a close relationship (Smith 1970; Hanks 1981). For white-tailed deer the percentage of fat in the body is related to the kidney fat index (KFI) by:

Percentage fat = 6.24 + 0.30KFI

(Finger *et al.* 1981). In mule deer both the weight of kidney fat and the KFI are correlated with total body fat (Anderson *et al.* 1969, 1972, 1990; Torbit *et al.* 1988).

Although a similar relationship was found for the brush-tailed possum (*Trichosurus vulpecula*) in South Island, New Zealand (Bamford 1970), a better correlation was found between total body fat and mesenteric fat.

Kidney fat index has been measured in two ways:

1 The kidney is pulled away from the body wall by hand and the surrounding connective tissue with embedded fat tears away along a natural line posterior to the kidney. A cut along the mid-line of the kidney allows the connective tissue to be peeled away cleanly. The KFI is the ratio of connective tissue plus fat weight to kidney weight summed for both kidneys.

2 The connective tissue is cut immediately anterior and posterior to the kidney, so that only the fat immediately surrounding the kidney is used. This has a small advantage of being more objective than (1), but it has the great disadvantage of discarding most of the tissue where fat is deposited, so much of the relevant variability in fat deposition is lost. Hence, the first method may be the more useful index.

4.9.3 *Bone marrow fat*

Bone marrow fat does not decline until after most of the kidney fat has been used (Fig. 4.11) in temperate and tropical ungulates, and in some marsupials (Ransom 1965; Bamford 1970; Hanks 1981). In mule deer, marrow fat changed most rapidly at very low levels of total body fat (Torbit *et al.* 1988). Consequently, a decline in bone marrow fat reflects a relatively severe depletion of energy reserves and therefore provides an index of severe nutritional stress, as was found for starving pronghorn antelope (Depperschmidt *et al.* 1987).

Mobilized marrow fat is replaced by water. Hence the ratio of dry weight to wet weight of marrow is a good measure of fat content. A number of studies on both temperate and tropical ruminants (Hanks 1981) indicate that as a close approximation:

Percentage marrow fat = percentage dry weight − 7

Dry weight of marrow is measured from the middle length of marrow in one of the long bones, avoiding the hemopoitic ends. The method has been used on wildebeest

Fig. 4.11 Kidney fat index (the ratio of kidney-plus-fat weight to kidney weight) is depleted almost entirely before bone marrow fat declines in ruminants. (After Hanks 1981.)

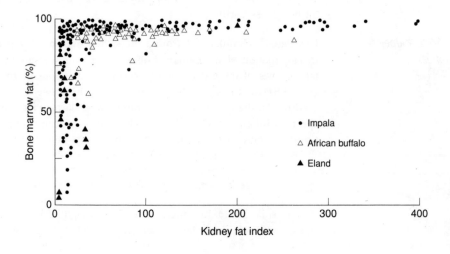

Fig. 4.12 Marrow fat of wildebeest (broken line) dying from natural causes in Serengeti is related to season and the percentage of crude protein in their diet (solid line, with 95% confidence limits). (After Sinclair 1977.)

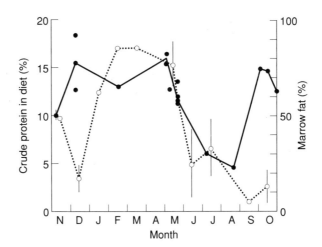

(Fig. 4.12) and deer (Klein and Olson 1960) to establish whether they had died from lack of food.

Broad categories of marrow fat content in ruminants are provided by the color and texture of the marrow (Cheatum 1949). This method is quick (it avoids collection of marrow) and it is sufficient to determine whether an animal has been suffering from undernutrition at death (Verme and Holland 1973; Kirkpatrick 1980; Sinclair and Arcese 1995). The categories with approximate fat values are:

1 Solid, white, and waxy: the marrow can stand on its own and contains 85–98% fat. Such animals are not suffering from undernutrition.

2 White or pink, opaque, gelatinous: the marrow cannot stand on its own, and covers a broad range of fat values (15–85%). It indicates that such animals have depleted fat reserves.

3 Yellow, translucent, gelatinous: the clear, gelatinous appearance is distinctive, and indicates that there is less than 15% fat and often only 1% fat. Such animals are starving.

4.9.4 *Bioelectrical impedance*

Bioelectrical impedance analysis uses an electrical current passed through anesthetized animals in a bioelectrical impedance plethysmograph. Resistance (R) and reactance (X_c) of the current are recorded, and these are related to impedance (Z) by:

$$Z = \sqrt{(R^2 + X_c^2)}$$

In wombats (*Lasiorhinus latifrons*) resistance is a good predictor of total body fat (Woolnough *et al.* 1997). The technique has also been applied to seals (Gales *et al.* 1994) and bears (Farley and Robbins 1994).

4.9.5 *Blood and urine indices of condition*

Blood parameters as indices of condition and food intake are potentially useful for living animals that are too large to be weighed easily. However, blood characteristics are not well known for most species. More work is needed. Different parameters have been examined in different studies. Plasma non-esterified fatty acid, protein-bound iodine, and serum total protein were all related to nutrition in Australian tropical cattle (O'Kelly 1973). Body condition of moose has been related to various sets of blood parameters ranked according to their sensitivity (Franzmann and LeResche 1978). Starting with the single best parameter, sensitivity increased by adding other measures: (i) packed cell volume (PCV); (ii) PCV plus hemoglobin content (Hb); (iii) PCV, Hb, Ca, P, and total blood protein (TP); and (iv) PCV, Hb, Ca, P, TP, glucose, albumin, and β-globulin.

Protein loss from the body was strongly correlated with body weight loss in white-tailed deer on experimentally restricted diets. Serum urea nitrogen and the ratio of urinary urea nitrogen to creatinine were the best blood and urine indicators of under-nutrition and protein loss (DelGiudice and Seal 1988; DelGiudice et al. 1990). For example, serum urea nitrogen is a good indicator of recent protein intake in white-tailed deer (Brown et al. 1995). Similarly, the ratio of urinary urea nitrogen to creatinine provided a reasonable predictor of physiological responses to nutrition in this species (DelGiudice et al. 1996).

4.9.6 *Problems with condition indices*

We have already noted that it is generally impractical to obtain measures of total body fat in large mammals. Various single indices such as kidney fat and bone marrow fat have been used but these are useful for specific purposes and cannot be used over the whole range of total body fat values. Kidney fat is more appropriate for estimating the upper range of body fat values and bone marrow fat represents the lower values. A combination of six indices of body fat deposits in carcasses has been proposed (Kistner et al. 1980). This method is useful for complete carcasses but it cannot be used for animals dying naturally because the soft parts are usually eaten by predators and scavengers, or they decompose. Under these conditions the only index that remains uniformly useful is that of bone marrow fat.

Bone marrow fat as an index is biased towards the low body fat values. It cannot reflect changes in the higher levels of body fat, so that very fatty bone marrow does not necessarily mean the animal is in good condition (Mech and DelGiudice 1985).

Many studies use some form of visual index of condition. However, studies where total body fat has been measured directly find poor correlations with body condition indices (Woolnough et al. 1997).

Although blood indices may be useful as a means of assessing condition and nutrition in living animals, they require careful calibration. Many of the blood characteristics are influenced by season, reproductive state, age, sex, and hormone levels. More importantly, they can be altered rapidly by the stress of capture and handling. All of these could act to obscure and confound changes in nutrition.

All estimates of body condition taken from a sample of the live population are poor indicators of the nutritional state of the population for two reasons. First, such samples are biased towards healthy animals because those in very poor condition are either dead or dying and not available for sampling. Second, the age groups that are most sensitive to density-dependent restriction in food supply – the very young and very old – form a small proportion of the live population. Thus, even a strictly random sample of the population will include a majority of healthy animals and consequently the mean value of condition will be very insensitive to changes in food supply. Therefore, it is unlikely that one can assess whether a population is regulated by food supply or by predators based solely on body condition samples of the live population. To make this assessment one should look at the condition of the animals that have died.

4.10 **Summary**   Nutrition and feeding behavior underlies many critical issues in wildlife ecology and management, such as determining the adequacy of food supplies for endangered species or determining the potential yield in response to harvesting. For carnivores, the nutrient composition of their prey is usually well balanced to a consumer's specific needs, whereas in herbivores the foods eaten may be deficient in key nutrients, such as protein or sodium. Many plant tissues defend themselves against herbivory using poisons, protective structures such as spines, or chemicals that bind to ingested proteins, making them unavailable for digestion. In herbivores, it can also prove difficult to assess food availability in a meaningful way, because the plant tissues eaten represent only a small fraction of the plant biomass present. Various animal-based measures, such as fecal nutrient composition, have been developed to assess food availability and body condition from the herbivore's point of view. Nutritional constraints often vary disproportionately with body size. Many aspects of the behavior and ecology of wildlife species are closely tied to seasonal and spatial variation in food availability, including social organization, spacing patterns, breeding synchrony, and mating system.

5 The ecology of behavior

5.1 Introduction

In this chapter we consider how ecological constraints shape the behavior of individual organisms and, conversely, the effect of individual behavior on the dynamics of populations and communities. This is part of the field known as **behavioral ecology**. We concentrate on foraging and social interactions because these characteristics often have important ecological ramifications that affect wildlife conservation and management. We start with a consideration of the many ways that organisms can choose what and where to eat, then move on to consider how ecological constraints affect social organization.

5.2 Diet selection

The range of mechanisms by which animals choose their diet is diverse. Some animals have a narrow range of preferences, sometimes even for a single species of plant. One of the best examples of this is the giant panda, which has evolved a special set of adaptations allowing it to feed largely on bamboo plants growing on the steep mountainsides of southern China (Schaller *et al.* 1985). Such species are termed **feeding specialists**. Other species tend to the opposite extreme – feeding relatively indiscriminately from a wide range of items. A good example of this would be the moose, which feeds from a wide array of plants, including grasses, woody plants, herbs, forbs, and even aquatic plants (Belovsky 1978). Such species are termed **feeding generalists**. Most wildlife species would fall between these extremes.

One can see an immediate advantage in having a broad diet: there is a much better chance of finding something to eat, no matter where the individual might find itself. There is also a disadvantage, however, to being a generalist: many of the possible items in the environment may be so nutritionally poor that they are scarcely worth pursuing, as we discuss in Chapter 4. For a herbivore, it may be impossible for an individual to survive on poor-quality items, even when supplies are unlimited. For carnivores, variation among prey derives less from differences in nutritional quality than from differences in size, visibility, ease of capture, or the associated risk of injury during capture. In both cases, choosing wisely (becoming a specialist) among a wide range of food items might prove advantageous. Much foraging theory relates to this question: how does an animal choose a diet that yields the highest rate of energy gain over time, energy that can be devoted to enhancing survival and reproduction? This question forms the core of **optimal foraging theory**, a set of mathematical models predicting the patterns of animal behavior that might be favored by natural selection (Stephens and Krebs 1986).

5.2.1 *Optimal diet selection: the contingency model*

The simplest way to consider optimal diet is to start with the functional response: the rate of consumption in relation to food availability (we discuss this further in Chapters 10 and 12). Most organisms have a decelerating functional response to

Fig. 5.1 Comparison of the expected rate of energy gain by a forager specializing on the most profitable prey (prey 1, solid curve) relative to the profitability of alternative prey 2 (broken line). An optimal forager would expand its diet whenever the abundance of prey 1 drops below the density at which these two curves intersect (slightly more than 20 per unit area in this hypothetical example).

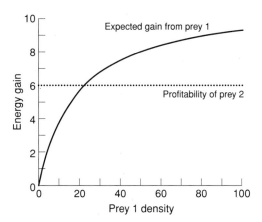

increasing food availability, often termed a **Type II** response (Holling 1959). The rate of energy gain $f(N_1)$ that an animal would experience as a consequence of the Type II functional response can be calculated as follows:

$$f(N_1) = \frac{e_1 a N_1}{1 + a h_1 N_1}$$

where e_1 is the energy content of each item of the more profitable prey type 1, a is the area searched per unit time by the consumer, h_1 is the time required to consume each item of prey type 1, and N_1 is the population density of prey type 1. The energy gain function $f(N_1)$ grows with increasing abundance of prey type 1, but there are diminishing returns to this relationship (Fig. 5.1). Indeed, there is an upper limit e_1/h_1 to the rate of energy gain, even when food is superabundant. This upper limit is set by the limited capacity of the animal to handle prey.

If a forager specialized by feeding only on prey type 1, it would realize a rate of energy return equivalent to $f(N_1)$. How would this compare with the energy gain if the forager generalized, by feeding on both prey types 1 and 2? If both prey types are mixed indiscriminately over the landscape traveled by our hypothetical forager, then the energy gain by a generalist would be calculated as follows:

$$g(N_1, N_2) = \frac{e_1 a N_1 + e_2 a N_2}{1 + a h_1 N_1 + a h_2 N_2}$$

This equation raises the following question: when does it pay to be a specialist and when to be a generalist? The answer is to specialize when $f(N_1) > g(N_1, N_2)$, but act like a generalist when $f(N_1) < g(N_1, N_2)$. Both strategies are equivalent when $f(N_1) = g(N_1, N_2)$. This special case occurs when $f(N_1) = e_2/h_2$. So, a forager that changed from being a specialist to a generalist whenever $f(N_1)$ fell below e_2/h_2 would do better than one that acted all the time as a specialist or as a generalist (Fig. 5.1). Such a foraging strategy is termed the "optimal" strategy, meaning that it yields the highest energetic returns over time. It is straightforward to extend this logic to any number of resource types. First, we rank prey in terms of profitability. Then we add prey to the diet so long as their profitability exceeds the expected rate of long-term intake obtained by specializing on all of the more profitable prey types.

Assuming that maximal acquisition of energy can improve the fitness of a forager (improve reproduction or survival), we might expect something like the optimal strategy to be favored by natural selection. Such selection does not necessarily mean that we would expect every animal to act like a little computer, perfectly assessing the implications of each behavioral decision it might make. Rather, a pattern of behavior that approximates the optimal strategy should be more successful at producing offspring than alternative patterns of behavior.

The optimal diet model makes a number of predictions that are testable by observation or, better still, experimental manipulation:

1 Foragers should rank food types in terms of their energetic profitability (energy content divided by handling time).

2 Foragers should always include the most profitable prey, then expand their diet to include less profitable prey when the expected rate of gain by specializing on more profitable prey matches the profitability of poorer prey.

3 The decision to specialize or generalize should depend on the abundance of highly profitable prey, but not on the abundance of less profitable prey.

4 An optimal forager should have an all-or-nothing response. By this we mean that the perfect forager would either always accept alternative prey or never accept them, depending on whether $f(N_1) > e_2/h_2$.

The optimal diet model has been tested in a variety of settings since it was first proposed (MacArthur and Pianka 1966; Schoener 1971; Pulliam 1974; Charnov 1976a). Despite its simplicity, the optimal diet model has proved remarkably successful in predicting foraging behavior (Stephens and Krebs 1986). Sih and Christensen (2001) reviewed the outcome of over 130 diet choice studies. They found that two out of three of the species studied, ranging from invertebrate herbivores to mammalian carnivores, showed foraging patterns qualitatively consistent with the optimal diet choice model. Optimal diet models tend to perform particularly well in situations where all of the relevant parameters have been accurately measured, enabling precise quantitative predictions.

A classic example is captive great tits (*Parus major*), trained to pick mealworms off a conveyer belt as it passes in front of them (Krebs *et al.* 1977). The birds became adept at choosing whether to specialize on one prey or to accept both prey indiscriminately, in accordance with predictions (1), (2), and (3) of the model (Fig. 5.2). However, the birds never mastered the all-or-nothing behavior that would be perfectly optimal. Instead, the foragers sometimes ate both prey types, and sometimes only the more profitable prey, a pattern termed **partial preference**. Such partial preferences are almost always observed, even in the most successful experiments, perhaps because foragers cannot discriminate perfectly amongst prey, or because foragers need continually to "test" alternative prey to assess their relative profitability.

Initial doubts that optimal diet choice theory could be applied to complex field situations with many prey species (Schluter 1981; Pierce and Ollason 1987) have been allayed by successful field studies (Sih and Christensen 2001), but the theory does require much effort in estimating many parameters. The optimal diet model has been less successful with predators which utilize mobile prey (e.g. weasels feeding on rodents) compared with those which utilize stationary prey (e.g. starfish feeding on mussels). There are a variety of reasons for this difference (Sih and Christensen 2001). In nature,

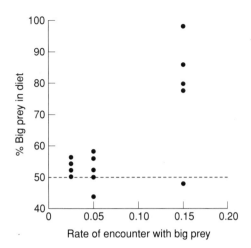

Fig. 5.2 Krebs *et al.*'s (1977) laboratory experiment on diet selection by *Parus major*. Two sizes of mealworms were delivered via conveyor belt in random order to the forager. The horizontal axis shows the rate of encounter with big prey. As the rate of encounter with big prey declined, the birds expanded their diet to include small prey.

food items are rarely mixed homogeneously across the landscape visited by the forager. This is particularly true of mobile prey which may be attempting to avoid predators. If alternative prey species tend to occur in different microhabitats, differ in their activity patterns, or have a different capacity for avoidance of predators, then simple frequencies of abundance may be a poor predictor of diet composition. In these cases, more complex models may be needed to predict optimal diet patterns.

In some cases, particularly herbivores, foragers need to maintain a balanced intake of particular nutrients, rather than simply maximizing energy gain in whatever way possible (see Chapter 4). For example, howler monkeys tend to choose a diet more heavily laden with leaves than should be optimal, perhaps because they must balance nutrients (Milton 1979). Similarly, the marine gastropod *Dolabella auricularia* grows several times faster on a mix of algal species than when fed on a single species of algae (Penning *et al.* 1993).

The optimal diet model represents a simplistic view of foraging. Most species have biological features that introduce additional elements into the decision-making process. For example, many species forage outwards from a central place, whether that place is a den, perch, or resting site. If the forager sallies forth, retrieves one or more prey items and then returns to the central place before feeding on the items, then this additional travel time and energetic expenditure must be accommodated to make useful predictions (Orians and Pearson 1979). Thus, beavers forage on a variety of woody plants on the shore surrounding the ponds or streams where they build their lodges. Several studies have shown that beavers feed more selectively the further out food items occur from the lodge, as predicted by central place foraging models (Jenkins 1980; McGinley and Witham 1985; Fryxell and Doucet 1993). The greater energetic cost of travel to more distant food requires that animals be more selective. As a consequence, the handling time of potential food items is dependent on distance, implying different patterns of diet selectivity.

5.2.2 *Optimal diet selection: effects on feeding rates*

Optimal patterns of foraging can have important effects on the rate of attack by consumers. The rate of consumption of prey type 1, $f(N_1)$, by an optimal forager is predicted by the following multispecies functional response:

$$f(N_1) = \frac{aN_1}{1 + ah_1N_1 + \beta(N_1)ah_2N_2}$$

where $\beta(N_1)$ depicts the probability of foraging on the poorer prey, which is a function of the density of preferred prey, calculated according to the optimal diet choice rule (Section 5.2.1).

This equation predicts that there would be a sharp drop in consumption of the more profitable prey at the point at which the forager expands its diet to include less profitable prey (Fig. 5.3). This drop implies that mortality risk for the more profitable species would decline accordingly. This process has been observed with beavers in large enclosures with either a single type of prey or a mix of prey species (Fryxell and Doucet 1993). When presented with a single species of prey, beavers had a Type II functional response, smoothly climbing with prey density, but at a decelerating rate (Fig. 5.4). When presented with a mix of prey, however, beavers expanded their diet as preferred prey declined in abundance, in accordance with an optimal diet model (Fryxell 1999). This led to a pronounced decline in predation risk to preferred prey as the diet expanded (Fig. 5.4).

Fig. 5.3 Predicted effect of optimal diet choice on the rate of intake of the preferred prey.

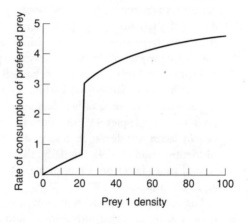

Fig. 5.4 Functional response of beavers presented with a preferred species of prey (trembling aspen saplings) or a mix of prey species (trembling aspen, red maple, and speckled alder saplings) in two separate trials. The rate of consumption of preferred prey (aspens) dropped precipitously as the animals expanded their diet to include less profitable prey (maples and alders) in the multispecies trial, causing the pronounced deviation between the two curves at low densities of aspen saplings. (After Fryxell 1999.)

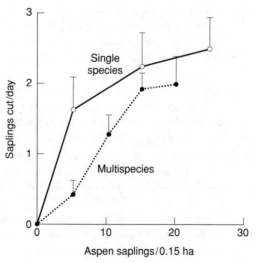

As a natural consequence of these multispecies effects on feeding rates, optimal patterns of diet choice can have important implications for predator–prey interactions. An adaptive predator would shift its attention away from some species as they become more and more rare. Such behavior can have a stabilizing influence on predator–prey dynamics, reducing the degree of variability over time of cyclical predator–prey systems (Gleeson and Wilson 1986; Fryxell and Lundberg 1994; Krivan 1996). This is especially likely when the growth rate of the predator is poor on alternative prey, when the forager exhibits partial preferences, or when alternative prey do not have overlapping spatial distributions (Fryxell and Lundberg 1997).

5.2.3 *Optimal diet selection: linear programming*

An alternative way to model diet choice employs a technique called **linear programming** to identify the optimal solution to a requirement influenced by several constraints (Belovsky 1978; Belovsky *et al.* 1989). When applied to optimal foraging, this allows researchers the means to explore more subtle hypotheses. Linear programming can be used to predict the optimal diet for a forager which is trying to maximize its intake of energy, while at the same ensuring that it obtains sufficient intake of a scarce nutrient to meet its metabolic requirements. When conducted for pairwise combinations of alternative foods, linear programming can be readily understood from simple graphs (Fig. 5.5).

Belovsky (1978) used linear programming to predict the optimal choice of aquatic versus terrestrial plants by moose, based on parameters for moose living on Isle Royale, a small island in Lake Superior. One constraint is that moose must obtain a daily intake of 2.57 g of sodium in order to meet their metabolic requirements. Terrestrial plants in this system are deficient in sodium, whereas aquatic plants have much higher concentrations. Like many other herbivores, moose have limits on the amount of food material that can be processed each day through the digestive tract. The total daily consumption of aquatic and terrestrial plants eaten by a moose cannot exceed this processing rate, which we call the **digestive constraint**. Moose also have limits on the number of hours that can be devoted to cropping food items. Finally, each food type has different profitability (ratio of digestible energy content to cropping time). Thus, time spent cropping energetically poor food items (such as aquatic plants) reduces the opportunity to look for energetically richer items (terrestrial plants). In other words, a moose might waste valuable time eating poor food that could be spent looking for better food.

Fig. 5.5 Linear programming model of diet selection in moose, based on Belovsky (1978). The lines represent constraints that must be met. The range of diets that fall within these constraints is enclosed in the triangle in the middle of the diagram.

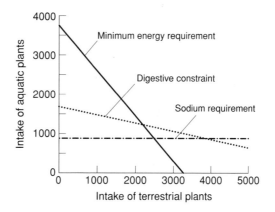

All of these constraints vary linearly with the proportion of each food type in the diet (Fig. 5.5). The optimal solution will occur at one of the intersections of the linear constraint lines. By multiplying the energy content of each prey type by the daily intake of that item at each of the intersection points, we can assess which intersection point offers the greatest energy returns, while guaranteeing that moose maintain a minimum acceptable level of sodium intake. The optimal solution in this case is to have a diet dominated by terrestrial plants, with a small fraction of aquatic plants.

Linear programming has been successfully applied to predict simple dietary preferences (e.g. forbs versus grasses) in a wide variety of species (Belovsky 1986). It has proven less successful at predicting the actual mix of species in herbivore diets. Like the contingency model, linear programming models are ultimately limited by the reliability of parameter estimates and the degree to which proper constraints have been identified. Nonetheless, it remains a very useful means of incorporating multiple constraints into dietary predictions.

5.3 Optimal patch or habitat use

Many resources naturally have patchy patterns of spatial distribution. This presents a number of problems for foragers, such as how to decide which patches or habitats to exploit, how long to stay in each patch once chosen, and how to adjust habitat preferences in light of choices made by other foragers. Optimality principles can be usefully applied to each of these problems.

5.3.1 Optimal patch residence time

We start by considering how long an animal should stay in a given patch. Let us take, for example, fig trees that are widely spaced throughout tropical rainforest. A toucan that wishes to eat figs is faced with deciding how long to feed at a particular fig tree before moving on to look for another. We have already seen that foragers must spend valuable time and energy looking for each food item that they might exploit. As a consequence, there are diminishing returns the longer the toucan stays at the tree because most foragers have a functional response that declines as resource density declines. After an initial period of rapid energy gain, the rate of accumulation of further energy by the animal begins to slow down as resource density drops lower and lower due to the animal's feeding (Fig. 5.6).

We can denote the cumulative energy gain by the function $G(t)$, meaning that cumulative gain depends on the time t spent in each patch. For simplicity, we assume that each patch is identical with respect to initial resource abundance and that the

Fig. 5.6 The cumulative gain of energy from a patch (solid curve) as a function of the time spent in the patch plus the average time spent traveling between patches. The broken line indicates the tangent to this gain curve that passes through the origin (0,0 on both axes). The slope of this tangent represents the long-term rate of energy gain relative to both travel and patch residence time. The point of intersection identifies the optimal patch residence time.

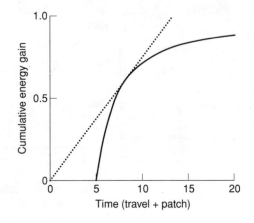

forager has no means of knowing exactly how long it will take to get to the next suitable patch, only how long it takes on average, based on its previous experience. The long-term rate of energy gain, $E(t)$, can be expressed as the total energy gained, $G(t)$, divided by the time spent within each patch (t) plus the average time it takes the forager to find a new patch ($1/\lambda$):

$$E(t) = \frac{G(t)}{\left(\dfrac{1}{\lambda}\right) + t} = \frac{\lambda G(t)}{1 + \lambda t}$$

Long-term intake is usually maximized at an intermediate amount of time spent within each patch. The optimal residence time can be found graphically by drawing the tangent to the gain curve that passes through the origin (Fig. 5.6). This tangent is known as the "marginal value" in economic jargon, so the optimal patch use model has come to be known as the **marginal value theorem** (Charnov 1976b).

The marginal value theorem makes a number of useful predictions:
1 Foragers should leave all patches when the rate of intake in those patches reaches a threshold value. This will typically occur at a particular density of prey.
2 Foragers should leave resource-poor patches much sooner than resource-rich patches.
3 The average distance among patches should influence the optimal time to leave a patch, the **giving-up time** (and by analogy the optimal **giving-up density** of prey). The optimal decision would be to stay in each patch longer when the distance among patches is long than when the distance is short.

Several studies have tested these predictions. Out of 45 published studies, 70% showed patterns of patch departure consistent with these predictions. In 25% of these studies, precise numerical predictions were upheld (Stephens and Krebs 1986). One of the most elegant examples is Cowie's (1977) study of patch use by great tits. Cowie built a series of perches in an aviary on which small containers with tight-fitting covers were attached. Several mealworms were placed in each container, and covered with sawdust. Birds learned to prise the lid off each container before searching for mealworms within it, the container being the "patch." By changing the tightness of lids, Cowie could control the time between cessation of foraging in one patch and the initiation of a bout of foraging in a new patch. He showed that birds were sensitive to travel time between patches, staying longer at patches when travel time was long than when it was short. Changes in departure time were well predicted by the marginal value theorem (Fig. 5.7).

5.3.2 *Patch use by herbivores*

For most large herbivores, food is continuously distributed across the landscape, rather than in definable patches. Nonetheless, local abundance of food still varies considerably from place to place. A slight modification to the marginal value theorem can readily accommodate this situation (Arditi and Dacorogna 1988). This model predicts that animals should feed whenever the cropping rate exceeds the average rate of cropping. A small herd of fallow deer (*Dama dama*) confined to a small pasture, grazed according to the marginal value rule, concentrating their feeding in sites where food abundance was higher than average (Focardi *et al.* 1996). However, a second deer herd which roamed over a much larger area showed little evidence of being sensitive to the marginal value of grazing. The marginal value rule seemed most applicable to

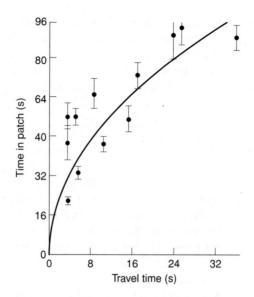

Fig. 5.7 Observed patch residence times by great tits foraging in an experimental aviary (shown by filled circles) versus the patch residence times predicted by the marginal value theorem (solid curve). (After Cowie 1977.)

the case where the deer had a much greater opportunity to develop detailed knowledge of the landscape. Similar patterns have been recorded in cattle (Laca *et al.* 1993; Distel *et al.* 1995) and dorcas gazelles (*Gazella dorcas*) (Ward and Salz 1994).

Large herbivores, particularly grazers, might also have good reason to avoid using patches of dense vegetation. The reason is that taller plants often have more cellulose and lignin than shorter plants to provide support for their height and weight. Consequently, a herbivore which grazed tall plants would obtain less nutritious and less digestible food than one which concentrated on younger growth forms. However, at very low plant sizes, the rate of cropping is very low, and this can also compromise rates of food intake. As a result, grazers should benefit best by feeding on intermediate height and biomass of grasses (Fig. 5.8). Several experimental studies have shown that large herbivores show grazing preference for swards of intermediate grass height and biomass, including cattle (de Vries *et al.* 1999), elk (Wilmshurst *et al.* 1995), bison (Bergman *et al.* 2001), red deer (Langvatn and Hanley 1993), and Thomson's gazelles (Fryxell *et al.* 2004). On the other hand, reindeer on

Fig. 5.8 Relative daily energy gain (○) and observed population densities of Thomson's gazelles (●) in relation to the biomass of grass in particular patches in Serengeti National Park. (After Fryxell *et al.* 2004.)

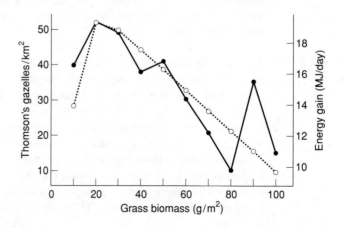

the arctic island of Svalbard prefer patches of tall vegetation, even though it is nutritionally inferior (Van der Wal *et al.* 2000), for reasons that are, as yet, unclear.

The marginal value theorem predicts that foragers should depart when the rate of food intake in a patch (i.e. the functional response) equals the average rate of food available elsewhere in the environment multiplied by a constant (proportional to the travel time between patches). This implies that foragers should concentrate in areas with above average prey abundance, ignoring areas with lower levels of prey abundance. Incorporation of this behavior into models of predators and prey in patches has a stabilizing influence on metapopulation dynamics (Chapter 7). Such behavior reduces the degree of variability in abundance over time of both predators and prey when averaged over all patches (Fryxell and Lundberg 1993; Krivan 1997). Average abundance in a collection of patches tends to be stable when abundance in a single patch at any given time tends to be independent of that in other patches (de Roos *et al.* 1991; McCauley *et al.* 1993). This is more likely when predators abandon patches with low prey abundance than when movement in and out of patches is unrelated to resource abundance.

5.4 Risk-sensitive habitat use

Many foragers are themselves at risk of being attacked by predators. Frequently such risk is highest when the forager is actively searching for food, rather than safely hidden away in a den or resting site. Incorporating predation risk is rarely straightforward in analyzing habitat use, yet we know from numerous empirical studies that it is important (Lima and Dill 1990). For example, risk-sensitive habitat use by the larvae of the aquatic insect *Notonecta* has been elegantly demonstrated in the laboratory (Sih 1980). Large *Notonecta* individuals often cannibalize smaller *Notonecta* individuals. Sih set up an experimental arena where individual *Notonecta* larvae could choose to feed in food-rich or food-poor patches. The larger *Notonecta* individuals selected food-rich patches, whereas smaller, more vulnerable, individuals foraged in the poor patches. This seems to be a logical way to reduce the risk of predation, at the cost of reduced food intake.

One of the most elegant examples of the complex effects of risk-sensitive foraging is the series of experiments conducted by Schmitz and co-workers in small caged populations of carnivorous spiders, herbivorous grasshoppers, and grasses and herbs (Schmitz *et al.* 1997). The grasshoppers suffer high rates of mortality from spiders under normal conditions. As a consequence, they tend to spend their time foraging on herbs, which are less nutritious than grasses, but offer better cover from predators. At the spatial scale of a grasshopper, a single plant is a patch, so dietary preferences are in fact habitat preferences. By gluing shut the mouthparts of spiders, researchers were able to assess the demographic impact of perceived risk of predation versus real predation. Results showed that grasshoppers subject to perceived risk of predation (but not actual predation) suffered mortality levels similar to those of grasshoppers subject to real predation. Both treatment groups suffered considerably higher mortality than grasshoppers in cages without predators, which quickly learned to forage on the more nutritious grasses rather than the safer, but less nutritious, herbs.

Bluegill sunfish (*Lepomis macrochirus*) have been shown to balance the risk of predation against foraging benefits in choosing habitats. Nearshore habitats offer dense protective cover, but relatively poor feeding. Open water offers better feeding, but more exposure. When predators are present, young bluegills tended to concentrate in the habitat offering the greatest cover, whereas older, invulnerable fish foraged in the open, where energy gain was highest (Werner *et al.* 1983).

Sensitivity to predation risk also underlies patterns of habitat use by many large herbivores. For example, white-tailed deer in the boreal forests of Wisconsin and Minnesota tend to concentrate in the no-man's land between wolf pack territories (Hoskinson and Mech 1976). Wolves tend to avoid going out of the area defended by their pack, because of a pronounced risk of being attacked by hostile neighbors (Lewis and Murray 1993; Mech 1994). This effectively creates refuges in between territories in which individual deer are relatively safe.

One of the major difficulties in testing for risk-sensitive habitat use is finding a sensitive way of measuring risk, ideally from the animal's point of view. Brown (1988) suggested that the giving-up density at feeding trays could be used as a field measure of habitat attractiveness, which should be sensitive to both predation risk and alternative foraging opportunities in the surrounding habitat. This technique has been successfully applied in a large number of field studies. For example, different species of granivorous rodents in the Negev desert have different assessments of risk in the same habitat, demonstrating interspecific differences in their perception of the risk of predation versus energetic gain (Brown *et al.* 1994).

A useful way to evaluate such decision-making, which balances trade-offs among competing risks and benefits to fitness, is known as **dynamic state variable modeling** (Mangel and Clark 1986; Clark and Mangel 2000). Although this approach is beyond what we can cover here, it offers a powerful means of evaluating the consequences of alternative behavioral activities that have complex trade-offs among energy gain, reproduction, and mortality risk. Indeed, it may be the only way to link complex sets of behaviors into a life-history framework. The monograph by Clark and Mangel (2000) offers an introduction to the techniques of dynamic state variable modeling, as well as describing a wide set of applications.

5.5 Quantifying habitat preference using resource selection functions

By now it should be apparent that there are good reasons for wildlife species to choose habitats carefully, to enhance the opportunities for feeding, while reducing the risk of being eaten. Moreover, most species have a suite of other needs to meet, including obtaining shelter from inclement weather, gaining access to water, or locating suitable breeding sites, such as cavities in dead trees or burrows. Quantification of specific habitat needs is known as **habitat assessment**, and this is an important area of wildlife ecology. Much of this interest derives from practical benefits: knowing precisely which wildlife habitats are essential allows appropriate management decisions regarding alternative forms of land use. Moreover, good understanding of habitat requirements can improve the odds of success when wildlife species are reintroduced to areas from which they were extirpated.

There are many ways to quantify wildlife habitat use. We shall focus on a recent approach, the **resource selection function** (Manly *et al.* 1993; Boyce and McDonald 1999). Resource selection functions offer a flexible means of quantifying the degree of habitat preference. Complex combinations of categorical and continuous variables can be readily accommodated using this method. Moreover, the method can use a Geographic Information System (GIS) to locate, manipulate, and analyze habitat data of interest.

GIS is a means of linking complex geographical information on physical structure, topographic relief, biological features, and human-made landscape elements into computerized databases. One important feature of GIS is rapid and simple construction of tailor-made "maps" that are readily accessible from a computer screen. This allows users to rapidly sift through complex spatial information in a visual context. Just as important, GIS allows the user to identify and measure spatial interrelationships among variables that would be exceedingly difficult to perform in the field. For example, one can rapidly calculate the size of forest stands of similar species composition, measure the distance of each of these stands from the nearest road, and calculate what fraction of the stands fall within the home range of a wildlife species of interest. From the point of view of assessing habitat selection, GIS also offers a convenient means of random sampling of geographic features across complex landscapes. GIS is clearly a technological breakthrough in the analysis of wildlife habitat needs that is transforming the way we think about conservation and management issues.

The logical basis of virtually all measures of selective use is comparison between the frequency of use of a particular resource (habitat) and its availability in the environment. We surmise that a resource (habitat) is preferred when its use by animals exceeds its availability and conversely that a resource (habitat) is avoided when its use is less than that expected from its availability in the environment. Note our purposeful intermingling of "resource" and "habitat." That is because we can use a similar analysis for determining whether animals preferentially choose diets as for determining whether animals show preferential habitat selection.

Perhaps the easiest way to understand the resource selection procedure is to walk through an example. The rufous bristlebird is a threatened passerine species living in coastal areas of Australia. Gibson *et al.* (2004) used GIS to evaluate critical habitat needs for bristlebirds in a site with competing land use interests (biodiversity values versus mining). Along a series of trails bisecting the study area, Gibson *et al.* recorded the presence (scored with a 1) or absence (0) of bristlebirds, noting the exact geographic coordinates of each positive identification made. They later transferred these sightings to a GIS, overlaying digitized topographic data on aspect, slope, and elevation as well as spatially explicit data on hydrology and vegetation complexity derived from multispectral remote sensing imagery. The probability that a habitat is used, $w(x)$, is given by the following logistic regression model:

$$w(x) = \frac{\exp(\beta_0 + \beta_1 X_1 + \ldots \beta_k X_k)}{1 + \exp(\beta_0 + \beta_1 X_1 + \ldots \beta_k X_k)}$$

where the logistic regression coefficients β_1 to β_k measure the strength of selection for the k habitat variables replicated over the full set of sample units. The function $w(x)$ is bounded between 0 and 1 and represents a probability of usage, given the set of habitat characteristics within a spatial unit. Given the descriptive nature of both the data on bristlebird presence or absence as well as habitat variables derived from the GIS, Gibson *et al.* elected to use model evaluation (Chapter 15) (Burnham and Anderson 1998) rather than classical hypothesis testing (Chapter 16). They found that there was a positive association between bristlebird presence and vegetation vertical complexity, but negative associations between bristlebird presence and "elevation," "distance to creek," "distance to the coast," and "sun incidence." This suggests that bristlebirds require densely vegetated stands in close proximity to coastal fringes and drainage lines. Such habitats composed approximately 16% of the study area, demonstrating how resource selection can help in the assessment of land use priorities for wildlife conservation in a planning context. There are many variations on this basic statistical design. For details, consult the comprehensive treatise by Manly *et al.* (1993).

Resource selection can be used to evaluate the potential success for reintroduction programs (Boyce and McDonald 1999). Mladenoff and co-workers (Mladenoff *et al.* 1995; Mladenoff and Sickley 1998) have used this approach to predict the potential for successful reintroduction of gray wolves to different parts of the USA. Data for existing wolf populations were first used to determine the suite of critical habitat variables for wolves and to relate local wolf densities to habitat features. GIS data were then fed into the resource selection models to predict the potential of different areas to support gray wolves. The model has been validated against data for an expanding wolf population in Wisconsin, demonstrating that this approach can be a useful planning tool.

Resource selection functions are also a powerful means of linking habitat characteristics with spatially realistic models of population viability. For example, Akçakaya and Atwood (1997) used logistic regression to develop a habitat suitability model for the threatened California gnatcatcher (*Polioptila c. californica*) in the highly urbanized environment of Orange County, California. Gnatcatcher distribution data were mapped onto a GIS map. Numerous geographical habitat features were then evaluated, and a resource selection probability function developed on the basis of the strongest suite of variables. Suitable habitat fragments were mapped onto the Orange County landscape and this spatial configuration was then modeled as a metapopulation to evaluate the long-term viability of gnatcatchers (see Chapters 7 and 17). This is a valuable way to evaluate the conservation needs of threatened populations. It is particularly appropriate for species utilizing fragmented landscapes, because it gives useful insights into the ecological implications of alternative land use policies and planning scenarios.

5.6 Social behavior and foraging

5.6.1 Density-dependent habitat selection

Given that there are differences in the intrinsic suitability of habitats, due to variation in resources, cover, and risk from predators, one might expect animals to concentrate in the most favorable habitats. The attractiveness of particular habitats is likely to depend, however, on the density of foragers already present. Birth rates tend to fall and mortality rates to climb as forager density increases (see Chapter 8). As a consequence, habitat suitability is often negatively associated with density. Density-dependent decline in habitat suitability could arise from a variety of causes, including resource depletion, direct interference among individuals, disease transmission, or elevated risk of predation on the foragers.

Density-dependent decline in habitat suitability can be extended to multiple habitats. Individuals should concentrate in the best habitat until the density in that habitat reduces its suitability to that of the next best alternative (Fig. 5.9). Thereafter, both habitats should receive equal use. The resulting pattern of distribution among alternative habitats is known as the **ideal free distribution** (Fretwell and Lucas 1970). It is free in the sense that every individual is presumed equal and capable of choosing the best option available. It is ideal in the sense that all individuals are presumed to have perfect knowledge about the relative suitability of each habitat on offer. Hence, it would not pay for any individual to deviate from the ideal pattern of distribution, because their fitness would be compromised. This is a prime example of an **evolutionarily stable strategy** (Maynard-Smith 1982). Once adopted by all the

Fig. 5.9 Schematic diagram of the ideal free distribution. As density in the preferred habitat 1 increases, suitability declines to a point indicated by the light broken line where it equals that in the poorer habitat 2 (60 units). At this point it pays some individuals to use habitat 2.

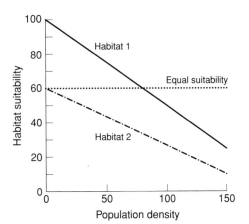

Fig. 5.10 Use of preferred habitats by three different bird species declines as population size increases. (After Sutherland 1996.)

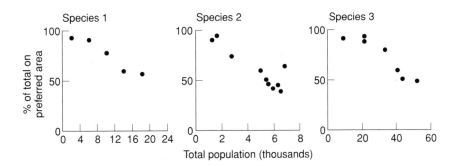

individuals in a population, no mutant or deviant strategy could do better. Hence, the evolutionarily stable strategy (often termed an ESS) will be favored by natural selection.

The ideal free hypothesis predicts that most individuals should be found in preferred habitats when forager population density is low, spilling over into less preferred habitats when forager density is high. This pattern has been demonstrated several times in different bird species (Fig. 5.10). One of the earliest examples was Brown's (1969) pioneering studies of great tits (*Parus major*) in the woodlands near Oxford, England. Brown showed that adult birds nested predominantly in woodland habitats in years with low bird abundance, expanding outwards into less attractive hedgerows only when densities were high. Krebs (1971) tested the assumption that this distribution stemmed from differences in fitness, by experimentally removing birds from woodland habitats, resulting in vacancies that were filled rapidly by former hedgerow "tenants."

A powerful way to test the ideal free hypothesis is to compare the feeding rates of individuals in different patches with different rates of food delivery. Milinski (1979) delivered food at differing rates to the two ends of an aquarium and measured the consequent pattern of distribution of sticklebacks. The ideal free hypothesis predicts that once they have determined the rate of food delivery at each end of the tank, the density of fish at each end should be proportional to the rate of food delivery. In other words, delivering twice as much food to one end of the tank should lead to two-thirds of the fish congregating in the food-rich patch. The sticklebacks redistributed themselves in precisely this manner (Milinski 1979). Similar results have been recorded in continuous food input experiments with numerous other species, including mallard ducks (Harper 1982), cichlid fish (Godin and Keenleyside 1984), and starlings (Inman 1990). Measurements in the field have been less supportive. However, animals in preferred habitats generally obtain higher rates of food intake than those relegated to poorer habitats (Sutherland 1996). Researchers frequently find that individuals vary in the quantity of food that they acquire, with more dominant or larger individuals securing more of the food delivered than lower-ranking individuals. This hierarchy suggests that although animals are capable of adjusting their behavior in predictable ways to accommodate the presence of other competitors for scarce resources, differences in dominance status tend to maintain differences in fitness (Sutherland 1996).

One way to accommodate these effects is through a modified model known as the **ideal despotic distribution** (Fretwell 1972). This model assumes that individuals choose the best habitat possible on the basis of their dominance status. The most dominant individuals choose first, followed by others in rank order of their dominance status. Under these conditions, individuals of similar status might well choose to split their time between two habitats offering similar levels of suitability, whereas high-ranking individuals invariably choose the best habitat. More importantly, the ideal despotic distribution predicts that there will be disparities in food intake, mortality risk, or reproductive success among individuals. These differences dissolve when we focus on individuals of similar rank.

5.6.2 *Interference among foragers*

The negative impact of other individuals on foraging success is sometimes termed **interference** by ecologists. It can result from direct aggression, stealing of food from other foragers, depletion of prey, or from uneaten prey scattering or hiding from other foragers. If we presume that aggression is the main cause of interference, we can predict how it will affect feeding rates. While searching for prey, individuals should encounter other predators at random. If each encounter between predators resulted in a wastage of w time units, then the foraging rate, $f(N, P)$, can be well approximated by a modified Type II functional response (Beddington 1975; Ruxton *et al.* 1992):

$$f(N, P) = \frac{aN}{1 + ahN + awP}$$

This formula predicts that interference should increase with predator forager density (Fig. 5.11). Similar logic can be used to develop an analogous model of interference arising from food thievery, also known as **kleptoparasitism** (Holmgren 1995).

Numerous field experiments have demonstrated such an increase in interference strength with forager abundance on organisms ranging from oystercatchers (*Haematopus ostralegus*) (Goss-Custard and Durrell 1987) to caribou (Manseau 1996). It is conventional to measure interference from plots of log intake versus log forager abundance. Thus, Fig. 5.12 illustrates changes in intake of cockles by oystercatchers in the Netherlands as a function of forager density (Sutherland 1996).

The ecological impact of interference can be profound (Beddington 1975; DeAngelis *et al.* 1975), adding a strong density-dependent effect to consumer–resource interactions that might otherwise be highly variable over time (see Chapters 12 and 19). Hence, interference can be a mechanism in the natural regulation of wildlife

Fig. 5.11 Changes in food intake as a function of predator (forager) density due to interference between predators.

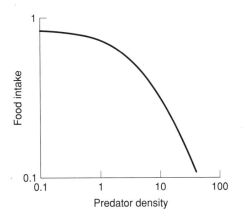

Fig. 5.12 Food intake by oystercatchers at two sites in the Netherlands declines as population density increases. (After Sutherland 1996.)

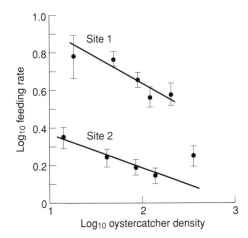

populations. An example of the effect of interference on carnivore population dynamics is the American marten, a mustelid carnivore in the forests of the USA and Canada (Fryxell *et al.* 1999).

5.6.3 *Territoriality*

Many wildlife species are territorial, meaning that they defend an area of (more or less) exclusive access from usage by other members of the population. Males, females, or both sexes may be territorial, depending on the ecological circumstances that apply. Territories may be defended solely during the breeding season, as in many birds, or throughout the year, as in many vertebrate carnivores; and they may be defended by individuals, such as in tigers, or by a pack of individuals, such as in gray wolves. The multitude of territorial forms mean that many different factors contribute to the adaptive significance of territoriality. Conversely, the consequences of territoriality at the population and community levels can also vary considerably.

Central to most arguments about the ecological basis of territoriality is the notion of **economic defendability** (Brown 1964; Dill 1978; Kodric-Brown and Brown 1978; Schoener 1983; Stephens and Dunbar 1993; Fryxell and Lundberg 1997). It would make little sense to try to defend any territory offering trivial benefits or whose costs are astronomical. Let us assume that the purpose of a territory is to gain access to

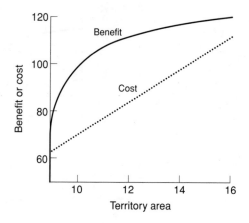

food supplies. The larger the territory, the greater is the abundance of food. However, we have already seen that there are diminishing returns, in terms of actual feeding rate, as prey abundance increases. As a consequence, food benefits decelerate with increasing territory size. Similarly, the time and energy needed to patrol the perimeter also rise with territory size. Moreover, the larger the territory, the greater the risk that other individuals will intrude. As a result, costs continue to rise steadily while benefits show diminishing returns with increasing territory size. The profit margin is clearly greatest for individuals which hold territories of intermediate size (Fig. 5.13). Provided that females are attracted to males which hold territories with sufficient resources successfully to rear offspring, the same sort of logic would predict that they favor intermediate-sized territories. In short, territory formation can be viewed as an economic decision, like many of the other behavioral processes described earlier in this chapter. Natural selection should favor the evolution of territoriality, if it enhances survival and long-term reproductive success.

Economic models of territory formation predict that territory size should be negatively related to both prey abundance (because it affects diminishing benefits) and forager abundance (because it affects the cost of defending the area). These predictions have been borne out in several field studies. For example, the size of rufous hummingbird (*Selasphorus rufus*) territories is inversely proportional to the abundance of flowers per unit area (Gass *et al.* 1976; Kodric-Brown and Brown 1978). Hence, larger territories hold approximately the same resource abundance as smaller, richer ones. Similar patterns have been observed in many other species, ranging from shorebirds (Myers *et al.* 1979) to roe deer (*Capreolus capreolus*) (Bobek 1977). Most convincing have been changes in territory size as a direct consequence of experimental alteration of either forager abundance (Bobek 1977) or resource levels (Myers *et al.* 1979). One difficulty with interpreting such experiments, however, is that experimental alteration of food levels often triggers changes in intruder pressure, so these factors tend to co-vary.

In many cases, individuals are faced with a choice of breeding (perhaps unsuccessfully) in a poor-quality territory or waiting for a vacancy in a better territory. Such is the case for many passerine birds, in which young birds are often relegated to poor breeding habitats. We discussed one such example earlier in the chapter: low-ranking great tits occupy hedgerow territories rather than prime territories in woodland (Krebs 1971). Removal of prime territory-holders leads to rapid replacement by members of the younger cohort. In other cases, younger individuals in poor habitat forgo breeding altogether, gambling instead on inheriting a good territory at a later date. A good example of this is the Seychelles brush warbler (*Bebrornis seychellensis*), in which many youngsters choose to stay at the nest and help their parents rear siblings rather than set off on their own (Komdeur 1992, 1993). Like all inheritances, this can be a risky proposition, because it depends on the probability that the helper survives and the (hopefully lower) probability that the current occupant does not. Economic models predict that such helping behavior will be selected for when there are pronounced disparities in the quality of potential breeding sites and where the probability of long life is reasonably good. The Seychelles warblers proved this point by abandoning helping behavior as soon as openings for good territories were created.

Territoriality can play a stabilizing role on population dynamics (Fryxell and Lundberg 1997). If there is an upper limit on the number of territories that can be supported, this can effectively cap breeding by the predator population, preventing large-scale predator–prey cycles of the sort described in Chapter 12. Since many top carnivores are territorial (e.g. wolves, weasels, lions, hyenas, and tigers), this suggests that a deeper understanding of carnivore territory formation and dynamics in relation to changes in abundance of both predators and prey is essential to adequate conservation and management efforts.

5.7 Summary

Foraging success is strongly affected by behavioral decisions of both predators and their prey. We consider a number of these decisions. First, foragers must decide which prey to attack and which to ignore. Optimal diet choice theory can be used to determine wise solutions to this problem, based on the opportunity cost of wasting time on poor prey in hand while better prey might yet still be found. For herbivores, foraging decisions are shaped by multiple constraints, such as balancing the need to meet requirements for a scarce nutrient with the objective of maximizing energy intake. Such problems can be approached using linear programming. Optimal patch departure theory is useful in considering how long foragers should stay in a particular area before moving on. The best theoretical solution is that animals be sensitive to the opportunity cost of wasting time in a poor patch, when a better patch lies close at hand. As for diet, herbivores choose patches in a manner that balances multiple constraints on their feeding.

Patch preferences can also be shaped by the need to trade off risk of predation against foraging gains. Risk-sensitive foraging demands complex approaches to decision evaluation, such as dynamic state variable modeling. Measurement of habitat-specific preferences demands special statistical tools, such as resource selection functions.

We also consider how social processes influence foraging decisions. Fitness is likely to decline as population density in preferred habitats grows, necessitating expansion of the population into poorer areas. Social interference plays an important role in this process, and we show how interference and territoriality can be viewed as adaptive responses to environmental conditions. All these behaviors have important consequences for the dynamics of wildlife populations.

6 Population growth

6.1 **Introduction**

In this chapter we deal with the internal workings of a population that result in a change of population size. The speed of that change is measured as rate of increase. Any such change alerts us that the fecundity rate, the mortality rate, or the age distribution, or more than one of these, has changed. Each of those parameters will be considered in turn and the relationships between them explained.

This chapter has two quite distinct functions. The first is to arm the reader with the theory of population dynamics. The second is to indicate which parts of that theory are immediately applicable to wildlife management and which parts are necessary only for a background understanding. The first function may appear to load a manager with unnecessary mental baggage, but without such knowledge mistakes are more than just possible, they are likely. Knowledge of atomic theory is not needed to mix a medicine, but without that knowledge a pharmacist will, sooner or later, make a critical mistake.

6.2 **Rate of increase**

If a population comprising 100 animals on (say) January 1 contained 200 animals on·the following January 1 then obviously it has doubled over 1 year. What will be its size on the next January 1 if it continues to grow at the same rate? The answer is not 300, as it would be if the growth increment (net number of animals added over the year) remained constant each year, but 400 because it is the growth rate (net number of animals added, divided by numbers present at the beginning of the interval) that remains constant. Thus the growth of a population is analogous to the growth of a sum of money deposited at interest with a bank. In both cases the growth increment each year is determined by the rate of growth and by the amount of money or the number of animals that are there to start with. Both grow according to the rules of compound interest and all calculations must therefore be governed by that branch of arithmetic.

Populations decrease as well as increase. The population of 100 animals on January 1 might have declined to 50 by the following January 1, in which case we say that the population has halved. If its decline continues at the same rate it will be down to 25 on the next January 1. Halving and doubling are the same process operating with equal force, the only difference being that the process is running in opposite directions. The terms by which we measure the magnitude of the process should reflect that equivalence. It is poorly achieved by simply giving the multiplier of the growth, 2 for a doubling and 0.5 for a halving, and it becomes even more confusing when these are given as percentages. We need a metric that gives exactly the same figure for a halving as for a doubling, but with the sign reversed. That would make it obvious that a decrease is simply a negative rate of increase.

It is achieved by expressing the rate of increase, positive or negative, as a geometric rate according to the following equation:

$$N_{t+1} = N_t \lambda = N_t e^r$$

in which N_t is population size at time t, N_{t+1} is the population size a unit of time later, e is the base of natural logs taking the value 2.7182817, and r is the exponential rate of increase. The **finite rate of increase** (λ) is the ratio of the two censuses:

$$\lambda = N_{t+1}/N_t$$

and therefore the **exponential rate of increase** is:

$$r = \log_e(N_{t+1}/N_t) = \log_e \lambda$$

We will try this out on a doubling and halving. With a doubling:

$$\lambda = 200/100 = 2$$

and so:

$$r = \log_e \lambda = 0.693$$

With a halving:

$$\lambda = 50/100 = 0.5$$

and so:

$$r = \log_e \lambda = -0.693$$

Thus a halving and a doubling both provide the same exponential rate of increase, 0.693, which in the case of a halving has the sign reversed (i.e. −0.693). It makes the point again that a rate of decrease is simply a negative rate of increase.

The finite rate of increase (i.e. the growth multiplier λ) and the exponential rate of increase r must each have a unit attached to them. In our example the unit was a year, and so we can say that the population is multiplied by λ per year. The exponential rate r is actually the growth multiplier of $\log_e$ numbers per year. That is something of a mouthful and so we say that the population increased at an exponential rate r on a yearly basis. Note that λ and r are simply different ways of presenting the same rate of change. They do not contain independent information.

Unlike the finite rate of increase, the exponential rate of increase can be changed from one unit of time to another by simple multiplication and division. If $r = -0.693$ on a yearly basis then $r = -0.693/365 = -0.0019$ on a daily basis. That simplicity is not available for λ.

The equations given above were simplified to embrace only one unit of time. They can be generalized to:

$$N_t = N_0 e^{rt}$$

where N_0 is population size at the beginning of the period of interest and N_t is the population size t units of time later. The average exponential rate of increase over the period is:

$$r = [\log_e(N_t/N_0)]/t$$

which can be written also as:

$$r = (\log_e N_t - \log_e N_0)/t$$

It would be of a waste of data to use only the population estimates at the beginning and end of the period to estimate the average rate of increase between those two dates. If intermediate estimates are available these can and should be included in the calculation to increase its precision. The appropriate technique is to take natural logarithms of the population estimates and then fit a linear regression to the data points each comprising $\log_e N$ and t. A linear regression takes the form $y = a + bx$ in which y is the dependent variable (in this case logged population size) and x is the independent variable (in this case time measured in units of choice). Our equation thus becomes:

$$\log_e N = a + bt$$

in which a is the fitted value of $\log_e N$ when time $t = 0$ and b is the increase in $\log_e N$ over one interval of time. Note that this is the definition of r, and so $r = b$. The equation for the linear regression may thus be rewritten:

$$\log_e N = a + rt$$

It can be converted back to the notation used in the example where rate of increase was measured between only two points by designating the start of the period as time 0:

$$\log_e N_t = \log_e N_0 + rt$$

which with a little rearranging converts to:

$$r = (\log_e N_t - \log_e N_0)/t$$

as before. Figure 6.1 shows such use of linear regression to estimate the rate of increase of the George River caribou herd in eastern Canada, yielding $r = 0.11$ (Messier *et al.* 1988).

6.2.1 *Intrinsic rate of increase*

The rate of increase of a population of vertebrates usually fluctuates gently for most of the time, around a mean of zero. If conditions suddenly become more favorable the population increases, the environmental improvement being reflected in a rise of fecundity and a decline in mortality. The environmental change might have been an increase in food supply, perhaps a flush of plant growth occasioned by a mild winter and a wet spring. The rate at which the population increases is

Fig. 6.1 Exponential population growth of the George River caribou herd, as discussed in the text. (After Messier *et al.* 1988.)

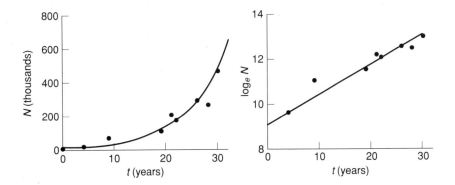

then determined by two things: on one hand the amount of food available and on the other the intrinsic ability of the species to convert that extra energy into enhanced fecundity and diminished mortality. Thus, it depends on an environmental effect and an intrinsic effect but neither is without limit. From the viewpoint of the animal both are constrained. There comes a point at which the animal has all the food it can eat, any further food having no additional effect on its reproductive rate and probability of survival. Similarly, an animal's reproductive rate is constrained at the upper limit by its physiology. Litters can be only so big and the interval between successive litters cannot be reduced below the gestation period. The potential rate of increase can never be very high, irrespective of how favorable the environmental conditions are, if the period of gestation is long (e.g. 22 months for the African elephant, *Loxodonta africana*). All species, therefore, have a maximum rate of increase, which is called their **intrinsic rate of increase** (Fisher 1930) and denoted r_m. It is a particularly important parameter in estimating sustainable yield (see Section 19.3).

Populations do not attain that maximum very often. It requires a very high availability of food and a low density of animals such that there is negligible competition for that food. These conditions are most closely approached when a population is in the early stage of active growth subsequent to the release of a nucleus of individuals into an area from which they were formerly absent. Figure 6.2 gives intrinsic rates of increase of several mammals, most of the data being gathered in that way. Alternatively the rate could be estimated from the initial stages of growth of a population recovering from overhunting. That would work for blue whales (*Balaenoptera musculus*) for example, which are presently recovering from intense over-harvesting between about 1925 and 1955 (Cherfas 1988).

Intrinsic rate of increase r_m tends to vary with body size. The relationship has been calculated (Caughley and Krebs 1983; Sinclair 1996) for herbivorous mammals as:

$$r_m = 1.5W^{-0.36}$$

where W is mean adult live weight in kilograms. Table 6.1 gives r_m calculated by that equation for a range of body weights. In the absence of other data it provides an approximation that can be used to make a first estimate of sustained yield (see Chapter 19).

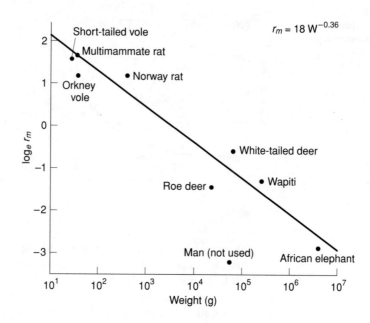

Fig. 6.2 Intrinsic rate of increase of mammals plotted against body weight. (After Caughley and Krebs 1983.)

$r_m = 18\,W^{-0.36}$

Short-tailed vole
Multimammate rat
Norway rat
Orkney vole
White-tailed deer
Wapiti
Roe deer
Man (not used)
African elephant

$\log_e r_m$

Weight (g)

Table 6.1 Expected intrinsic rates of increase r_m on a yearly basis for herbivorous mammals as estimated from mean adult live weight.

Weight (kg)	r_m
1	1.50
10	0.65
100	0.29
1000	0.08

6.3 Fecundity rate

A population's rate of increase is determined by its size, by how many animals are born, and by how many die during a year. Hence, birth rate is an important component of population dynamics that can be measured in a number of ways. Of these the most useful is fecundity rate.

We measure fecundity rate as the number of female live births per female per unit of time (usually 1 year). That figure is often broken down into age classes to give a fecundity schedule as in Table 6.2, and each value is denoted m_x, female births per female in the age interval x to $x + 1$.

6.4 Mortality rate

The number of animals that die over a year is another important determinant of rate of increase, and again it can be measured in a number of ways. We measure it as the mortality rate, the number of animals that die during a unit of time (usually 1 year) divided by the number alive at the beginning of the time unit. As with fecundity, the rate is often given for each interval of age.

The pattern of mortality with age is summarized as a life table, which has a number of columns as in Table 6.3. The first is the age interval labeled by the age at the beginning of the interval and denoted x. The second is survivorship l_x, the probability at birth of surviving to age x. The third is mortality d_x, the probability at birth of dying in the age interval x, $x + 1$. The fourth, the most useful, is mortality rate q_x, the probability of an animal age x dying before the age of $x + 1$. The fifth,

Table 6.2 A fecundity schedule calculated for chamois.

Age (years) (x)	Sampled number (f_x)	Number pregnant or lactating (B_x)	Female births per female ($B_x/2f_x$) (m_x)
0	–	–	0.000
1	60	2	0.017
2	36	14	0.194
3	70	52	0.371
4	48	45	0.469
5	26	19	0.365
6	19	16	0.421
7	6	5	0.417
> 7	10	7	0.350

From Caughley (1970).

Table 6.3 Construction of a partial life table.

Age (years) (x)	Survival frequency (f_x)	Survivorship (l_x)	Mortality (d_x)	Mortality rate (q_x)
0	1200	1.00	0.58	0.58
1	500	0.42	0.17	0.40
2	300	0.25	0.08	0.32
3	200	0.17	–	–
.	.	.	.	.
.	.	.	.	.
.	.	.	.	.

age-specific survivorship p_x, is the probability at age x that an animal will still be alive on its next birthday.

Probabilities are estimated from proportions. The probability of a bird surviving to age x can be estimated for example by banding 1200 fledglings and recording the number still alive 1 year later, 2 years later, 3 years later, and so on. Let us say those frequencies were 500, 300, 200. Survivorship at age 0 (i.e. at birth) is 1200/1200 = 1, by age 1 year it has dropped to 500/1200 = 0.42, further to 300/1200 = 0.25 at age 2 years, and further still to 200/1200 = 0.17 at age 3 years.

No further data are needed to fill in the other columns corresponding to these values of l_x because each is a mathematical manipulation of the l_x column. Mortality d_x is calculated as $l_x - l_{x+1}$ (1 − 0.42 = 0.58 for $x = 0$ and 0.42 − 0.25 = 0.17 for $x = 1$). Mortality rate q_x is calculated as $(l_x - l_{x+1})/l_x$ or d_x/l_x (0.58/1 = 0.58 for $x = 1$ and 0.17/0.42 = 0.40 for $x = 2$). Table 6.3 shows the table fully constructed up to age 2 years, that for age 3 years being partial because data for age 4 years are needed to complete it. The subsequent rows would be filled in each year as the data became available.

So, constructing a life table is straightforward when the appropriate data are available. Pause for a moment to contemplate the difficulty of obtaining those data. Banding 1200 fledglings, or whatever number, poses no more than a problem in logistics. The difficulty comes in estimating what proportion of those birds are still alive at the end of the year. Nonetheless, there have been a number of direct studies of vital rates in wildlife species, based on mark–recapture methods (Lebreton *et al.* 1992; Gaillard *et al.* 2000).

Approximation methods are also available, based on age structure. If one can age a sample of the living population, or alternatively establish the ages at death of a sample of deaths from that population, an approximate life table can, in some circumstances, be constructed from those age frequencies.

6.5 Direct estimation of life-table parameters

There are basically two different ways in which life-table data can be directly estimated. The first, and rarest, method is to monitor the fates of all individuals in a relatively small population that is carefully studied over a long time. For example, virtually every young lion born to the population inhabiting the ecotone between the Serengeti plains and adjacent woodlands has been carefully monitored over the past three decades (Packer *et al.* 2005). The unique combination of facial spots, scars, and other features make it possible to visually recognize every individual and keep track of their fate. By collating data for each specific cohort, one can readily calculate the probability that any member born to this group survives to age x (the l_x series), by simply dividing the number of survivors at age x by the initial group size.

Even in this ideal situation, however, there are thorny problems associated with the estimation of life-table parameters. The difficult issue is that survival is like a game of chance: the outcome can vary considerably from one replicate to another (see Chapter 17). For example, a 0.5 probability of survival for an initial group of four individuals can lead to no survivors (expected 6.3% of the time), one survivor (expected 25% of the time), two survivors (expected 37.5% of the time), three survivors (expected 25% of the time), or even four individuals (expected 6.3% of the time). So the fact that two out of four individuals in a cohort survive over a given year does not conclusively demonstrate that the probability of survival really is 0.5, nor does the observation of no individuals surviving provide conclusive evidence against such a rate. As a result of the inherently variable nature of demographic processes, it is difficult to ascribe a particular risk of mortality with high likelihood, unless very large numbers of individuals are involved or such observations are repeated over many years.

The second way to estimate life-table parameters directly is to mark a large number of individuals at time t (A_t), then recover some of those individuals (b_{t+1}) in a subsequent sampling session, say a year later, to estimate the probability of survival. Marked individuals might be equipped with leg bands (as in many bird studies), ear-tags (as in many studies of small mammals), or even radiotransmitters (as in many studies of large mammals). If the true number of survivors is B_{t+1}, then the number of marked animals in the sample (b_{t+1}) depends on the detectability of individuals in each sample (c), typically under the presumption that $b_{t+1} = cB_{t+1}$. In this situation, not only is there stochastic variation to contend with, but also sample variation associated with detectability of individuals in the population. By chance, we might detect a relatively large number of marked individuals in a subsequent sample, for reasons wholly unrelated to survival probability.

The confidence we ascribe to survival probabilities estimated using these mark–recapture techniques depends critically on sample size, probability of recapture if an animal is still alive, mobility of marked animals and their loyalty to the site at which originally caught, the number of replicate sampling intervals, and whether or not newly marked animals have been repeatedly added to the population or not (Lebreton *et al.* 1992; Nichols 1992). Over the past two decades, there has been a revolution of sorts in the analysis of mark–recapture data, using sophisticated computer programs, such as SURGE (Lebreton *et al.* 1992) or MARK (White and Burnham 1999). Many of these programs are available free of charge from the World Wide Web. We point interested readers to the encyclopedic review of demographic methodology by Williams *et al.* (2002) for an insightful discussion of different mark–recapture approaches and their statistical analysis.

Sophisticated mark–recapture experiments to estimate demographic parameters often involve comparisons among a large number of competing models (does survival vary among sexes, over time, across age groups, or between sites?). As we discuss in Chapter 15, such comparisons often require use of information–theoretic approaches to identify the most parsimonious model to explain a given data set. Recent versions of demographic analysis software, such as MARK, commonly include formal means of making choices among competing models (commonly either via Akaike's information criterion (AIC) optimization evaluation or likelihood-ratio testing).

6.6 Indirect estimation of life-table parameters

If certain conditions (see the end of this section) are met, the age distribution of the living population can be used as a surrogate for the survival frequency f_x of Table 6.3 to produce an approximate life table. Many of the bovids can be aged from annual growth rings on the horns; some species of deer, seals, and possums produce growth layers in the teeth; and fish form growth lines on the scales. Unbiased samples of animals which have died from natural causes or from the live population yield data that may be amenable to life-table analysis.

It is sometimes possible to estimate the life table from a sample of individuals taken indiscriminately from the live population. This is most often derived from hunting statistics, although the reliability of such measures is often questionable, given the tendency for most hunters to select bigger or older animals. It is better to rely on catastrophic events that indiscriminately sample a cross-section of individuals in the population.

Flash floods during the autumn of 1984 killed thousands of woodland caribou from the George River herd in northern Quebec as they were migrating to their winter range. A large number of carcasses from this freak event washed up on the banks of the Caniapiscau River, where wildlife biologists working with the Quebec government retrieved them (Messier *et al.* 1988). The resulting sample of 875 female caribou 1 year of age and older was assumed to reflect the standing age composition of the living population. The frequency of newborns was estimated from calf–mother counts on the calving grounds.

If any study population is unchanging (termed "stationary"), the standing age distribution reflects survival frequencies by age. In the case of the George River caribou herd, however, a series of censuses were available showing strong evidence of exponential increase over the previous two decades, with $r = 0.11$ (Fig. 6.1). This introduces a bias into life-table parameter estimation, because older animals were born into a much smaller population than were younger individuals. The appropriate way to cope with this bias is to transform the age frequency data before deriving the life table. Table 6.4 demonstrates how to transform the age structure data, by multiplying the observed frequency at age x (f_x) by a coefficient (e^{rx}), that corrects for the bias in observed age frequencies caused by population growth.

One often needs to further smooth the age frequency data, especially when the data come from a relatively small sample of animals, to guarantee a continual decline in frequency with each successive age group. This is usually done by fitting a

Table 6.4 Life table for female caribou in the George River herd. Column 2 gives the original data from dead animals. Column 4 corrects column 2 by multiplying by e^{rx}, and column 4 smooths column 3.

Age	Frequency	Corrected frequency	Smoothed frequency	l_x	d_x	p_x	q_x	m_x
0	236.1	236.1	236.1	1.000	0.286	0.286	0.714	0
1	138	154.0	168.5	0.714	0.007	0.010	0.990	0
2	156	194.4	167.0	0.707	0.017	0.024	0.976	0.06
3	113	157.2	163.0	0.690	0.027	0.039	0.961	0.35
4	94	145.9	156.6	0.663	0.037	0.056	0.944	0.4
5	83	143.9	147.9	0.626	0.044	0.070	0.930	0.4
6	65	125.8	137.3	0.582	0.053	0.091	0.909	0.4
7	63	136.1	125.0	0.529	0.057	0.108	0.892	0.4
8	57	137.4	111.4	0.472	0.063	0.133	0.867	0.4
9	40	107.6	96.6	0.409	0.065	0.159	0.841	0.4
10	24	72.1	81.2	0.344	0.067	0.195	0.805	0.4
11	18	60.4	65.4	0.277	0.067	0.242	0.758	0.4
12	12	44.9	49.5	0.210	0.066	0.314	0.686	0.4
13	7	29.2	33.9	0.144	0.064	0.444	0.556	0.4
14	1	4.7	18.8	0.080	0.061	0.763	0.238	0.4
15	4	20.8	4.4	0.019	0.019	1.000	0.000	0.4

From Messier *et al.* (1988).

quadratic or cubic curve to the age distribution, using the values derived from the curve in place of the actual observations, as demonstrated for the George River caribou in Table 6.4. The survivorship series is then constructed by dividing each age frequency by 236, the d_x series as $l_x - l_{x+1}$, and the q_x series as d_x/l_x. If the age frequency data had not been smoothed, there would have been instances in which the observed frequency of an older age group exceeded that in the next youngest age group, implying survival rates exceeding 100%, an obvious impossibility.

An unbiased sample of ages at death due to natural causes, as might be obtained by a picked-up collection of skulls, may in some circumstances be treated as a multiple of the d_x series. Table 6.5 gives an example from African buffalo (Sinclair 1977). Only those skulls aged 2 years or older were counted because skulls from younger animals disintegrate quickly. These age frequencies are given in the second column of the table and total 183 skulls. The third column corrects for the missing younger frequencies: sample counts of juveniles in the field showed that the mortality rate over the first year of life was 48.5% and that 12.9% of the original cohort died in the second year. Hence, if the original cohort is taken as 1000, 485 of these would die in the first year of life and 129 in the second year. These values are tabled. They account for 614 of the original cohort, leaving 386 to die at older ages. The age frequencies of the 183 animals in the second column are thus each multiplied by 386/183 to complete the third column. The fourth column, d_x, is formed by dividing the fd_x frequencies by 1000 so that they sum to unity. Survivorship at age 0 (i.e. birth) is then set at one and the subsequent l_x values calculated by subtracting the corresponding d_x from each. Mortality rates q_x are calculated as before, as $q_x = d_x/l_x$.

The reliability of any life table developed indirectly from either a sample from the live population or a sample of animals that die of natural causes depends on how closely the data meet the underlying assumptions of the analysis:

Table 6.5 Construction of a life table from a pick-up sample of African buffalo skulls. The table is not corrected for rate of increase.

Age (x)	Mortality frequency (f_x)	Mortality corrected (fd_x)	Mortality (d_x)	Survivorship (l_x)	Mortality rate (q_x)
0	–	485	0.485	1.000	0.485
1	–	129	0.129	0.515	0.250
2	2	4	0.004	0.387	0.010
3	5	11	0.011	0.383	0.029
4	5	11	0.011	0.372	0.030
5	6	13	0.013	0.361	0.036
6	18	38	0.038	0.348	0.109
7	17	36	0.036	0.310	0.116
8	20	42	0.042	0.274	0.153
9	17	36	0.036	0.232	0.155
10	15	32	0.032	0.196	0.163
11	16	34	0.034	0.164	0.207
12	18	38	0.038	0.130	0.292
13	15	32	0.032	0.092	0.348
14	14	29	0.029	0.060	0.483
15	8	17	0.017	0.031	0.548
16	5	10	0.010	0.014	0.714
17	1	2	0.002	0.004	0.500
18	0	0	0.000	0.002	0.000
19	1	2	0.002	0.002	1.000
	183	1001	1.001		

1 The sample is an unbiased representation of the living age distribution in the first case or of the true frequency of ages at death in the second. The exercise would have to control the usual biases implicit in hunting activities if the sample of the living age distribution were obtained by shooting. One would be unlikely to use a sample obtained by sporting hunters, for example. The first age class is usually underestimated in a picked-up sample of ages at death because the skulls of young animals disintegrate much faster than do those of adults, thereby significantly biasing the table.

2 Age-specific fecundity and mortality must have remained essentially unchanged for a couple of generations.

3 Whether the sample is of the living population or of the ages at death, the population from which it came must have a rate of increase very close to zero, or else the data must be transformed to accommodate the observed rate of population change over the past two generations. Major fluctuations in recent rates of growth invalidate virtually all such indirect methods. This can limit the usefulness of such exercises in wildlife management.

6.7 Relationship between parameters

We restrict the following discussion to females for simplicity, but the points made apply also to the male segment of the population.

Remember that l_x is survivorship to age x, m_x is production of daughters per female at age x, and r is the exponential rate at which the population increases. Then:

$$\sum l_x m_x e^{-rx} = 1$$

which is the basic equation of population dynamics. If the survivorship and fecundity schedules hold constant, the population's age distribution will converge to the constant form of:

$$S_x = l_x e^{-rx}$$

which is called the **stable age distribution**. S_x is the number of females in a particular age class divided by the number of females in the first age class. The basic equation may thus be written $\sum S_x m_x = 1$. In the special case of rate of increase being zero, the stable age distribution, now called the **stationary age distribution**, is $S_x = l_x$ by virtue of $e^{-0x} = 1$. That is the justification for using such an age distribution to construct a life table. The stationary age distribution is the special case of the stable age distribution that obtains when $r = 0$. It has been argued that, since fecundity and mortality schedules seldom remain constant for long, the stable age distribution is little more than a mathematical abstraction, although a useful one. Although the stable distribution can be attained fairly quickly (roughly two generations) after mortality and fecundity patterns stabilize, most wildlife species that have been adequately studied have mortality and fecundity schedules that fluctuate, sometimes substantially, from year to year.

6.8 Geometric or exponential population growth

Thomas Malthus in 1798 recognized that populations have an intrinsic tendency towards exponential or geometric growth, just as a bank account at fixed interest grows geometrically with the amount of money in the account. The growth of such populations can be calculated as either a continuous or a discrete process. For simplicity, we will concentrate on discrete time representations of population growth. Strictly speaking, such models are most applicable to organisms whose patterns of deaths and births follow a seasonal or annual cycle of events, which includes most wildlife species. Consider, for example, a population whose finite growth rate (λ) is 0.61 and whose initial density (N_0) is 1.5. The geometric growth model predicts subsequent changes in density over time according to $N_t = N_0 \lambda^t$. The outcome depends on whether λ is larger or smaller than 1. When $\lambda < 1$ (Fig. 6.3) there is a decelerating pattern, while the outcome is changed to an accelerating pattern of growth when $\lambda > 1$ (Fig. 6.4).

Fig. 6.3 Population changes according to the geometric model with $\lambda = 0.61$.

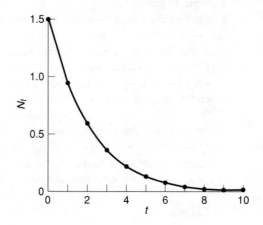

Fig. 6.4 Population changes according to the geometric model with $\lambda = 1.65$.

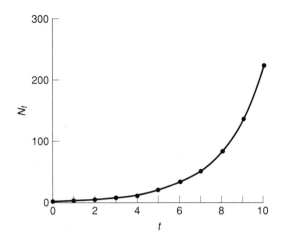

As we saw earlier in this chapter, the geometric model can be readily translated into the exponential model:

$$N_t = N_0 e^{rt}$$

Hence, it is straightforward to shift between representation of population dynamics in continuous time and discrete time. Such simple models are most appropriate for small populations introduced into a new environment or for a short period follow-ing a perturbation. For example, the George River caribou herd in eastern Canada grew exponentially at a rate of $r = 0.11$ during a 30-year period following recovery from overharvesting (Messier *et al.* 1988).

6.9 Summary

The dynamic behavior of a population – whether it increases, decreases, or remains stable – is determined by its age- or stage-specific mortality and fecundity rates inter-acting with the underlying distribution of ages or stages in the population. A wide variety of techniques are available for estimating age-specific parameters, summarized in the life table. When age-specific rates of fecundity and survivorship remain con-stant, the population's age distribution assumes a stable form, even though its size may be changing. These demographic parameters determine the rate of population change over time, forming the logical basis for many conservation and management decisions.

7 Dispersal, dispersion, and distribution

This chapter explores some of the reasons why populations are found where they are. We describe the finer-scaled pattern as the dispersion and the broader scale as the distribution. We offer examples of how different factors such as temperature and seasonality limit the distribution of wildlife. We then discuss the causes for dispersal, and finally methods of modeling rates of dispersal of populations.

Dispersal is the movement an individual animal makes from its place of birth to the place where it reproduces. Dispersal is not to be confused with **migration** (movement backward and forward between summer and winter home ranges) or with **local movement** (movement within a home range). The terms immigration and emigration are used in mark–recapture studies to mean movement into and out of a study area of arbitrary size and location. Migration is used by population geneticists to mean "the movement of alleles between semi-isolated subpopulations, a process that by definition involves gene flow between subpopulations" (Chepko-Sade *et al.* 1987). Although these uses differ from their ecological uses, the difference is usually obvious from context and causes little confusion.

Dispersion is the pattern of spatial distribution taken up by the animals of an area. Dispersions may be fixed if the animals are sessile but more commonly they change with time under the influence of a changing dispersion of resources. A dispersion at a given time may be changed by dispersal, or local movement, or both.

The **distribution** of a population or species is the area occupied by that population or species. It is depicted as the line drawn around the dispersion. The distribution can be subdivided into gross range and breeding range, and it can be mapped at different scales.

7.2 Dispersal

Dispersal is an action performed by an individual (Johnson and Gaines 1990). An animal disperses or it remains within its maternal home range. If it disperses, it may move only that distance sufficient to bring it to the nearest unoccupied and suitable area within which to establish its own home range, or it might move a considerable distance, crossing many areas that look suitable enough, before settling down.

The mechanism of dispersal may also vary. The individual may be pushed out of the maternal home range by a parent or it may move without any prompting save for that supplied by its genes. The young of some species never meet their parents (e.g. frogs, reptiles, the mound-building birds of the family Megapodidae) and so must provide their own motivation. In mammals, at least, there are two forms of dispersal that have been recognized (Stenseth and Lidicker 1992). **Presaturation** dispersal is seen in some species of small mammals where juveniles leave their natal range even when the density of the population is low. The mechanism is either that

Fig. 7.1 The frequency distributions of distances dispersed by juvenile rat kangaroos. (Data from Jones 1987.)

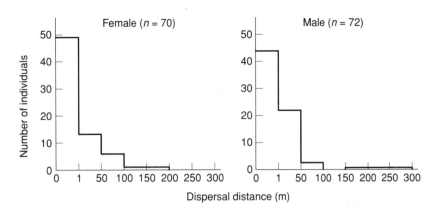

the juveniles leave voluntarily, their behavior being innately determined by their genes (e.g. in Belding's ground squirrels (*Spermophilus beldingi*); Holekamp 1986); or that adults forcibly exclude juveniles. **Saturation** dispersal is seen in many large mammals (Sinclair 1992). In this case dispersal occurs when a population reaches a threshold density determined by food limitation. Dispersal then is density dependent (see Chapter 8 for an explanation of this mechanism) so that population density remains the same on the initial area. Examples of this have been described for Himalayan tahr (*Hemitragus jemlahicus*) as they spread through the Southern Alps of New Zealand (Parkes and Tustin 1985), and for wood bison (*Bison bison*) as their population increased through their former range in the boreal forest of Canada (Larter *et al*. 2000) (see Section 7.6 for modeling range expansion).

The likelihood of dispersal differs markedly between individuals of a population. Figure 7.1 shows a sample of distances dispersed by juvenile kangaroo rats (*Dipodomys spectabilis*) (Jones 1987), a solitary, nocturnal, grain-eating desert rodent. Females averaged 29 m and males 66 m, but the majority of individuals did not disperse at all. Jones (1987) reported that adults of this species do not disperse much: 70% of adult males and 61% of adult females remained in one mound for the rest of their lives. Juvenile females of red deer (*Cervus elaphus*) seldom disperse but adopt home ranges that overlap those of their mothers. In contrast, males leave the natal home range between the ages of 2 and 3 years, mostly joining stag groups in the vicinity (Clutton-Brock *et al*. 1982).

Patterns of dispersal are related to the type of mating system (Greenwood 1980, 1983; Dobson 1982; Greenwood and Harvey 1982). Thus, in mammals, females are concerned with obtaining resources while males compete for mates. In general, males disperse in promiscuous and polygynous species because they are more likely to find new mates by doing so, while females are philopatric (i.e. remain at their birth site) because they are more likely to find food in areas they know well. Both sexes disperse in monogamous species. Amongst higher vertebrates, one sex is more prone to dispersal than the other. Thus, in mammals males are the dispersers whereas in birds it is the females which disperse, although there are exceptions for both groups. For example, in mammals females are the dispersers in wild dogs (*Lycaon pictus*) and zebra (*Equus burchelli*). In fishers (*Martes pennanti*) and wolves, both sexes disperse equally (Arthur *et al*. 1993; Boyd and Pletscher 1999).

The causes of dispersal fall into three broad categories: competition for mates, avoidance of inbreeding, and competition for resources (Johnson and Gaines 1990). In polygynous species, females invest more in each offspring than do males, and so their reproductive success is determined by resource competition. Male reproductive success is limited by the number of mates they can find, so competition for mates is important.

Inbreeding avoidance is often cited as a cause of dispersal on theoretical grounds (reviewed in Thornhill (1993); see Section 17.3.5 for an explanation of the genetics of inbreeding depression). Inbreeding depression was observed in a captive wolf (*Canis lupus*) population (Laikre and Ryman 1991). In contrast, there was no evidence of inbreeding depression or avoidance in a social carnivore, the dwarf mongoose (*Helogale parvula*) (Keane *et al.* 1996). In general, the occurrence of inbreeding depression depends on the species (Waser 1996). There are some instances where inbreeding avoidance has been found, as in some species of birds (Pusey 1987; Keller *et al.* 1994), primates (Pusey 1992), rodents (Hoogland 1982), and marsupials (Cockburn *et al.* 1985). However, there are many instances where populations occur in small numbers, inbreeding is not avoided, and there is no deleterious effect of inbreeding (Keane *et al.* 1996). In other cases there are multiple causes of dispersal (Dobson and Jones 1985).

Dispersers tend to have lower survival than those that remain in their natal area. In arctic ground squirrels (*Spermophilus parryii*) survival of philopatric juveniles was 73%, whereas survival of dispersing squirrels was in the range 25–40%. Also, survival declines with the distance of dispersal due to the increasing probability of being caught by predators (Byrom and Krebs 1999). The survival of dispersing ferrets (*Mustela furo*) in New Zealand was 100% where predators had been removed experimentally compared with only 19–71% in areas where predators were present (Byrom 2002). However, survival of dispersing male San Joachin kit foxes (*Vulpes macrotis mutica*) was higher than that for philopatric males (Koopman *et al.* 2000), indicating exceptions to the rule.

7.3 Dispersion

Dispersions may be random, clumped, or spaced. The most common is a **clumped dispersion** (sometimes called a **contagious dispersion**). If the area is divided into quadrats and the frequency distribution of animals per quadrat is recorded, the variance of that distribution will equal its mean if the animals are randomly distributed (a Poisson distribution), the variance will be greater than the mean if the animals are clumped at that scale, and the variance will be less than the mean if the animals space themselves.

Scale is important when dispersions are considered because two or more orders of dispersion may be imposed upon each other: randomly distributed clumps of animals for example. In these circumstances a quadrat in a grid of small quadrats will include either part of a group or it will miss a group: its count will be of many animals or of no animals. When the grid comprises large quadrats, an average quadrat will contain several groups of animals and the variation in counts between quadrats will be less marked. The dispersion is the same whether the quadrats used to sample it are large or small, but in this case the clumping as measured by the variance/mean ratio will appear to be more intense when quadrats are small.

An alternative to characterizing dispersion in terms of the frequency distribution of quadrats containing 0, 1, 2, etc., animals per quadrat is instead to record the frequency distribution of nearest-neighbor distances or of the distances between randomly chosen points and the nearest animal to each. The problem of quadrat size does not arise because no quadrats are involved, but no simple measure is presently available for distributions of distances that clearly differentiates classes of dispersions, one from the other, given the wealth of possible dispersions. However, J.M. Cullen and M. Bulmer (in Patterson 1965) provide a formula for calculating the random distribution of inter-individual (or intergroup) distances in a known area. Given the same number of individuals N, distributed randomly with respect to each other in the same area A, then the proportion (p) of individuals having their nearest neighbor at a distance x is given by the expression:

$$p_x = \exp[(-\pi N/A)(x - 0.5a)^2] - \exp[(-\pi N/A)(x + 0.5a)^2]$$

where a is the unit of measurement used. The number at distance x is Np_x. Thus, if one observes 200 birds in an area of 2 km radius ($A = 12.57 \times 10^6$ m^2), and observations are in units of 50 m ($= a$), then the expected frequency of distances at the nearest interval ($x_1 = 25$ m) is 23.5, that at the next interval ($x_2 = 75$ m) is 55.2, and so on until the sum of Np_x equals 200. We see that the increments of x must start with the first one equal to $0.5a$ (midpoint of the first interval) and then increase in increments of a (thus 25, 75, 125, 175, etc.). By comparing this frequency of distances with the observed frequency one can identify clumped or overdispersed distributions.

Dispersion is affected by the **home range** of individuals, that is the area used during the normal daily activities. Traditionally, home ranges are estimated from radiotelemetry locations (usually > 30 locations are required) using computer software packages. Habitat type affects range area (Relyea et al. 2000), as does the gender of the individual (McCullough et al. 2000). Some species have tight habitat preferences, their dispersion reflecting where that habitat is to be found. Others are more catholic in their requirements and will therefore be distributed more evenly across the landscape. The ecology of the dispersion is important. Dispersion can be measured more directly, however, by the average distance between locations (Conner and Leopold 2001). We considered the concept of home range in Chapter 5, in which we outline methods of determining the key determinants of home range use.

When we design surveys to count wildlife (see Chapter 13) we have to pay attention to its dispersion and allocate our sampling units accordingly. We explore this practical aspect of dispersion more fully in Section 13.4.

7.4 Distribution

Krebs (2001) considered that "the simplest ecological question one can ask is simply: Why are organisms of a particular species present in some places and absent in others?" There are several interesting ways that this question can be answered. We start with a consideration of the ultimate limits of a species' range, before going on to consider the distribution of introduced or invading species and finally to consider patterns of occupancy in spatially subdivided populations (metapopulations).

Figure 7.2 shows three hypothetical distributions, not as a map but as a plot within a range of mean annual temperature and mean annual rainfall. For species A, temperature and rainfall act independently of each other in setting limits to distribution. A single mean temperature and a single mean annual rainfall is all one needs to predict whether or not the species will be in a given area.

The distribution of species B is also determined by temperature and rainfall but this time in an asymmetric interactive manner. Distribution is determined absolutely

Fig. 7.2 Three hypothetical envelopes of adaptability of a species to temperature and moisture: A, the two factors act independently; B, the level of one factor influences the effect of the other; C, the effect of each factor varies according to the level of the other. (After Caughley *et al.* 1988.)

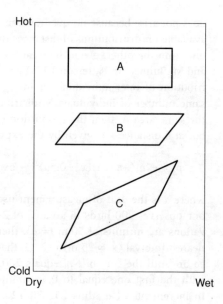

by an upper and lower limiting temperature but it is demarcated within those bounds by rainfall whose effect varies with temperature. High rainfall is tolerated only in hot areas and low rainfall only in colder areas where evaporation is reduced.

The distribution of species C is controlled by a symmetric interaction of rainfall and temperature. The species' tolerance of high temperatures increases with increasing annual rainfall, and the tolerance of the species to rainfall increases with temperature. This is a two-way interaction.

A known range of tolerance to one or more factors like temperature and rainfall does not translate directly into a map of distribution because the factors may interact as in example B and C of Fig. 7.2, the level of one factor determining the effect of another. Whether distribution is determined by one or several factors depends critically on the geographic dispersion of the levels of each factor.

7.4.1 *Range limited by temperature*

Temperature can limit the distribution of animals through direct effects on their physiology and indirectly by affecting resources. Some distributions can be described empirically by temperature contours (isotherms). Thus, the southern limit to northern hemisphere seals is set by sea surface temperatures never exceeding 20°C (Lavigne *et al.* 1989). The reason is unclear, but most seals breed in regions of high marine productivity and these are largely restricted to high latitudes. Similarly, the penguins of the southern hemisphere inhabit seas with temperatures lower than 23°C. Most penguin species inhabit latitudes between 45°S and 58°S where marine productivity is high (Stonehouse 1967). They reach the equator at the Galapagos Islands off the Pacific coast of South America, but only because those shores are bathed by the cold Humboldt current.

The northern limit for rabbits (*Oryctolagus cuniculus*) in Australia is marked by the 27°C isotherm. These temperatures coincide with high humidity, and the combination of the two causes resorption of embryos so that the animals cannot breed.

Cold is clearly an important factor limiting species in the Arctic and subarctic. Although the Arctic is an important breeding ground for birds, most leave during winter. Only four North American species can withstand the cold to reside year round in the Arctic: the raven (*Corvus corax*), the rock ptarmigan (*Lagopus mutus*), the snowy owl (*Nyctea scandiaca*), and the hoary redpoll (*Acanthis hornimanni*) (Lavigne *et al.* 1989). Amphibians and reptiles are particularly affected by temperature. The American alligator (*Alligator mississipiensis*) cannot tolerate temperatures below 5°C. Although several species of amphibians and reptiles tolerate freezing temperatures, in general there is a negative relationship between the number of species and the latitude. The direct effect of cold in limiting the distribution of these groups is probably less important than the availability of hibernation sites remaining above lethal temperatures (Lavigne *et al.* 1989).

Movements of large mammals can be affected by temperature. In the Rocky Mountains several ungulates, such as moose (*Alces alces*), elk, and deer, move down hill for the winter. Sometimes a temperature inversion in winter positions a warmer air layer above a colder one, and in these conditions Dall sheep (*Ovis dalli*) in the Yukon climb higher rather than lower.

The limiting effects of temperature are demonstrated by changes in the range of several species during historic times. Temperatures increased in the northern hemisphere between 1880 and 1950. The breeding ranges of herring and black-headed gulls (*Larus argentatus*, *L. ridibundus*) moved north into Iceland, and that of green woodpeckers (*Picus viridis*) extended into Scotland. Temperatures have declined since 1950 and the breeding ranges of snowy owls and ospreys (*Pandion haliaetus*) have moved south (Davis 1986). On the American prairies the warming period was associated with severe droughts in the 1930s. As a result the cotton rat (*Sigmodon hispidus*) has spread north (Davis 1986). Further changes in the distribution of these and many other wildlife species are expected in the future, as a result of global warming.

Cold temperatures themselves may be less important than the consequent changes in snow pack. Caribou must expend greater amounts of energy in exposing ground lichens when snow develops a crust (Fancy and White 1985). Even further north on Canada's High Arctic Islands the warming temperatures of spring melt the surface snow. As the water trickles through the snow pack it freezes when it hits the frozen ground and forms an impenetrable layer. The caribou abandon feeding in those areas and may migrate across the sea ice to areas where the wind has blown the shallow snow away (Miller *et al.* 1982).

Deep snow limits other species also. North American mountain sheep (*Ovis canadensis*, *O. dalli*) are usually found in winter on cold windswept ledges where there is little snow. Deer (*Odocoileus* species) are limited by snow cover of moderate depths (< 60 cm) whereas moose can walk through meter-deep snow (Kelsall and Prescott 1971). Both move to coniferous forest in late winter because the snow is less deep there (Telfer 1970; Rolley and Keith 1980).

The stress of cold temperature has resulted in various adaptations to conserve energy, the most notable being the hibernation of ground squirrels during winter and the dormancy and lowering of body temperature of bears. Hummingbirds also lower body temperature overnight to about 15°C or when resting in cold conditions, a state called **torpor**. The limiting effect of temperature on ground squirrels operates indirectly through soil type, slope, and aspect. Squirrels need to dig burrows deep enough to avoid the cold and this requires sandy, friable soil. They also need to avoid being swamped by melt water in spring, so burrows are situated on slopes where water can drain away. Similarly, in Australia, the distribution of rabbits within the 27°C isotherm is influenced by soil type, soil fertility, vegetation cover, and distribution of water (Parer 1987).

7.4.2 Range limited by water loss and heat stress

High temperatures are often combined with high solar radiation and restricted water supplies. In high-rainfall areas the last factor is important for restricting distribution; in arid regions all three have interrelated effects on animals. These effects are expressed as heat loads built up in the body, and there are various adaptations to overcome them.

Adaptations to high temperatures include behavioral responses such as using shade in the middle of the day and restricting feeding to the hours of darkness. Both eland (*Taurotragus oryx*) and impala (*Aepyceros melampus*) reduce heat stress by feeding at night in East Africa (Taylor 1968a). At the driest times of year both species boost water intake by switching from grazing grasses and forbs to browsing on succulent shrubs (Taylor 1969; Jarman 1973).

Solar radiation restricts the movements of animals that are large and that have dark coats. Elephant and buffalo are examples where they seek shade in the heat of the day to cool off (Sinclair 1977). Coat color and structure can reduce heat loads. The lighter tan-colored coat of hartebeest (*Alcelaphus buselaphus*) reflects 42% of short-wave solar radiation as against only 22% for the darker coat of eland. In both species re-radiation of long-wave thermal radiation is greater than that absorbed, and this represented 75% of total heat loss (Finch 1972).

High heat loads can be avoided by sweating when water is abundant. African buffalo, eland, and waterbuck use sweating for evaporative cooling (Taylor 1968a; Taylor *et al.* 1969b). Buffalo keep body temperature in the range 37.4–39.3°C and allow body temperature to rise to 40°C only when water is restricted. They cannot reduce water loss from sweating when water is restricted (Taylor 1970a,b). Waterbuck show similar physiological adaptations. When water is restricted for 12 hours at 40°C ambient (environmental) temperature they lose 12% of their body weight compared with the 2% for beisa oryx (*Oryx beisa*) which is a desert-adapted species (Taylor *et al.* 1969b). As a consequence both buffalo and waterbuck must remain within a day's walk of surface water.

Large animals can afford to lose water by sweating but smaller animals such as the gazelles cannot. They employ panting instead, as do species in arid areas (e.g. the beisa oryx) or species on open plains with high solar radiation, such as wildebeest (Robertshaw and Taylor 1969; Taylor *et al.* 1969a; Maloiy 1973).

Some species can adapt to extreme arid conditions by allowing their body temperature to rise before they start panting: up to 43°C for Thomson's gazelle (*Gazella thomsonii*) and 46°C for Grant's gazelle (*G. granti*) (Taylor 1972). Other adaptations for water conservation include restriction of urine output, concentrating the urine, and reabsorbing water from the feces. Dikdik, a very small antelope that lives in semi-arid scrub away from water, had the lowest fecal water content and the highest urine concentration of all antelopes, followed by hartebeest, impala, and eland (Maloiy 1973).

Grazing ungulates in Africa are restricted to areas within reach of surface water and all show behavioral adaptations such as night feeding or migration (Sinclair 1983). Those that can do without water are all browsers (Western 1975). Beisa oryx and Grant's gazelle select hygroscopic shrubs (*Disperma* species). They eat them at night because these shrubs contain only 1% free water in the day but absorb water from the air at night to boost the water content of the leaves to 43% (Taylor 1968b).

Perhaps not apparent at first sight is the restricted availability of water for wildlife in cold regions. Not only are many of those regions deserts, as their rainfall is low, but during winter the moisture is available only as snow, and valuable energy is needed to melt it. Arctic mammals go to some lengths to conserve water. Caribou recycle nitrogen to reduce the formation of urine, thereby conserving water.

7.4.3 *Range limited by day length and seasonality*

The distribution of many North American birds is limited at northern latitudes by season length, the number of days available for breeding above a certain temperature. This is another aspect of temperature limitation. However, the southern boundary is limited by day length, the number of hours available for feeding themselves and their young (Emlen *et al.* 1986; Root 1988).

Seasons are highly predictable in the northern temperate latitudes of North America and Eurasia, and many birds and mammals have evolved a response to **proximate factors** (i.e. the immediate factors affecting an animal), particularly day length (photoperiod), which trigger conception and result in the production of young during optimum conditions. Such conditions are the **ultimate factors** (i.e. the underlying selection pressure) to which an animal is adapted by breeding seasonally (Baker 1938).

Increasing photoperiod determines the start of the breeding season in many bird species (Perrins 1970), while declining photoperiod triggers the rut in caribou; the rut is so synchronized that most conceptions occur in a mere 10-day period starting around the first day of November (Leader-Williams 1988). Moose and elk also have highly synchronized birth seasons (Houston 1982), which suggests photoperiodic control of reproduction.

Among tropical ungulates only the wildebeest is known to respond to photoperiod. In southern Africa it uses solar photoperiod to synchronize conceptions, but near the equator where solar photoperiod varies by only 20 minutes in the year it is cued by a combination of lunar and solar photoperiod (Spinage 1973; Sinclair 1977).

In variable environments with less predictable seasons, as in the tropics and arid regions, animals tend not to use photoperiod to anticipate conditions but rather adjust their reproductive behavior to the current conditions. Thus, tropical birds begin breeding when the rainy season starts, responding to the increase in insect food supply and the growth spurt of the vegetation (Sinclair 1978). In some arid areas such as Western Australia the seasonality of rain is relatively predictable but its location is not. Emus (*Dromaius novaehollandiae*) there travel long distances searching for areas that have received rain (Davies 1976), as do male red kangaroos (*Macropus rufus*) (Norbury *et al.* 1994).

Most ungulates produce their young during the wet season in Africa and South America, but put on fat before giving birth. This fat is then used during lactation, the period when the energy demands on the female are highest (Ojasti 1983; Sinclair 1983). Therefore, nutrition in the seasonal tropics becomes both the proximate and ultimate factor determining the timing of births. An example is provided by the lechwe (*Kobus lechee*), an African antelope that lives on seasonally flooded riverine grasslands (Fig. 7.3). During the peak of the floods animals are confined to the less preferred surrounding woodlands. The greatest area of flood plain is exposed at the low point in the flood cycle and it is at this time, corresponding with greatest availability of food, that births take place. In Zambia the peak of births occurs in the dry season 3 months after the rains; in the Okavango swamp of Botswana it occurs in

Fig. 7.3 The numbers of lechwe, a flood plains antelope of southern Africa (●, left axis), increase on the flood plains as water recedes in the Chobe River exposing the greatest area of high-quality food. The recruitment of newborn per 100 females (○, right axis) shows that births occur at this time. (After Sinclair 1983, which is after Child and von Richter 1969.)

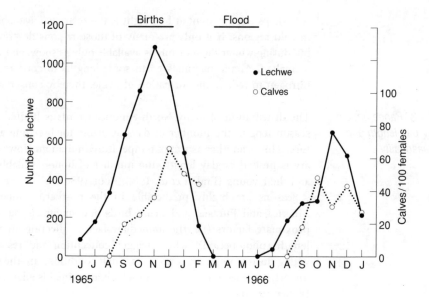

the middle of the wet season 9 months after the previous rains; but both occur when the swamp grasslands are most available.

7.4.4 *Range limited by biotic factors*

So far we have discussed range limitation by abiotic environmental factors. However, these abiotic factors can interact with biotic processes such as predation and competition to further limit a species' range. For example, the geographic distribution of arctic fox (*Alopex lagopus*) is largely in the tundra of the Holarctic and is separate from that of the more southerly red fox (*Vulpes vulpes*). However, their ranges overlap in some areas of North America and Eurasia. The northern limit of the red fox's range is determined directly by resource availability, which is determined by climate. The southern limit of the arctic fox's range is determined by interspecific competition with the more dominant red fox (Hersteinsson and Macdonald 1992).

7.5 **Distribution, abundance, and range collapse**

A major pattern in ecology is the positive relationship between the range of a species and its abundance. In general, locally abundant species have wide ranges whereas rare species have narrow ranges (Brown 1995; Gaston and Blackburn 2000). This observation has led to Rapoport's rule, namely that the latitudinal extent of a species' range increases towards the poles (Rapoport 1982). This general pattern is modified by species richness, rainfall, vegetation, and land surface as seen in studies of birds (Gentilli 1992) and mammals (Pagel *et al.* 1991; Letcher and Harvey 1994; Smith *et al.* 1994).

Of great importance in conservation management is what happens to a species' range when the population declines. One expects that population densities tend to be higher at the center of a population's range than at the periphery. Geographic ranges should collapse from the outside, with the center being the last to go (Brown 1995). Analyses of range contractions in a wide variety of animals and plants suggest that populations often collapse first in the center, leaving isolated fragments on the periphery (Lomolino and Channell 1995; Channell and Lomolino 2000). These collapses were due to the variety of causes outlined by Caughley (1994). Thus, peripheral populations not only provide a refuge for endangered species but also represent genetic and morphological varieties that differ from central populations (Lesica and Allendorf 1995).

7.6 Species reintroductions or invasions

Many species of wildlife have been eliminated from their traditional range, for one reason or another. This can even happen to common species, like the plains bison (*Bison bison*). Europeans arriving in North America encountered millions of bison on the Great Plains. In remarkably short order, this massive population was nearly extirpated, through a combination of commercial hunting by Europeans and subsistence hunting by aboriginal groups, competition with livestock, and fencing off of migration routes (Isenberg 2000). Since the turn of the century, the plains bison has been re-established by wildlife authorities to parts of its former range, though in nothing like its former abundance. Such reintroductions are becoming ever more common.

In other cases, species have naturally recovered from catastrophic decline, expanding into their former range. A well-documented example is the California sea otter *Enhydra lutris* (Lubina and Levin 1988). This species was nearly exterminated throughout its Pacific coast range through overharvesting by fur traders in the late nineteenth century, before a moratorium on harvesting was signed in 1911. A small relict population of otters survived in an inaccessible part of the California coast south of Monterey Bay. This small population provided the nucleus for gradual spread of the population both northwards and southwards along the coast.

Whether intentional or accidental, such reintroductions have some fascinating characteristics that have important bearing on their successful conservation. Key among these is the interplay between demography and patterns of movement.

7.6.1 Diffusive spread of reintroduced species

Although there are many elegant ways to model patterns of movement by invasive or reintroduced species (Turchin 1998), simple random walk models can often predict the pattern of spread surprisingly well. We first consider what is meant by a random walk, then use this algorithm to develop a simple model of population distribution.

What pattern would emerge over time, for a single individual that moves randomly every day of its life? We will assume that this hypothetical animal can only move forwards, backwards, or sideways, one step at a time. We further assume that each of these events is as probable as remaining where it is. To model this, we need to sample randomly from a uniform probability distribution (see Box 7.1).

Box 7.1 Modeling a random walk in space.

We first need to randomly sample a large number of values uniformly distributed between 0 and 1, assigning each random number on this interval to either the variable *p* or the variable *q*. We then use these probabilities to mimic forwards versus backwards movement using the variable *x* and side-to-side movements using the variable *y*, using the logic shown below. As a result of this logic, an animal would move left one-third of the time, right one-third of the time, and stay on track one-third of the time. Similar probabilities correspond to forwards, backwards, and stationary outcomes.

$$x_{t+1} = \begin{cases} x_t - 1 & \text{if} \quad P_t < 0.333 \\ x_t + 1 & \text{if} \quad P_t > 0.333 \wedge P_t < 0.667 \\ x_t & \text{otherwise} \end{cases}$$

$$y_{t+1} = \begin{cases} y_t - 1 & \text{if} \quad q_t < 0.333 \\ y_t + 1 & \text{if} \quad q_t > 0.333 \wedge q_t < 0.667 \\ y_t & \text{otherwise} \end{cases}$$

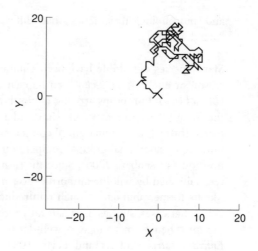

Fig. 7.4 Hypothetical trajectory over 100 time steps for a single individual following a random walk, starting from the origin (0,0).

For this kind of random walk model, most trajectories tend to find their way back to a position not far from the initial starting point (Fig. 7.4). In other words, walking randomly is not a very effective means of getting anywhere new. This is a useful null model, however, that sets an extreme standard against which we might evaluate the movements of real organisms. The random walk model is perhaps most plausible at large spatial scales, such as for dispersing juveniles, in which animals have no past experience with local conditions.

We can readily expand this kind of model to a group of individuals (Case 2000). To keep it simple, we will concentrate on only one spatial dimension, such as for sea otters dispersing up and down the coast of California. Let's say that there are 100 individuals released at a central position "0" and that each individual has a 20% probability of moving left and a 20% probability of moving right, with position along this axis indicated by the variable x. This probability we will term "d" for dispersal. Local changes in density of individuals can be modeled in the following manner:

$$N_{x,t+1} = N_{x,t} - 2dN_{x,t} + dN_{x-1,t} + dN_{x+1,t}$$

The local population loses $2 \times d \times N$ individuals due to movement in either direction, but gains $d \times N$ individuals from each adjacent site. We need to repeat this exercise over the full range of distance intervals.

The output of this model demonstrates two important features (Fig. 7.5). First, the spatial distribution of individuals in the population begins to take on a bell-shaped or normal distribution over time. Second, the rate of spread is initially fast, but slows over time. This is because movement away from the release point is balanced to a considerable degree by movement backwards. This slower movement away becomes more pronounced over time because the distribution is getting flatter. When dynamics are driven purely by random motion the population range spreads at a rate proportional to √time. If we repeat this simulation with a larger fraction of dispersers (say $d = 0.3$), the rate of spread will increase accordingly. The rate of spread is proportional to √d.

We can use our random walk model to derive the differential equation that defines diffusive changes in local density, over a continuous gradient of space and time:

Fig. 7.5 Variation in the population density of individuals over time, when those individuals redistribute themselves every time step (t) according to an unbiased random walk.

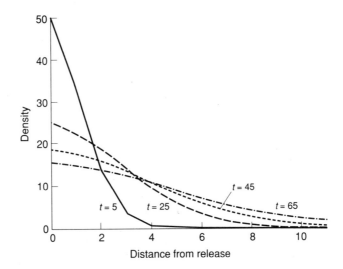

$$\Delta N_x = N_{x,t+1} - N_{x,t} = -2dN_{x,t} + dN_{x-1,t} + dN_{x+1,t}$$

We then rearrange terms on the right-hand side of the equation:

$$\Delta N_x = d[(N_{x-1,t} - N_{x,t}) - (N_{x,t} - N_{x+1,t})]$$

The rate that individuals accumulate at site x depends on the degree of difference between the density gradient below the site and the density gradient above the site. In other words, it is not the gradient itself, but the rate of change of the density gradient over space that dictates the rate of diffusive movement. Mathematicians refer to the rate of change of the density gradient as the second derivative. If this occurs over short enough intervals of time and space, then the result is the following differential equation (called the diffusion equation in one dimension):

$$\frac{dN(x, t)}{dt} = D\frac{\partial^2}{\partial x^2}N(x, t)$$

The solution to this equation is the normal distribution:

$$N(x, t) = \frac{N_0}{\sqrt{4\pi Dt}}\exp\left[\frac{-(x - \mu)^2}{4Dt}\right]$$

where t is the time since the animals were released, μ is the initial position (usually 0), and D is the diffusion coefficient. It reflects how fast individuals tend to diffuse away from an initial point of release. We discuss how to calculate it below. This equation may look familiar – it is closely related to the normal (sometimes called Gaussian) probability distribution. The variance in spatial locations is given by $\sigma^2 = 2Dt$.

The easiest way to estimate the diffusion coefficient D is to estimate the mean-squared displacement of the individuals in the population over time. One simply measures the distance of a given individual from its original release point, squares that displacement to get rid of positive versus negative values, sums the squared displacements for all individuals, and divides this sum by the total sample size to estimate mean-squared displacement. D is then calculated by dividing mean-squared displacement by $2t$.

In the more typical case of diffusion in two dimensions (x and y, centered at the release point), these equations are slightly altered:

$$\frac{dN(x,\ y,\ t)}{dt} = D\left[\frac{\partial^2}{\partial x^2}N(x,\ y,\ t) + \frac{\partial^2}{\partial y^2}N(x,\ y,\ t)\right]$$

$$N(r,\ t) = \frac{N_0}{4\pi Dt}\exp\left(\frac{-r^2}{4Dt}\right)$$

where r is the distance (i.e. radius) from the release point. Despite the slight change in formula, this equation also predicts that the range occupied by the population is proportional to $\sqrt{\text{time}}$. This is a very useful prediction that differs from other models of population spread, as we shall shortly see.

7.6.2 Spread of reintroduced species: diffusion + exponential growth

As we discussed in Chapter 6, a newly reintroduced population is likely to have plenty of resources with which to grow and multiply. This logically leads to geometric or exponential growth, at least in the initial period following release. Unrestricted population growth can be readily incorporated in our random walk model of population spread. We simply multiply the local population by the finite rate of growth λ (in this case, let us say that $\lambda = 1.05$):

$$N_{x,t+1} = \lambda N_{x,t} - 2dN_{x,t} + dN_{x-1,t} + dN_{x+1,t}$$

The rate of spread now seems to be much more consistent over time (Fig. 7.6) than was the case for diffusive movement alone (Fig. 7.5). In fact, the population range now spreads at a rate proportional to t and d because the population grows fastest where density is highest. This relationship tends to create a rapid rate of change in

Fig. 7.6 Variation in the population density of individuals over time, when those individuals redistribute themselves every time step (t) according to an unbiased random walk. Unlike Fig. 7.5, the population is also growing at an annual rate of $\lambda = 1.05$.

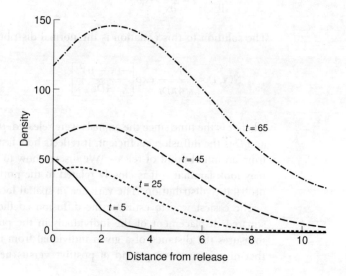

the density gradient, which we have already suggested fuels a high rate of diffusion. The net result is a population that explodes over both time and space.

Another interesting feature of the diffusion + exponential growth model is that a standing wave of animals spreads over time across the landscape (Fig. 7.6), rather than the gradually eroding "mountain" seen in the pure diffusion model (Fig. 7.5). This wavelike form of spread is echoed in most models that incorporate population growth as well as diffusive movement, such as those with logistic growth or predator–prey models. The velocity with which this wave rolls across the landscape is identical in virtually all such models: $v = 2\sqrt{(\lambda D)}$.

7.6.3 *Empirical tests of diffusion theory*

We should be able to discriminate between alternative models of population spread by looking at population range versus time. If the rate of increase of the radius of population distribution becomes less over time, then this deceleration would be consistent with a pure diffusion process, in which population growth is not involved. On the other hand, constant increase in radial spread of the population would be most consistent with the diffusion + exponential growth model.

Skellam (1951) made this comparison using data on a population starting with five muskrats (*Ondatra zebithica*) translocated into the countryside near Prague in 1905 (Fig. 7.7). Skellam's analysis, supported by more rigorous analysis by Andow *et al.* (1990), clearly demonstrated that the radial spread of muskrats increased linearly over time, at a rate of 11 km/year (Fig. 7.8), thus supporting the exponential diffusion model.

Similar analysis of the naturally recovering population of California sea otters also supports the diffusion + exponential growth model (Lubina and Levin 1988), although the pattern is more complex. Radial spread to the north was slower than that to the southern California coast. Moreover, there seemed to be a dramatic jump in the distance dispersed per year as the otters moved into sandy coastal areas with less of their preferred rocky habitat.

Since those early days, there has been considerable development of alternative models of population spread. These recognize directional bias on the part of the

Fig. 7.7 Spatial spread over time of a small population of muskrats introduced into the countryside near Prague. (After Elton 1958.)

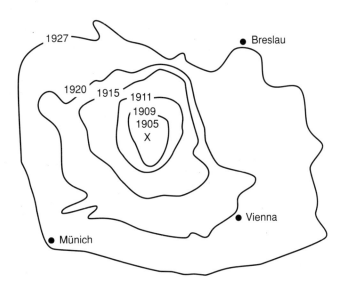

Fig. 7.8 Radial spread (measured as the square root of area) of the population of muskrats introduced into the countryside near Prague. (After Skellam 1951.)

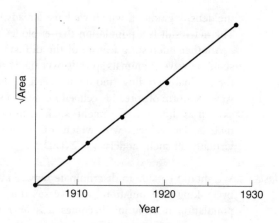

disperser, changes in disperser motivation, or heterogeneous environmental effects on dispersal tendency (Turchin 1998). Nonetheless, the simple model of diffusive spread combined with exponential growth often does a tolerably good job of predicting patterns of population spread over time. These successful predictions suggest that both rapid population growth at the wave front and some degree of randomness in the pattern of movement contribute heavily to observed patterns of spread in many wildlife species. Both the theory and empirical mechanisms underlying animal movement across complex ecological landscapes are now developing rapidly, because both have important conservation implications.

7.7 Dispersal and the sustainability of metapopulations

Dispersal also plays a key role in understanding the dynamics of species that are subdivided, for one reason or another, into discrete subpopulations. Provided that there is some degree of dispersal amongst subpopulations, ecologists refer to the larger aggregate as a **metapopulation**, or population of populations (Hanski and Gilpin 1997). Metapopulations can occur in a variety of contexts. Bird species on continental islands are an obvious example (Sæther *et al.* 1999). However, it is just as valid to think of butterflies inhabiting grassy glades in a matrix of boreal forest as a metapopulation (Hanski *et al.* 1994). Since 1980 there has been a surge in interest in metapopulation dynamics, fueled in part by the recognition that human environmental impacts often lead to fragmentation of natural areas, creating effective metapopulations from populations that were continuously distributed in the not-so-distant past. Here we outline some of the basic principles of metapopulation dynamics, particularly with relation to the impact of further habitat loss.

7.7.1 *Metapopulation dynamics of a single species*

There are many ways one can represent metapopulations, but the Levins model (Levins 1969) and its subsequent modifications (reviewed by Gyllenberg *et al.* 1997) have perhaps been the most influential. Let p be the proportion of occupied sites, c the probability of successful colonization, and e the probability of extinction of an occupied site. The rate of change in the number of occupied sites is calculated in the following manner:

$$\frac{dp}{dt} = cp(1 - p) - ep$$

Fig. 7.9 Dynamics over time of a metapopulation with colonization rate $c = 0.90$ and extinction rate $e = 0.45$.

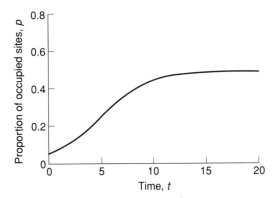

The first term represents colonization of new sites, and the second term represents extinction. Provided that $c > e$, this model predicts that the proportion of occupied sites will converge over time on the following equilibrium:

$$p_{eq} = 1 - \frac{e}{c}$$

This can be clearly seen in the simulation shown in Fig. 7.9. The stability of the metapopulation at equilibrium belies a constant turnover of subpopulations. A substantial fraction of sites (45% in fact) go extinct per unit time. This extinction rate does not get translated into a dangerous collapse of population because of a stream of colonists from the remaining occupied sites. As we shall discuss in Chapter 17, local extinction is expected to become common when subpopulations have been reduced to low numbers, simply due to chance demographic events or rapid genetic loss. Hence, fragmentation of the environment into numerous small patches creates a situation in which local extinction risk is a very real possibility.

Empirical data consistent with the metapopulation scenario are accumulating. One of the best-documented examples is Hanski and co-workers' studies of the Glanville fritillary (*Melitaea cinxia*), an endangered butterfly inhabiting a number of forested islands in the eastern Baltic Sea, off the coast of Finland (Hanski *et al.* 1994; Saccheri *et al.* 1998). Field studies have concentrated on one island, Åaland, in particular.

The spatial distribution of butterflies on Åaland is quite patchy, in keeping with the patchy distribution of the larval food plants. The Finnish team repeatedly censused the number of butterfly larvae at each of several hundred locales. As the larvae are colonial and quite conspicuous, it is relatively straightforward to ascertain whether local extinction has taken place in the small grassy meadows. Results of the repeated censusing demonstrated that extinction was common amongst these local subpopulations, in accordance with metapopulation theory (Fig. 7.10).

As we might expect, many factors influenced the risk of extinction, including size of the local subpopulation, degree of genetic variability, and the degree of isolation from neighboring sites (Hanski *et al.* 1994; Saccheri *et al.* 1998). A high degree of turnover of local populations was normal, with the overall prevalence determined by the probabilities of colonization versus extinction. Unfortunately, recent population trends suggest that the Glanville fritillary may be fighting a losing battle against extinction.

Fig. 7.10 A map of Åaland, showing sites of local subpopulations of the Glanville fritillary butterfly. Sites with suitable host plants occupied by larvae in 1995 are shown by filled symbols, whereas open circles depict unoccupied sites. A subset of 42 occupied sites were studied in detail, shown by large circles. Seven of these subpopulations, depicted by triangles, had gone extinct by the next sampling period. (After Saccheri *et al.* 1998.)

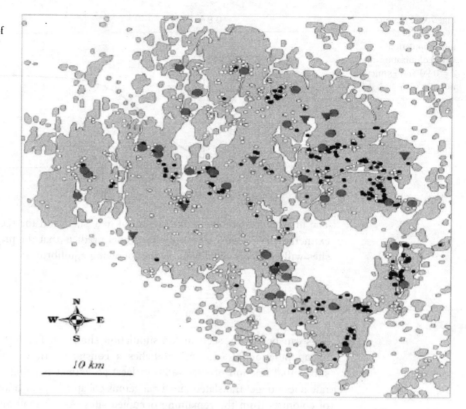

Well-studied examples of metapopulation dynamics in vertebrates are less common. Long-term studies of pool frogs along the coast of Sweden demonstrate a steady pattern of subpopulation turnover (Sjögren Gulve 1994). Similar patterns of extinction and recolonization have been shown in a number of other systems: cougars inhabiting chapparal shrub patches in urban southern California (Beier 1996), pikas living on mine tailings in the Sierra Nevada mountains of California (Smith and Gilpin 1997), sparrows on windswept islands off the coast of Norway (Sæther *et al.* 1999), and beavers inhabiting isolated ponds in Canada (Fryxell 2001). There is no doubt that the preconditions for metapopulation dynamics exist. The unresolved question is how common they might be.

7.7.2 Habitat loss and metapopulation collapse

Levin's (1969) simple metapopulation model can be readily modified to predict the effect of habitat loss. Let H reflect the proportion of sites destroyed by man, so that any propagule that lands on a degraded site cannot persist. The dynamics of this degraded environment are depicted as follows:

$$\frac{dp}{dt} = cp(1 - H - p) - ep$$

As a result of habitat loss, the equilibrium level of occupancy is reduced (Fig. 7.11). If H is large enough, the metapopulation may not be able to persist at all. This is a simple, but graphic, way to look at the potential costs of habitat degradation. Empirical examples of habitat degradation leading to extinction are largely anecdotal, but nonetheless

Fig. 7.11
Metapopulation
dynamics over time for
the same parameters as
in Fig. 7.9 ($c = 0.90$,
$e = 0.45$), now with
substantial habitat loss
($H = 0.3$) among the
previously inhabitable
sites. This leads to
a reduction in the
metapopulation
equilibrium from
50% to 20%.

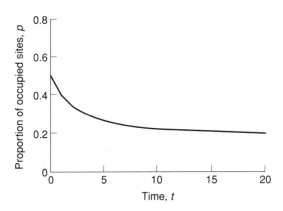

abound in the natural history literature, as we discuss in Chapter 18. The importance of habitat loss may be magnified in the future, if man does not learn to limit habitat fragmentation and prevent further alienation of subdivided patches of habitat.

7.7.3 Fragmented territorial systems

Through natural or human-influenced patterns of disturbance, suitable patches of habitat for breeding can become widely separated. Individuals setting up breeding territories must locate these suitable patches in an unfavorable matrix before they can set up a territory to attract mates (Lande 1987). Under many circumstances, the dynamics of territory occupancy by breeding pairs and their offspring are logically similar to those of a classic metapopulation (Noon and McKelvey 1996). Extinction is amplified when the probability of territory (patch) colonization is low relative to the probability of territory vacancy (local extinction) arising due to mortality. A key difference, however, is that successful breeding requires that both a male and a female independently discover a suitable territory site. At low probabilities of discovery, some individuals may never find mates. This **Allee effect** (after Allee (1938) who defined the process) could cause extinction if the overall level of territory occupancy falls below a critical level (Lande 1987, 1988; Courchamp *et al.* 2000a).

It is possible to imagine a metapopulation of local protected areas, each of whose internal dynamics are determined by the fragmented territory model (Lamberson *et al.* 1992, 1994). Under these circumstances, similar levels of species persistence could be obtained by a few large reserves, even if they are widely spaced, or a large number of small reserves, each of which is much more vulnerable in isolation to extinction. Depending on which circumstances prevail, the management priority would switch from maintaining territory quality within specific reserves to enhancing dispersal across a large network of reserves (Noon and McKelvey 1996).

A good example of this kind of situation involves the northern spotted owl (*Strix occidentalis*) of the western USA (Lande 1988). Spotted owls require substantial tracts of old growth forest for their breeding territories, but 80% or more of the mature forest in the northwestern USA has been logged over the past half century. As a consequence, local populations of owls are increasingly isolated from each other by large areas of clear-cutting. Moreover, variation in forest structure at the local level can influence territory occupancy. Concern about the long-term viability of northern spotted owl populations led to debates about appropriate management schemes for the public forest lands in the Pacific Northwest (Doak 1989; Lamberson *et al.* 1992, 1994; Doak and Mills 1994). The solution was both to control the future loss of mature forest and to manage the spatial pattern of forest utilization in such a way as to maintain an effective metapopulation structure (Murphy and Noon 1992; Noon and McKelvey 1996). This kind of management controversy will only become more common with further fragmentation of existing wild lands.

7.7.4 Source–sink and island–mainland systems

Special kinds of metapopulation dynamics occur when some patches are large enough or productive enough to sustain permanent subpopulations, whereas other patches are small enough or unproductive enough that local extinction is common. If both the permanent patches and the transient patches can support positive population growth, such an arrangement is termed a **mainland–island system**. Examples include the checkerspot butterfly (*Euphydryas editha bayensis*), which inhabits scattered patches of serpentine soil in coastal California (Harrison *et al.* 1988) and spiders in the Bahama Islands (Schoener and Spiller 1987).

In other cases, only a fraction of patches can sustain positive subpopulation growth, whereas individuals in other patches always experience higher rates of mortality than birth. Such an arrangement is referred to as a **source–sink system**, with source sites supplying a steady stream of dispersers that fan out to surrounding sinks (Pulliam 1988; Pulliam and Danielson 1991). Despite the fact that sinks are incapable of supporting viable local populations, through immigration from source patches they can have substantial numbers of individuals. Beavers (*Castor canadensis*) inhabiting shallow lakes in the mixed deciduous and boreal forest of southern Ontario provide a good example of a mammalian species with source–sink dynamics (Fryxell 2001). Beavers at a small fraction of colonies have sufficient food supplies to support substantial production of offspring year after year. These populate the surrounding area when they disperse. Most of the other colonies rarely produce viable young.

Clearly, the conservation needs of mainland–island and source–sink systems differ from those of classic metapopulations. Mainland or source sites take on disproportionate importance in sustaining viable populations over the larger landscape. Loss of even small amounts of these critical source or mainland habitats could be unsustainable.

7.8 Summary

The distribution is the area occupied by a population or species, the dispersion is the pattern of spacing of the animals within it, and dispersal, migration, and local movement are the actions that modify dispersion and distribution. Dispersion and distribution are states; dispersal, migration, and local movement are processes. The edge of the distribution is that point at which, on average, an individual just fails to replace itself in the next generation. Its position may be set by climate, substrate, food supply, habitat, predators, or pathogens. The limiting factor can often be identified by the trend in density from the range boundary inward.

Dispersal plays a key role in dictating the rate of spread of a species reintroduced into a new area or one recovering from catastrophic decline. Diffusion models are often an effective means of modeling the spread of reintroduced species, particularly if they incorporate both demographic and random walk processes. We demonstrate the logical basis for the simplest random walk and diffusion models. Dispersal is also integral to the dynamics of organisms occupying spatially subdivided habitats forming metapopulations. Simple models demonstrate that the long-term persistence of metapopulations depends on the relative probabilities of extinction versus dispersive colonization. There is some empirical evidence for regular turnover of colonies and high rates of extinction and colonization. Variations on the metapopulation theme include source–sink systems, island–mainland systems, and metapopulations with internal territory structure.

8 Population regulation, fluctuation, and competition within species

8.1 Introduction

In this chapter we first describe the theory and evidence for the stability of populations through **regulation**. We then analyze the processes that can cause fluctuations and population cycles, using models to develop our understanding of the processes. Finally, we examine one of the major causes of regulation, namely competition between individuals for resources, or **intraspecific competition**. Other causes of regulation such as predation will be dealt with in Chapter 10. Chapter 12 outlines an alternative approach to analyzing resource use.

8.2 Stability of populations

If we look at long-term records of animal populations we see that some populations remain quite constant in size for long periods of time. Records of mute swans (*Cygnus olor*) in England from 1823 to 1872 (Fig. 8.1) illustrate that although the population fluctuates, it remains within certain limits (190–1150). Other populations, such as those of insects or house mice (*Mus domesticus*) in Australia (Fig. 8.2), fluctuate to a much greater extent and furnish no suggestion of an equilibrium population size. Nevertheless such populations do not always go extinct and they remain in the community for long periods. Occasionally one finds unusual situations where populations show regular cycles. The snowshoe hare (*Lepus americanus*) in northern

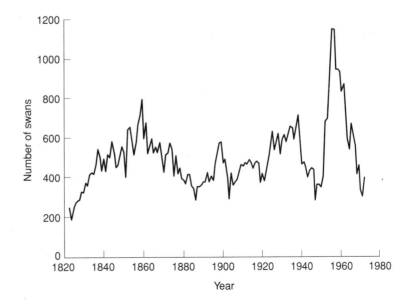

Fig. 8.1 Some populations remain within relatively close bounds over long time periods. The mute swan population of part of the river Thames, England (estimated by total counts) shows a steady level or gentle increase despite some perturbations due to severe winters, for example in 1946–47 and 1963–64. (Data from Cramp 1972.)

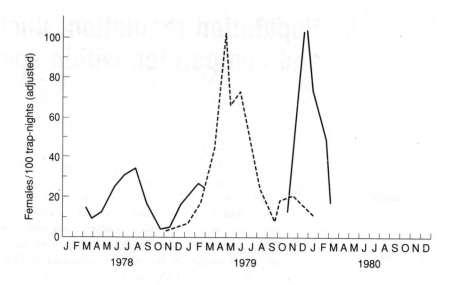

Fig. 8.2 Density indices for old female house mice on contour banks and in stubble fields of rice crops in southeastern Australia. Broken lines distinguish the crop cycle cohort of 1978–79 from that of 1977–78 and 1979–80. The extent of the peak in January 1980 is unknown due to a poisoning campaign. (After Redhead 1982.)

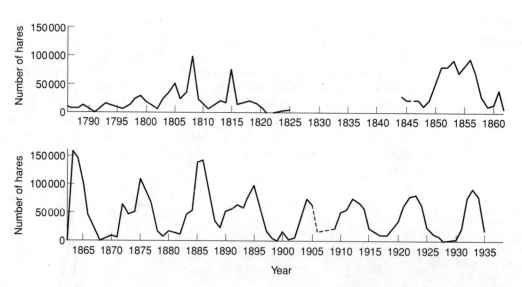

Fig. 8.3 Snowshoe hares in the boreal forest of Canada show regular fluctuations in numbers with a 10-year periodicity. Data are from the Hudson Bay Company fur records up to 1903 and questionnaires thereafter. (After MacLulich 1937.)

Canada shows the clearest (Fig. 8.3), as indicated by the furs collected by trappers for the Hudson Bay Company over the past two centuries (MacLulich 1937).

This relative constancy of population size, or at least fluctuation within limits, is in contrast to the intrinsic ability of populations to increase rapidly. The fact that population increase is limited suggests that there is a mechanism in the population that slows down the rate of increase and so regulates the population. We discuss first the theory for how populations might be **limited** and **regulated**.

8.3 **The theory of population limitation and regulation**

8.3.1 *Density dependence*

Populations have inputs of births and immigrants and outputs of deaths and emigrants. For simplicity we will confine discussion to a self-contained population having only births (B) and deaths (D) per unit time.

If either the proportion of the population dying increases or the proportion entering as births decreases as population density increases then we define these changes in proportions as being **density dependent**. The underlying causes for the changes in these rates are called **density-dependent factors**.

Births and deaths as a proportion of the population $(B/N_t, D/N_t)$ can be related to the instantaneous birth (b) and death (d) rates in the following way.

The change in population per unit time is:

$$N_{t+1} - N_t = B - D$$

the instantaneous rate of increase (r) is given by:

$$r = b - d$$

and the finite rate of increase (λ) is given by:

$$\lambda = N_{t+1}/N_t = e^r$$

Therefore:

$$e^{b-d} = (N_{t+1}/N_t) = (B - D + N_t)/N_t$$

If $d = 0$, $D = 0$ then:

$$e^b = (B + N_t)/N_t = [1 + (B/N_t)]$$

and

$$b = \log_e[1 + (B/N_t)]$$

Similarly if $b = 0$, $B = 0$, and D/N_t is much less than 1, then:

$$d = \log_e[1 + (D/N_t)]$$

If B and D fall in the range of 0–20% of the population then b and d are nearly linear on N, and they remain approximately linear even if B and D are 20–40% of N. This range covers most of the examples we see in nature, so for our purposes we can say that D/N_t and B/N_t change with density in the same way as do b and d, and both go through the origin.

In Fig. 8.4a we plot b against density (or population size) N as a constant so that it is a horizontal line. If we now plot d as an increasing function of density, we see that where the two lines cross, $b = d$, and the population is stationary, at the equilibrium point K. The difference between the b and d lines represents r, and this declines linearly as density increases, in the same way as it does for the logistic curve (see Section 8.6). In Fig. 8.4a the decline in r is due solely to d being density dependent.

Fig. 8.4 Model of density-
dependent and density-
independent processes.
(a) Birth rate, *b*, is held
constant over all densities
while mortality, *d*, is density
dependent. The population
returns to the equilibrium
point, *K*, if disturbed. The
instantaneous rate of increase,
r, is the difference between *b*
and *d*. (b) As in (a) but *b* is
density dependent and *d* is
density independent. (c) Both
b and *d* are density dependent.
(d) *d* is curvilinear so that the
density dependence is stronger
at higher population densities.

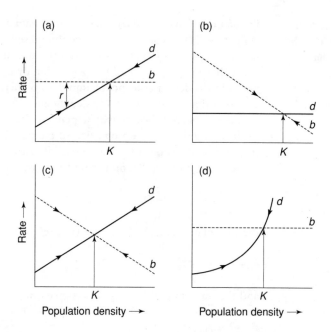

Since *b* (or B/N_t) is constant in this case we describe it as **density independent** (i.e.
it is unrelated to density). In real populations density-independent factors such as
weather may affect birth and death rates randomly. Rainfall acted in this way on greater
kudu (*Tragelaphus strepsiceros*) in Kruger National Park, South Africa, causing mortality of juveniles (Owen-Smith 1990).

We can apply the same arguments if we assume that *b* is density dependent and
d is density independent (Fig. 8.4b) or if both are density dependent (Fig. 8.4c). So
far we have assumed that the density-dependent factor has a linear effect on rate of
increase as in the logistic curve. However, density-dependent mortality is more likely
to be curvilinear, as in Fig. 8.4d.

8.3.2 *Limitation and limiting factors*

In Fig. 8.5 we take the argument a little further. Let us assume a constant (density-
independent) birth rate *b*. Shortly after birth a density-independent mortality d_1 (depicted
here as a constant) kills some of the babies so that inputs are reduced to b_1. There
follows a density-dependent mortality d_2, and the population reaches an equilibrium
at K_3. If mortality d_1 had not occurred (or was smaller), the equilibrium population
would be at K_1. Therefore, the presence or absence of the density-independent factor causing d_1 alters the size of the equilibrium population.

The strength or severity of the density-dependent factor is indicated by the slope
of d_2. If the density-dependent factor becomes stronger such as to produce d_3 instead
of d_2, the slope becomes steeper and the equilibrium population drops from K_3 to K_4
(or K_1 to K_2 if d_1 is absent). Thus, altering the strength of density-dependent factors
also alters the size of the equilibrium population.

We define the process determining the size of the equilibrium population as **limitation**, and the factors producing this are **limiting factors**. We can see, therefore,
that both density-dependent and density-independent factors affect the equilibrium
population size and so they are all limiting factors. Any factor that causes mortality
or affects birth rates is a limiting factor.

Fig. 8.5 Model showing that the equilibrium point, K, can vary with both density-dependent and density-independent processes. Birth rate, b, is held constant over all densities. In sequence, a density-independent mortality d_1 reduces the input to the population to b_1. There follows a density-dependent mortality d_2 or d_3. The intercept of b or b_1 with d_2 or d_3 determines the equilibrium (K_1–K_4).

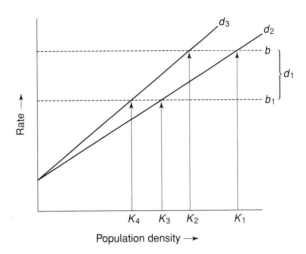

8.3.3 *Regulation*

Populations are often disturbed from their equilibrium, K, by temporary changes in limiting factors (a severe winter or drought or influx of predators might reduce the population; a mild winter or good rains might increase it). The subsequent tendency to return to K is largely due to the effect of density-dependent factors, and this process is called **regulation**. Therefore, regulation is the process whereby a density-dependent factor tends to return a population to its equilibrium. We say "tends to return" because the population may be continually disturbed so that it rarely reaches the equilibrium. Nevertheless this tendency to return to equilibrium results in the population remaining within a certain range of population sizes. Superficially it appears as if the population has a boundary to its size, and it fluctuates randomly within this boundary. However, it is more constructive to picture random fluctuations in both the density-independent (d_1) and density-dependent (d_2) mortalities as the shaded range in Fig. 8.6a. This results in a fluctuation of the equilibrium population

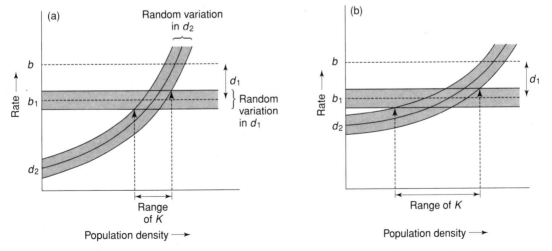

Fig. 8.6 Random variation in the mortalities d_1 and d_2 (indicated by the shaded area) are the same in (a) and (b). In (a) there is stronger density dependence at the intercept of b_1 and d_2 than in (b), and this difference results in a smaller range of equilibria, K, in (a) than in (b).

indicated by the range of K. Figure 8.6a shows that this range of K is relatively small when the density-dependent mortality is strong (steep part of the curve). Figure 8.6b shows the range of K when the density-dependent mortality is weak. We can see that the range of K (which we see in nature as fluctuations in numbers) is very much greater when the density-dependent mortality is weak than when it is strong. Note in Figs 8.6a and 8.6b that differences in amplitude of fluctuations are due to changes in the strength of the density-dependent mortality because we have held density-independent (random) mortality constant in this case.

8.3.4 *Delayed and inverse density dependence*

Some mortality factors do not respond immediately to a change in density but act after a delay. Such **delayed density-dependent factors** can be predators whose populations lag behind those of their prey, and food supply where the lag is caused by the delayed action of starvation. Both causes can have a density-dependent effect on the population but the effect is related to density at some previous time period rather than the current one. For example, a 34-year study of white-tailed deer in Canada indicated that both the population rate of change and the rate of growth of juvenile animals are dependent on population size several years previously, rather than current population size (Fryxell *et al.* 1991). A similar relationship was found with winter mortality of red grouse (*Lagopus lagopus*) in Scotland (Fig. 8.11). Delayed density dependence is indicated when mortality is plotted against current density and the points show an anticlockwise spiral if they are joined in temporal sequence (Fig. 8.11). These delayed mortalities usually cause fluctuations in population size, as we will demonstrate later in this chapter.

Predators can also have the opposite effect to density dependence, called an **inverse density-dependent** or **depensatory** effect. In this case predators have a destabilizing effect because they take a decreasing proportion of the prey population as it increases, thus allowing the prey to increase faster as it becomes larger. Conversely, if a prey population is declining for some reason, predators would take an increasing proportion and so drive the prey population down even faster towards extinction. In either case we do not see a predator–prey equilibrium. We explore this further in Chapter 10.

8.3.5 *Carrying capacity*

The term **carrying capacity** is one of the most common phrases in wildlife management. It does, however, cover a variety of meanings and unless we are careful and define the term, we may merely cause confusion (Caughley 1976, 1981). Some of the more common uses of the term are discussed below.

Ecological carrying capacity

This can be thought of abstractly as the K of the logistic equation, which we derive later in this chapter (Section 8.6). In reality it is the natural limit of a population set by resources in a particular environment. It is one of the equilibrium points that a population tends towards through density-dependent effects from lack of food, space (e.g. territoriality), cover, or other resources. As we discussed earlier, if the environment changes briefly it deflects the population from achieving its equilibrium and so produces random fluctuations about that equilibrium. A long-term environmental change can affect resources, which in turn alters K. Again the population changes by following or tracking the environmental trend.

There are other possible equilibria that a population might experience through regulation by predators, parasites, or disease. Superficially they appear similar to that equilibrium produced through lack of resources because if the population is disturbed through culling or weather events it may return to the same population size. To distinguish the equilibria produced by predation, by resource limitation, and by a combination of the two, we need to know whether predators or resources or both are affecting b and d.

Economic carrying capacity

This is the population level that produces the maximum offtake (or maximum sustained yield) for culling or cropping purposes. It is this meaning that is implied when animal production scientists and range managers refer to livestock carrying capacity. We should note that this population level is well below the ecological carrying capacity. For a population growing logistically its level is $\frac{1}{2}K$ (Caughley 1976).

Other senses of carrying capacity

We can define carrying capacity according to our particular land use requirements. At one extreme we can rate the carrying capacity for lions on a Kenya farm or wolves on a Wyoming ranch as zero (i.e. farmers cannot tolerate large predators killing their livestock).

A less extreme example is seen where the aesthetic requirements of tourism require reducing the impact of animals on the vegetation. Large umbrella-shaped *Acacia tortilis* trees make a picturesque backdrop to the tourist hotels in the Serengeti National Park, Tanzania. In the early 1970s, elephants began to knock over these trees. Whereas elephants could be tolerated at ecological carrying capacity in the rest of the park, in the immediate vicinity of the hotels the carrying capacity for elephants was much lower and determined by human requirements for scenery.

8.3.6 *Measurements of birth and death rates*

Birth rates are inputs to the population. Ideally we would like to measure conception rates (**fecundity**), pregnancy rates in mammals (**fertility**), and births or egg production. In some cases it is possible to take these measurements, as in the Soay sheep of Hirta (Clutton-Brock *et al.* 1991). Pregnancies can be monitored by a variety of methods including ultrasound, X-rays, blood protein levels, urine hormone levels, and rectal palpation of the uterus (in large ungulates). In many cases, however, these are not practical for large samples from wild populations.

Births can be measured reasonably accurately for seal species where the babies remain on the breeding grounds throughout the birth season. Egg production, egg hatching success, and fledgling success can also be measured accurately in many bird populations. However, in the majority of mammal species birth rates cannot be measured accurately, either because newborn animals are rarely seen (as in many rodents, rabbits, and carnivores) or because many newborn animals die shortly after birth and are not recorded in censuses (as in most ungulates). In these cases we are obliged to use an approximation to the real birth rate, such as the proportion of the population consisting of juveniles first entering live traps for rodents and rabbits, or juveniles entering their first winter for carnivores and ungulates. These are valid measures of recruitment.

Death rates are losses to the population. Ideally they should be measured at different stages of the life cycle to produce a life table (see Section 6.4). Once sexual maturity is reached, age classes often cannot be identified and all mortality after that age is therefore lumped as "adult" mortality. Mortality can be measured directly by using mortality radios which indicate when an animal has died, as was done by Boutin *et al.* (1986) and Trostel *et al.* (1987) for snowshoe hares in northern Canada.

Survivorship can be calculated over varying time periods by the method of Pollock *et al.* (1989).

Mortality caused by predators can also be measured directly if the number of predators (numerical response) and the amount eaten per predator (functional response) are known (see also Chapters 5, 10, and 12). Such measurements are possible for those birds of prey that regurgitate each day a single pellet containing the bones of their prey. With appropriate sampling, the number of pellets indicates the number of predators, and prey per pellet shows the amount they eat. This method was used for raptors (in particular the black-shouldered kite, *Elanus notatus*) eating house mice during mouse outbreaks in Australia (Sinclair *et al.* 1990).

8.3.7 *Implications*

We should be aware of a number of problems associated with the subject of population limitation and regulation:

1 Much of the literature uses the terms limitation and regulation in different ways. In many cases the terms are used synonymously, but the meanings differ between authors. Since any factor, whether density dependent or density independent, can determine the equilibrium point for a population, any factor affecting *b* or *d* is a limiting factor. It is, therefore, a trivial question to ask whether a certain cause of mortality limits a population – it has to. The more profound question is in what way do mortality or fecundity factors affect the equilibrium.

2 Regulation requires, by our definition, the action of density-dependent factors. Density dependence is necessary for regulation but may not be sufficient. First, the particular density-dependent factor that we have measured, such as predation, may be too weak, and other regulating factors may be operating. Second, some density-dependent factors have too strong an effect, and consequently cause fluctuations rather than a tendency towards equilibrium (see Section 8.7).

3 The demonstration of density dependence at some stage in the life cycle does not indicate the cause of the regulation. For example, if we find that a deer population is regulated through density-dependent juvenile mortality, we do not have any indication from this information alone as to the cause of the mortality. Correlation with population size is merely a convenient abbreviation that hides underlying causes. Density itself is not causing the regulation; the possible underlying factors related to density are competition for resources, competition for space through territoriality, or an effect of predators, parasites, and diseases (see Section 8.7).

8.4 Evidence for regulation

There are three ways of detecting whether populations are regulated. First, as we have seen in Section 8.3.3, regulation causes a population to return to its equilibrium after a perturbation. Perturbation experiments should therefore detect the return towards equilibrium. Similarly, natural variation in population density, provided it is of sufficient magnitude, can be used to test whether per capita growth rates decline with density (Chapter 15). Second, if we plot separate and independent populations at their natural carrying capacity against some index of resource (often a weather factor) there should be a relationship. Third, we can try to detect density dependence in the life cycle.

8.4.1 *Perturbation experiments*

If a population is moved experimentally either to below or above its original density and then returns to this same level we can conclude that regulation is occurring. An example of downward perturbation is provided by the northern elk herd of

Fig. 8.7 The wildebeest population in Serengeti increased to a new level determined by intraspecific competition for food, after the disease rinderpest was removed in 1963. (After Mduma *et al.* 1999 and unpublished data.)

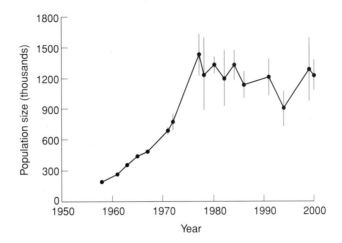

Yellowstone National Park (Houston 1982). Before 1930, the population estimates ranged between 15,000 and 25,000. Between 1933 and 1968 culling reduced the population to 4000 animals. Culling then ceased and the population rebounded to around 20,000 (Coughenour and Singer 1996). This result is consistent with regulation through intraspecific competition for winter food (Houston 1982), since there were no natural predators of elk in Yellowstone until the return of wolves in the early 1990s.

Density is usually recorded as numbers per unit area. If space is the limiting resource (as it might be in territorial animals), or if space is a good indicator of some other resource such as food supply, numbers per unit area will suffice in an investigation of regulation. However, space may not be a suitable measure if density-independent environment effects (e.g. temperature, rainfall) cause fluctuations in food supply. It may be better to record density as animals per unit of available food or per unit of some other resource.

The Serengeti migratory wildebeest experienced a perturbation (Fig. 8.7) when an exotic virus, rinderpest, was removed. The population increased fivefold from 250,000 in 1963 to 1.3 million in 1977 and then leveled out (Mduma *et al.* 1999). This example is less persuasive than that of the Yellowstone elk because the pre-rinderpest density (before 1890) was unknown, but evidence on reproduction and body condition suggests that rinderpest held the population below the level allowed by food supply, a necessary condition for a perturbation experiment implicating a disease.

A case of a population perturbed above equilibrium is provided by elephants in Tsavo National Park, Kenya (Laws 1969; Corfield 1973). From 1949 until 1970, the population had been increasing due in part to immigration from surrounding areas where human cultivation had displaced the animals. A consequence of this artificial increase in density was depletion of the food supply within reach of water. In 1971, the food supply ran out and there was starvation of females and young around the water holes. After this readjustment of density, the vegetation regenerated and starvation mortality ceased.

8.4.2 *Mean density and environmental factors*

A population uninfluenced by dispersal and unregulated (i.e. it has no density-dependent factors affecting it) will fluctuate randomly under the influence of weather and will eventually drift to extinction (DeAngelis and Waterhouse 1987).

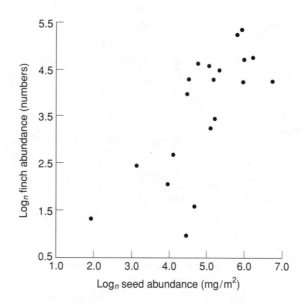

Fig. 8.8 The total abundance of seed-eating finches in savanna habitats of Kenya is related to the abundance of the food supply. Such a positive relationship in unconnected populations may demonstrate regulation. (After Schluter 1988.)

Just by chance there may be for a time a correlation between density and environmental factors. However, if we take many separate populations, the probability that all of them are simultaneously correlated with an environmental factor by chance alone is very small. Therefore, if we find a correlation between mean densities from independent populations with an environmental factor, there is a strong inference that weather is influencing some resource for which animals are competing, and which results in regulation about some equilibrium point.

An example of this approach is shown in Schluter's (1988) study of seed-eating finches in Kenya (Fig. 8.8): finch abundance from various populations is correlated with seed abundance. Other examples of density correlated with weather factors are given in Sinclair (1989).

8.4.3 *Examples of density dependence*

As we discussed in Section 8.3.7, density dependence is a necessary but not sufficient requirement to demonstrate regulation. There are an increasing number of studies in the bird and mammal literature demonstrating density-dependent stages in the life cycle. For birds (Fig. 8.9a), the long-term study on great tits (*Parus major*) at Oxford, England has shown that winter mortality of juveniles was related to the number of juveniles entering the winter (McCleery and Perrins 1985). In contrast (Fig. 8.9b), it was early chick mortality in summer that was density dependent for the English partridge (*Perdix perdix*) (Blank *et al.* 1967).

For mammals, density-dependent juvenile mortality has been recorded for red deer on the island of Rhum, Scotland (Clutton-Brock *et al.* 1985) (Fig. 8.10a), for reindeer in Norway (Skogland 1985) (Fig. 8.10b), for feral donkeys (*Equus asinus*) in Australia (Choquenot 1991), and for greater kudu in South Africa (Owen-Smith 1990). Adult mortality was density dependent for African buffalo in Serengeti (Sinclair 1977). In each case, the cause was lack of food at critical times of year. Reproduction is known to be density dependent in both birds (Arcese *et al.* 1992) and mammals (Clutton-Brock *et al.* 1991). Figure 8.10c shows that the proportion of Soay sheep that give birth at 12 months of age declines with density. Fowler (1987) reports over

Fig. 8.9 Examples of density-dependent mortality in birds. (a) Great tit (*Parus major*) overwinter mortality (log of [juveniles in winter/first year breeding population]) plotted against log juvenile density in winter. (After McCleery and Perrins 1985.) (b) Chick mortality of European partridge (*Perdix perdix*) (measured as log hatching population/log population at 6 weeks) plotted against log hatching population, in Hampshire, England. (After Blank *et al.* 1967.)

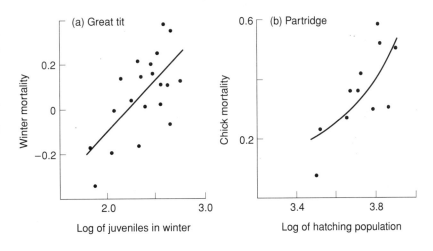

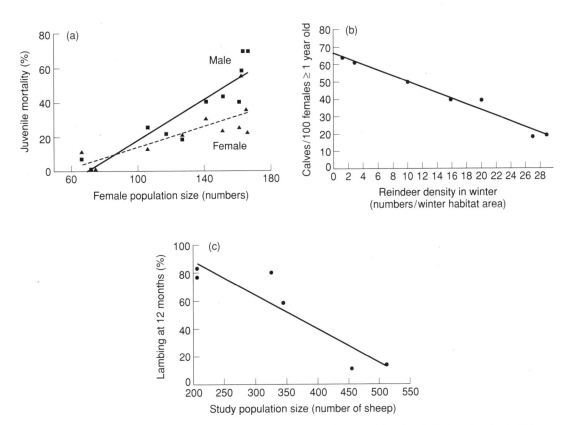

Fig. 8.10 Density dependence in large mammals. (a) Juvenile mortality of male and female red deer on the island of Rhum, Scotland. (After Clutton-Brock *et al.* 1985.) (b) Juvenile recruitment per 100 female reindeer older than 1 year in Norway. (After Skogland 1985.) (c) The fertility rate of 1-year-old Soay sheep on St Kilda island. (After Clutton-Brock *et al.* 1991.)

Fig. 8.11 The proportion of a red grouse population in Scotland which disappears over winter (August–April) is related to population density in the previous August in a complex way. Mortality varied according to whether the population was increasing or decreasing. By joining the points sequentially an anticlockwise cycle is produced, indicating a delayed density-dependent effect in the cause of the mortality. By plotting the percentage disappearance against density 1 year earlier, a closer fit can be obtained for a regression line. Thus the delay is 1 year. Numbers at the points are years. (After Watson and Moss 1971.)

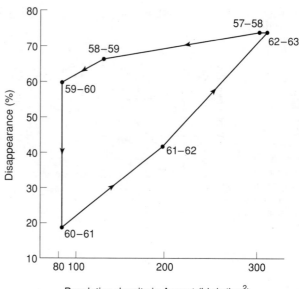

100 studies of terrestrial and marine mammal populations where density dependence was detected.

Delayed density dependence has been recorded in winter mortality of snowshoe hares in the Yukon and in overwinter mortality of red grouse in Scotland (Watson and Moss 1971) (Fig. 8.11). For the hares the delay appears to have been due to a lag of 1–2 years in the response of predator populations to changing hare numbers (Trostel *et al.* 1987), while for the grouse the delay came from density responding to food conditions in the previous year (see Section 8.8.3).

8.5 Applications of regulation

Causes of population change can be divided into (i) those that disrupt the population and often result in "outbreaks," and which can be either density dependent or density independent; and (ii) those that regulate and therefore return the population to original density after a disturbance (Leirs *et al.* 1997). These are always density dependent.

Knowledge of regulation may be useful for management of house mice (*Mus domesticus*) plagues in Australia. In one experimental study (Barker *et al.* 1991), mice in open-air enclosures were contained by special mouse-proof fences. The objective was to create high densities, thus mimicking plague populations, in order to test the regulatory effect of a nematode parasite (*Capillaria hepatica*). It turned out that they could not test the effect of the parasite because other factors regulated the population and thus obscured any parasite effect. The replicated populations declined simultaneously. Why did this happen? By dividing up the life cycle into stages they found that late juvenile and adult mortality was strongly density dependent but that other stages, including fertility and newborn mortality, were not. This allowed them to discount causes that would affect reproduction and focus more closely on what was happening amongst adults, in particular the social interactions of mice.

Other studies suggest that mouse populations in Australia may be regulated by predators, disease, and juvenile dispersal (Redhead 1982; Sinclair *et al.* 1990). Under conditions of superabundant food following good rains, the reproductive rate of females increased faster than the predation rate, and an outbreak of mice occurred. The implication of these results for management is that if reproduction could be reduced, for example through infections of the *Capillaria* parasite, then predation may be able to prevent outbreaks even in the presence of abundant food for the mice.

8.6 Logistic model of population regulation

In Chapter 6, we derived geometric and exponential growth models. In 1838, Pierre-Francois Verhulst published a paper (Verhulst 1838) that challenged the assumption of unlimited growth implicit in these models. Verhulst argued that the per capita rate of change (dN/N dt) should decline proportionately with population density, simply due to a finite supply of resources being shared equally among individuals. If each individual in the population gets a smaller slice of the energy "pie" as N increases, then this would prevent them from devoting as much energy to growth, reproduction, and survival than would be possible under ideal conditions. As we saw in Chapter 6, changes in demographic parameters lead to corresponding changes in the finite rate of population growth λ_t or its equivalent exponential rate r_t, where t denotes a specific point in time. Other factors, such as risk of disease, shortage of denning sites, or aggressive interactions among population members, might also cause the rate of population growth to decline with population size. The simplest mathematical depictions of such phenomena are commonly termed "logistic" models.

There are numerous ways to represent logistic growth. For simplicity, we will focus on population growth modeled in discrete time, which is often a reasonable approximation for species that live in a seasonal environment. One of the most commonly used forms is called the Ricker equation, in honor of the Canadian fisheries biologist, Bill Ricker, who first suggested its application to salmon stocks (Ricker 1954):

$$N_{t+1} = N_t e^{r_{max}\left(1-\frac{N_t}{K}\right)}$$

The Ricker logistic equation represents the exponential rate of increase under ideal conditions as r_{max}, with a proportionately slower rate of increase with each additional individual added to the population. When the rate of increase has slowed to the point that births equal deaths, then the population has reached its carrying capacity K. These two population parameters (r_{max} and K) dictate how fast the population recovers from any perturbation to abundance.

A population growing according to the logistic equation would have slow growth when N is small, grow most rapidly when N is of intermediate abundance, and grow slowly again as N approaches carrying capacity K (Fig. 8.12). This kind of sigmoid or S-shaped pattern is often termed **logistic growth**.

At first, it may seem somewhat counterintuitive that a proportional decline in per capita demographic rates could produce the non-linear growth pattern seen in Fig. 8.12. The answer lies in the fact that population changes are dependent on both population size and the per capita growth rate, in much the same way that growth of a bank account depends both on the money already in the account and the interest rate. When a population is small, the per capita rate of change will tend to be large, in fact close to r_{max}, because either birth rates are high or mortality rates are low. Nonetheless, the population will still display a slight change from one year to the next because the population is small. At the other end of the spectrum, despite

Fig. 8.12 Population growth according to the logistic equation, with $r_{max} = 0.5$, initial population density $N_0 = 1.5$, and carrying capacity $K = 100$.

Fig. 8.13 Net recruitment $(N_{t+1} - N_t)$ as a function of population density N_t, according to the Ricker logistic growth model, with $r_{max} = 0.5$ and $K = 100$.

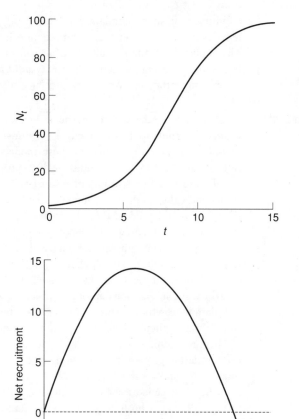

the fact that N is enormous, the population will similarly display modest change from year to year. This is because the per capita rate of growth is small, because either birth rates are low or mortality rates are high. It is only when the population is of intermediate size and growing at an intermediate per capita rate that growth is maximized (Fig. 8.13).

Population data displaying the classic sigmoid pattern of change are rare. It will only be seen when a population is reduced to very low initial density and then monitored closely over an extended period. So, logistic growth will not be obvious in most populations that we might see around us in nature, which are presumably close to their carrying capacity. In some cases, however, populations have been perturbed (reduced) to low densities, and give us a rare glimpse of logistic growth in the field. For example, as we discussed earlier, the Yellowstone elk herd has been aggressively culled at various times in the past, particularly in the late 1960s. Cessation of culling operations, stimulated by a new policy of natural regulation in US National Parks, led to a subsequent pattern of elk recovery reminiscent of the sigmoid pattern predicted by the logistic model (Fig. 8.14). Similarly, release of the Serengeti wildebeest population from the exotic disease rinderpest led to a subsequent sigmoid pattern

Fig. 8.14 Population dynamics of Northern Yellowstone elk between 1968 and 1989. (Data from Coughenour and Singer 1996.)

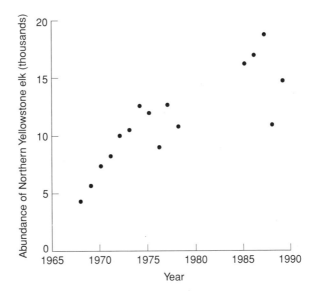

of change (Fig. 8.7) reminiscent of the logistic model. Indeed, perturbation is an important ingredient in detecting natural regulation and logistic growth, because it gives us evidence to work with, unlike populations kept close to their ecological carrying capacity. We demonstrate how to estimate the parameters for the Ricker logistic model, and compare it with other possible population growth models, in Chapter 15.

All environments show some degree of variability in conditions from year to year. Such stochastic or random variation can have a strong influence on the dynamics of even tightly regulated species. We can explore this by applying the Ricker logistic model to some typical empirical data. Figure 8.14 shows records of elk censused in Northern Yellowstone National Park between 1968 and 1989 (Coughenour and Singer 1996). We see that there is the barest hint of a sigmoid pattern in these data. Nonetheless, exponential growth rates r_t calculated over this two-decade period show a strong density-dependent decline in growth rates when the population is large (Fig. 8.15).

The scatter around the regression line (termed "residual" variation) in Fig. 8.15 shows that natural regulation explains only part of the demographic response by a wild population to changes in density. Even when the population is tightly regulated, as is obviously the case here, there can be considerable variation in growth rates from year to year that is not explained by density dependence. Some of this variability is due to stochastic climatic variation that characterizes every natural environment, some places more than others. In the case of Northern Yellowstone elk, for example, precipitation in the preceding 2 years is probably responsible for much of the residual variation shown in Fig. 8.15, judging from its effect on offspring production and survival rates (Coughenour and Singer 1996). This probably stems from a strong linkage between precipitation and forage availability to elk.

Variability in population growth rates can also stem from "demographic stochasticity." This term refers to variation in the numbers of individuals born or dying per unit time, simply due to chance (Chapter 17). The principle is familiar to anyone who has played a game of cards or spun a roulette wheel. For a given probability of survival, say 0.25, we do not necessarily expect exactly a quarter of the population

Fig. 8.15 Exponential growth rates for Northern Yellowstone elk between 1968 and 1989 in relation to population density at the beginning of each yearly interval. (Data from Coughenour and Singer 1996.)

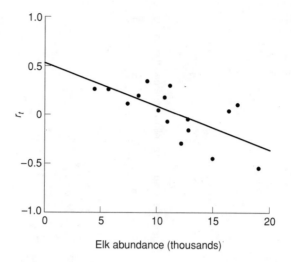

to survive, but rather anticipate that by chance sometimes a larger fraction will survive, sometimes a smaller fraction. We consider this process in more exact mathematical detail in Chapter 17, when we discuss population viability analysis. In wildlife management we need to disentangle demographic from environmental sources of stochasticity (Sæther *et al.* 2000; Bjørnstad and Grenfell 2001).

We should include in our population models the variability in growth rates due to environmental and demographic stochasticity. We do this by simulating natural stochastic variation and adding this variation to the exponential growth rate r_t predicted by population density. We first need to calculate the residual variation in growth from the data in Fig. 8.15:

$$\sigma = \sqrt{\sum_{t=0}^{15} \frac{[r_t - (0.518 - 0.00004404N_t)]^2}{16}}$$

where 0.518 is the intercept (r_{max}) of the regression line drawn through the observed values of r_t versus N_t, and -0.00004404 is the slope. We calculate the deviation between each observation of r and the value predicted by the regression line at that population density, square each deviation to standardize positive versus negative values, sum the squared deviations, and divide by the sample size (16 in this case) to estimate the mean-squared deviation. This is the residual variability, denoted by σ^2. For the Northern Yellowstone elk, $\sigma^2 = 0.0361$.

Once equipped with an estimate of the residual variation based on the observed data, we draw values of the random variable ε from a bell-shaped (i.e. normal) probability distribution with the same magnitude of residual variation σ_t. In MATHCAD, this normal probability distribution is the function called **rnorm**, which also requires the user to input the required number of random values, and mean and standard deviation of the normal distribution from which these values will be drawn. For the elk example:

$$\varepsilon = \text{rnorm}(51, \mu, \sigma)$$

Fig. 8.16 Simulated dynamics of elk, based on the Yellowstone National Park population. (Data from Coughenour and Singer 1996.)

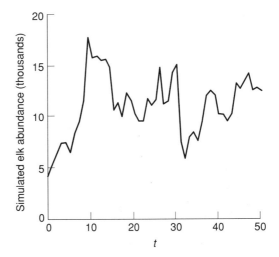

where $\mu = 0$ and $\sigma = 0.19$. We then combine the random normal deviate at any point in time (ε_t) with the rate of increase predicted by the Ricker logistic equation, $r_{max}(1 - N_t/K)$, to predict changes in abundance. We use a different symbol (n_t) for the simulated density:

$$n_{t+1} = n_t e^{r_{max}\left(1 - \frac{n_t}{K}\right) + \varepsilon_t}$$

We plot the simulated elk data (n_t) in Fig. 8.16. We see that the trends in the simulated population are completely different from those of the real population (Fig. 8.14), but the overall magnitude of variability is similar. This similarity occurs because we have included both the stochastic (environmental and demographic) processes that tend to perturb the population away from its carrying capacity and the natural regulatory processes that tend to restore the population, once perturbed. Both processes are common in the natural world, and therefore we need to accommodate them in our management planning.

Such stochastic simulations, sometimes termed **Monte Carlo models**, offer useful insights into the degree of variation that wildlife biologists and managers might expect to see over a long time. Monte Carlo simulation is central to the procedure known as population viability analysis, which we describe in Chapter 17.

8.7 Stability, cycles, and chaos

Paradoxically, the same density-dependent processes that are responsible for natural regulation can also induce population fluctuations, at least under special circumstances. One way that this can happen is when the maximum rate of growth is particularly high. For example, consider the dynamics of a hypothetical population whose maximum rate of increase $r_{max} = 3.3$ and carrying capacity $K = 100$ (Fig. 8.17). In this case the population does not increase smoothly over time and level off at the carrying capacity, but rather the population fluctuates erratically over time, with no apparent repeated pattern. Such a pattern of population change is known as **deterministic chaos** (May 1976). It arises because the population grows so fast that it tends to overshoot the carrying capacity, a process known as **overcompensation** (May and Oster 1976). Once above the carrying capacity the net recruitment is negative

Fig. 8.17 Simulated dynamics over time of two different populations growing according to the Ricker logistic equation, with $r_{max} = 3.3$ and $K = 100$. The first population was initiated at a density of 2.0 individuals per unit area, whereas the second population was initiated at a slightly higher density of 2.1 individuals per unit area. The rapid divergence in population dynamics due to slight changes in starting conditions is typical of deterministic chaos.

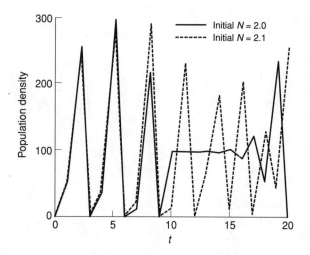

Fig. 8.18 Plot of predicted recruitment (N_{t+1}) relative to N_t (the heavy curve), equilibrium line at which $N_{t+1} = N_t$ (thin broken line), and trajectory of population dynamics over time for a simulated population following the Ricker logistic model, with $r_{max} = 1.3$ and $K = 100$ (thin solid line).

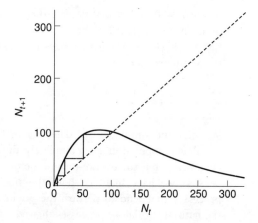

(Fig. 8.13), so the population declines rapidly. Repetition of this boom–bust pattern of overshooting the carrying capacity and subsequent decline to levels below the carrying capacity results in the erratic fluctuations of deterministic chaos seen in Fig. 8.17. For lower rates of increase ($2.0 < r_{max} < 2.7$) the pattern of fluctuation would be regular cycles, rather than deterministic chaos, but the underlying cause is still overcompensation.

The underlying cause of instability due to overcompensatory density dependence can be appreciated better by plotting the population dynamics over time on a graph with N_t on the horizontal axis and N_{t+1} on the vertical axis (Fig. 8.18). The diagonal identifies potential points of equilibria, at which $N_{t+1} = N_t$. We will also plot the recruitment curve. Dynamics are plotted by starting at a particular value of N_0, projecting upwards to the recruitment curve, that identifies the next year's population density. Then we project horizontally to the broken equilibrium line, before repeating the process. At modest values of r_{max}, the recruitment curve is low and has a shallow angle of incidence as it intersects the equilibrium line. The result is that the population trajectory becomes pinched between the recruitment curve and the equilibrium line as it converges on K. This leads to stability.

Fig. 8.19 Plot of predicted recruitment (N_{t+1}) relative to N_t (the heavy curve), equilibrium line at which $N_{t+1} = N_t$ (thin broken line), and trajectory of population dynamics over time for a simulated population following the Ricker logistic model, with $r_{max} = 3.3$ and $K = 100$ (thin solid line).

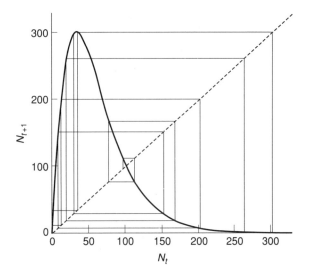

Now, let us consider the pattern arising when $r_{max} = 3.3$ (Fig. 8.19). The recruitment curve has a pronounced hump and intersects the equilibrium line at a sharp angle (> 90°). The recruitment curve is so sharply peaked that recruitment events tend to overshoot the carrying capacity. This leads to the population collapsing to well below the carrying capacity, where the boom–bust cycle begins anew. In this way, the population never reaches an equilibrium, despite the fact that there is strong density dependence. This example demonstrates overcompensation, and it occurs when the angle of incidence of the recruitment curve exceeds 90° as it approaches the equilibrium line (May 1976; May and Oster 1976).

A diagnostic feature of deterministic chaos is that slight changes in starting conditions lead to quite different population dynamics over time. In Fig. 8.17, the simulated dynamics of the two hypothetical populations, started at slightly different densities, became quite different later on, illustrating their sensitivity to initial conditions. Both populations go through similar changes in the first few years but rapidly diverge thereafter, displaying different patterns of fluctuation.

Chaotic growth and fluctuation is unlikely for large wildlife species, which tend to have values of r_{max} that are less than 0.5, well outside the parameter range in which cycles or chaos could arise through the simple mechanism we have described. Cycles or chaos can also arise, however, in other ways that *are* quite feasible for large wildlife species.

We have thus far limited our discussion to the simplest pattern of density dependence: linear changes in per capita rates of reproduction or survival. We saw earlier in the chapter (Figs 8.4 and 8.6) that there is no reason to expect natural regulation to be linearly density dependent. Some wildlife biologists have even argued that it may be the exception rather than the rule (Fowler 1981), and adult mortality in Serengeti wildebeest is a good example (Mduma *et al.* 1999).

Another example of non-linear demographic responses is seen in the feral Soay sheep on the St Kilda archipelago off the coast of Scotland. These sheep, similar in many ways to the ancestral sheep first domesticated by man, were initially introduced during the second millennium BC. They have roamed wild for several decades on several of the St Kilda islands, the best known of which is the small island of Hirta.

Fig. 8.20 Survival of
Soay sheep lambs on
the island of Hirta in
relation to adult
population density:
females (solid line);
males (broken line).
(After Clutton-Brock
et al. 1997.)

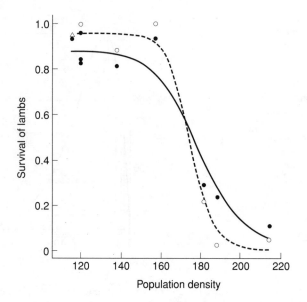

A fraction of the Hirta population uses an open, grassy area once occupied by people. Here the pregnant females use abandoned stone huts for shelter during birth. As a consequence, sheep numbers can be counted accurately and a large fraction of the newly born lambs can be caught and marked. Tracking these known individuals over the subsequent years has allowed unusually detailed calculations of age-specific reproduction and survival (Clutton-Brock *et al.* 1997).

The demographic pattern that has emerged from these field studies shows pronounced threshold effects of population density on sheep survival (Fig. 8.20). When the population is less than 200 adult sheep, survival of lambs, yearlings, and adults tends to be high: typically more than 90% in adults and yearlings and more than 80% in lambs. Increase in sheep abundance beyond the threshold tends to be accompanied by a precipitous decline in survival to low levels, sometimes as low as 10% (Fig. 8.20). Simulation models constructed with threshold survival and fecundity effects generate regular fluctuations of Soay sheep at 6-year intervals (Grenfell *et al.* 1992), qualitatively similar to the patterns seen in the real population (Figs 8.21–8.23). We show how to construct such a model in Box 8.1.

The model does not capture all of the variability in sheep abundance observed on St Kilda. Like all models, our age-structured model leaves out many important features. The model has no direct link with food supply or disease, both of which are important in shaping dynamics. Catastrophic mortality is largely caused by starvation, and vulnerability to starvation is exacerbated by high nematode infestation in the intestinal tract of individual sheep (Gulland 1992; Clutton-Brock *et al.* 1997). Perhaps more importantly, the model has no demographic or environmental stochasticity, which, as we have already seen, can considerably influence long-term dynamics. Using the Monte Carlo approach we outlined before, we could add such stochasticity.

Strong effects of weather variation can influence the population dynamics of Soay sheep (Grenfell *et al.* 1998; Coulson *et al.* 2001a). Populations of sheep on adjacent, isolated islands tend to be loosely synchronized, because they share a common climate (Grenfell *et al.* 1998). Although density-dependent processes regulate Soay

Fig. 8.21 Sigmoid survival functions in relation to the population density of adult+yearling Soay sheep, for adults (dashed-dotted line), yearlings (dotted line), and lambs (solid line), estimated by Clutton-Brock *et al.* (1997).

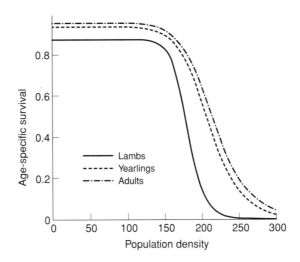

Fig. 8.22 Simulated dynamics of an age-structured population with sigmoid survival and fecundity functions with the same parameters as those of the Soay sheep population on the island of Hirta. (After Clutton-Brock *et al.* 1997.)

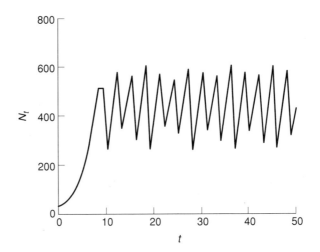

Fig. 8.23 Observed variation in the population of Soay sheep over the entire St Kilda archipelago. (After Clutton-Brock *et al.* 1997.)

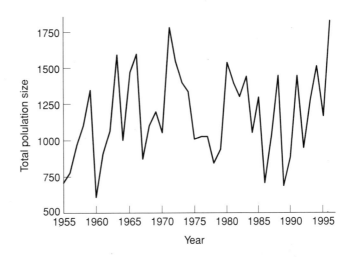

Box 8.1 Model of the
Soay sheep population
on St Kilda.

Threshold effects on mortality can be well described by a sigmoid function:

$$\Psi(i, N) = \frac{p_{max_i}}{1 + (\alpha_i N)^{\beta_i}}$$

where i refers to the age group (from 0 for newborns up to 2 for adults), N is population density of yearlings and adults, p_{max} is maximum survival rate, and α and β are parameters determining the shape of the sigmoid survival function. Clutton-Brock $et\ al.$ (1997) estimated the parameters of the Ψ function, from several years of data. These values are shown below:

$$p_{max} = \begin{pmatrix} 0.88 \\ 0.94 \\ 0.96 \end{pmatrix}$$

$$\alpha = \begin{pmatrix} 0.00562 \\ 0.00484 \\ 0.00467 \end{pmatrix}$$

$$\beta = \begin{pmatrix} 15.3 \\ 9.46 \\ 8.93 \end{pmatrix}$$

By applying these sigmoid functions, we can mimic the threshold effect (Fig. 8.21). Similar sigmoid functions can be fitted to age-specific fecundity rates of females:

$$\Omega(i, N) = \frac{m_{max_i}}{1 + (\alpha\alpha_i N)^{\beta\beta_i}}$$

$$m_{max} = \begin{pmatrix} 0.335 \\ 0.643 \\ 0.643 \end{pmatrix}$$

$$\alpha\alpha = \begin{pmatrix} 0.00629 \\ 0.00589 \\ 0.00589 \end{pmatrix}$$

$$\beta\beta = \begin{pmatrix} 24.1 \\ 14.1 \\ 14.1 \end{pmatrix}$$

By applying these density-dependent survival and fecundity rates to specific age classes, we can estimate changes in abundance over time:

$$\begin{pmatrix} n_0, t+1 \\ n_1, t+1 \\ n_2, t+1 \end{pmatrix} = \begin{pmatrix} \sum_i n_{j,t}\, \Omega\left(i, \sum_j n_{j,t}\right) \Psi\left(i, \sum_j n_{j,t}\right) \\ n_{0,t}\, \Psi\left(0, \sum_j n_{j,t}\right) \\ n_{1,t}\, \Psi\left(1, \sum_j n_{j,t}\right) + n_{2,t}\, \Psi\left(2, \sum_j n_{j,t}\right) \end{pmatrix}$$

sheep, the precise pattern of regulation is strongly affected by age structure. The mix of age groups on any of the islands is highly changeable, and slight modifications in age structure alter the dynamic consequences of density-dependent processes (Coulson *et al.* 2001a). The Soay sheep example illustrates how populations can fluctuate through a combination of (i) stochastic environmental effects; (ii) non-linear demographic responses; and (iii) delays that arise through a complex age structure.

8.8 Intraspecific competition

Regulation can occur by a number of mechanisms such as predation or parasitism, but a more common cause is competition between individuals for resources. Such resources can be food, shelter from weather or from predators, nesting sites, and space to set up territories. We have seen some examples already in Figs 8.9 and 8.10.

8.8.1 *Definition*

Intraspecific competition occurs when individuals of the same species utilize common resources that are in short supply; or, if the resources are not in short supply, competition occurs when the organisms seeking that resource nevertheless harm one or other in the process (Birch 1957).

8.8.2 *Types of competition*

When individuals use a resource so that less of it is available to others, we call this type of competition **exploitation**. This includes both removal of resource (consumptive use) when food is consumed and occupation of a resource (pre-emptive use) when resources such as nesting sites are used (see Section 12.3). Individuals competing for food need not be present at the same time: an ungulate can reduce the food supply of another that arrives later.

Another type of competition involves the direct interaction of individuals through various types of behavior. This is called **interference** competition. One example of behavioral interference is the exclusion of some individuals from territories. Another is the displacement of subordinate individuals by dominants in a behavioral hierarchy.

8.8.3 *Intraspecific competition for food*

Experimental alteration of food supply

Food addition experiments provide the best evidence for intraspecific competition. Krebs *et al.* (1986) supplied extra food to snowshoe hares in winter from 1977 to 1985. This raised the mean winter density fourfold at the peak of the 10-year population cycle. Similarly, Taitt and Krebs (1981) increased the density of vole populations (Fig. 8.24) by giving them extra food. The elk population at Jackson Hole, Wyoming, is kept at a higher level than otherwise would be the case by supplementary feeding in winter (Boyce 1989). These examples show that food is one of the factors limiting density.

The dense shrubland (chaparral) of northern California contains two shrubs, chamise (*Adenostema taxiculatum*) and oak (*Quercus wislizenii*), that are preferred food for black-tailed deer (*Odocoileus hemionus*). These shrubs resprout from root stocks after burning to provide the new shoots which are the preferred food. Taber (1956) showed that on plots thinned by experimental burning, herbaceous food supply increased to 78 kg/ha from the 4.5 kg/ha found on control plots; and the shrub component increased to 460 kg/ha from 165 kg/ha. Deer densities consequently increased from 9.5/km^2 on the experimental controls to 22.9/km^2 on the treatment plots, while fertility of adult females increased from 0.77 to 1.65 young per female.

Red grouse (*Lagopus lagopus*) live year round on heather (*Calluna vulgaris*) moors in Scotland. Their diet consists almost entirely of heather shoots. Watson and Moss

Fig. 8.24 The numbers of Townsend's voles on trapping grids increase in proportion to the amount of food that is provided, indicating that intraspecific competition regulates the population. Control (dashed-dotted line); low food addition (dashed line); high food addition (solid line); shaded area indicates winter. (After Taitt and Krebs 1981.)

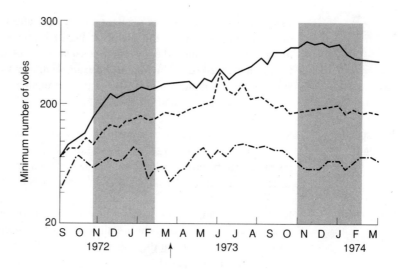

(1971) described experiments where some areas were cleared of grouse, fertilized with nitrogen in early summer, and then were left to be recolonized. Fertilizing increased the growth and nutrient content of heather. The size of their territories did not differ between fertilized and control areas when grouse set up their territories in fall. However, territorial grouse that had been present all winter reared larger broods on the fertilized than on the control areas, indicating that reproduction was affected by overwinter nutrition. Territory sizes did decline in the following fall and densities increased, showing the 1-year lag of density responding to nutrition. On other areas, old heather was burned every 3 years, creating a higher food supply of young regenerating heather. Territory size on these plots decreased (as density increased) in the same year as the treatment, so there was a more immediate response than on the fertilized plots.

Direct measures of food

Snowshoe hare populations in the boreal forests of Canada and Alaska reach high numbers every 10 years or so. Measurements of known food plants, and feeding experiments, suggested that the animals ran short of food at peak numbers (Pease *et al.* 1979). Other measures such as the amount of body fat (Keith *et al.* 1984) and fecal protein levels (Sinclair *et al.* 1988) also identified food shortage at this time (see Section 4.9).

African buffalo graze the tropical montane meadows of Mt Meru in northern Tanzania, keeping the grass short. Grass growth rates and grazing offtake were measured by use of temporary exclosure plots. Growth in the rainy season was more than sufficient for the animals, but in the dry season available food fell below maintenance requirements (Sinclair 1977).

Murton *et al.* (1966) measured the impact of wood-pigeons (*Columba palumbus*) on their clover (*Trifolium repens*) food supply. Food supply was measured directly by counting clover leaves in plots. Pigeons consumed over 50% of the food supply during winter. They feed in flocks, those at the front of the flock obtaining more food than those in the middle or at the back. The proportion of underweight birds (< 450 g) was related directly to the overwinter change in numbers (Fig. 8.25) and

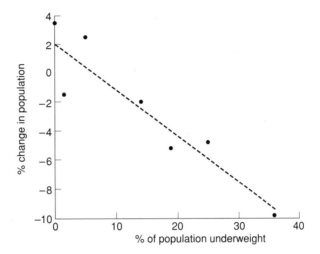

Fig. 8.25 The percentage change in a wood pigeon population in England is related to the proportion of the population that is underweight. (Data from Murton *et al.* 1966.)

inversely related to the midwinter food. Thus competition within flocks resulted in some animals starving, and the change in numbers was related to the proportion that starved.

Indirect measures of food shortage

Indirect evidence for competition for food comes from indices of body condition (see Section 4.9). The last stores of body fat that are used by ungulates during food shortages are in the marrow of long bones such as the femur. Bone marrow fat can be measured directly by extraction with solvents. However, since there is an almost linear relationship between fat content and dry weight (Hanks 1981) (see Section 4.9.3), it is easier to collect a sample of marrow from carcasses found in the field and oven dry it. A cruder but still effective method is to describe the color and consistency of the marrow, a method introduced by Cheatum (1949).

Other fat stores such as those around the heart, mesentery, and kidney are used up before the bone marrow fat starts to decline (see Section 4.9). The relationship between kidney and marrow fat holds for many ungulate species (see Fig. 4.11). If both kidney and marrow fat can be collected, a range of body conditions can be recorded. However, often the marrow fat is all that is found in carcasses because scavengers have eaten the internal organs.

Klein and Olson (1960) used bone marrow condition indices to conclude that deer in Alaska died from winter food shortage, as did Dasmann (1956) for deer in California. Similarly, migratory wildebeest in Serengeti that died in the dry season were almost always in poor condition, as judged by the bone marrow, and this was correlated with the protein level in their food (see Fig. 4.12). This dry season mortality was density dependent and was sufficiently strong to allow the population to level out (Sinclair *et al.* 1985; Mduma *et al.* 1999).

Problems with measurement of food supply

To determine whether competition for resources such as food is the cause of regulation we need to know what type of food is eaten, how much is needed, and how much is available. What is needed must exceed what is available for competition to occur. The types of food eaten form the basis for many studies on diet selection, sometimes called **food habit studies**. These in themselves do not tell us what is needed in terms of digestible dry matter, protein, and energy. We should note that such requirements are unknown for most wild species and we have to use approximations from other, often domestic, species. The amount of food available to animals is particularly difficult to assess because we are unlikely to measure potential food in the same way as does an animal. For example, animals are likely to be far more selective than our crude sampling and so we are likely to record more "food" than the animal sees. Our measures of food supply are often seriously flawed, and this is one of the reasons why direct evidence for intraspecific competition for food is rare. There is far more indirect evidence for competition provided by indicators such as body condition.

8.9 Interactions of food, predators, and disease

The effect of limited food on population demography can go beyond the direct effects of undernutrition. There can also be synergistic interactions with predation and disease. Animals may alter their behavior when food becomes difficult to find in safe areas, searching increasingly in areas where they are at risk of predation in order to avoid eventual starvation (McNamara and Houston 1987; Lima and Dill 1990). This is called **predator-sensitive foraging** and has been observed in snowshoe hares (Hik 1995; Hodges and Sinclair 2003). Such behavior can result in increased predation well before starvation takes effect, as seen in wildebeest (Sinclair and Arcese 1995).

Disease can also interact synergistically with food, pathological effects suddenly becoming apparent at a certain, sometimes early, stage of undernutrition (see Chapter 11). Sometimes food, disease, and predators all interact. Wood bison numbers in the Wood Buffalo National Park, Canada, switch suddenly from a high-density food-regulated state to a low-density predator-regulated state when diseases, such as tuberculosis and brucellosis, affect the population (Joly and Messier 2004).

8.10 Summary

Regulation is a biotic process which counteracts abiotic disturbances affecting an animal population. Two common biotic feedback processes are predation and intraspecific competition for food. These are called density-dependent factors if they act as negative feedbacks. Negative feedback imparts stability to the population. Disturbances are provided by fluctuating weather or other environmental conditions (termed environmental stochasticity) or chance effects on reproduction and survival (termed demographic stochasticity). They are called density-independent factors and will cause populations to drift to extinction if there are no counteracting density-dependent processes operating. For wildlife management it is necessary to know (i) what are the causes of the density-dependent processes that stabilize the population, and what are the causes of fluctuations and instability; and (ii) which age and sex groups are most influenced by these stabilizing or destabilizing processes.

One way to understand such effects is to model density-dependent changes in population growth rate, using logistic models. Application of such models shows that whereas density dependence is often stabilizing, overcompensatory density dependence can itself encourage population fluctuation, beyond the degree we would expect due to demographic or environmental stochasticity. A common cause of regulation is intraspecific competition for food.

Competition occurs if the needs of the population exceed availability. To measure such competition we need to know how much food is available and how much is needed, and whether it is density dependent. Food can also interact with predation and disease to regulate populations.

9 Competition and facilitation between species

9.1 Introduction

Species do not exist alone. They live in a community of several other species and some of these will interact. There are various forms of interaction between species; competition, commensalism (facilitation), mutualism (symbiosis), predation, and parasitism are the main ones. These are defined by the way each species affects the other, as is shown in Table 9.1. In competition each species suffers from the presence of the other, although the interaction need not be balanced. With commensalism or facilitation one species benefits without affecting the other, while in mutualism both benefit. These can be thought of as the converse of interspecific competition. With predation and parasitism one species benefits to the disadvantage of the other. We shall discuss predation in Chapter 10 and parasitism in Chapter 11 and will confine ourselves here to interspecific competition and mutualism.

9.1.1 Definition

Interspecific competition is similar to intraspecific competition. It occurs when individuals of different species utilize common resources that are in short supply; or, if the resources are not in short supply, competition occurs when the organisms seeking that resource nevertheless harm one or other in the process (Birch 1957).

9.1.2 Implications

Interspecific competition deals with the cases when there are two or more species present, and we should be aware of a number of implications arising from this definition.
1 Competition must have some effect on the fitness of the individuals. In other words, resource shortage must affect reproduction, growth, or survival, and hence the ability of individuals to get copies of their genes into the next generation.
2 Although it is necessary for species to require common resources (i.e. overlap in their requirements), we cannot conclude there is competition unless it is also known that the resource is in short supply, or that they affect each other.
3 The amount of resource such as food that is available to each individual must be affected by what is consumed by other individuals. Thus two species cannot compete if they are unable to influence the amount of resource available to the other species, or to interfere with that species obtaining the resource.

Table 9.1 Types of interaction.

Species 2	Species 1		
	+	0	−
+	Mutualism	Commensalism	Predation/parasitism
0	Commensalism		(Amensalism) Competition
−	Predation/parasitism	Competition	Competition

4 Both exploitation and interference competition (see Section 8.8.2) can occur between species, although interference between species is relatively uncommon.

9.2 Theoretical aspects of interspecific competition

To obtain an understanding of what might be the expected outcome from a simple and idealized interspecific competition we return to the logistic equation:

$$dN_1/dt = r_{m1} \times N_1 \times (1 - N_1/K_1) \tag{9.1}$$

9.2.1 Graphical models

The term in parentheses $(1 - N_1/K_1)$ describes the impact of individuals upon other individuals of the same species and on the population growth rate dN_1/dt. We must now add a term representing the impact of the second species N_2 on species 1. The equation for the effect of species 2 on population growth of species 1 is:

$$dN_1/dt = r_{m1} \times N_1 \times (1 - N_1/K_1) \times (a_{12} \times N_2/K_1) \tag{9.2}$$

where r_{m1} is the intrinsic rate of increase for species 1.

The ratio N_2/K_1 represents the abundance of species 2 relative to the carrying capacity (K_1) of species 1. It is a measure of how much of the resource is used by species 2 that would have been used by species 1. The **coefficient of competition** a_{12} measures the competitive effect of species 2 on species 1. If we define the competitive effect of one individual of species 1 upon the resource use of an individual of its own population as unity, then the coefficient for the effect of other species is expected to be less than unity. We expect this because individuals will compete more strongly with those similar to themselves than with the dissimilar individuals of other species. This does not always occur: when two species differ greatly in size an individual of the larger species (l) may consume far more of a resource than one of the smaller species (s) and in this case the a_{sl} could be greater than unity. The converse effect of species 1 on species 2 is denoted by the coefficient a_{21} in the equation for the other species:

$$dN_2/dt = r_{m2} \times N_2 \times (1 - N_2/K_2) \times (a_{21} \times N_1/K_2) \tag{9.3}$$

These two equations (9.2, 9.3) are called the Lotka–Volterra equations, after the two authors who produced them (Lotka 1925; Volterra 1926a). We can examine the implications of the equations graphically by plotting the numbers of species 2 against those of species 1, as in Fig. 9.1a. First we plot the conditions for species 1 when dN_1/dt is zero. There are the two extreme points when N_1 is at K_1 so that N_2 is zero, and when N_1 is zero because species 2 has taken all the resource. This latter point can be found from eqn. 9.2 by setting dN_1/dt to zero and rearranging so that it simplifies to:

$$N_1 = K_1 - a_{12} \times N_2$$

If the resource is taken entirely by species 2, then:

$$N_1 = 0, \text{ and } N_2 = K_1/a_{12}$$

Of course there can be any combination of N_1 and N_2 so that dN_1/dt is zero; this is seen from the diagonal line joining these two extreme points. To the left of this line

Fig. 9.1 Isoclines for the Lotka–Volterra equations. (a) At any point on the isocline $dN_1/dt = 0$. This indicates where the number of species 1 is held constant for different population sizes of species 2. Species 1 increases to the left of the isocline, but decreases right of it. (b) The isocline where $dN_2/dt = 0$. This shows where the population of species 2 is held constant at different values of species 1. Species 2 increases below the line, but decreases above it.

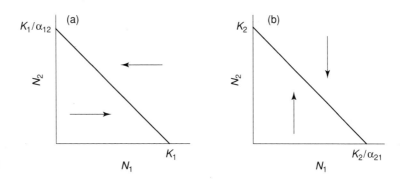

dN_1/dt is positive so that N_1 increases, and to the right it is negative and N_1 decreases as indicated by the arrows. At all points on the line (called an **isocline**) the population is stationary. Exactly similar reasoning produces the equivalent diagram for species 2 (Fig. 9.1b). Below the line (isocline) N_2 increases, and above it N_2 decreases.

With these two diagrams describing the competitive abilities of the two species independently we can now predict the outcome of competition between them. If we put the two diagrams in Fig. 9.1 together, as in Fig. 9.2a, we see that K_1 is larger than K_2/a_{21}. The latter term is the number of species 1 required to drive species 2 to extinction, and since it is possible for species 1 to exist at higher numbers than this level (i.e. at K_1), species 1 will drive species 2 down. On the other axis we see that

Fig. 9.2 The relationship of the two species' isoclines determines the outcome of competition. (a) Species 1 increases at all values of species 2 so that species 1 wins. (b) The converse of (a) such that species 2 wins. (c) In the region where the isocline of species 1 is outside that of species 2, species 1 wins and vice versa, so that either can win. (d) A stable equilibrium occurs because all combinations tend towards the intersection of the isoclines.

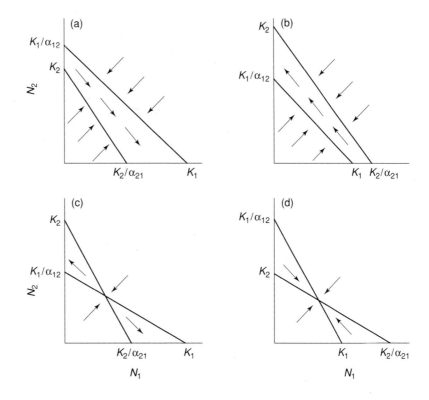

K_2, which is the maximum number of species 2 that the environment can hold, is less than that necessary to drive down species 1. Therefore species 2 always loses when the two species occur together, as can be seen by the resultant arrows and by the fact that the species 1 isocline is always outside that of species 2.

The above outcome is not the only possible solution, for this depends on the relative positions of the two isoclines that are shown in Figs 9.2b–d. Figure 9.2b is the converse to that of Fig. 9.2a so that species 2 always wins. In Fig. 9.2c we see that $K_2 > K_1/a_{12}$ and $K_1 > K_2/a_{21}$ so that, depending on the exact combination of the two population sizes, either can win. Where the two isoclines cross there is an equilibrium point but this is unstable in the sense that any slight change in the populations will cause the system to move to either K_1 or K_2 and the extinction of one of the species. In nature we would never see such an equilibrium.

Figure 9.2d also shows the two isoclines crossing, but in this case $K_2 < K_1/a_{12}$ and $K_1 < K_2/a_{21}$ (i.e. individuals of the same species affect each other more than do individuals of the other species, and neither is capable of excluding the other). This also means that intraspecific competition is always greater than interspecific competition. Hence, whatever the combination of the two populations, the arrows show that the system moves to the equilibrium point, which is therefore stable. This situation can occur only if there is some form of separation in the resources that they use, which we call niche partitioning (see Section 9.6).

9.2.2 Implications and assumptions

1 We can see from the figures that the outcome of competition depends upon the carrying capacities (K_1 and K_2) and the competition coefficients (a_{12} and a_{21}) according to the Lotka–Volterra model. The intrinsic rate of increase has no influence on which species will be the eventual winner.

2 Coexistence occurs when intraspecific competition within both species is greater than interspecific competition between them.

3 These equations can be expanded to include the effects of several species on species 1 by summing the $a \times N$ terms. This assumes that each species acts independently on species 1.

4 There are several other assumptions underpinning the logistic equation, for example constant environmental conditions leading to constant r and K, and no lags in competing species' responses to each other. Furthermore, the competition coefficients are constant: the intensity of competition does not change with size, age, or density of the competing species.

These assumptions mean that the Lotka–Volterra equations, like the logistic one, are simplistic and idealized. It is unlikely that the assumptions hold, although they may be approximated in some cases. The real value of these models is that they show how it is possible for coexistence to occur in the presence of competition, and that exclusion is not necessarily predetermined but may depend on the relative densities of the competing species.

9.3 Experimental demonstrations of competition

9.3.1 Perturbation experiments

Much of the work in ecology has assumed that competition has occurred and is necessary for the coexistence of species, and competition is one of the major assumptions in Darwin's theory of natural selection. Nevertheless, it is necessary to demonstrate that interspecific competition does actually take place. One of the most direct approaches is to carry out a **removal experiment** whereby one of the species is removed, or reduced in number, and the responses of the other species are then

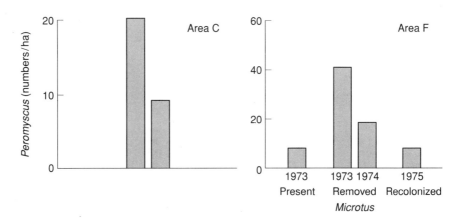

Fig. 9.3 Population densities of deermice (*Peromyscus*) on two areas from which voles (*Microtus*) were removed in the years 1973 and 1974. On area F only deermice were monitored after voles were allowed to recolonize. Deermice were absent or in low numbers before and after the vole removal, but high during the removal. (Data from Redfield *et al.* 1977.)

recorded. If competition has been operating we would expect that either the population, or reproductive rate, or growth rate of the other species would increase.

Forsyth and Hickling (1998) showed from an incidental removal experiment through hunting that Himalayan tahr (*Hemitragus jemlahicus*) are associated with declining populations of chamois (*Rupicapra rupicapra*). Competition appears to occur through behavioral interference, with the larger tahr excluding chamois. Another experiment, illustrated in Fig. 9.3, examined the competitive effect of voles (*Microtus townsendii*) on deermice (*Peromyscus maniculatus*). Deermice normally live in forests but one race on the west coast of Canada can also live in grassland, the normal habitat of voles. Redfield *et al.* (1977) removed voles from three plots and compared the population response of the deermice there with that on two control areas. On one control there were no deermice, on the other 4.7/ha. All the removal areas showed increases in deermouse numbers, one going from 7.8/ha before removal to 62.5/ha 2 years later. At the end of the study, when the workers stopped removing voles, these animals recolonized, reaching densities of 109/ha, while deermice numbers dropped to 9.4/ha. In another experiment, instead of removing voles, Redfield disrupted the social organization of the voles by altering the sex ratio so that there was a shortage of females, but the density remained similar to the controls. In this area deermice numbers increased from nearly zero to 34/ha. This result suggests that it was **interference competition** due to aggression from female voles that excluded the deermice because the density and food supply remained the same.

A similar type of experiment was conducted on desert rodents in Arizona by Munger and Brown (1981). They excluded larger species from experimental plots while smaller species were allowed to enter. Plots were surrounded by a fence, and access was controlled by holes cut to allow only the smaller species to enter. There were two types of small rodents, those that ate seeds (granivores) and those that ate a variety of other foods as well (omnivores). Munger and Brown predicted that if there was exploitation competition for seeds between the large and small granivores then the latter should increase in number in the experimental plots, while the omnivore populations should stay the same; if, however, the increased density of granivores was an artifact of the experiment (e.g. by excluding predators) then the number of small omnivores should also increase. Figure 9.4 shows that after a 1-year delay small granivores reached and maintained densities that averaged 3.5 times higher on the removal plots than on the controls, but the small omnivores did not show any significant

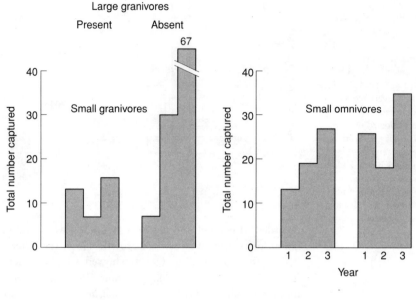

Fig. 9.4 Exclusion of large granivorous rodents resulted in an increase in the small granivorous rodent population relative to control areas, indicating competition. Small omnivorous rodent numbers did not increase significantly, indicating lack of competition. (Data from Munger and Brown 1981.)

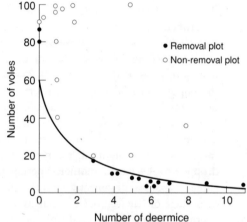

Fig. 9.5 Number of deermice (*Peromyscus maniculatus*) known to be alive at time t plotted against number of voles (*Microtus ochrogaster*) known to be alive at time $t - 1$ ($t - 1$ is the sample period 2 weeks earlier than period t). (After Abramsky *et al.* 1979.)

increase. These results are consistent with the interpretation that there was competition between large and small granivorous rodents.

Although the above examples produced results consistent with the predictions of interspecific competition, there was no attempt to measure the competition coefficients. However, Abramsky *et al.* (1979) carried out a similar removal experiment on the shortgrass prairie in Colorado in which a competition coefficient was measured. In this case voles (*M. ochrogaster*) were removed and the response of deermice (*P. maniculatus*) recorded. Figure 9.5 shows the negative relationship between the number of deermice present in the removal plot and the number of voles present in the previous sampling period 2 weeks earlier, as expected if competition were acting. To measure the competition effect (a) of voles on deermice, the Lotka–Volterra equation was used. At equilibrium $dN_1/dt = 0$, and so:

$$K_1 = N_1 + a \times N_2 \times (2^{0.75})$$

where K_1 is the carrying capacity of the environment for individuals of deermice when alone, N_1 is the number of deermice, N_2 is the number of voles, and $2^{0.75}$ is the conversion factor, and standardizes the species in terms of their metabolic rates.

The body weight (W) of voles is about two times that of deermice, and the basal metabolic rate (M) is taken as $M = W^{0.75}$ (see Section 4.5.2). Using various combinations of N_1 and N_2 an average estimate of $a = 0.06$ was obtained.

Properly designed removal experiments are difficult to carry out for practical reasons, so it is not surprising that they have not yet been performed with large mammals.

9.3.2 Natural experiments

An easier approach uses natural absences or combinations of species to observe responses that would be predicted from interspecific competition. For example, mallard ducks (*Anas platyrhynchos*) breed on oligotrophic (low nutrient) lakes in Sweden (Pehrsson 1984). Some of the lakes contained fish while others did not. In lakes with fish, the density of mallards was lower, mean invertebrate food size was lower, and emerging insects were significantly smaller. In an experiment where ducklings were released, their intake rate was higher on lakes without fish (Table 9.2). These results imply competition between ducks and fish.

Another type of natural experiment is illustrated by the distributions of two gerbilline rodent species in Israel (Abramsky and Sellah 1982). One species, *Gerbillus allenbyi*, lives in coastal sand dunes and is bounded in the north by Mt Carmel. In the same region the other species, *Meriones tristrami*, is restricted to non-sandy habitats. In the coastal area north of Mt Carmel, *M. tristrami* occurs alone and inhabits several soil types including the sand dunes. Abramsky and Sellah suggested that *M. tristrami* colonized from the north and was able to bypass Mt Carmel, whereas *G. allenbyi* colonized from the south and could not pass the Mt Carmel barrier. In the region of overlap, south of the barrier, interspecific competition had excluded *M. tristrami* from the sand dunes. They tested this hypothesis by removing *G. allenbyi* from habitats where the two species overlapped, and found that there was no significant increase in *M. tristrami*. They concluded that there was no present-day competition occurring. Instead they suggested that competition in the past had resulted in a shift in habitat choice so that there was no longer any detectable competition.

Islands are sometimes used to look at the distributions of overlapping species, because on some islands a species can occur alone while on others it overlaps with related species. The theory of interspecific competition would predict that when alone a species would expand the range of habitats it uses (a process we call **competitive release**), while on islands where there are several species the range of habitats contracts

Table 9.2 Mean dry weight of subaquatic invertebrates available to mallard ducklings and the rate of duckling food intake in Calnes with and without fish in Sweden.

	Year	Lakes without fish	Lakes with fish
Mean dry weight (May–June)	1977	119.8 (21.0)	45.3 (13.7)**
	1978	159.0 (9.9)	26.5 (4.8)**
Duckling feeding (food items/min)	1977	12.4 (0.6)	9.5 (0.5)***
	1978	20.4 (5.1)	7.9 (0.7)**

$**P < 0.01$; $***P < 0.001$.
From Pehrsson (1984).

Fig. 9.6 Habitats of ground doves (*Chalcophaps* and *Gallicolumba*) on islands off New Guinea demonstrate "competitive release." (After Diamond 1975.)

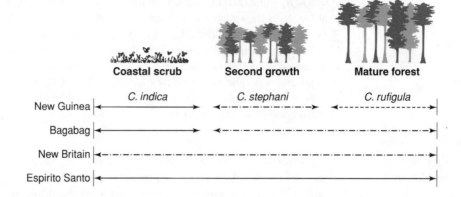

(**competitive exclusion**). A good example of this is seen in ground doves on New Guinea and surrounding islands (Diamond 1975). On the larger island of New Guinea there are three species each with its own habitat (Fig. 9.6): *Chalcophaps indica* in coastal scrub, *C. stephani* in second growth forest further inland, and *Gallicolumba rufigula* in the interior rainforest. On the island of Bagabag *G. rufigula* is absent and *C. stephani* expands into the mature forest. On some islands (New Britain, Karkar, Tolokiwa) only *C. stephani* occurs and it uses all habitats, while on the island of Espirito Santo only *C. indica* occurs and this species also expands into all habitats. It is assumed that this habitat expansion has been due to competitive release through the absence of the other potential competitors.

9.3.3 Interpreting perturbation experiments

Perturbation experiments are designed to measure responses of populations that would be predicted from interspecific competition theory. We should be aware, however, that there are two types of perturbations (Bender *et al.* 1984). One, called a **pulse** experiment, involves a one-time removal of a species. We then measure the rate of return by the various species to the original equilibrium. This requires accurate measurements of rates of population increase, which in practice is not easy and in fluctuating environments very difficult. As a consequence few of these experiments are carried out.

The other type of perturbation is the continuous removal, or **press**, experiment. Let us assume that species 1 is reduced to a new level and kept there. Other species are allowed to reach a new equilibrium and it is this level that is observed. This type of perturbation avoids having to measure rates but there are other problems. If there are more than two species in a community, which in most cases there are, an increase in another species' population is neither a necessary nor sufficient demonstration of competition. First, species 1 and species 2 may not overlap, and so not compete, but they may affect each other through interactions with other competing species: this is **indirect competition**. Second, the two species could be alternative prey for a food-limited predator. Changes in the population of species 1 could affect that of species 2 by influencing the predator population: this has been termed **apparent competition** (Holt 1977) and we will discuss it again below (see Section 9.8).

All of the examples we have discussed above are press experiments and strictly speaking, in order to demonstrate competition unequivocally, we would need to know that: (i) resources were limiting; (ii) there was overlap in the use of the resources; (iii) other potential competitors were having a negligible effect; and (iv) predator populations were not responding to the experiment. In few cases have all these conditions been met. Because of these difficulties an entirely different approach to the study of interspecific competition has measured the pattern of overlap in the use of resources. We now consider this approach.

9.4 The concept of the niche

In Chapter 3 we saw that different species on different continents appeared to adopt the same role in the community and often these species have evolved similar morphological and behavioral adaptations. This place in the community is called the **niche**, defined by Elton (1927) as the functional role and position of the organism in its community. (We provide the modern definition later.)

For practical reasons the niche has come to be associated with use of resources. Thus, we can plot the range and frequency of seed sizes eaten by different bird species, as a hypothetical example, in Fig. 9.7a. Species that exploit the outer parts of the

Fig. 9.7 (a) Hypothetical frequency distribution of seeds of different sizes indicating the range and overlap of potential niches for granivores. (After Pianka *et al.* 1979.) (b) Range of seed sizes eaten by British finches feeding on herbaceous plants. Seeds are in five size categories A to E. The finches are redpoll (*Carduelis flammea*), linnet (*C. cannabina*), greenfinch (*Chloris chloris*), and hawfinch (*Coccothraustes coccothraustes*). (After Newton 1972.)

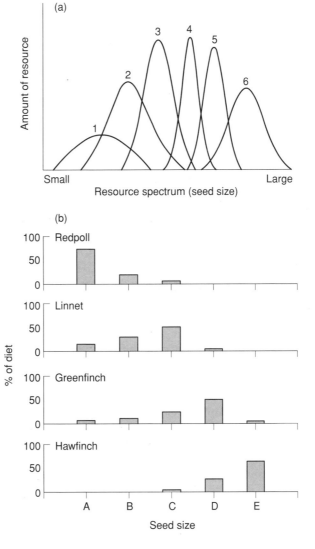

Fig. 9.8 Hypothetical frequency distribution of species 1 and species 2 along two parameter gradients: (a) moisture; (b) temperature. Outline of the species distributions when considering the two parameters simultaneously shows niches that can be either distinct (c), or overlapping (d).

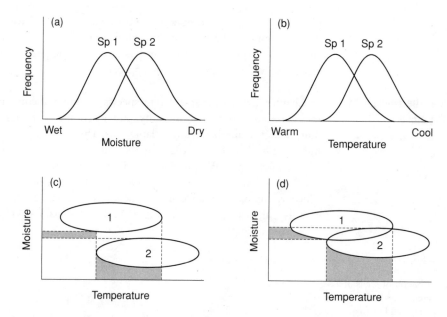

resource spectrum use a broader range of resources because they are less abundant. Some species, for example 2, 3, and 4, overlap while others such as 2 and 5 do not. Overlap is necessary (but not sufficient) to demonstrate competition. An example (Fig. 9.7b) is provided by the range of seed sizes eaten by finches in Britain (Newton 1972). In this case we see that, contrary to the theoretical distribution proposed in Fig. 9.7a, there is a broader range of seed sizes eaten by these finches in the middle range than by birds eating seeds at the extremes.

So far we have considered only one resource axis, that is, one variable such as seed size. When we consider two or more axes the picture becomes less clear cut in terms of overlap. Take two species, 1 and 2, which overlap along two axes, for example moisture and temperature as in Figs 9.8a and 9.8b. If we plot the outline of the two species distributions by considering the two axes simultaneously we see that it is possible for the two distributions to be distinct (Fig. 9.8c) or to overlap (Fig. 9.8d). Which one occurs depends on whether individuals show **complementarity** (i.e. individuals that overlap on one axis do not do so on the other one (Fig. 9.8c), or overlap simultaneously on both axes (Fig. 9.8d)).

An example of complementarity is shown in Fig. 9.9. DuBowy (1988) examined the resource overlap patterns in a community of seven North American dabbling ducks all of the genus *Anas*, by plotting habitat overlap against food overlap for pairs of species. In winter, when it is assumed that resources were limiting, points for pairs were below the diagonal line (Fig. 9.9a), indicating complementarity: pairs with high overlap in one dimension had low overlap in the other. In contrast, during summer species pairs showed high resource overlap in both dimensions (several points are outside the line), indicating that species fed on the same food at the same place. In summary, the change in niche of these duck species from summer to winter results in lower overlap and by implication lower competition at a time when we would expect that resources would be limiting. Note, however, that neither the lack of resources nor interspecific competition was demonstrated, merely that the results conform to what we would predict if competition had been acting.

Fig. 9.9 Resource overlaps in seven species of dabbling ducks. Below the broken line there is complementarity in overlaps. (a) In winter, high habitat overlap between pairs of species tends to be associated with low food overlap, demonstrating complementarity. (b) In summer there is simultaneous overlap in both dimensions. (After DuBowy 1988.)

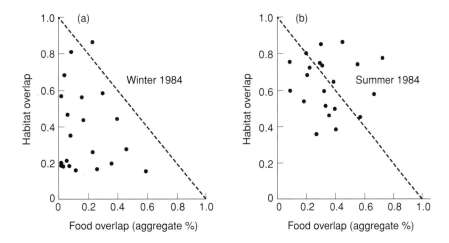

Green (1998) found complementarity in ducks along habitat and feeding behavior axes. He found in dabbling and diving ducks in Turkey that pairs with similar habitat had dissimilar feeding mechanisms.

We have considered only two dimensions of a niche so far, but clearly the niche must include every aspect of the environment that would limit the distribution of the species. We cannot draw all these dimensions on a graph but we could perhaps imagine a sort of sphere or volume with many dimensions, which could theoretically describe the complete niche. Hutchinson (1957) described this as the **n-dimensional hypervolume**. This is the fundamental niche of the species and is defined by the set of resources and environmental conditions that allow a single species to persist in a particular region (Schoener 1989; Leibold 1995). This suggests that the niche is in some way discrete. However, resource measures are usually continuous so the discreteness does not come from these. Rather it comes from the constraints of the species in terms of their morphology, physiology, and behavior – a species is more efficient at using some combinations of a resource than other combinations, while other species have different combinations where they are most efficient. These peaks of efficiency, then, are the adaptive peaks exhibited by a species (Schluter 2000).

The fundamental niche is rarely if ever seen in nature because the presence of competing species restricts a given species to a narrower range of conditions. This range is the observed or **realized niche** of the species in the community. It emphasizes that interspecific competition excludes a species from certain areas of its fundamental niche. In terms of the Lotka–Volterra diagrams (see Fig. 9.2) the weaker competitor has no realized niche in Figs 9.2a and 9.2b, and for Fig. 9.2d parts of the fundamental niche are not used.

The difference between the two types of niches can be seen in a study by Orians and Willson (1964) of red-winged blackbirds (*Agelaius phoeniceus*) and yellow-headed blackbirds (*Xanthocephalus xanthocephalus*). Both species make their nests among reeds in freshwater marshes of North America, and, if alone, both will use the deep-water parts of the marsh (there is greater protection from mammalian predators here). However, when the two species occur together, the yellow-heads exclude the red-winged blackbirds, which are then restricted to nesting in the shallow parts. Thus, the fundamental niche for nesting red-winged blackbirds is the whole marsh, but the realized niche is the shallow-water reedbed. Coexistence occurs from the partitioning of the resource (nesting habitat), and the divergence of realized niches.

9.5 The competitive exclusion principle

In 1934 Gause stated that "as a result of competition two species hardly ever occupy similar niches, but displace each other in such a manner that each takes possession of certain kinds of food and modes of life in which it has an advantage over its competitor" (Gause 1934). In short, two species cannot live in the same niche, and if they try one will be excluded; second, coexisting species live in different niches. This is known as the **principle of competitive exclusion**, or Gause's principle (Hardin 1960), and has become one of the fundamental tenets of ecology. It proposes that species can coexist if adaptations arise to effectively partition resources. Examples of such adaptations include the use of different microhabitats, different components of prey, different ways of feeding, different life stages of the same prey, different time periods in the same habitat, or taking advantage of disturbance, and interference competition (Richards *et al.* 2000). Therefore, Gause's principle has become the basis for studies of resource partitioning and overlap as a way of measuring interspecific competition.

There are, however, two serious problems with Gause's statement. The first is that it is a trivial truism, because we have already identified the two coexisting populations as being different by calling them different species, and, therefore, if we look hard enough we are likely to find differences in their ecology as well. This is called a tautology: having defined the species as being different, it should be no surprise to find they are different.

The second problem is that the principle is untestable. It cannot be disproved because either result (exclusion or coexistence) can be attributed to the principle. To disprove the principle it is necessary to demonstrate that the niches of two species are identical. Yet, as we can see from Fig. 9.8, what appears to be overlap, even complete overlap, may not be so when an additional axis is taken into account. Since we can never be sure that we have measured all relevant axes in describing the niches of two species, we can never be sure that the two niches are the same. Hence, we cannot disprove the principle.

9.6 Resource partitioning and habitat selection

9.6.1 Habitat partitioning

Despite these problems with the competitive exclusion principle, it underlies the numerous studies of habitat partitioning amongst groups of coexisting species. Lamprey (1963) described the partitioning of habitats by species of savanna antelopes in eastern Africa. A similar study by Ferrar and Walker (1974) showed how various antelopes in Zimbabwe used the three habitat types of grassland, savanna, and woodland (Fig. 9.10). In both cases there was partitioning as well as overlap.

Similar studies by Wydeven and Dahlgren (1985) show partitioning of both habitat and food in North American ungulates (Fig. 9.11). In Wind Cave National Park elk and mule deer have similar winter habitat choices, as do pronghorn and bison, but these pairs have very different diets. For example, the diet of bison contains 96% grass as against 4% for pronghorn.

Interspecific overlap in the diet niches of two sibling bat species (*Myotis myotis*, *M. blythii*) of Switzerland shows niche partitioning: *M. myotis* feeds largely on ground insects (carbid beetles) whereas *M. blythii* feeds mostly on grass-dwelling insects (bush crickets). This allows coexistence within the same habitats (Arlettaz *et al.* 1997). MacArthur (1958), in a now classic paper, described the different feeding positions

Fig. 9.10 Habitat partitioning and overlap by ungulates in Kyle National Park, Zimbabwe. The width of the boxes reflects the degree of preference. (After Ferrar and Walker 1974.)

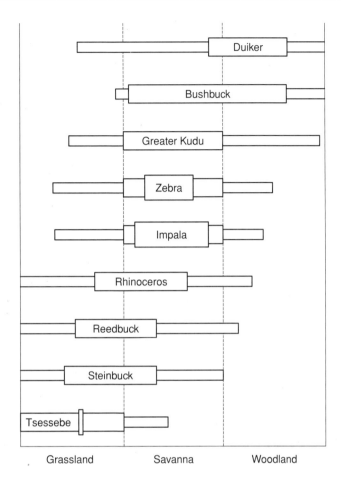

Grassland   Savanna   Woodland

of five species of warblers within conifer trees in the northeastern USA. They varied in both height in the tree and use of inner or outer branches. Nudds *et al.* (1994) found habitat partitioning in dabbling ducks in both Europe and North America. Species with a high density of filtering lamellae in their bills (fine filter feeders) tended to live in deep water with short, sparse vegetation compared with those species with few lamellae which lived in shallow water with tall, dense vegetation.

9.6.2 *Limiting similarity*

As we have mentioned above it should not be surprising that species divide up the resource available to them. However, Gause's principle implies that there should be a limit to the similarity of niches allowing coexistence of two species. Earlier studies predicted values of limiting similarity based on theoretical arguments (MacArthur and Levins 1967). If the distance between the midpoints of species distributions along the resource axis is d and the standard deviation of the curves (such as those in Fig. 9.7a) is w (the relative width) then limiting similarity can be predicted from the ratio d/w. However, various assumptions, such as the curves must be similar, normally distributed, and along only one resource axis, make this approach unrealistic.

Pianka *et al.* (1979) asked: how much would niches overlap if resources were allocated randomly among species in a community? A frequency distribution of niche overlaps generated from randomly constructed communities is shown in Fig. 9.12.

Fig. 9.11 (a) Diet and
(b) winter habitat use
of elk, mule deer,
pronghorn, and bison
in Wind Cave National
Park, South Dakota.
Where habitat choice is
similar there are major
differences in diet.
(Data from Wydeven
and Dahlgren 1985.)

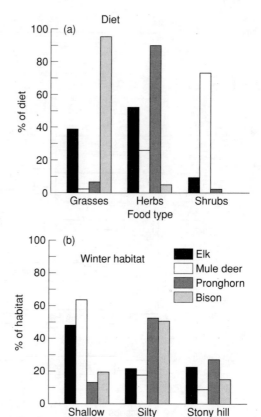

Fig. 9.12 Frequency
distribution of niche
overlaps in 100
randomly constructed
communities with 15
species and five equally
abundant resource
states. (After Pianka
et al. 1979.)

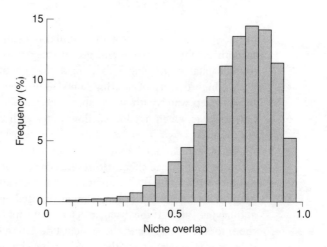

This can be compared with distributions of observed overlap in diets of desert lizards
from 28 sites on three continents (Fig. 9.13). Those in the Kalahari desert of southern Africa showed the greatest degree of overlap (because one food type, termites,
comprised a large amount of the diet), and those in Australia the least. In no case
were observed distributions similar to the random distributions: there were far more

Fig. 9.13 Distributions of observed overlap in the diets of desert lizards. Dietary overlap is higher in Kalahari lizards, where termites comprised 41% of the diet. (After Pianka *et al.* 1979.)

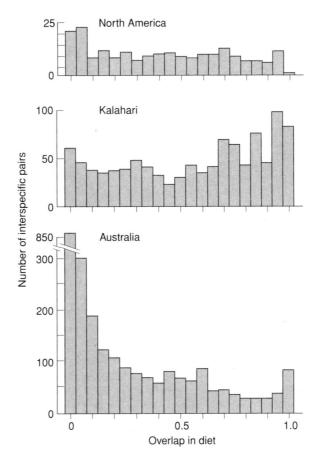

species pairs with small overlaps than would be predicted by chance, implying that interspecific competition was causing niche segregation.

9.6.3 Habitat selection from field data

As we have seen, species usually differ from each other by choosing different resources such as food types, habitats, etc. We call this choice **habitat selection**. One approach to measuring the competition coefficients has been to look at the variation of species density in different habitats. First, the variation in density due to habitats, and other resources, is estimated by statistical procedures such as multiple regression (Crowell and Pimm 1976). Then the remaining variance should be attributable to interspecific competition with another identifiable species. An example of this approach is given by Hallett (1982) in a study of 10 desert rodent species in New Mexico. He measured habitat variables related to the common plants such as number of individuals, plant height, distance to nearest plant from trap, and percent cover. Regression analysis was used to partition the variance in capture frequency at trap stations due to habitat variables and competitors. Competition was observed within one group of three species, *Perognathus intermedius*, *Perognathus penicillatus*, and *Peromyscus eremicus* (Table 9.3). Although the competitive effects differed from year to year, they were not random. Also the inhibitory effects were not symmetrical: thus *P. eremicus* always had a greater effect on the other two species than the reverse, and similarly *P. intermedius* had a greater effect on *P. penicillatus* than vice versa.

Table 9.3 Matrix of competition coefficients for the *Perognathus–Peromyscus* guild for each year. Entries are the partial regression coefficient after removal of the effects of the habitat variables. The coefficients are the effects of the column species (independent variable) on the row species (dependent variable).

	1971		1972			1973		
	PP	PI	PP	PI	PE	PP	PI	PE
PP	...	−0.43*	...	−0.17	−0.42*	...	−0.12	−0.82*
PI	−0.17	...	−0.09	...	0.05	−0.05	...	−0.39*
PE	NI	NI	−0.09	0.19	...	−0.12*	−0.10	...

*$P < 0.05$.
PP, *Perognathus penicillatus*; PI, *Perognathus intermedius*; PE, *Peromyscus eremicus*.
NI, not included in the analysis.

Fig. 9.14 The negative correlation in numbers of two woodmice species (*Apodemus mystacinus* and *A. sylvaticus*) in Israel. (After Abramsky 1981.)

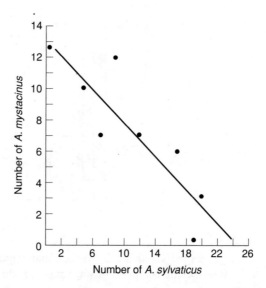

Abramsky (1981) used a similar regression method to look at interspecific competition and habitat selection in two sympatric rodents, *Apodemus mystacinus* and *A. sylvaticus*, in Israel. Plotting the densities of the two species in different habitats against each other (Fig. 9.14) indicated a negative relationship and suggested that there may be interspecific competition operating. However, he found that species abundances could have been the result of habitat differences alone; the effect of the presence of the other species was negligible in this case, implying no competition.

There are problems with the regression method, some of which are outlined by Abramsky *et al.* (1986). One is that if sympatric populations of different species differ greatly in average abundance, then estimates of their variance and regression coefficients are distorted. In turn estimates of competition are unreliable. A second problem lies in the assumption of constant competition coefficients; if competition is weak when populations are close to equilibrium (which we assume is when regressions from field populations are estimated), but strong when disturbed from equilibrium (the situation in perturbation experiments), then regression analysis is likely to miss competitive effects while experiments will indicate their presence. A third problem is that we can never be sure that we have accounted for all the variability in density at various sites from environmental factors; there may be some factor that has been overlooked to account for the remaining variability instead of attributing it to interspecific competition.

9.6.4 *The theory of habitat selection*

Since species prefer to use some habitats over others, we ask how does this choice change when resources become limiting? There are two hypotheses that we should consider. We start with the **theory of optimal foraging** which predicts that when resources are not limiting, species should concentrate their feeding on the best types of food or the best types of habitats and ignore the others no matter how abundant they are. When resources are limiting, a species should expand its niche to include other types of food, habitat, etc. This is the expected response under intraspecific competition.

When two species are present one might expect both to respond to declining resources by expanding their niches and so increase the overlap. However, Rosenzweig's (1981) **theory of habitat selection** introduces a second hypothesis. This predicts that when resources are limiting, species should contract their niches as a result of inter-specific competition. He considered two different situations: we start by assuming that there are two species, 1 and 2, and two habitats, A and B. In the first case, called the **distinct preference** case, both species use both habitats but each prefers to use a different one (i.e. species 1 treats A as the better and B as the poorer habitat, while species 2 does the reverse). In the second case, the **shared preference** case, again both species use both habitats, but now both treat the same habitat A as the better and B as the poorer one.

In either case we should first consider the habitat choice of a species when no competitors are present. Under conditions of abundant resources, such as food, a species should confine itself to its preferred habitat. As density increases and resources become limiting through the feeding of other individuals, the species will continue to remain in the better habitat (A) so long as the food intake rate is greater than what it would be in the poorer habitat (B). At some point, density in habitat A increases so that intake rate falls until it equals that in habitat B: at this stage the species should not confine itself to A but should expand its habitat use in such a way that densities keep intake rates similar in the two habitats. The intake rate at which the species changes from one to two habitats is called the **marginal value**.

Now we consider what happens when there is a competitor present and resources are limiting. The outcome depends on which of the two preference cases occurs. First, in the distinct preference case each species will confine itself to its preferred habitat rather than expand into the other habitat. Therefore, when resources are limiting, species will specialize, contract their niches and reduce overlap. When resources are abundant they should use either one or both habitats depending on their intake rates in the two habitats. Second, in the shared preference case we have to assume that one of the species (1) is dominant and can exclude by behavior, or other means such as scent marking, the second species from the preferred habitat A. If species 2, the subordinate, is to coexist with species 1 then it must be more efficient at using the less preferred habitat B than species 1. If the dominant is more efficient in both habitats then it will exclude species 2. Therefore, when resources are limiting, one species – the dominant – will not change its habitat choice. In contrast, the subordinate will change its choice from A to B: the competitive effect is asymmetrical, with the dominant having a large effect on the subordinate while the reverse effect is small.

Fig. 9.15
Demonstration of
distinct and shared
preferences in habitat
selection by three
species of honeycreepers
in Hawaii. The two
main flowering trees
are (a) *Sophora* and
(b) *Metrosideros*. At low
flower numbers *Loxops*
(O) fed on *Sophora*, and
Himatione (■) fed on
Metrosideros, showing
"distinct preference."
Vestaria (△) fed on
both trees, excluding
the other species, but
only at high flower
numbers, indicating
"shared preference."
Note the reverse scale
on the *x*-axis. (After
Pimm and Pimm 1982.)

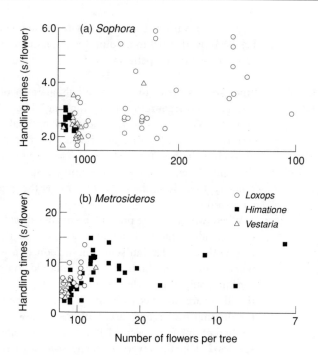

In a test of these hypotheses Pimm and Pimm (1982) recorded the feeding choices of three nectar-feeding bird species (*Himatione*, *Loxops*, *Vestiaria*) on the island of Hawaii. There were two main tree species, *Metrosideros* and *Sophora*, which came into flower at different times of the year. The evidence for the distinct preference case is seen in Fig. 9.15. When the number of flowers is high, all three species feed on both trees. When flowers per tree are low (and assuming that this indicated limiting resource) only *Loxops* feed on *Sophora*, and only *Himatione* feed on *Metrosideros*. Thus, both species reduce their niche width and specialize. There was also evidence of shared preference. *Vestaria* feed on both tree species but only at high flower numbers, and physically exclude the other species by visual and vocal displays. In contrast, both *Himatione* and *Loxops* spend much of their feeding time on trees with few flowers. Thus, these two species are confined to poorer feeding areas during times when resources are low, as predicted by the theory.

Rosenzweig's theory predicts that niches contract when resources are limiting and there is interspecific competition. We have seen that the Hawaiian honeycreepers may conform to the predictions, but what about other species? Information from wildlife both agrees and disagrees with the predictions. The overlap in diet of sympatric mountain goats (*Oreamnos americanus*) and bighorn sheep (*Ovis canadensis*) is high in summer but reduced in winter (Dailey *et al.* 1984), as predicted by the theory. In ducks we have already seen that during winter there is a decrease in overlap (Fig. 9.9). Burning grasslands increases the nutrient content of regenerating plants and may produce locally abundant food. Under these conditions mountain goats and mule deer (*Odocoileus hemionus*) actually increase dietary overlap (Spowart and Thompson Hobbs 1985). In contrast, elk and deer in natural forests increased dietary overlap in winter when resources were assumed to be least available, contrary to expectation (Leslie *et al.* 1984).

We should note that we do not have actual measures of the food supply in these examples, so we cannot be sure that we are seeing competition. In Serengeti, Tanzania, wildebeest are regulated by lack of food in the dry season (Mduma *et al.* 1999), so that overlap with this species at this time should result in competition. However, overlap in both diet and habitat between wildebeest and several other ungulate species increases or does not change between wet and dry seasons (Hansen *et al.* 1985; Sinclair 1985). One interpretation could be that interspecific competition is asymmetrical, with the impact of the rarer ungulates on the numerous wildebeest being real but very slight, while the reverse does not occur because these other ungulates are kept at low density by predation (Sinclair *et al.* 2003).

9.7 Competition in variable environments

So far we have discussed the patterns of occupancy and utilization of habitats as if they were constants for a species, or that they changed only seasonally. However, longer-term studies are now showing that species densities vary in the same habitat and they also change over a longer time scale measured in years. Thus populations may go through periods when there are abundant resources and, although there is overlap with other species, even at the supposedly difficult time of year there is no competition (Fig. 9.16). Occasionally there are periods of resource restriction and it is only at these times that one sees competition and niche separation (Wiens 1977).

9.8 Apparent competition

9.8.1 *Shared predators*

Some of the predicted outcomes from interspecific competition include the reduction of populations, the contraction of niches, and exclusion of species from communities. However, these predictions are also to be expected when species have non-overlapping resource requirements but share predators, especially when predators can increase their numbers fast.

Let us suppose there is a predator that is food limited, and which feeds on two prey species. The prey are both limited by predation and not by their own food supplies. If species 1 increases in number then this should lead to more predators, which in turn will depress species 2 numbers. This result is called **apparent competition**

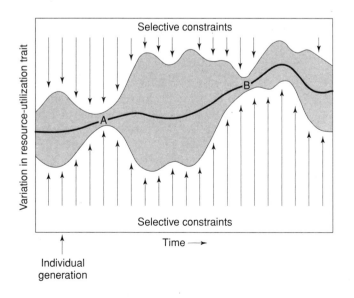

Fig. 9.16 Changes over time in the mean (thick line) and variance (shaded area) of a selective constraint such as resource availability. At times A and B there are "bottlenecks" when competition is more likely. (After Wiens 1977.)

because it produces the same changes in prey populations as would be predicted from interspecific competition (Holt 1977, 1984). Examples of apparent competition are given in Section 9.10.2 and Chapter 21, where predators are causing the demise of secondary prey, the rare roan antelope in Kruger National Park, South Africa (Harrington *et al.* 1999; McLoughlin and Owen-Smith 2003), and the wildebeest in Manyara National Park, Tanzania, as a result of a high abundance of buffalo, the primary prey.

If two prey species live in the same habitat, as in the wildebeest and buffalo example in Manyara, then at high intensities of predation coexistence is unlikely. On the other hand, coexistence is promoted if the two species select different habitats, that is, niche partitioning occurs.

Another version of apparent competition can occur through shared parasites. One species can be a superior competitor if it supports a parasite which it transmits to a more vulnerable species. For example, when gray squirrels (*Sciurus carolinensis*) were introduced to Britain, they brought a parapox virus that reduced the competitive ability of the indigenous red squirrel (*S. vulgaris*) (Hudson and Greenman 1998). The latter has largely been displaced, occurring now in only a few small locations of its former range. Gray squirrels are displacing red squirrels through competition in Italy, and could be spreading through Europe (Wauters and Gurnell 1999).

9.8.2 *Implications*

Since the observed responses of prey populations to changes in predator numbers are similar to those from interspecific competition, we cannot infer such competition simply from observations or even experiments that show either changes in species population size or niche shifts. We need to know (i) whether resources are limiting; and (ii) the predation rates and predator numbers.

9.9 **Facilitation**

9.9.1 *Examples of facilitation*

Facilitation is the process whereby one species benefits from the activities of another. In some cases the relationship is **obligatory** as in the classic example of the nereid worm (*Nereis fucata*), which lives only in the shell of hermit crabs (*Eupagurus bernhardus*). The crabs are messy feeders and scraps of food float away from the carcass that is being fed upon; these scraps are filtered out of the water by the worm. While the worm benefits, the crab appears not to suffer any disadvantage (Brightwell 1951). In other cases the relationship may be **facultative**, by which we mean that the dependent species does not have to associate with the other in order to survive, but does so if the opportunity arises. Thus, cattle egrets (*Ardea ibis*) often follow grazing cattle in order to catch insects disturbed by these large herbivores. Although the birds increase their prey capture rate by feeding with cattle, as they probably do by following water buffalo (*Bubalus bubalus*) in Asia and elephants and other large ungulates in Africa, they are quite capable of surviving without large mammals (McKilligan 1984). The European starling (*Sturnus vulgaris*) also follows cattle on occasions. In contrast, its relative in Africa, the wattled starling (*Creatophora cinerea*), seems always to follow large mammals and in Serengeti they migrate with the wildebeest like camp followers.

Vesey-Fitzgerald (1960) suggested that there was grazing facilitation amongst African large mammals. Lake Rukwa in Tanzania is shallow and has extensive reedbeds around the edges. The grasses, sedges, and rushes can grow to several meters in height, and in this state only elephants can feed upon the vegetation. As the elephants feed and trample the tall grass they create openings where there is lush

Fig. 9.17 The proportion of the population of different ungulate species using short grass areas on ridge tops (upper catena) in Serengeti. The larger species leave before the smaller at the start of the dry season. (After Bell 1970.)

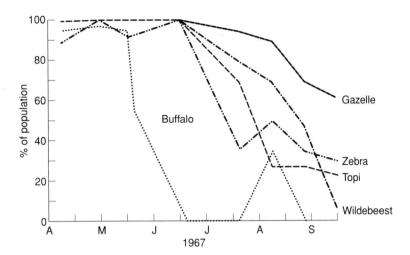

regenerating vegetation. This provides a habitat for African buffalo, which in turn provide short grass patches that can be used by the smaller antelopes such as topi. In this case elephants are creating a habitat for buffalo and topi that would not otherwise be able to live there. Therefore, the presence of elephants increases the number of herbivores that can live in the Lake Rukwa ecosystem. Vesey-Fitzgerald called this sequence of habitat change in the grasslands a **grazing succession**.

Bell (1971) has described a similar grazing succession amongst the large mammals of Serengeti. In certain areas of Serengeti there is a series of low ridges bounded by shallow drainage lines. The ridges have sandy, thin soils and support short, palatable grasses. The drainage lines have fine silt or clay soils that retain water longer than those on the ridges and so support dense but coarse grasses, which remain green long into the dry season. Between these two extremes there are intermediate soil types on the slopes. The whole soil sequence from top to bottom is called a **catena**. In the wet season when all areas are green, all five non-migratory species (wildebeest, zebra, buffalo, topi, and Thomson's gazelle) feed on the ridge tops. Once the dry season starts the different species move down the soil catena into the longer grass in sequence, with the larger species going first (Fig. 9.17). Thus, zebra is one of the first species to move because it can eat the tough tall grass stems. By removing the stems, the zebra make the basal leaves in these tussock grasses more available to wildebeest and topi, and these in turn prepare the grass sward for the small Thomson's gazelle. Thus, there is a grazing succession.

Zebra, wildebeest, and Thomson's gazelle also have much larger migratory populations in Serengeti separate from the smaller resident populations discussed above. It is tempting to think that the movements of these migrants follow the same pattern as those of the resident populations. Indeed, McNaughton (1976) has shown that migrating Thomson's gazelle prefer to feed in areas already grazed by wildebeest because these areas produce young green regrowth not found in ungrazed areas. The gazelle take advantage of this growth, which was stimulated by the grazing, and so benefit from the wildebeest.

The relationship between the migrating zebra and wildebeest is more complex. Although zebra usually move first from the short grass plains to the long grass dry-season areas, the wildebeest population (1.3 million), which is much larger than

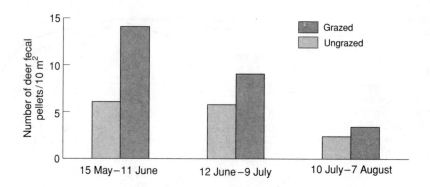

Fig. 9.18 Facilitation of deer grazing by cattle is demonstrated by deer fecal-pellet groups on cattle-grazed plots during each of the 3 months of study. Deer preferred to graze plots used by cattle the previous winter. (After Gordon 1988.)

that of zebra (200,000), often do not follow the zebra but take their own route and eat the long dry grass. Therefore, most migrant wildebeest obtain no benefit from the zebra. In contrast, zebra may be benefiting from the wildebeest for a completely different reason. In the wet season, when there is abundant food, many zebra graze very close to the wildebeest, and by doing so they can avoid predation because most predators (lions and spotted hyenas (*Crocuta crocuta*)) prefer to eat wildebeest. Only if there are no wildebeest within range will predators turn their attention to zebra. Therefore, it pays zebra to make sure there are wildebeest nearby. In the dry season, however, zebra compete with wildebeest for food. Zebra, therefore, show habitat partitioning and avoid the wildebeest. However, by doing so they probably make themselves more vulnerable to predators again (Sinclair 1985). Thus, zebra have to balance the disadvantages of predation if they avoid wildebeest with competition if they stay with the wildebeest. We can see a seasonal change from facilitation in the wet season when there is abundant food, to competition in the dry season when food is regulating the wildebeest.

An example of facilitation has been recorded on the island of Rhum, Scotland. There cattle were removed in 1957 and reintroduced to a part of the island in 1970 where they grazed areas used by red deer. Pasture used by cattle in winter results in a greater biomass of green grass in spring compared with ungrazed areas. Gordon (1988) found that deer preferentially grazed areas in spring that had previously been used by cattle (Fig. 9.18), and subsequently there were more calves per female deer.

On the North American prairies both black-tailed prairie dogs (*Cynomys ludovicianus*) and jackrabbits (*Lepus californicus*) benefit from grazing by cattle. If grazing is prevented, then the long grass causes prairie dogs to abandon their burrows. At a site in South Dakota where cattle were fenced out, there were half as many burrows as on adjacent areas where grazing was continued. Snell and Hlavachick (1980) showed that a large prairie dog site of 44 ha could be reduced to a mere 5 ha by the elimination of cattle grazing in summer to allow the grass to grow. Presumably under natural conditions when American bison grazed the prairies there was facilitation by bison allowing prairie dogs to live in the long grass prairies. Facilitation could be mutual because both pronghorn and bison respond to the vegetation changes caused by prairie dogs, both species using prairie dog sites (Coppock *et al.* 1983; Wydeven and Dahlgren 1985; Huntly and Inouye 1988; Miller *et al.* 1994).

This example illustrates two management points which follow from the understanding of the interaction (facilitation) between large mammal grazers and prairie dogs:

(i) a simple management program (through grazing manipulation) could be devised to control the prairie dogs, without the use of harmful poisons which could affect other species; and (ii) in many areas prairie dogs are becoming very scarce and their colonies need to be protected. In addition, the black-footed ferret (*Mustela nigripes*), one of the rarest mammals in the world, depends entirely on prairie dogs and it is thought that their very low population has resulted from the decline in prairie dog populations. The conservation of both species would benefit from the manipulation of grazing practices.

In another example where facilitation improved management for wildlife, Anderson and Scherzinger (1975) showed that ungrazed grassland resulted in tall, low-quality food in winter for elk. Cattle grazing in spring maintained the grass in a growing state for longer. If cattle were removed before the end of the growing season, the grasses could regrow sufficiently to produce a shorter, high-quality stand for elk; their population increased from 320 to 1190 after grazing management was introduced.

In Australia, rabbits prefer very short grass. Rangelands that have been overgrazed by sheep benefit rabbits, through facilitation, and rabbit numbers increase. When sheep are removed and long grass returns, rabbit numbers decline. In Inner Mongolia, China, the substantial increase in livestock numbers since 1950 has produced a grass height and growth rate that favors Brandt's vole (*Microtus brandti*) populations so that there has been an increase in the frequency of population outbreaks of this species (Zhang et al. 2003).

The saltmarsh pastures of Hudson Bay in northern Canada are grazed by lesser snow geese (*Chen caerulescens*) during their summer breeding (Bazely and Jefferies 1989; Hik and Jefferies 1990; Wilson and Jefferies 1996). The marshes are dominated by the stoloniferous grass *Puccinellia phryganodes* and the rhizomatous sedge *Carex subspathacea*. At La Perouse Bay some 7000 adults and 15,000 juvenile geese graze the marsh, taking 95% of *Puccinellia* leaves. These are nutritious, with high amounts of soluble amino acids. From exclosure plots it was found that natural grazing by geese increased productivity by a factor of 1.3–2.0. Experimental plots with different levels of grazing by captive goslings showed that above-ground productivity of *Puccinellia* was 30–100% greater than that of ungrazed marsh. In addition, the biomass (standing crop) of the grass was higher if allowed to regrow for more than 35 days following clipping. Immediately after the experimental grazing the biomass was less than the ungrazed plots, so that at some point between then and the eventual measurements the biomass on the treated and untreated plots was the same. Even so, the production rate of shoots was higher on the grazed sites. Other experimental plots where grazing was allowed but from which goose feces were removed showed that biomass returned to the level of control plots, but no further. Thus, it appears that goose droppings, which are nitrogen rich and easily decomposed by bacteria, stimulate growth of *Puccinellia*. Geese, therefore, benefit each other from their grazing by fertilizing the grass, a form of intraspecific facilitation.

In summary, facilitation occurs when one species alters a habitat, or creates a new habitat, which allows the same or other species to benefit. We have discussed grazing systems, in particular, but the concept applies in many other cases. For example, many hole-nesting birds and mammals in North America such as wood ducks (*Aix sponsa*) and flying squirrels (*Glaucomys sabrinus*) depend on woodpeckers to excavate the holes, a form of facilitation. Knowledge of such interactions is important for the proper management and conservation of ecosystems.

9.9.2 Do grasses benefit from grazing?

If a species such as Thomson's gazelle benefits from the grazing effects of wildebeest due to the increased productivity of the plants, then do the plants themselves benefit? In other words, what benefits do the plants receive from being grazed and growing more? In evolutionary terms (see Chapter 3) we have to rephrase this as, "Does herbivory increase the fitness of individual plants?" In ecological terms one may ask, "Does the plant grow more after herbivory?"

The studies of lesser snow geese on the saltmarshes of Hudson Bay, which we have discussed above, are now showing that the grass *Puccinellia* comes in different genotypes (Jefferies and Gottlieb 1983). Nineteen grazing experiments have shown that under grazing there is selection for those genotypes that are fast growing. These types have the ability to take up the extra nitrogen from the goose feces and seem to outcompete slower-growing genotypes. This is, therefore, an evolutionary benefit from grazing. Plots where grazing is prevented show that, after 5 years, change to slower-growing genotypes was only just beginning. The more immediate ecological benefit from grazing again comes from the addition of nutrients resulting in a 30–50% increase in biomass.

In general there are few studies that show plants increasing their fitness as a result of herbivory (Belsky 1987). In contrast, we can look at communities of plants and see that if the majority of plants, such as grasses, can tolerate grazing (i.e. survive despite herbivory) a few other intolerant species in that community may not survive due to inadvertent feeding or trampling by large mammals (i.e. apparent competition rather than true competition between plants). This may be simply a consequence of grazing and not necessarily an evolutionary advantage for the grass species. Nevertheless, McNaughton (1986) has argued in opposition to Belsky that grasses and grasslands have evolved in conjunction with their large mammal herbivores, especially in Africa. From an evolutionary point of view a grass individual that by chance evolved an antiherbivore strategy (such as the production of distasteful chemicals) should be able to spread through the grassland. We have to surmise at this stage that antiherbivore adaptations are constrained in some way; for example, it could be that production of distasteful chemicals results in the plant being less successful, in root competition, or in the uptake of nutrients, as in the example of lesser snow geese grazing.

On the surface it appears disadvantageous for a grass to grow more as a response to grazing because it would provide more food and invite further grazing. However, growth could also be viewed as a damage repair mechanism that is making the best of a bad situation (i.e. the grass may lose fitness less by growing than by not doing so).

In summary, we know too little about both the ecological and evolutionary consequences of herbivory on plants. We are left with many questions and opposing views, and more work is needed.

9.9.3 Complex interactions

Competition, parasitism, and predation are all processes that have negative effects on a species. However, when they act together they may end up having a beneficial effect. For example, acorns of English oak (*Quercus robur*) are parasitized by weevils and gall wasps, and are eaten by small mammals. Very high mortality rates are imposed on healthy acorns by small mammals, but parasitized acorns are left alone. While most of the parasitized acorns also die, some survive and are avoided by the small mammals. Thus, higher survivorship and hence fitness occurs when the plant is parasitized (Crawley 1987; Semel and Andersen 1988).

9.10 **Applied aspects of competition**

9.10.1 *Applications*

It is important that we should understand the underlying concepts of interspecific competition if we are to comprehend how species might or do actually interact in the field. There are several applications where we need to be aware of potential competition: (i) in conservation where we might have to protect an endangered species from competition with another dominant species; (ii) in managed systems such as rangelands and forests where there could be competition between domestic species and wildlife – for example an increase in livestock or the expansion of rangeland might cause the extinction of wildlife species, or wildlife might eat food set aside for the domestic animals; and (iii) if we want to introduce a new species to a system, for example a new game bird for hunting, and there could be competition from other resident species.

9.10.2 *Conservation*

Let us imagine a situation where we want to conserve a rare species but we are concerned about possible competition from a common species. For example, roan antelope (*Hippotragus equinus*), a fairly rare species, were released in Kruger National Park, South Africa, as part of a conservation program. There were concerns that the numerous wildebeest would exclude the roan antelope. In this case the management response was to cull the wildebeest (Smuts 1978). More recent evidence indicates that predators, supported by an abundant zebra population, are limiting and even excluding this rare species (Harrington *et al.* 1999; McLoughlin and Owen-Smith 2003). Thus, apparent competition is the dominant process here. A similar example involved the extremely rare Arabian oryx (*Oryx leucoryx*). A few of the last remaining individuals of this species were captured in Arabia in the early 1960s and taken to the San Diego zoo. There numbers increased and some have been successfully reintroduced to Oman (Stanley Price 1989).

In both of these examples it would be important to detect whether there was competition with resident species. We have seen that simple measures of overlap or even changes of overlap with season may not be good indicators of competition. Similarly, observations that an increase in a common species is correlated with a decrease in the rare species does not mean that competition is the cause because of the problem of apparent competition. These measures are necessary but not sufficient. In addition we would need a measure of: (i) resource requirement; (ii) availability of limiting resources and demonstration that one is in short supply; and (iii) the predation rates on both the target species and alternative prey.

A second kind of problem comes from changes in habitat. Assume there is coexistence and habitat partitioning between two species along the lines of Rosenzweig's shared preference hypothesis described above. Since studies of diet and habitat selection would show that both species prefer the same habitat, one may be tempted to manage an area by increasing the preferred habitat at the expense of the other habitats. In this case, however, only one species, the dominant, would benefit and the other would decline. The breeding habitats of the yellow-headed and red-winged blackbirds (Orians and Willson 1964) may be a case in point. Both species prefer deeper-water marshes but one may predict that increasing the depth of a marsh where both occur, thereby leaving little shallow water, may well result in the exclusion of the red-winged blackbird.

In Lake Manyara National Park, Tanzania, there is a habitat consisting of open grassland on the lake shore which is used by wildebeest. Adjacent to this shoreline is savanna consisting of longer grass with scattered trees and shrubs preferred by African

Fig. 9.19 The
percentage of the diet of
white-tailed deer (open
bars) and cattle (shaded
bars) made up of
shrubs, grasses, and
herbs. Diet overlap
increased in winter
when food was limiting.
(Data from Thill 1984.)

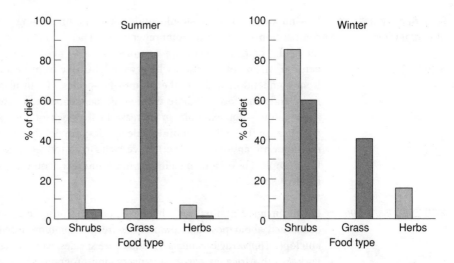

buffalo. In 1961 heavy rains caused the lake levels to rise and flood the open grass-lands, a situation which remained for the rest of the 1960s. The wildebeest were forced to use the savanna habitat, which they did not prefer, and after 4 years the population went extinct. On the surface this appeared to be due to competition with buffalo. Closer inspection of the situation showed that lions, whose densities were high because of the high buffalo population, had eradicated the wildebeest. Wildebeest normally escape predation by running, which they can do in habitats with short grass and little cover for ambush by predators. Once wildebeest were confined to the savanna they were less capable of avoiding lions. Buffalo, on the other hand, avoid predators by hiding in thickets and defending themselves with their horns, and this they could do in savanna but not on the open grassland. Thus, each prey species had its own specialized antipredator habitat that allowed coexistence between the prey species, as predicted by Holt's (1977) "apparent competition" hypothesis. Once this habitat partitioning broke down, the predator was able to eliminate one of the species. The process of apparent competition explains these observations better than true competition.

9.10.3 *Competition between domestic species and wildlife*

There are a number of studies designed to detect whether there is competition between livestock and wildlife. Thill (1984) recorded the seasonal diets of cattle and white-tailed deer in three forested and two clear-cut sites in Louisiana pine forests. Woody plants made up more than 85% of the diet of deer on the forest sites throughout the year (Fig. 9.19). For cattle diets these plants made up less than 16% in summer and fall but rose to 60% in winter and to 48% in spring. The overlap between the two species in overall diet was highest in winter at 46% and lowest in summer at 12%. In contrast, on cleared sites deer continued to eat mainly woody plants but cattle ate more than 80% grass year round. Diet overlap was only 17% in summer and fell to 10% in winter. Since the two species were in the same habitat and there did not appear to be predators, there could be a real possibility for interspecific competition if cattle were confined to forest sites; in fact most of them stayed on the open sites. It is possible that because cattle and deer have not evolved together we do not see the expected decrease in overlap in winter, so that competition is increased rather than avoided at this time.

Thill (1984) points out the advantages for multiple use management derived from the diet partitioning. As forest practices intensify, forest ages decrease and the young stands become impenetrable without artificial clearing. They are also poor areas for deer forage. If cattle were used to graze these sites they could be kept open and so benefit deer by improving accessibility and increasing production of the second growth deer food plants. This presumes that increasing deer numbers is the management objective. We should recognize that deer can have negative impacts, particularly on rare plants and birds (McShea *et al.* 1997), so management of deer needs to be carefully evaluated.

Hobbs *et al.* (1996) manipulated elk densities in randomized block experiments in sagebrush grassland to study the effects of competition with cattle. The effects of elk on cattle exhibited a threshold where low densities of elk had no effect but above a certain density there were both competition and facilitation effects. Food intake declined in direct proportion to elk density because elk reduced the biomass of standing dead grass in winter. There were some weak facilitatory effects of elk grazing through an increase in digestibility and nitrogen content of the remaining grass available to cattle.

Cattle can also have indirect competitive effects by altering habitat structure. In a study of bird communities using the riverine shrub willow habitats in Colorado, Knopf *et al.* (1988) found that cattle grazing altered the structure of the shrubs but not the plant composition. Areas with only summer grazing contained larger bushes widely spaced and with few lower branches when compared with those areas that experienced only winter grazing. The difference in structure affected migratory bird species according to how specific their habitat preferences were. Densities of those with wide habitat preference (e.g. yellow warbler (*Dendroica petechia*), song sparrow (*Melospiza melodia*)), did not change between the sites. Those with moderate niche width (American robin (*Turdus migratorius*), red-winged blackbird (*Agelaius phoeniceus*)), were three times more numerous on the winter-grazed sites; while those with narrow niches (willow flycatcher (*Empidonax traillii*), white-crowned sparrow (*Zonotrichia leucophrys*)), occurred only on the winter-grazed sites.

Hayward *et al.* (1997) found from a 10-year exclosure of cattle in riparian habitats of arid zones in New Mexico that small mammals were 50% more abundant in areas where cattle were excluded. Similarly, kangaroo rats (*Dipodomys merriami*) were more abundant in semi-desert shrubland where cattle grazing was reduced (Heske and Campbell 1991), and reptiles were also more abundant (Bock *et al.* 1990).

9.10.4 *Introduction of exotic species*

Exotic species, those that do not normally live in a country, are introduced for a variety of reasons, and very often they become competitors with the native wildlife. Rabbits in Australia are perhaps the most conspicuous example of this, for they have been implicated in the decline of native herbivores through either direct competition or apparent competition by supporting exotic predators such as foxes (Short and Smith 1994; Short *et al.* 2000; Robley *et al.* 2001). Dawson and Ellis (1979) measured the dietary overlap between the rare yellow-footed rock-wallaby (*Petrogale xanthopus*), the euro (*Macropus robustus*), which is a kangaroo, and two introduced feral species, the domestic goat and European rabbit. During periods of high rainfall the rock-wallaby's diet was mostly of forbs (42–52%) but the proportion of forbs in the ground cover was only 14%. Under drought conditions they still preferred forbs (there was 13% forbs in the diet when forbs were hardly detectable in the vegetation) but trees and shrubs formed the largest dietary component (44% browse). At this season major components for the other species were: euros, 83% grass; goats, 65% browse; rabbits, 25% browse. The rock-wallabies overall diet overlap was 75% with goats, 53% with rabbits, and 39% with euros. In good conditions dietary overlap was still substantial but lower than when drought prevailed. At that time the overlap was 47% and 45% with goats and euros, respectively. Thus, potential competition was greatest with goats and rabbits and least with the indigenous euro.

In North America the introduced starlings (*Sturnus vulgaris*) and house sparrows (*Passer domesticus*) have competed with the native bluebird (*Sialia sialis*) for nesting sites, with the result that bluebird numbers have declined considerably (Zeleny 1976).

Not all introductions result in competition. Chukar partridge (*Alectoris chukar*) have been introduced to North America as a game bird. Their habitat includes semi-arid mountainous terrain with a mixture of grasses, forbs, and shrubs. In particular, they like the exotic cheatgrass (*Bromus tectorum*). Chukar introductions succeeded only where cheatgrass occurred. These habitat requirements are unlike those of native game birds such as sage grouse (*Centrocercus urophasianus*), and thus little competition has taken place (Gullion 1965). Robley *et al.* (2001) showed that the endangered burrowing bettong (*Bettongia lesurur*) in Australia was able to cope with drought stress much better than rabbits because they could eat a variety of herbs and shrubs that rabbits could not eat. If anything these bettongs could outcompete rabbits. Thus, the decline of these marsupials was due to apparent competition from foxes (Short *et al.* 2000).

9.11 Summary

Interaction between species can be competitive or beneficial. Competition occurs when two species use a resource that is in short supply, but a perceived shortage in itself should not be used as unsupported evidence of competition. Instead, the relationships must be determined by manipulative experiments reducing the density of one to determine whether this leads to an increase of the other. Care should be taken to eliminate other factors such as predation that may cause the response. Facilitation is the process by which one species benefits from the activities of the other. It often takes the form of one species modifying a less suitable food supply to make it more suitable for another species, and where one species modifies a habitat making it more favorable for another.

These two effects – competition and facilitation – can often be manipulated by management to increase the density of a favored species.

10 Predation

10.1 **Introduction**

We start by describing the behavior of predators with respect to prey. With this knowledge we explore some theoretical models for predator–prey interactions. Finally, we examine how the behavior of prey can influence the rate of predation. This chapter complements the approach given in Chapter 12 for analyzing interactions between trophic levels.

10.2 **Predation and management**

The previous two chapters have dealt with interactions between individuals on the same trophic level. Predation usually involves interactions between trophic levels where one species negatively affects another. With respect to our three issues of management – conservation, control, and harvesting – predators and predation are of great interest. For rare prey species, the presence of a predator can make the difference between survival or extinction of the prey, especially if the predator is an introduced (exotic) species. This type of problem is particularly important on small islands, but also on isolated larger land areas such as New Zealand and Australia. In contrast, where prey are pests, predators may be useful as biological control agents. Ironically, it was for just this purpose that the small Indian mongoose (*Herpestes auropunctatus*) was introduced to Hawaii and the stoat (*Mustela erminea*) to New Zealand. Unfortunately, in these cases the predators found the indigenous birds and small marsupial mammals easier to catch, so that the predators themselves became the problem. Finally, where harvesting of a prey species (by sport hunters, for example) is the objective, the offtake by natural predators must be taken into account, or one runs the risk of overharvesting and causing a collapse of the prey population.

10.3 **Definitions**

Predation can be defined as occurring when individuals eat all or part of other live individuals. This excludes detritivores and scavengers which eat dead material.

There are four types of predation.

1 *Herbivory*. This occurs when animals prey on green plants (grazing, defoliation) or their seeds and fruits. It is not necessary that the plants are killed; in most cases they are not. Seed predators (granivores) and fruit eaters (frugivores) often kill the seed, but some seeds require digestion to germinate. We discuss herbivory in Chapter 12.

2 *Parasitism*. This is similar to herbivory in that one species, the parasite, feeds on another, the host, and often does not kill the host. It differs from herbivory in that the parasite is usually much smaller than the host and is usually confined to a single individual host. The behavior of nomadic herders in Africa who live entirely on the blood and milk of their cattle would also fit the definition of parasitism. Insect parasites (parasitoids) lay their eggs on or near their host insects which are later killed and eaten by the next generation. We discuss parasitism further in Chapter 11.

163

3 *Carnivory*. This is the classical concept of predation where the predator kills and eats the animal prey.

4 *Cannibalism*. This is a special case of predation in which the predator and prey are of the same species.

10.4 The effect of predators on prey density

Table 10.1 compares caribou and reindeer populations in areas with different levels of wolf predation (Seip 1991). Densities vary by two orders of magnitude, the highest densities being in areas with few or no predators. The lowest caribou densities are in areas subject to high and constant predation. Conversely, Fig. 10.1 shows, first, that wolf densities are positively related to moose densities in Alaska and Yukon (i.e. the highest wolf densities are in areas with the highest moose density). This suggests that wolves are regulated by their food supply. Second, when wolves are removed

Table 10.1 Density of caribou and reindeer populations in relation to the level of predation.

Category	Location	Density/km^2
Major predators rare or absent	Slate islands	4–8
	Norway	3–4
	Newfoundland (winter range)	8–9
	South Georgia	2.0
Migratory Arctic herds	George River	1.1
	Porcupine	0.6
	Northwest Territories	0.6
Mountain-dwelling herds	Finlayson	0.15
	Little Rancheria	0.1
	Central Alaska	0.2
Forest-dwelling herds	Quesnel Lake	0.03
	Ontario	0.03
	Saskatchewan	0.03

After Seip (1991).

Fig. 10.1 Wolf density is related to moose density in Alaska and Yukon. In areas where wolves are culled (●) moose can reach higher densities than in areas where there is little culling of predators (○). (After Gasaway *et al.* 1992.)

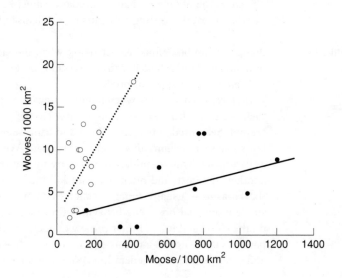

moose densities go up (Gasaway *et al.* 1992). Other studies in Alaska and Yukon have repeated these wolf removals and show similar increasing moose and caribou populations (Boertje *et al.* 1996; Hayes and Harestad 2000). In a natural experiment where red foxes were removed for several years by an epizootic of sarcoptic mange, prey numbers increased, particularly those of hares and several grouse species (Lindstrom *et al.* 1994). In general, predator removal experiments show that the prey population increases or that some index such as calf or fledgling survival increases.

The observations in Table 10.1 and Fig. 10.1 appear to go in opposite directions. No interpretation of cause and effect is possible because they represent correlations. We cannot tell, for example, whether the predators are truly regulating the prey at levels well below that allowed by the food supply or whether predators are catching those that are suffering from malnutrition (so that predators are not regulating), or whether both processes are occurring. Prey availability is influenced by a number of factors: (i) whether there are alternative, more preferred, prey in the area; (ii) the size of this and other prey populations; (iii) the vulnerability of different age and sex classes; (iv) whether the predators specialize on particular prey; and (v) how the environment effects the efficiency of the predators in catching prey. To understand these processes we need to understand the behavior of predators.

10.5 The behavior of predators

In order to interpret predator–prey interactions we must first understand how predators respond to their prey. We ask three questions. How do predators respond to: (i) changes in prey density; (ii) changes in predator density; and (iii) differences in the degree of clumping of prey? We look at these in the following three sections.

10.5.1 *The functional response of predators to prey density*

The response of predators to different prey densities depends on: (i) the feeding behavior of individual predators, which is called the **functional response** (see also Sections 12.4 and 12.5); and (ii) the response of the predator population through reproduction, immigration, and emigration, which is called the **numerical response** (see also Section 12.5) (Solomon 1949). We deal with the functional response first.

Understanding of the functional response was developed by Holling (1959). If we imagine a predator that: (i) searches randomly for its prey; (ii) has an unlimited appetite; and (iii) spends a constant amount of time searching for its prey, then the number of prey found will increase directly with prey density as shown in Fig. 10.2a. This is called a **Type I** response. For the lower range of prey densities some predators may show an approximation to a Type I response, such as reindeer feeding on lichens (Fig. 10.3), but for the larger range of densities these assumptions are unrealistic. For one thing, no animal has an unlimited appetite. Furthermore, a constant search time is also unlikely. Each time a prey is encountered, time is taken to subdue, kill, eat, and digest it (handling time, h). The more prey that are eaten per unit time (N_a), the more total time (T_t) is taken up with handling time (T_h) and the less time there is available for searching (T_s) (i.e. search time declines with prey density, N).

Thus, handling time is given by:

$$T_h = hN_a \tag{10.1}$$

and total time is:

$$T_t = T_h + T_s \tag{10.2}$$

Fig. 10.2 (a) Types of functional response shown as the number of prey eaten per predator per unit time relative to prey density. (b) As for (a) but plotted as the percentage of the prey population eaten.

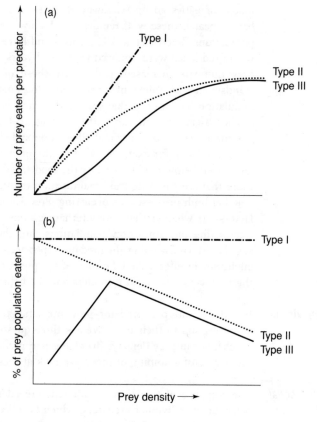

Fig. 10.3 The linear (Type I) functional response of reindeer feeding (dry matter intake) on lichen. (After White *et al.* 1981.)

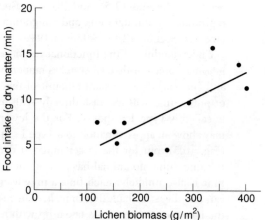

The searching efficiency or attack rate of the predator, a, depends on the area searched per unit time, a', and the probability of successful attack, p_c, so that:

$$a = a'p_c \qquad (10.3)$$

The number of prey eaten per predator per unit time (N_a) increases with search time, search efficiency, and prey density, so that:

$$N_a = aT_s N \tag{10.4}$$

Substituting eqns 10.1 and 10.2 into 10.4 we get:

$$N_a = a(T_t - hN_a)N \tag{10.5}$$

or:

$$N_a = (aT_t N)/(1 + ahN) \tag{10.6}$$

This is Holling's (1959) "disc equation" which describes a **Type II** functional response, where N_a increases to an asymptote as prey density increases (Fig. 10.2a).

When there are several prey types (species, sex or age classes), the multispecies disc equation for prey type i eaten per predator is then:

$$N_{ai} = (a_i T_t N_i)/(1 + \sum_j a_j h_j N_j) \tag{10.7}$$

where the sum is across all prey types eaten.

The Type II functional response can be constructed from the parameters of the disc equation estimated from observations. Searching efficiency is the product of p_c and a'. The probability of capture is usually low, about 0.1–0.3 in most wildlife cases (Walters 1986). The area of search, a', can be approximated from (distance moved) × (width of reaction field or detection distance). Handling time per prey item, h, can be obtained from direct observation or from maximum feeding rates because the maximum rate is $1/h$. Examples of such calculations are given in Clark et al. (1979) and Walters (1986).

The important effect of the Type II response is seen when numbers eaten per predator are re-expressed as a proportion of the living prey population alive (Fig. 10.2b). The Type II curve shows a decreasing proportion of prey eaten as prey density rises. Figure 10.4a illustrates the Type II response of European kestrels (*Falco tinnunculus*) feeding on voles (*Microtus* species) in Finland (Korpimäki and Norrdahl 1991). The functional responses of herbivores are not as well known as those of carnivores but where measured they appear to be Type II as in Fig. 10.4b for bank voles (*Clethrionomys glareolus*) feeding on willow shoots (Lundberg 1988). Deer and elk show Type II functional responses to their food supply. Dale et al. (1994) report Type II responses of wolves preying on caribou.

Holling found a third type of functional response (**Type III**; Fig. 10.2). The number of prey caught per predator per unit time increases slowly at low prey densities, but fast at intermediate densities before leveling off at high densities, producing an S-shaped curve. When those eaten are expressed as a proportion of the live population, the proportion consumed increases first, then declines. Hen harriers (*Circus cyaneus*) in the UK show a Type III functional response (Fig. 10.5) to changes in red grouse (*Lagopus l. scoticus*) populations (Redpath and Thirgood 1999).

The S shape of this curve is attributed to a behavioral characteristic of predators called **switching**. If there are two prey types, A being rare and B being common, the predators will concentrate on B and ignore A. Predators may switch their search from B to A, thus producing an upswing in the number of A killed when A becomes more common. There is often a sudden switch at a characteristic density of A. Birds have

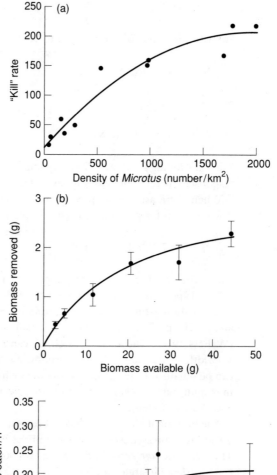

Fig. 10.4 The Type II functional response of: (a) European kestrel feeding at different densities of voles (*Microtus* species). "Kill" rate is voles eaten per predator per breeding season. (After Korpimäki and Norrdahl 1991.) (b) Bank voles feeding on willow shoots. (After Lundberg 1988.)

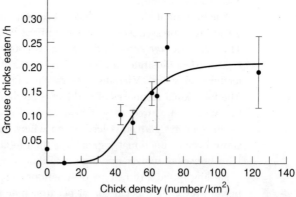

Fig. 10.5 The Type III functional response of hen harriers feeding on red grouse chicks in Britain. (After Redpath and Thirgood 1999, with permission.)

a **search image** of a prey species such that they concentrate on one prey type while ignoring another. As the rare prey (A) becomes more common, birds (such as chickadees (*Parus* species) searching for insects in conifers) will accidentally come across A often enough to learn a new search image and switch their searching to this species.

In practice, it is often difficult to determine whether there is a Type II or III response because the differences occur at low densities of prey and measurements are usually imprecise. The most robust evidence comes from determining whether predators ignore prey until there is a sizeable prey density available: that would indicate a Type III response.

10.5.2 *Predator searching*

The success with which predators catch prey depends upon the density of the predator population. Predators usually react to the presence of other individuals of their own kind by dispersing. Mammal and bird predators are usually territorial and evict other individuals once the space has been fully occupied. These examples are forms of "interference," as discussed in Section 8.8.2.

Interference progressively reduces the searching efficiency of the predator as predator density increases. The drop in searching efficiency caused by crowding lowers the asymptote of the functional response curve. Interference also has a stabilizing influence on predator numbers because it causes dispersal once predators become too numerous. Both interference behavior of predators and antipredator behavior of prey result in non-linear predation rates as predator populations increase. Sometimes this can result in a decrease in predation rate at higher predator levels (Abrams 1993).

10.5.3 *Predator searching and prey distribution*

Prey usually live in small patches of high density with larger areas of low density in between; in short, prey normally have a clumped distribution. This can be seen in the patchiness of krill preyed upon by whales, of insects in conifers searched for by chickadees, of seeds on the floor of a forest eaten by deermice, of caribou herds preyed upon by wolves, and of impala herds hunted by leopards (*Panthera pardus*).

Searching behavior of predators is such that they concentrate on the patches of high density. By concentrating on these patches, predators have a regulating effect on the prey because of the numerical increase of predators by immigration (see Section 10.6).

10.6 **Numerical response of predators to prey density**

We define the numerical response of predators as the trend of predator numbers against prey density (see also Section 12.5 for other ways of looking at this). As prey density increases, more predators survive and reproduce. These two effects, survival and fecundity, result in an increase of the predator population, which in turn eats more prey. An example of this is Buckner and Turnock's (1965) study of birds preying on larch sawfly (*Pristiphora erichsonii*) (Table 10.2). As prey populations increased, the number of birds eating them also increased by reproduction and immigration. When plotted against prey density, predator numbers increased to an asymptote determined by interference behavior such as territoriality (Fig. 10.6). Territoriality results in dispersal so that resident numbers stabilize. Wolves at high density have high dispersal rates, around 20% for adults and 50% for juveniles (Ballard *et al.* 1987; Fuller 1989). In New Zealand, the response of feral ferrets (*Mustela furo*) and cats to an experimental reduction of their primary prey (European rabbits) was a rapid long-distance dispersal (Norbury *et al.* 1998). Extreme long-distance dispersal (800 km) of lynx

Table 10.2 The predation rate on larch sawfly in areas of tamarack (*Larix laricina*) (high density) and mixed conifers (low density). Bird predators include new world warblers and sparrows, cedar waxwing (*Bombycilla cedrorum*), and American robin (*Turdus migratorius*).

	High density (N/km^2)	Low density (N/km^2)
Sawfly larvae	528×10^4	9.88×10^4
Sawfly adults	50.75×10^4	1.16×10^4
Birds	58.1	31.1
Predation of larvae (%)	0.5	5.9
Predation of adults (%)	5.6	64.9

After Buckner and Turnock (1965).

Fig. 10.6 The numerical response may be depicted as the trend of predator numbers against prey density.

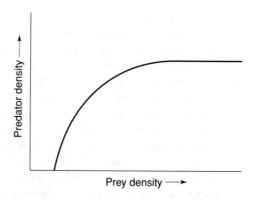

has been observed in northern Canada when numbers of their primary prey, snowshoe hares, collapse.

The initial increase in numerical response may or may not be density dependent. However, because of the asymptote, the numerical response at higher prey densities can only be depensatory (inversely density dependent). This means it has a destabilizing effect on the prey population, by either driving the prey to extinction or allowing it to erupt. This is an important characteristic of populations and it is illustrated in Buckner and Turnock's (1965) study: the proportion of sawfly eaten by birds in the area of high prey density was lower than that in the low-density area (i.e. predation was depensatory and, therefore, could not keep the sawfly population down). The conditions when regulation can or cannot occur are discussed in Section 10.7.

10.7 The total response

We can now multiply the number of prey eaten by one predator (N_a, the functional response) with the number of predators (P, numerical response) to give a total mortality, M, where:

$$M = N_a P \tag{10.8}$$

The instantaneous change in prey numbers is:

$$dN/dt = -N_a P \tag{10.9}$$

and an approximation for changes in prey number, over short intervals when prey populations do not change too much (< 50%), is given by:

$$N_{t+1} = N_t + N_t e^{-N_a P/N_t} \tag{10.10}$$

where $N_t = N$ in eqn. 10.6 (Walters 1986).

If we express this total mortality, M, as a proportion of the living prey population, N, we can get a family of curves, as shown in Fig. 10.7, which depend on whether or not there is density dependence in the functional and numerical responses. If there is density dependence (for example from a Type III functional response) then we have a curve with an increasing (regulatory) part followed by a decreasing (depensatory) part. These are called the **total response curves**, and examples are shown for

Fig. 10.7 Theoretical total response curves (solid lines) of a predator population, measured as the percentage of prey population eaten, in relation to prey at different densities, (a) when there is density dependence in either the functional or numerical response, and (b) when there is no density dependence. The broken line represents the per capita net recruitment of prey $(dN/dt)(1/N)$ assuming logistic growth (i.e. after effects of competition for food have been accounted for). K is a stable point with no predators; A, C, and C^1 are stable with predators, and B is an unstable boundary point.

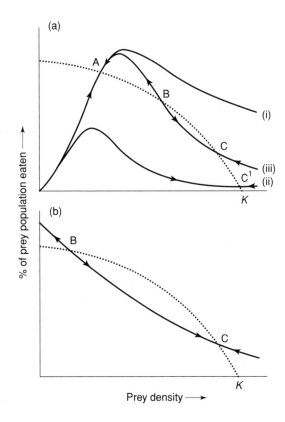

some of Holling's small mammals (Fig. 10.8a) and for wolves eating moose (Fig. 10.8b) (Boutin 1992).

10.7.1 Regulatory effects of predation

In Fig. 10.7 these total response curves have been superimposed on the per capita net recruitment rate of prey $(dN/dt)(1/N)$. For the case where we have density dependence (Fig. 10.7a) there are several stable equilibria (A, C, C^1) where prey net recruitment is balanced by total predation mortality. The point B is an unstable equilibrium where any perturbation to the system (from weather for example) will result in the prey declining to A or increasing to C. In practice B is never seen and is regarded as a boundary between domains of attraction towards A or C.

Curve (i) illustrates the case where predators can regulate the prey population under the complete range of prey densities and hold the prey at a low density A. One possible example of this occurs where both wolves and grizzly bears (*Ursus arctos*) prey upon moose in Alaska (Ballard *et al.* 1987; Gasaway *et al.* 1992). Wolves appear to keep moose densities at low levels (< 0.4/km²). When wolves were removed in a culling operation, the mortality of juvenile moose caused by bears increased so that moose numbers remained at the low level. Moose are kept at similar low levels by the density-dependent predation from wolves in Quebec (Messier and Crete 1985). Red foxes can regulate some small marsupials in desert regions of Australia and some medium-sized marsupials in large eucalypt forests of Western Australia (Sinclair *et al.* 1998). The combined predation of two raptor species, the hen harrier (*Circus cyaneus*) and peregrine falcon (*Falco peregrinus*), on red grouse in Scotland and England

Fig. 10.8 Total response curves of predators at different prey densities. (a) Two shrews (*Blarina*, *Sorex*) and the deermouse (*Peromyscus*) eating European sawfly cocoons. (After Holling 1959.) (b) The proportion of moose populations killed by wolves in different areas of North America. (After Boutin 1992.)

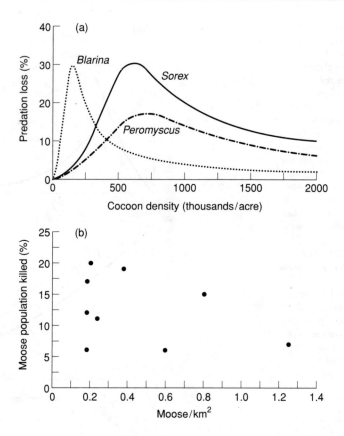

was density dependent in winter and was probably regulating the prey (Thirgood *et al.* 2000). Regulation was due at least partly to the Type III functional response referred to earlier (Redpath and Thirgood 1999).

Curve (ii) can occur when prey are regulated by intraspecific competition for food. Predators then kill malnourished animals and the effect on the prey population is depensatory rather than regulatory. This may be occurring on Isle Royale in Lake Superior where wolves cannot increase sufficiently to regulate the moose population (Fig. 10.9). Moose appear to be regulated by food (Peterson and Vucetich 2003) and wolf predation is merely depensatory (Fig. 10.9). Similar depensatory predation is exhibited in the total response of wolves depredating moose in the Findlayson valley (Hayes and Harestad 2000).

Curve (iii) is the special case where both A and C are present and we have multiple stable states. This situation has been suggested for a few predator–prey systems. One example is that of foxes feeding on rabbits in Australia (Pech *et al.* 1992). Foxes were experimentally removed from two areas and the rabbit populations increased in both, as would be expected from any of the curves in Fig. 10.7, and so by itself the increase in prey tells us little about the nature of predation. However, when foxes were allowed to return to the removal areas there was some evidence that rabbits continued to stay in high numbers rather than return to their original low densities. This result suggests that we have curve (iii) and not (i) or (ii): the interpretation is that rabbit populations, originally at A, were allowed to increase above the boundary

Fig. 10.9 Wolf (broken line) and moose (solid line) numbers on Isle Royale, during 1959–2003, show that the wolf population follows the fluctuations of moose, which are limited by food. (After Peterson and Vucetich 2003, with permission.)

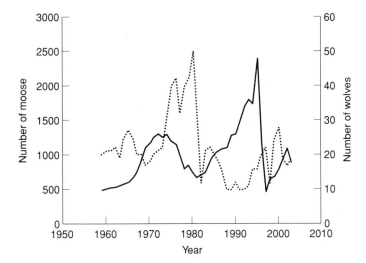

density, B, so that when foxes reinvaded the experimental area rabbit numbers continued towards C.

The "forty-mile" caribou herd of Yukon may have exhibited behavior characteristic of multiple stable states (Urquhart and Farnell 1986). Traditionally, this herd, whose range is on the Yukon–Alaska boundary, numbered in the hundreds of thousands – one estimate by O. Murie in 1920 was 568,000. In the 1920s and 1930s goldminers and hunters killed tens of thousands. After the Second World War, when the Alaska Highway and associated roads were built, hunting increased further. By 1953 numbers were estimated at 55,000 and by 1973 there were only 5000 animals left. Although wolf numbers declined along with their prey, as one might expect, the proportional effect of predation was thought to be high. After 1973 hunting of caribou was restricted and during 1981–83 wolf numbers were reduced from 125 to 60. Thereafter, wolf numbers returned to pre-reduction levels. Although caribou numbers increased marginally to 14,000 during the wolf reductions, they have remained at approximately this level since the early 1980s. Despite the lack of accurate population estimates, the density changes shown by the "forty-mile" herd are so great (almost two orders of magnitude) that it is reasonably clear there has been a change in state from a high level determined by food to a low level determined by predators. The wolves may have been able to take over regulation because hunting could have reduced the caribou population size below the boundary level, B.

Another example of two states may be seen in the wildebeest of Kruger National Park, South Africa (Smuts 1978; Walker *et al.* 1981). In this case high numbers of wildebeest were reduced by culling. When the culling was stopped numbers continued to decline through lion predation, suggesting the system had been reduced below point B. A herbivore–plant interaction with two stable states is seen in Serengeti woodlands (Dublin *et al.* 1990; Sinclair and Krebs 2002). Woodland changed from high to low density in the 1950s and 1960s by severe disturbance from fires. In the 1970s elephant browsing was able to hold woodlands at low density despite a low incidence of fires. Then, poaching removed elephants in the 1980s and trees have regenerated in the 1990s. Elephant numbers are rebounding in the 2000s but they cannot reduce tree density.

Figure 10.7b shows the case where predators have no regulatory effect but can cause the extinction of the prey species if prey numbers are allowed to drop below B. Predation mortality is greater than prey net recruitment below B so that the prey population will decline to extinction. The conditions for this situation occur when there is no switching by predators (i.e. there is a Type II functional response), there is no refuge for the prey at low densities, and predators have an alternative prey source (their primary prey) to maintain their population when this (secondary) prey species is in low numbers.

Various mechanisms have been modeled by Gascoigne and Lipcius (2004). The inverse density-dependent effect of predation on secondary prey (i.e. greater proportional predation as numbers decline) was shown experimentally using hens eggs in sooty shearwater (*Puffinus griseus*) nests in New Zealand. Smaller colonies of nests experienced higher proportional egg predation from rats and mustelids that were dependent on other primary prey such as European rabbits (Jones 2003).

Low densities of the secondary prey could be produced by reduction of habitat (as has occurred with many endangered bird species), or hunting. For example, wolves prey upon mountain caribou in Wells Gray Park, British Columbia during winter, but not in summer when caribou migrate beyond the range of wolves (Seip 1992). Recruitment for this herd in March is 24–39 calves/100 females. In contrast, caribou in the adjacent Quesnel Lake area experience predation year round, and the average recruitment is 6.9 calves/100 females. This population suffers an adult mortality of 29% (most of which is caused by wolves), well above the recruitment rate, and so the population is declining. Wittmer *et al.* (2005) have shown that the predation rate increases as caribou density declines, causing the populations to decline even faster (Fig. 10.10a), as predicted in Fig. 10.7b. Moose are now the primary prey in this system and maintain the wolf population. However, moose have only recently entered this ecosystem, having spread through British Columbia since the 1900s as a result of logging practices, so that previously wolves would not have had this species to maintain their populations at low caribou numbers. One interpretation, therefore, is that before the arrival of moose, caribou were probably the primary prey of wolves and the system was stable at either A or C. Moose have now become the primary prey, caribou have become the secondary prey, and they may be vulnerable to local extinction (Hayes *et al.* 2000). Similar caribou declines have been recorded in central Canada (Rettie and Messier 1998).

Habitat fragmentation for passerine birds breeding in deciduous forests of North America is thought to be the primary reason for the major decline in their populations (Wilcove 1985; Terborgh 1989, 1992). The interior of large patches of forest provides a refuge against nest predation from raccoons (*Procyon lotor*), opossums (*Didelphis virginiana*), and striped skunks (*Mephitis mephitis*), and parasitism from brown-headed cowbirds (*Molothrus ater*). Fragmentation of the forests reduces this refuge because nests are now closer to the edge of the forest where there are more predators and nest parasites. Predation rates are inversely related to forest patch size which must be related to total prey population (Fig. 10.10b). In large forest tracts nest predation is only 2%, in small suburban patches it is close to 100% and well above the recruitment rate. Since small fragmented forest patches are the norm in much of North America, many populations of bird species may be in the situation shown in Fig. 10.7b where the density is left of the boundary B and declining to extinction.

Fig. 10.10 Depensatory total responses. (a) Wolf predation on different woodland caribou herds in British Columbia. Predation rate increases as caribou density declines, causing the populations to decline even faster. (After Wittmer *et al.* 2005.) (b) Various mammal and bird predators on passerine bird nests as a function of forest patch size. These patches are an index of prey population size. (After Wilcove 1985.)

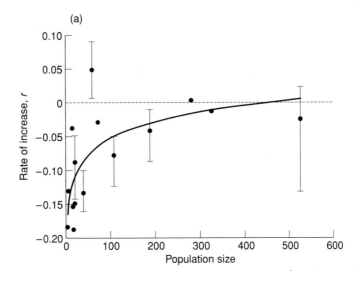

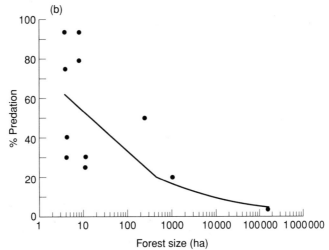

A similar result was observed in southern Sweden with forest fragments embedded in an agricultural landscape. Andren (1992) recorded the impact of various species of the crow family as predators of artificial nests placed in the forest. Two species, the European jay (*Garrulus glandarius*) and the raven (*Corvus corax*), were confined to forest and were absent from small fragments, so their impact declined with fragmentation. Jackdaws (*C. monedula*) and black-billed magpies (*Pica pica*) were largely in agriculture. The hooded crow (*C. corone*) lived in agriculture but invaded forest patches, causing increased predation along the forest edge and within small fragments.

In Kruger National Park, the expansion of zebra (*Equus burchelli*) populations into dry habitats when water holes were constructed in the middle of last century allowed lions to move into those areas. Consequently, rare secondary prey, such as roan antelope (*Hippotragus equinus*) and tsessebe (*Damaliscus lunatus*), have been driven towards extinction (Harrington *et al.* 1999).

Cougar (*Puma concolor*) appear to be having an inverse density-dependent effect, destabilizing bighorn sheep (*Ovis canadensis*) populations in the Sierra Ladron of New Mexico, USA. These effects occur because cougar prey primarily on domestic cattle, which therefore subsidize the cougar population in this area (Rominger *et al.* 2004). The introduction of exotic predators and their exotic primary prey in Australia and New Zealand has caused declines and extinctions of endemic marsupials and birds. Thus, red foxes that depend on European rabbits and sheep carrion are able to drive black-footed rock-wallabies (*Petrogale penicillata*) and other marsupials to extinction in Australia (Kinnear *et al.* 1998; Sinclair *et al.* 1998). In New Zealand, stoats (*Mustela erminea*), black rats (*Rattus rattus*), and brush-tailed possums (*Trichosurus vulpecula*) that depend upon exotic house mice (*Mus domesticus*), a variety of exotic passerine birds, and fruits are driving endemic birds such as kokako (*Callaeas cinerea*) and yellowheads (*Mohoua ochrocephala*) to extinction (King 1983; Murphy and Dowding 1995); experimental reductions of these predators have allowed an increase in the endemic birds (Elliott 1996; Innes *et al.* 1999).

10.8 Behavior of the prey

We have seen how the behavior of predators can influence the nature and degree of predation. We will now examine how the behavior of prey affects predation rates.

10.8.1 Migration

If a prey species can migrate beyond the range of its main predators, then their populations can escape predator regulation (Fryxell and Sinclair 1988a). This has been shown theoretically (Fryxell *et al.* 1988a) and there are some examples supporting this idea. The explanation for this escape from predator regulation is that predators, with slow-growing, non-precocial young, are obliged to stay within a small area to breed. In contrast, ungulate prey, with precocial young, do not need to stay in one place because the young can follow the mother within an hour or so of birth. Thus, the prey can follow a changing food supply while the predators cannot. For example, the wildebeest migrations in Serengeti can follow seasonal changes in food and are regulated by food abundance; meanwhile their lion and hyena predators, although commuting up to 50 km from their territories, cannot move nearly as far as the wildebeest. Other examples from Africa are reported for wildebeest migrations in Kruger National Park, South Africa (Smuts 1978), and white-eared kob (*Kobus kob*) in Sudan (Fryxell and Sinclair 1988b). In North America a similar escape from predation is suggested for migrating caribou herds – the George River herd in Quebec (Messier *et al.* 1988), the barren-ground caribou (Heard and Williams 1991), the Wells Gray Park mountain caribou through altitudinal migration (Seip 1992), and possibly the "forty-mile" caribou before hunting reduced the herd (Urquhart and Farnell 1986).

10.8.2 Herding and spacing

Theoretical studies propose that animals can reduce their risk of predation by forming groups, herds, or flocks (Hamilton 1971), and that group sizes should increase with increasing predator densities (Alexander 1974). The benefit from avoiding predators, however, is counteracted by the cost of intraspecific competition within the group. There should be some group size where the benefit–cost ratio is optimized (Terborgh and Janson 1986).

The effect of predators on herding behavior is illustrated in Fig. 10.11. The relationship between muskox (*Ovibos moschatus*) group size and wolf density suggests that predators are the most likely explanation for differences in group size in different populations (Heard 1992).

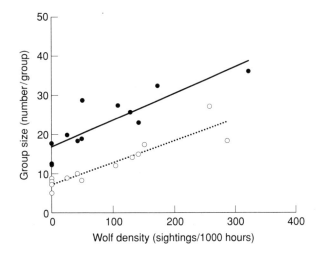

Fig. 10.11 Muskox group size on arctic islands in Northwest Territories, Canada, is related to wolf density in both summer (●) and winter (○). (After Heard 1992.)

The opposite behavior to herding is shown by many female ungulates when they give birth. At that time they leave the herd and become solitary. This is seen in impala and other antelopes in Africa, and cervids, mountain sheep, and forest caribou in North America (Bergerud and Page 1987). This behavior relies on predators spending most of their search time in areas of high prey density, so that solitary prey at low density experience a partial refuge and hence lower predation rates.

Another form of refuge is used by deer in winter when they congregate in loose groups in the small areas between wolf home ranges (Rogers *et al.* 1980; Nelson and Mech 1981). These areas appear to be unused by the wolves, a sort of "no-man's-land," and hence they act as a refuge for the prey.

10.8.3 Birth synchrony

Some prey species synchronize their reproduction to lower predation rate on their young, a behavior called **predator swamping**. This synchrony is over and above that imposed by the seasons. For example, moose and caribou have highly synchronized birth periods (Leader-Williams 1988), as do many ungulates in Africa (Sinclair *et al.* 2000). Other examples of breeding synchrony are seen in lesser snow geese (Findlay and Cooke 1982) and colonial seabirds (Gochfeld 1980).

Experimental studies, such as that of foxes feeding in breeding colonies of black-headed gulls (*Larus ridibundus*) in England, have shown that those gulls breeding outside the main nesting period are more likely to lose their nests to predators (Patterson 1965). However, synchrony may or may not be adaptive depending on the abilities of the predator and the type of synchrony (Ims *et al.* 1988). Thus, if prey form groups and all groups are synchronized together, then predator swamping can occur. However, if reproduction is synchronized within groups but not between them, then predation rate on juveniles could be increased rather than decreased by this behavior, and this depends on the type of predator (Ims 1990). In general, breeding synchrony should be evaluated not just in terms of predation. Other aspects such as seasonality of the environment should be considered. These aspects are important for conservation because species that rely on grouping behavior and synchrony of breeding will be vulnerable to excessive predation if human disturbances alter either aspect.

10.9 **Summary**

Some of the important points for conservation and management that we can derive from this discussion of predation are as follows.

1 Predator and prey populations usually coexist. Prey may be held at low density by predator regulation or at high density by intraspecific competition for food or other resources, and here predators are depensatory.

2 It is possible that both systems may operate in the same area, leading to multiple stable states. This may be generated by a Type III functional response or by a density-dependent numerical response at low prey densities. The system may move from one state to another as a result of disturbance. Such dynamics may occasionally underlie the outbreak of pest species and the decline of species subject to hunting.

3 Conversely, there are situations where the prey population could go extinct, particularly with a Type II predator functional response, no refuge for prey, and alternative food sources for the predators. This is important in conservation where habitat changes may reduce refuges; introduced pests such as rats may provide alternative prey for predators of rare endemic species; or invading prey such as moose or white-tailed deer assume the role of primary prey and so cause the original prey to become vulnerable to extinction as secondary prey.

4 Which of the above occurs depends on the ability of the predator to catch prey and the ability of the prey to escape either by using a refuge or by reproducing fast enough to make up the losses. A very efficient predator defined by a high predator/prey ratio will hold the prey at low density.

11 Parasites and pathogens

11.1 Introduction and definitions

This chapter introduces parasitism and disease within wildlife populations. It addresses how an infection affects a population's dynamics and how it spreads through a population. The veterinary aspects of infection, special to each parasite and host, are not dealt with here. Instead we look at examples of how parasites and disease regulate populations, structure communities and affect conservation of endangered species, reduce the potential yield of harvested populations, or are of use in controlling pests.

Parasites feed on living hosts and (unlike predators) do not always kill the hosts. Some parasites have many hosts, others are species specific. Parasites and pathogens can best be divided into two classes: **microparasites**, which include viruses, fungi, and bacteria, and **macroparasites**, such as arthropods (e.g. fleas, ticks), nematodes, and cestodes (e.g. tapeworms). Microparasites and macroparasites have a roughly equivalent kind of effect upon their hosts and so can be lumped together as parasites. The debilitating effect of the parasite upon the host is termed **disease**. (At the end of the book we include a glossary of terms most often used in parasitology and epidemiology.)

11.2 Effects of parasites

All animals support many species of parasites. For example, the American robin (*Turdus migratorius*) has at least 62 macroparasite species, the European starling (*Sturnus vulgaris*) has 126 helminth species alone, the African buffalo over 60 species, and we, ourselves (*Homo sapiens*), as many as 149 species (Windsor 1998). Many of these species live with their hosts through a substantial portion of the life of the host, causing some minor debilitation. These parasite species are adapted to their hosts, and the hosts are adapted to the presence of the parasite. Such parasites are said to be **endemic**. The disease caused by this type of parasite is called **enzootic**. (Note the special use of the term endemic in this context. In another context a species is endemic when it is confined naturally to one location such as an island or a habitat.)

Endemic parasites cause **chronic** impacts on a host, that is, low-level, persistent, non-lethal debilities or diseases. In contrast to endemic parasites, there are others that cause **epizootic** disease (in animals) or **epidemic** disease (in humans). These parasites cause relatively short-term, major, and often fatal debilities. As a result of human impacts and global climate change on ecosystems, we are experiencing the appearance of new diseases, sometimes termed **emerging infectious diseases**. Enzootic and epizootic diseases have different effects on ecosystems, endangered species, and introduced pests. Parasites may lower the standing biomass of a host population. Hence, they are disadvantageous if the host population is to be conserved or harvested, and advantageous if the host population is to be controlled.

179

The role of disease in mammals can be generalized to all vertebrates (Yuill 1987). Parasites can be expected in all wildlife species in every ecosystem. Death of the host is unusual and occurs only if (i) serious illness facilitates transmission, as in rabies; (ii) the parasite does not depend on the infected host for survival and can complete its life cycle after the host dies; and (iii) the pathogen moves through host populations over a wide geographic area and over a long period of time. Disease may have a drastic effect on survival of wildlife but more commonly its effects are subtle. It can adversely affect natality or normal movement. Brucellosis in caribou has both effects. A caribou cow infected with brucellosis may abort her fetus, and the same disease may also cause lameness from degenerative arthritis in the leg joints. Infective agents can also affect the host's energy balance by reducing energy intake or increasing energy costs through higher body temperature and metabolic rate.

11.3 **The basic parameters of epidemiology**

11.3.1 *Simple compartment models of parasite–host interactions*

Simple models for describing the way a disease establishes and spreads through a population start by assuming a constant host population size. This assumption allows us to understand transmission processes over short time intervals. More complex models can also account for changes in parasite and host populations.

For directly transmitted infections of microparasites such as rinderpest we can divide the host population (N) into three groups: **susceptibles** (S), **infected** (I), and **recovered** (R). The dynamic relationships are illustrated in a simple compartment model, called the SIR model (Fig. 11.1) (Anderson and May 1979). Host population size is determined by birth and death rates. Death rates arise from disease and other causes. The effects of disease are described by: (i) the per capita rate of mortality due to disease (α); (ii) the per capita rate of recovery (γ); (iii) the transmission rate or coefficient (β); and (iv) the per capita rate of loss of immunity.

The rate of change of the susceptible population is given by the rate of transmission of disease from infected to susceptibles. Thus:

$$\frac{dS}{dt} = -\frac{\beta SI}{N} \tag{11.1}$$

where N and β are assumed to be constant.

The rate of change of the infected population is given by the rate of transmission from infected to susceptibles minus the rate of recovery of infected animals. Thus:

Fig. 11.1 The SIR model showing the relationships between susceptibles, infected, and recovered components of the population.

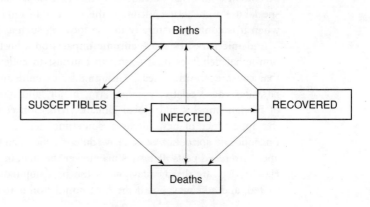

$$\frac{dI}{dt} = \frac{\beta SI}{N} - \gamma I \tag{11.2}$$

Here we assume γ is a constant and that transmission is directly related to the proportion of infected individuals in the population (I/N) times the size of the susceptible population (S).

The rate of change of the recovered population is given by the rate of recovered infected individuals. Thus:

$$\frac{dK}{dt} = \gamma I \tag{11.3}$$

11.3.2 *Thresholds of infection and the transmission coefficient*

When does a disease become epidemic, that is, start to spread through a population? The answer to this question depends on the net reproductive rate, R_0, of the pathogen. For microparasites R_0 is the average number of secondary infections produced by one infected individual, and for macroparasites it is the average number of offspring per parasite that grow to maturity. If the parasite has two sexes it can also be defined as the average number of daughters reaching maturity per adult female.

If R_0 is less than unity the initial inoculum of parasites will decay to extinction. R_0 is not a constant for a parasitic species but is determined by the varying characteristics of both the parasite and the host populations, particularly the density of the host. The conditions leading to persistence of the infection are given by Anderson and May (1986) and Anderson (1991) as the ratio of the rate at which new infectives are generated (β) to the rate at which they are lost:

$$R_0 = \frac{\beta}{\gamma} \tag{11.4}$$

An epidemic occurs if $R_0 > 1$, meaning that more infectives are generated than are lost. An epidemic stops when $R_0 = 1$. We can stop an epidemic by vaccinating a proportion, C, of the susceptible individuals. We can then reduce R_0 by

$$R_0 = \frac{(1 - C)\beta}{\gamma} \tag{11.5}$$

The proportion to be vaccinated to prevent an epidemic (i.e. to keep $R_0 = 1$) is:

$$C > 1 - \frac{\gamma}{\beta} = 1 - \frac{1}{R_0} \tag{11.6}$$

Thus, the proportion to be vaccinated is critically dependent on R_0. If $R_0 = 2$ then 50% of susceptibles must be vaccinated, if $R_0 = 10$ then 90% must be vaccinated (Krebs 2001).

The relationship can be expressed also in terms of a threshold host density N_T below which the infection will die out:

$$N_T = (\alpha + b + \gamma)/\beta > N \tag{11.7}$$

where b is the mortality rate of uninfected hosts.

This makes the point that R_0 is dependent upon host density. Note that if the parasite is highly virulent (large α), if recovery is rapid (large γ), or if the parasite transmits poorly between hosts (small β), then a dense population (large N_T) is needed to stop the infection dying out. Equation 11.7 can be elaborated to take in the effect of an incubation period and post-infection immunity (both of which increase N_T) and "vertical" transmission of the infection whereby a fraction of the offspring of an infected female are born infected (which lowers N_T).

These equations encapsulate two important concepts of epidemiology:

1 that persistence or extinction of an infection is determined by only a few traits of the host and parasite;

2 that the density of the hosts must exceed some critical threshold to allow the infection to persist and spread.

We examine two examples of disease persistence in wildlife populations.

Swine fever

An example of the study of epidemiology involves classical swine fever (CSF) in wild pigs of Pakistan (Hone *et al.* 1992). This is a viral disease of pigs spread primarily by close proximity of hosts. The disease is widespread in Europe, Asia, and Central and South America. Understanding of its epidemiology is relevant in efforts to keep it out of North America and Australia.

Classical swine fever was introduced to a population from wild boar (*Sus scrofa*) in a 45 km² forest plantation in Pakistan. The known starting population (all of which were susceptibles) was 465. One infected animal was released into this population. After 69 days, 77 deaths had been recorded and it was assumed there were no deaths of uninfected animals. The regression of cumulative mortality over time provided a deterministic estimate of the transmission variable β as 0.00099/day. The threshold population of pigs (N_T) below which the disease cannot persist was estimated by

$$N_T = \frac{\alpha + \gamma}{\beta}$$

where α is the mortality rate from infection and γ is the recovery rate. Animals were infective for 15 days over this period. The mortality rate was 0.2/day and the recovery rate was 1/15 or 0.067/day. Thus N_T was $(0.2 + 0.067)/0.00099 = 270$ animals.

The number of secondary infections (R_D) is the ratio of the number of susceptibles (S) (in this case the starting population of 465) to the threshold population N_T. Thus:

$$R_D = S/N_T = 465/270 = 1.7$$

A disease establishes when R_D is unity or greater, but this is valid only for the initial population and not a prediction for persistence.

In general, six pieces of information are required from an epizootic to make predictions about the transmission of a disease:

1 the initial abundance of hosts;

2 the number of infectives initially involved;

3 the number of deaths during the epizootic;

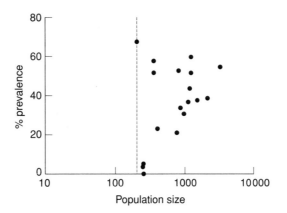

Fig. 11.2 The threshold for establishment of brucellosis in bison of Yellowstone National Park is around 200 animals. (Data from Dobson and Meagher 1996.)

4 the incubation period;

5 the recovery rate;

6 the disease-induced mortality rate.

Brucellosis in Yellowstone National Park

Brucella abortus is a bacterium of the reproductive tract. It causes abortions and is transmitted by animals licking aborted fetuses and grazing contaminated forage. It is common in many ungulates of Africa and has been present in the elk and bison of Yellowstone National Park since the introduction of domestic stock to North America. There are species-specific differences in the effects of the disease on hosts. In elk over 50% of females abort their first fetus, whereas in bison few if any do so (Thorne *et al.* 1978; Meyer and Meagher 1995).

Bison can acquire brucellosis from elk where the two species mix. Initially healthy bison in Grand Teton National Park acquired the disease from elk on the adjacent National Elk Refuge when the two species fed together in winter at Jackson Hole (Boyce 1989). Modeling of the epidemiology (Dobson and Meagher 1996) shows a threshold population for establishment in bison of around 200 animals (Fig. 11.2), and the proportion of the host population infected increases directly with population density. The threshold population, however, is so low that it is very difficult to eradicate the disease – the population would need to be reduced below 200, a cull deemed to be unacceptable in a national park.

11.4 Determinants of spread

Rate of spread (c) of an infection is determined, as is persistence, by traits of both the parasite and the host, particularly the rate of mortality (α) caused by the disease and the net reproductive rate (R_0) of the pathogen. Källén *et al.* (1985) give the relationship as:

$$c = 2[D\alpha(R_0 - 1)]^{0.5} \tag{11.8}$$

where D is a diffusion coefficient more or less measuring the area covered by the wandering of an infected animal over a given period of time. Dobson and May (1986a) calculated the constants of that equation for rinderpest in Africa from the observed radial spread of 1.4 km per day. Pech and McIlroy (1990) used a more elaborate version the other way around, estimating from a knowledge of the equation's constants a potential spread of foot and mouth disease of 2.8 km per day through a population of feral pigs in Australia.

11.5 Endemic pathogens

11.5.1 Birth rates

Parasites usually take some of the energy and protein eaten by the host and so the host suffers some loss. Such losses, if severe enough, can affect the reproductive ability of the host. The nematode *Capillaria hepatica*, experimentally introduced to laboratory mice, resulted in a reduced number of live young born and higher mortality of young before weaning. Such a reduction of natality and early survival might prevent the plagues of mice that are a feature of Australian wheatlands (Singleton and Spratt 1986; Spratt and Singleton 1986).

The bacterium *Brucella abortus* can reduce both conceptions and births in some ungulate species. In birds, parasites can reduce reproduction through forced desertion of nest sites, as in cliff swallows (*Hirundo pyrrhonota*) and many seabirds, or reduction of clutch size (barn swallow, *H. rustica*), delays in mating (great tit, *Parus major*), and lower body condition (house wren, *Troglodytes aedon*) (references in Loye and Carroll 1995). Red grouse (*Lagopus lagopus*) in northern England produced larger clutches of eggs and showed higher hatching success when the nematode *Trichostrongylus tenuis* was reduced with anthelmintic drugs (Hudson 1986). In general, there are still few data on the effect of parasites on host birth rates.

11.5.2 Mortality rates

Laboratory mice infected with the nematode *Heligmosomoides polygyrus* exhibited mortality rates in proportion to the intensity of infection (Scott and Lewis 1987). Soay sheep (*Ovis aries*) on the North Atlantic island group of St Kilda exhibit population crashes every 3 or 4 years. Mortality is highest towards the end of winter, and dead animals had high nematode worm burdens. Live animals that were experimentally treated with anthelmintic drugs had higher survival rates (Gulland 1992). Other studies of rodents and hares show that mortality is often associated with high parasite burdens, for example helminths in snowshoe hares (Keith *et al.* 1984) and botflies in *Microtus* voles (Boonstra *et al.* 1980).

11.6 Endemic pathogens: synergistic interactions with food and predators

The great majority of parasites and diseases coexist with their hosts over long periods, and their prevalence does not exhibit wide fluctuations over time. Direct mortality from these parasites is usually low. In contrast, they can have important indirect effects by (i) responding to the nutritional state of the host and becoming pathogenic or otherwise increasing vulnerability to predation; and (ii) altering the behavior of hosts (Poulin 1995).

11.6.1 Interactions of parasites with food supply

There is much evidence that the pathogenicity of parasites is influenced by the nutritional status of the host. In one experimental study Keymer and Dobson (1987) repeatedly infected mice every 2 weeks for 12 weeks with larvae of the helminth *Heligmosomoides polygyrus*. Mice on a low-protein diet accumulated parasites in direct proportion to the infective dose. In contrast, those on high-protein diets had worm burdens that reached a plateau and even declined over time, and overall the worm burdens were lower for the same dose.

In a field study of snowshoe hares in Manitoba, Murray *et al.* (1997) reduced natural burdens of sublethal nematodes using anthelmintic drugs. On three of six study areas hares were provided with extra high-quality food during the winter when food is normally limiting. They found that survivorship of hares depended on a synergistic interaction of food and parasites. Overwinter survival was 56% in control animals (unfed and normal worm burdens). In unfed but parasite-reduced animals survival was 60%, while survival was 73% in untreated but fed animals. However, in fed and parasite-treated animals survival reached 90%.

Field experimental studies of the effects of parasites are rare, and most information comes from descriptive studies where animals dying in poor nutritional condition also have high parasite burdens. Studies of the periodic mortality of Soay sheep on St Kilda indicate that animals were emaciated and malnutrition was the cause of death. However, dead animals also had high nematode counts, indicating an interaction between food and parasites (Gulland 1992).

11.6.2 Interaction of parasites with predators

Parasites and pathogens can increase a host's vulnerability to predation by changing its ability to escape the predator. Snowshoe hares with high nematode burdens in spring were more likely to be caught in live traps than those with lower worm burdens (Murray *et al.* 1997). Wood bison (*Bison bison*) populations may be held at low densities by predators only when there is a high prevalence of diseases such as tuberculosis and brucellosis (Joly and Messier 2004).

In the red grouse (*Lagopus lagopus*) there is a complex interaction between the nematode *Trichostrongylus tenuis* and predators such as red fox (*Vulpes vulpes*). These game birds, being ground nesters, are vulnerable to predation while incubating eggs. Normally grouse emit scent in the feces that can be detected by trained dogs (and presumably by foxes) up to 50 m away. However, during incubation female grouse stop producing cecal feces and dogs cannot locate the birds more than 0.5 m away. The parasite *T. tenuis* burrows into the cecal mucosa and disrupts its function so that the bird cannot control its scent (Dobson and Hudson 1994). Hudson *et al.* (1992) demonstrated experimentally the effect of these worms on the detectability of incubating red grouse by dogs. They treated some birds with anthelmintic drugs to reduce their worm burdens. Trained dogs found many fewer treated birds than untreated birds with naturally high worm burdens (Table 11.1). Thus, parasites increased the susceptibility of grouse to predation.

Parasites may also increase predation on hosts by altering the behavior of the host either as an incidental consequence of debility or as a specific adaptation to enhance transmission; the latter occurs when the predator is the final host in the life cycle of the parasite. In the former case a disease that causes debility of the host makes it more conspicuous to predators through abnormal behavior and especially flight

Table 11.1 Red grouse nests found by dogs (scent) and random search (researchers) with respect to treatment of the female with an anthelmintic to reduce burdens.

Year	Treatment	Number found	
		Dog scenting	Human search
1983			
Treated	Low worm burden	6	7
Untreated	High worm burden	37	10
1984			
Treated	Low worm burden	9	7
Untreated	High worm burden	29	7

From Hudson *et al.* (1992).

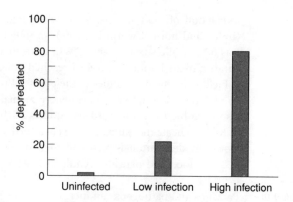

Fig. 11.3 Parasitized killifish are more heavily depredated by birds than are unparasitized fish (adapted from Lafferty and Morris 1996.

behaviors. Other general responses are unusual levels of activity, disorientation, and altered responses to stimuli.

Three lines of evidence support the hypothesis that modified behavior of hosts is a strategy adapted to increase transmission (Lafferty and Morris 1996). First, hosts infected by transmissible stages of parasites often behave differently. Second, experimentally infected prey are more readily eaten by predators in laboratory experiments; and third, infected prey are eaten by predators more frequently than expected in field studies. We mentioned above that parasitized snowshoe hares in spring were more likely to be eaten by predators (Murray *et al.* 1997). Conspicuous behaviors exhibited by killifish (*Fundulus parvipinnis*) were linked to parasitism by larval trematodes. Field experiments showed that parasitized fish were heavily depredated by birds, the final hosts (Fig. 11.3).

11.7 Epizootic diseases

Unlike enzootic diseases, epizootic ones have intermittent effects on host populations. There are outbreak phases with rapid spread and high mortality followed by periods of quiescence when they lie dormant in host species. The great majority of the pathogens causing epizootics are microparasites such as viruses and bacteria. Some case studies illustrate their behavior.

11.7.1 *Rinderpest*

Rinderpest is a virus from the Morbillivirus group (genus *Morbillivirus*, family Paramyxoviridae) that produces measles in humans and canine distemper in dogs, cats, and hyenas. It is probably the oldest member of the group, from which others evolved. Predators can develop cross-immunity to distemper by feeding on herbivores infected with rinderpest (Rossiter 2001). Its natural host is cattle and it is endemic to Asia. It is highly contagious through droplet infection via licking and sneezing. It causes high fever, and inflammation and lesions of the alimentary and respiratory passages.

Rinderpest, as far as evidence goes, was absent from Africa until it was introduced during the 1880s from Egypt to southern Sudan and Ethiopia. By 1889 it was causing epidemics in eastern Africa where it killed 95% of the cattle and similar proportions of closely related wildlife, especially African buffalo but also wildebeest, and less closely related giraffe, warthog, greater and lesser kudu, and other antelope species (Rossiter 2001). By 1896 the epidemic had reached the tip of South Africa and the West African coast causing similar mortality. Thereafter rinderpest reappeared at roughly 20-year intervals producing slightly less virulent epizootics. Mortality of susceptible animals was at least 50%.

From 1961 to 1976 an Africa-wide cattle vaccination campaign (called JP-15) aimed to eradicate the disease from cattle and thereby from wildlife. The latter, being unnatural hosts, could not maintain the disease by themselves. They obtained it by contact with cattle (Sinclair 1977, 1979a, 1995; Plowright 1982). Although the campaign largely succeeded, a few foci of infection remained in the remote regions of southern Sudan and Mali. By 1979 new outbreaks had appeared in Mali, Mauritania, and Senegal, and in 1981 it appeared in East Africa, dying out in 1984. Lax vaccination programs contributed to the spread, and it now appears that vaccination of cattle is required indefinitely (Walsh 1987; Rossiter 2001).

11.7.2 *Myxomatosis*

The *Myxoma* virus is endemic to rabbit species in South America. It was deliberately introduced to Australia in 1950 as a biological control agent for the European rabbit that had become a serious exotic pest. The initial spread from the source infection on the upper Murray River was via mosquito vectors that had increased as a result of recent floods. The first wave of the epidemic took 6 months to cross Australia and mortality was 99%. The virus has remained in the population since 1950 and every 2 years or so outbreaks occur, although mortality has declined to about 87%. Initial abundance of rabbits dropped considerably, but over the decades rabbit numbers have increased as they have become resistant to the virus, and virulence has declined. Rabbit fleas (*Spilopsyllus cuniculus*) were introduced to augment the spread of the disease in wetter regions of Australia, and they now act as major vectors of the virus (Fenner and Fantini 1999).

11.7.3 *Rabbit hemorrhagic disease*

Rabbit hemorrhagic disease (RHD) is caused by a virus (*Lagovirus*, family Caliciviridae) that first appeared in domestic rabbits in China during the 1980s. Subsequently it has caused heavy mortality of wild rabbits throughout Europe. It is closely related to a disease killing European hares (*Lepus europaeus*). It was being tested on Wardang Island, South Australia as a possible biological control agent for rabbits in Australia when it escaped from the island and established in wild rabbits on the mainland in 1996 (Mutze *et al.* 1998). Although mechanisms of transmission and spread are not fully understood, blowflies (*Calliphora* species), a psychodid fly, the rabbit flea (*S. cuniculus*) and culicine mosquitos are carriers of the virus. Initial mortality of rabbits was high (about 90%) and rabbit numbers have remained depressed since its introduction (Kovaliski 1998; Fenner and Fantini 1999).

11.8 Emerging infectious diseases of wildlife

The rinderpest in Africa is an early example of a number of diseases that have recently appeared in wildlife and human populations. The phocine distemper virus of gray seals (*Halichoerus grypus*) has spread along the coast of Europe (Kennedy 1990). In Australia two orbiviruses cause blindness of eastern gray kangaroos (*Macropus giganteus*) (Hooper *et al.* 1999), and the *Chlamydia* bacterium causes blindness and urogenital disease in koala (*Phascolarctos cinereus*). The chytrid fungus (*Batrachochytrium dendrobatidis*) causes mortality and decline in amphibian populations in many parts of the world (Berger *et al.* 1998). These are some wildlife cases; acquired immunodeficiency syndrome (AIDS), the ebola virus, tick-borne spirochetal bacteria causing Lyme disease (*Borrelia burgdorferi*), and the virus causing severe acute respiratory syndrome (SARS) are human examples.

These emerging infectious diseases (EIDs) are associated with a range of underlying causal factors. They can be classified on the basis of three main pathways of infection: (i) EIDs associated with a jump from domestic to wildlife populations living nearby; (ii) those connected with direct human intervention through translocation of host or parasite; and (iii) those with no human or domestic animal associations.

1 The rinderpest is a clear case of the transfer of a virus from cattle to susceptible wildlife hosts that had not met the disease before. Similarly, canine distemper has spread into wild dog (*Lycaon pictus*) populations of Serengeti causing the local extinction of that species (Ginsberg *et al.* 1995), into lions causing a 40% mortality, and also into hyenas (Roelke-Parker *et al.* 1996). Most likely the rapidly expanding human population surrounding the Serengeti ecosystem, with its associated domestic dogs that carry the disease, are the source of these new outbreaks. Another example is brucellosis (Meagher and Meyer 1994). This was introduced to North America with the import of cattle, and the disease then jumped to elk and bison in Yellowstone National Park, USA and Wood Buffalo National Park in Canada.

2 The translocation of wildlife for agriculture, hunting, and conservation has increased exposure of wildlife to novel diseases. Translocation of fish and amphibians may have caused the ranavirus epizootics now threatening many amphibian populations (Daszak *et al.* 2000). Rabies epizootics in the eastern USA developed from translocations of infected raccoons (*Procyon lotor*) from the southern USA where the disease was enzootic (Rupprecht *et al.* 1995).

Zoos and captive feeding programs may inadvertently expose animals to novel diseases due to the close proximity of neighboring hosts. Asian elephants (*Elephas maximus*) in zoos have contracted a lethal herpes virus from neighboring African elephants (*Loxodonta africana*) (Richman *et al.* 1999). There is considerable concern that the agent for bovine spongiform encephalopathy (BSE) could be transferred to zoo-held wildlife through contaminated food, and thereby to free-living wildlife (Daszak *et al.* 2000).

3 Climate change may be having an effect on the emergence, frequency, and intensity of epizootics. For example, African horse sickness in South Africa (Bayliss *et al.* 1999) and various aquatic diseases (Marcogliese 2001) have been affected by climatic events. The fungal disease cutaneous chytridiomycosis is the cause of amphibian mortality in Central American and Australian rainforests (Berger *et al.* 1998; Morell 1999). The synchronous emergence of this novel disease in widely spaced sites affecting a wide range of species is thought to be the result of global climate change (Pounds *et al.* 1999).

In general, the causes of EIDs are largely ecological. These are (i) movement and migration of hosts and pathogens to new environments; (ii) the change of environment *in situ* through global climate change; and (iii) a change in agricultural and forestry practices that brings species into contact. Changes in genetic characteristics of the pathogens play little if any part in EIDs except perhaps in their ability to jump to new hosts (Krause 1992; Schrag and Wiener 1995).

11.9 Parasites and the regulation of host populations

As we have seen, most endemic parasites interact with other factors such as food and predators to reduce host population numbers. There are few examples where parasites, on their own, regulate the host population, that is act in a density-dependent way. One clear example comes not from an endemic parasite, but from an emerging epizootic disease. The poultry pathogen *Mycoplasma gallisepticum* has entered a previously unknown host, the house finch (*Carpodacus mexicanus*) in North America.

Fig. 11.4 The rate of change of a house finch population due to mortality by the pathogen *Mycoplasma gallisepticum* is density dependent. (Data from Hochachka and Dhondt 2000.)

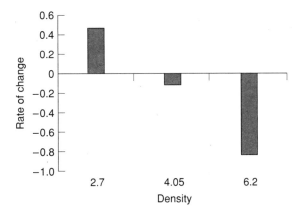

The decline in finch population caused by the disease was proportional to the initial density of finches, with the result that 3 years after the start of the epizootic most finch populations had stabilized at similar densities (Fig. 11.4). Thus, the mortality was density dependent and the disease has regulated the finch population (Hochachka and Dhondt 2000).

We have already mentioned the emerging epizootic, the rabbit hemorrhagic disease, which was released in Australia in 1996 and has caused major declines in European rabbit numbers (Mutze *et al.* 1997). Subsequently, the disease appears to be keeping rabbit numbers at very low levels. Bighorn sheep (*Ovis canadensis*) populations regularly experience pneumonia outbreaks caused by the bacterium *Pasteurella* (Miller 2001). This has caused declines of bighorn sheep throughout western North America. It can regulate bighorn sheep numbers, particularly in the Idaho area of the USA, keeping populations well below those determined by food resources. The source of the disease is domestic sheep, which are less susceptible to mortality from the pathogen than are the bighorns (Monello *et al.* 2001). More anecdotally, the rinderpest virus was probably regulating the African buffalo and wildebeest populations of Serengeti, Tanzania before its removal through vaccination of cattle in 1963 (Sinclair 1977).

The role of endemic pathogens, particularly macroparasites, in regulating hosts is not clear. The nematode *Heligmosomoides polygyrus* regulated laboratory mouse populations (Scott 1987; Scott and Lewis 1987; Scott and Dobson 1989). We have yet to find examples from the field. However, recent studies suggest that macroparasites may, at least, be causing population cycles. For example, red grouse populations in Britain exhibit 7-year cycles and it appears that these could be produced by the nematode *Trichostrongylus tenuis* (Hudson and Dobson 1988). Winter mortality was the major factor determining changes in grouse numbers, although breeding losses were also important. Both winter loss and breeding loss were correlated with the intensity of parasite infection. Cycles could be resulting from time delays in the recruitment of parasites so that they are partly out of phase with host numbers (Dobson and Hudson 1992; Hudson *et al.* 1992). This idea was tested experimentally by reducing parasite burdens with anthelmintic drugs. Treatment of the grouse population prevented the normal decline in numbers, demonstrating that parasites were the cause of the decline phase of the cycle (Hudson *et al.* 1998).

11.10 **Parasites and host communities**

As more research on parasites is carried out we are becoming aware of the role of parasites in structuring the diversity and abundance of host communities. This is a new area, and much remains to be done (Minchella and Scott 1991; Poulin 1999). Most parasites have shorter life cycles and much faster rates of increase than their hosts. These features are the opposite to those of predators, and therefore parasites can have different impacts on the structure of host communities.

11.10.1 *Altering species interactions*

Parasites can have three types of impact on host communities (Poulin 1999).
1 *Competition.* Parasites can affect competitive interactions between two species by having a greater effect on one of the pair. A superior competitor may become an inferior competitor in the presence of the parasite. The northward spread of white-tailed deer in the hardwood forests of North America was accompanied by its meningeal nematode parasite *Parelaphostrongylus tenuis.* This worm is lethal to both the moose and caribou that were the original inhabitants of the forest, and populations of these species have declined (Anderson 1972; Nudds 1990; Schmitz and Nudds 1994). Thus, the parasite has altered the relative abundance of the three host species by affecting one less than the others. Schall (1992) shows that competition in *Anolis* lizards is altered by the presence of the malaria parasite (*Plasmodium azurophilum*). On the Caribbean island of St Maarten the normally dominant *A. gingvivinus* excludes the subordinate *A. wattsi* that is found only in the central hills. However, the parasite is common in *A. gingvivinus*, and rarely so in *A. wattsi*. In the presence of the malaria the two coexist.
2 *Reducing predation.* Parasites may reduce the efficiency of predators or herbivores in obtaining prey so that the prey increase at the expense of their competitors. In other words, parasites could alter the effect of "apparent competition" mentioned in Chapter 9. Little has been documented at the carnivore trophic level. In herbivores, reduced food intake in reindeer is induced by gastrointestinal nematodes (Arneberg *et al.* 1996) and so heavily grazed, palatable plants could increase in abundance. The presence of rinderpest in the Serengeti ecosystem (Section 11.7.1) reduced the dominant herbivore, wildebeest, by some 80%. One consequence of this reduction of wildebeest was to increase the biomass of grasses on the Serengeti plains and decrease both the diversity and abundance of small dicot species which are overshadowed and outcompeted by the grasses.
3 *Increasing prey susceptibility.* Parasites can increase the availability of prey for a predator and so alter the competitive relationships between predators. We have already mentioned that parasites alter prey behavior to the benefit of their predators (Section 11.6.2). There are no data on how altered prey behavior affects the community of predators.

11.10.2 *Complex ecosystem effects*

Red grouse in northern England have declined in numbers due to an increase in prevalence of the tick-borne louping ill virus, which affects the central nervous system. The increase of the disease was produced by a change in the relative abundance of two plant species of the heath communities inhabited by grouse: heather (*Calluna vulgaris*), which is the major food for grouse, and bracken (*Pteridium aquilinum*), which produces a humid mat layer, the habitat for ticks. Bracken is increasing at the expense of heather because it can invade when heather is burned. Sheep ticks (*Ixodes ricinus*) are maintained by domestic sheep and mountain hares (*Lepus timidus*). The spread of bracken has increased the exposure of grouse to ticks and hence louping ill virus (Dobson and Hudson 1986; Hudson *et al.* 1995).

Myxomatosis was introduced to rabbits in England in 1953. It caused severe mortality of rabbits and resulted in several indirect effects on the ecosystem (Ross 1982a,b). The normally closely grazed grass lawns on the chalk downs changed to tussock grassland with *Festuca rubra* and heather, *Calluna vulgaris*, invading. Indeed, rabbits, which were introduced to Britain from Europe a thousand years ago, had maintained the species composition of these grasslands for so long that no one knew of any alternative state. There was an initial increase in diversity of flowering herbs followed by dominance of tussock grasses, and eventually some areas turned to woodlands. Plant succession affected animal diversity: European hares (*L. europaeus*), voles (*Microtus agrestis*), and ants increased, while the sand lizard (*Lacerta agilis*) decreased. Predators that depended on rabbits, such as stoats (*Mustela erminea*) and buzzards (*Buteo buteo*) also declined. Similar changes were recorded in South Australia after rabbit hemorrhagic disease reduced rabbits there in 1996.

Dutch elm disease, caused by the fungus *Ceratocystis ulmi*, decimated elm trees (*Ulmus* species) in the early 1970s in Britain. These were amongst the most abundant trees in agricultural areas, and their removal changed the physical structure of the habitats for birds. Death of the trees had less effect than the removal of the trees, because these provided nesting and feeding sites. Bird diversity was reduced by eight species (from 36 to 28) as a consequence. Later, as the dead trees disappeared, increased light levels changed the herbaceous plant community (Osborne 1985).

These examples illustrate how the presence or absence of a disease can have complex indirect effects that filter down even to the plant community.

11.11 Parasites and conservation

Parasites and pathogens can be important in all three components of wildlife management. They can cause conservation problems by reducing the densities of species targeted for conservation, they can reduce the potential yield of harvested populations, but, on the positive side, they can be used to control pest species. The following sections provide examples of each to give a feel for the range of effects.

The long period of natural selection over which a parasite and its obligate host sort out an accommodation with each other ensures that a persistent infection has little influence on the density of the host. If, however, the specific characters of the host and parasite are such that usually $R_0 < 1$ then the infection is likely to be sporadic and may have a large but temporary depressing effect on the density of the host. Bubonic plague and (until recently) smallpox acted in this way against humans.

As we have mentioned earlier, parasites can reduce both birth and survival rates, and hence affect population size. Therefore, they are relevant to the conservation of small populations and can be a cause of population decline (see Chapters 17 and 18). There are several ways in which threatened species may be exposed to parasites.

11.11.1 Introduction of domestic or exotic species

Microparasitic diseases are now implicated in the decline and extinction of several wildlife species, particularly in carnivores (Ginsberg *et al.* 1995; Kat *et al.* 1995; Tompkins and Wilson 1998; Murray *et al.* 1999). Thus, African wild dog (*Lycaon pictus*), Ethiopian wolf (*Canis simensis*), and Blanford's fox (*Vulpes cana*) in Israel have all been decimated by rabies or canine distemper contracted from domestic dogs. The arctic fox (*Alopex lagopus semenovi*) on the Aleutian islands has contracted mange also from dogs.

Parasites, particularly microparasites, have their greatest effect when they jump from one species of host to another. That process is also a major source of evolutionary opportunities for parasites. For example, the knee worm (*Pelecitus roemeri*) has been "captured" by the wallabies and kangaroos from the parrot family. The effect of the worm is unknown in parrots, but in macropods the worm induces a fibrous capsule up to the size of a cricket ball on the animal's knee. Trans-specifics are the parasites and pathogens to watch out for. They can cause significant conservation problems (see Section 11.11.3) but they can also sometimes be used to control pest species (see Section 11.12). Other trans-species parasites must be guarded against because they cause considerable additional mortality. We have already mentioned the myxoviral rinderpest epidemic that swept the length of Africa in the 1890s and killed large numbers of wild ungulates, particularly African buffalo. Asian cattle were its original host but it jumped across to wild ungulates when it reached Africa. The decline of moose populations in Nova Scotia and New Brunswick is associated with infestation by the nematode brainworm *Parelaphostrongylus tenuis* that jumped from its original host, white-tailed deer. The infestation in moose is fatal but there is little evidence that the parasite can maintain itself in the moose except by reinfection from white-tailed deer (Anderson 1972). However, the relationship between meningeal worm, white-tailed deer, and moose has not been studied experimentally, and not all the circumstantial evidence is consistent (Samuel *et al.* 1992).

The translocation of domestic or exotic wildlife may lead to parasites and pathogens jumping to a new suite of species. In Australia, native animals such as kangaroos and wombats became infected with common liver flukes (*Fasciola hepatica*) acquired from sheep and cattle. The liver flukes cause severe lesions in the liver of the wombat (Spratt and Presidente 1981).

We have already seen how the presence of sheep brought in the sheep tick to English moorlands and the louping ill virus to red grouse, causing their populations to decline. Similarly, high mortality of monkeys (*Presbytis entellies, Macaca radiata*) in India occurred after cattle were introduced and increased the numbers of the tick *Haemophysalis spinigera*. This tick carried the flavivirus causing Kysanur forest disease (Hudson and Dobson 1991).

Perhaps the most well-known example of parasites introduced by wild exotic species is that of avian malaria in the Hawaiian islands (Dobson and Hudson 1986; van Riper *et al.* 1986; McCallum and Dobson 1995; Cann and Douglas 1999). Early extinction of lowland native bird species (twelfth to nineteenth centuries) resulted from agricultural clearing of forests, and later the introduction of rats and Indian mongooses (*Herpestes auropunctatus*) that depredated their highly vulnerable nests. The mosquito, *Culex quinquefasciatus*, was introduced in 1826. It spread across all the islands but it did not carry avian malaria – no cases were detected from blood samples in the early 1900s.

Many species of exotic birds were introduced to the islands between 1900 and 1930 as a response to the damage caused by the clearing of forests for agriculture. As in New Zealand and Australia previously, there was an organization (the Hui Manu) committed to introducing exotics. It is now clear that these exotics were responsible for bringing in the avian malaria (*Plasmodium relictum capistranoae*). Native birds were highly susceptible to the parasite, and many species became extinct because of it. Now native species are restricted to habitats above 600 m where both the mosquito and exotic birds are at low density. Reintroduction of native species in the lowlands is not feasible in the presence of the parasite.

The species jumping process may also operate the other way, with wildlife acting as reservoirs of parasites and pathogens transmitted to domestic stock. The controversies over brucellosis in bison, and its transmission to domestic stock both in the USA and Canada, and the transfer of tuberculosis (*Mycobacterium bovis*) from European badgers to cattle are obvious examples (Peterson *et al.* 1991; Clifton-Hadley *et al.* 2001). The appearance of SARS in humans in 2002 is thought to have arisen because people in southern China keep civets (*Viverra zibetha*) in captivity and eat them.

11.11.2 *The alteration of habitat*

We explore in Chapter 18 the general consequences of degraded habitats and fragmentation for conservation. One particular effect of habitat fragmentation is that it increases exposure to parasites. In birds, and perhaps in other animal groups, Loye and Carroll (1995) suggest three mechanisms:

1 Increased edge habitat due to fragmentation increases the contact rate between species in adjacent habitats and exposes those in fragments to new vectors and new parasites to which they are more susceptible.

2 Loss of habitat could force birds to reuse old nests, exposing them to higher numbers of fleas, ticks, or other nest-living parasites.

3 As a special case in birds, fragments expose birds not only to predators commuting from the surrounding agriculture, but also to brood-parasite birds such as the brown-headed cowbird (*Molothrus ater*) of North America.

Some of these mechanisms are illustrated by Loye and Carroll (1995) for the Puerto Rican parrot (*Amazona vittata*). This species is restricted to a single fragment of high-elevation forest. Habitat degradation, harvesting for the pet trade, and novel parasites have been factors in its decline. In particular, fatal parasitism of nestlings by muscid botflies became a problem after the pearly eyed thrasher (*Margarops fuscatus*) invaded the forest fragment in about 1950. The thrasher nestlings are host to abundant blood-feeding botfly maggots. The introduced native thrasher, therefore, brought in its endemic parasite that then spread to a new but rare host, the parrot.

11.11.3 *Captive breeding and reintroductions*

Parasites and pathogens can be a factor driving the decline of an endangered species (see Section 17.2.8) and can become an issue in the recovery of endangered species. Parasites and pathogens can hinder or thwart attempts to establish captive breeding populations. Thorne and Williams (1988) review the well-known example of the first attempts to establish a captive breeding colony of black-footed ferrets (*Mustela nigripes*) in the USA. A previous attempt to establish a breeding colony in the early 1970s failed because canine distemper virus (CDV) killed the only two litters. The source colony also disappeared. The extreme susceptibility of the black-footed ferrets to CDV became apparent when four of six black-footed ferrets died after being vaccinated for CDV in the 1970s. The vaccine had been previously shown to be safe in domestic ferrets. In 1981 the species was rediscovered in Wyoming, and the colony's vulnerability to disease was quickly realized. Precautions were taken to minimize human introductions of disease, especially CDV and influenza. The population declined from an estimated peak population of 128 in 1984 to only 16 in 1985. The decline spurred an attempt to start a captive breeding colony, but the first six ferrets captured rapidly succumbed to canine distemper. Despite all precautions CDV infected the colony and most of its members died from it. The few surviving eventually formed a breeding population. Nonetheless, as Thorne and Williams (1988) note, "The captive breeding program went from a carefully planned approach with ideally selected, unrelated founder animals to a crisis situation with related animals, a poor sex ratio, and few mature, experienced breeder males."

Captive-bred animals released into the wild may spread disease or pick up parasites and pathogens from endemic wildlife. A potential example of the former is Jones's (1982) report of the release of Arabian oryx captive-raised in the USA for a national park in Oman, which was delayed when the animals tested positive for antibodies to bluetongue disease. The failure of reintroduced animals of woodland caribou to an island within their historic range in Ontario, Canada is an example of lethal trans-species parasites and the problems that can be encountered with reintroduced animals becoming infected with a disease from the endemic wildlife. The area had been colonized by white-tailed deer and the caribou became infected with meningeal worm from the deer via a gastropod secondary host (Anderson 1972).

Another example comes from the captive breeding of whooping cranes (*Grus americana*). An eastern equine encephalitis (EEE) virus fatally infected 7 of the 39 captive-bred population at the Patuxent Wildlife Research Center in Maryland, USA. At that time, in 1985, the captive population accounted for about 25% of the world's population. EEE virus causes sporadic outbreaks of disease in mammals and birds in the eastern USA and is spread by mosquitos. No deaths are usually seen in endemic hosts, but introduced game birds such as ringneck pheasants (*Phasianus colchicus*) are vulnerable. Among the some 200 sandhill cranes in neighboring pens to the whooping cranes some birds were serum positive for EEE virus but no clinical signs were found. The discovery of the vulnerability of the whooping cranes to a common pathogen was seen as an unrecognized risk and an obstacle to the species' recovery (Carpenter *et al.* 1989).

11.12 Parasites and control of pests

Spratt (1990), in reviewing the possible use of helminths for controlling vertebrate pest species, pointed out the marked contrast between the numerous successes in biological control of insects and the almost universal failure of such methods to control vertebrates. The one unequivocal success has been the use of the *Myxoma* virus to control European rabbits in Australia (Fig. 11.5).

Myxomatosis is a benign disease in *Syvilagus* (cottontail) rabbits in South America which is transmitted mechanically by mosquitoes. In the European rabbit (*Oryctolagus*), which is a pest in Australia and England, the virus from *Sylvilagus*

Fig. 11.5 Numbers of rabbits trapped in Australia (million kg of skins) shows a rapid decline after 1950 when the *Myxoma* virus was introduced to control rabbits. (Data from Fenner 1983.)

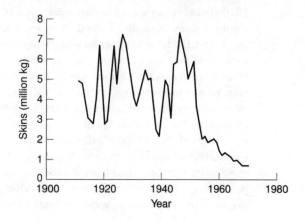

produced a generalized disease that is almost always lethal. Myxomatosis was deliberately introduced into Australia in 1950 and into Europe in 1952. It was first spectacularly successful in controlling the rabbit pest, but biological adjustments occurred in the virulence of the virus and the genetic resistances of the rabbits. After 30 years of interaction, natural selection has resulted in a balance at a fairly high level of viral virulence. (Fenner 1983)

The initial annual mortality rates were very high in Australia, over 95%, but these dropped progressively over the next few years. There is a widespread perception that the rabbits and the disease accommodated to each other and, therefore, that myxomatosis provided only a temporary respite. This is not so. The rabbit density at equilibrium with the disease is considerably lower than the mean density in the absence of myxomatosis.

Parer *et al.* (1985) demonstrated the controlling effect of this virus. They used a relatively benign strain of *Myxoma* to immunize rabbit populations against the more virulent field strains that swept through the study area in most years. Rabbit densities increased by a factor of 10 under this treatment. Even after the rabbits and the virus had reached an accommodation with each other, the disease was apparently holding mean density of rabbits to about 10% of the densities prevalent before introduction of the *Myxoma* virus.

11.13 **Summary**

Most parasites and pathogens have little effect on their hosts. When a parasite jumps from one host species to another, it is the "naivety" of the new host to the parasite or pathogen which is responsible for the reaction of the new host individual or the new host population to the parasite. In both meningeal worm in cervids, other than white-tailed deer, and liver fluke infection in wombats it is the dramatic host immunological response to the new parasite which is responsible for the debilitation in the animals. Such a response has been dampened down over time as the meningeal worm evolved in white-tailed deer and liver fluke evolved in sheep, and so we do not see the same level of debilitation in the "normal" host species.

The key points from the epidemiology of parasites and pathogens are that the fate of an infection is determined by only a few traits of the host and parasite, and there is a critical density of the host that allows the infection to persist and spread. Efforts to reduce the effects of parasites and pathogens can be at their most important in the management of small populations of endangered species, be they in the wild or in captivity. Diseases of harvested wildlife are more rarely controlled unless they present a potential hazard to people. Few attempts to use parasites and pathogens to control pest wildlife have been successful.

12 Consumer–resource dynamics

12.1 Introduction

In this chapter we explore those things an animal needs to eat to survive and reproduce: resources. This leads to a description of the structure and dynamics of consumer–resource systems, where both the consumers and their resources may interact in complex ways. We show how to analyze such systems by breaking them down into their dynamic components. This approach is used to compare several different systems: kangaroos and plants in Australia, trees, moose, and wolves in North America, small mammals in northern Europe, and snowshoe hares and lynx in Canada.

12.2 Quality and quantity of a resource

A **resource** is defined as something that an animal needs, whose consumption by one individual makes the resource unavailable to another individual. The most obvious example is food, and to that may be added shelter, water, or nesting sites. By definition, a resource is beneficial. As the availability of the resources rises, the fecundity and probability of survival of an individual is enhanced.

Food resources are often characterized by two attributes: the amount of food available to an animal and the suitability of that food to the animal's requirements. For example, quality may be described as the percentage of digestible protein in the food, whereas quantity may be measured as dry mass of food per hectare. This often leads to a discussion on whether quality or quantity of the food is the most important to the animal. In most cases the distinction is meaningless. It indicates that the resource is being measured in the wrong units. If the resource is in fact digestible protein, then that is what should be measured. The availability of the resource should be expressed as dry weight of digestible protein per hectare. Its measurement may entail measuring dry weight of herbage as an intermediate step, but that does not make herbage the resource.

12.3 Kinds of resources

It is necessary at this stage to give a classification of resources because the interaction between the resources and the animals that depend upon them can take several forms. These in turn influence the dynamics of the population in different ways.

The use of a resource may be **pre-emptive**. An example is the use of nesting holes by parrots. Individuals are either winners or losers. On the other hand the use of a resource may be **consumptive**. All individuals have access to the resource and each individual's use of it reduces the level of the resource available to other individuals. An example is the use of plants by herbivores. We see that both pre-emptive and consumptive use of a resource removes a component of the resource from use by other individuals. Consumptive use removes the component permanently whereas pre-emptive use removes it temporarily.

To complete the classification, there may be an **interactive relationship** between the population and the resource in that the level of the resource influences the rate of increase of the population, and reciprocally the level of the population's density influences the rate of increase of the resource. The dynamics of the animals interact with the dynamics of the resource, that is the relationship between a herbivore and its food supply and between a predator and its prey resource. In a **reactive relationship**, however, the rate of increase of the animal population reacts to the level of the resource (as before) but the density of the animals has no reciprocal influence on the rate of renewal of the resource. The relationships between a scavenger and its food supply or between a herbivore and salt licks are examples of reactive relationships.

We start by developing a general theoretical framework that applies in principle to all consumer–resource relationships, regardless of whether they focus on plants and herbivores, carnivores and their herbivorous prey, or all three.

12.4 Consumer–resource dynamics: general theory

The origin of consumer–resource theory can be traced directly to the contributions of two early ecologists: Alfred Lotka and Vito Volterra (Kingsland 1985). Starting from very different backgrounds, these two men simultaneously developed a similar framework for thinking about interactions between consumers and their resources (Lotka 1925; Volterra 1926b), a general framework that is still in common use, albeit with considerable change in biological details. The framework is a set of mathematical expressions for simultaneous changes in the density of consumers (denoted here by N) and their resources (denoted V):

$$\frac{dV}{dt} = \text{growth of resource} - \text{mortality due to consumption}$$

$$\frac{dN}{dt} = \text{growth of consumer due to consumption} - \text{mortality}$$

In the case of resources, mortality is largely due to consumption, whereas consumers experience a constant background level of mortality. For example, one might model reproduction and mortality by the following equations (Rosenzweig and MacArthur 1963):

$$\frac{dV}{dt} = r_{max}V\left(1 - \frac{V}{K}\right) - \frac{aVN}{1 + ahV}$$

$$\frac{dN}{dt} = \frac{acVN}{1 + ahV} - dN$$

where r_{max} is maximum per capita rate of resource recruitment, K is resource carrying capacity in the absence of consumers, a is area searched per unit time by consumers, h is handling time for each resource item, c is a coefficient for converting resource consumption into offspring, and d is consumer per capita mortality rate. The particular form of the equations used in this example is based on the most commonly observed patterns. In the absence of consumption (e.g. when $N = 0$), the resource population has a logistic pattern of growth (see Chapter 8). In other words, the resource population is self-regulating. Consumption rates and per capita rates of growth by

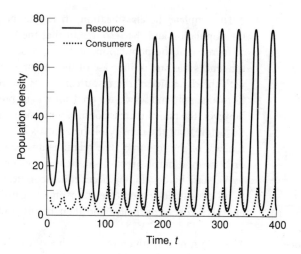

Fig. 12.1 Cyclic dynamics over time for the general consumer–resource model with the following parameter values: $a = 0.1$, $h = 0.2$, $c = 0.2$, $d = 0.3$, $r_{max} = 0.4$, and $K = 120$.

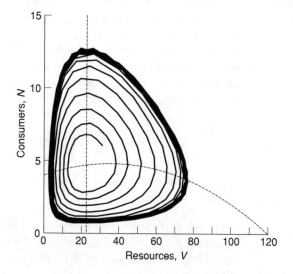

Fig. 12.2 Data from Fig. 12.1 re-plotted as a phase-plane diagram of consumers versus resources.

predators are presumed to increase and level off according to Holling's (1959) Type II functional response (see Chapter 10).

Depending on the parameter values that one uses, this model is capable of a variety of dynamics. In Fig. 12.1, we demonstrate one possible outcome: cyclic fluctuations of both consumers and resources over time.

Rather than plot densities of both resource and consumer populations against time, ecologists often plot the density of consumers against that of resource. This is known as a phase-plane diagram. For example, in Fig. 12.2 we re-plot the data shown in Fig. 12.1 as such a phase-plane diagram.

The phase-plane trajectory shown in Fig. 12.2 displays a pattern spiraling outwards from the starting point until it converges on a repetitive pattern known as a **stable limit cycle**. For most realistic consumer–resource models, this is a common outcome (Rosenzweig 1971; May 1972, 1973). If we had started at values outside the stable limit cycle, we would observe a spiral inwards until the trajectory converged once again on the stable limit cycle.

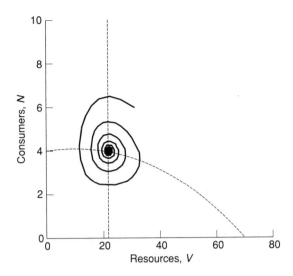

Fig. 12.3 Phase-plane diagram of the dynamics over time for a stable form of the consumer–resource model with the following parameter values: $a = 0.1$, $h = 0.2$, $c = 0.1$, $d = 0.3$, $r_{max} = 0.4$, and $K = 70$.

There are some useful additional lines displayed in the phase-plane diagram shown in Fig. 12.2: the null isoclines (sometimes termed a **nullcline**) for consumers (the vertical line) and resources (the hump-shaped curve). A null isocline identifies combinations of consumer and resource densities at which one of the populations is unchanging. In other words, at any of the consumer and resource combinations lying on the hump-shaped isocline, consumption exactly matches the rate of resource production, so resource density would be unchanging. Similarly, at the resource density shown by the vertical broken line, the consumer population acquires just enough resources to allow it to balance mortality by offspring production. In this case, there is only one sustainable combination of consumer and resource densities at which both are unchanging – the point of intersection of the two null isoclines. If we somehow set both populations to this equilibrium point, they would stay there. Slight deviation from the equilibrium leads to spiraling outwards of the consumer–resource trajectory until the stable limit cycle is reached (Fig. 12.2). Hence, the coexistent equilibrium is dynamically unstable, at least for these parameter values. Other trivial equilibria are also present: both N and $V = 0$, or $N = 0$ and $V = K$. These equilibria are also unstable for the parameter combination shown in Fig. 12.2.

For other parameter values, a second sustainable outcome is possible: a stable equilibrium for both consumers and resources (Fig. 12.3). The only difference between the models plotted in Figs 12.2 and 12.3 is the carrying capacity of resources. Decrease in the resource carrying capacity tends to be stabilizing, whereas enrichment of the carrying capacity of resources tends to be destabilizing. This has been termed the "paradox of enrichment," whereby provision of a better resource environment only leads to destabilization of consumers (Rosenzweig 1971).

Although the complete explanation for this phenomenon is complex, the system can be usefully viewed as reflecting a dynamic tension between stabilizing influences (such as self-regulation by resources) and destabilizing influences (such as consumption of resources). The reason that consumption tends to be destabilizing is that the per capita risk of resource mortality for a given consumer density is inversely related to resource density (see Chapter 10). Hence, an increase in resource levels leads to diminished risk of death, which imparts a positive, rather than negative, feedback on population dynamics. When the carrying capacity is small, the consumer null isocline lies close to the resource carrying capacity, to the right of the hump in the resource null isocline. This is a region where stabilizing influences are stronger than destabilizing influences. In contrast, when the carrying capacity is large, the consumer null isocline lies far away from the carrying capacity, where destabilizing influences hold sway.

At yet other parameter combinations, consumers would be unable to persist, simply because the intake of resources at any resource level is unable to compensate for mortality. The possible outcomes of this consumer–resource model depend entirely on the parameter values. Predicting the outcome of even this highly simplified representation requires detailed knowledge of the magnitude of ecological parameters. We now go on to illustrate how this approach can be applied to a well-studied system: red kangaroos and their food plants in Australia.

12.5 Kangaroos and their food plants in semi-arid Australian savannas

12.5.1 Plant dynamics

The dynamics of a renewable resource can be quite complicated, containing elements of seasonality, intrinsic growth pattern, and the modification of those two by the animals using the resource. To clarify some general issues, we shall consider in some detail a well-studied example: the growth of the herbage layer fed upon by kangaroos in the arid zone of Australia.

Figure 12.4 shows Robertson's (1987) estimate of the **plant growth response**, growth by ungrazed herbaceous plants in response to rainfall. He sampled growth rates on a kilometer grid over 440 km² of the arid zone of Australia. The measurements were repeated every 3 months for 3.5 years and rainfall was recorded for each 3-month interval. Look at the curve labeled 60 mm. It indicates that the higher the biomass at the start of the 3-month period the lower is the increment of further biomass added over the next 3 months. That is to be expected because plants compete for space, water, light, and nutrients. The 60 mm and 40 mm curves shown in Fig. 12.4 are part of a family of curves each representing that trend for a given rainfall over 3 months. We can summarize the figure by saying that the higher the rainfall the higher the growth increment, but for a given rainfall the higher the starting biomass the lower

Fig. 12.4 Plant growth rate as a function of plant biomass at the beginning of the interval and rainfall during the interval, for pastures in Kinchega National Park. (After Robertson 1987.)

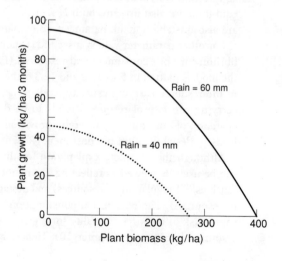

the growth increment. Hence, the rate of plant growth is influenced by both rainfall and plant biomass at the beginning of the period.

Figure 12.4 is a graphical representation of a regression analysis that estimated the relationship between growth increment in kg/ha over 3 months (Δ) on the one hand and starting biomass (V) and rainfall in mm (R) on the other:

$$\Delta = -55.12 - 0.01535V - 0.00056V^2 + 3.946R$$

Unlike the logistic model, plant growth in the Australian study was highest at low levels of abundance, rather than at intermediate levels of abundance (see Chapter 8). This is probably due to there being an ungrazeable plant reserve below ground. At low levels of plant abundance, rapid regrowth is enabled by translocation from these below-ground tissues. Such an ungrazeable refuge tends to lend a stabilizing influence to the interaction, as we shall shortly see.

12.5.2 The functional response of kangaroos to plant abundance

Having established how fast the resource grows in the absence of grazing and browsing, we now need to know what happens to it when a herbivore is present. The amount a herbivore eats per unit time is a constant only when it is faced by an *ad libitum* supply. Herbivores are seldom so lucky. The trend of intake against food availability is therefore curved, being zero when the level of food is zero and rising with increasing food to a plateau of intake. From there on no increase in food supply has any effect on the rate of intake because the animal is already eating at its maximum rate. Such a curve is called a **functional response** or **feeding response**, the trend of intake per individual against the level of the resource (see also Chapter 10). It can be represented symbolically by an equation such as:

$$I = c[1 - \exp(-bV)]$$

where I is plant consumption, c is the maximum (satiating) intake, V is the level of the resource, and b is the slope of the curve, a measure of grazing efficiency. The last has another meaning. Its reciprocal $1/b$ is the level of the resource V at which 0.63 (i.e. $1 - e^{-1}$) of the satiating intake is consumed.

Figure 12.5 shows the dry weight food intake (I) by a red kangaroo at various levels of pasture biomass when it is grazing annual grasses and forbs interspersed with scattered shrubs (Short 1987). The equation for a 35 kg kangaroo is:

$$I = 86[1 - \exp(-0.029V)]$$

The satiating intake is 86 kg/3 months, occurring when pasture biomass exceeds 300 kg/ha.

Short (1987) estimated these two functional responses by allowing high densities of kangaroos and rabbits to graze down pasture in enclosures, the offtake per day being estimated as the difference between successive daily estimates of vegetation biomass corrected for trampling. Daily intake could be estimated for progressively lower levels of standing biomass because the vegetation was progressively defoliated during the experiment. We scale up this daily intake rate to intake per 3 months to maintain a similar time frame as for the plant growth data.

Although the functional response has been discussed here in the context of a plant–herbivore system, all of that discussion carries over to prey–predator systems.

Fig. 12.5 Food intake
per individual red
kangaroo per day at
varying levels of food
availability. (After Short
1987.)

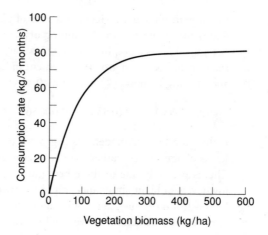

They are exactly analogous. The only difference lies in the difficulty of measuring a predator's food intake. An ability to measure intake by way of radioactive tracers has greatly simplified that problem. A good example is Green's (1978) use of radio-sodium to estimate how much meat a dingo eats in a day.

12.5.3 *The numerical response of kangaroos to plant abundance*

The functional response gives the effect of the animal upon a consumable resource. In contrast, the **numerical response** gives the effect of the resource on the change in animal numbers. If the resource is used in a pre-emptive rather than a consumptive way (e.g. nesting holes used by parrots), then it may be adequate to represent the numerical response by consumer density of the animals against the level of the resource (e.g. nesting holes per hectare). If the animals' use of the resource is consumptive, however, then the relationship between the animals and the resource is best portrayed as the instantaneous rate of population increase against the level of the resource.

Figure 12.6 shows the numerical response relationship between rate of increase of red kangaroos and the biomass of pasture. Bayliss (1987) estimated rates of increase from successive aerial surveys, and pasture biomass from ground surveys. As with the functional response, the numerical response has an asymptote: there is an upper limit to how fast a population can increase and no extra ration of a resource will

Fig. 12.6 Rate of
increase of red
kangaroos on a 3-
monthly basis in
relation to food
availability. (After
Bayliss 1987.)

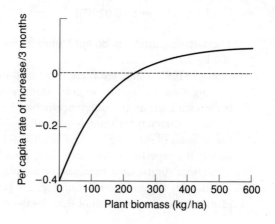

force that rate higher. The numerical response differs from the functional response in that negative values are both possible and logically necessary. If not the population would increase to infinity.

The numerical response can usually be described by an equation of the form:

$$r = -d + a[1 - \exp(-fV)]$$

where r is the exponential rate of increase of the animals, d is the maximum rate of decrease, and a is the maximum extent to which that rate of decrease can be alleviated. Hence $a - d = r_{max}$ is the maximum rate of increase. Demographic efficiency, the ability of the population to increase when resources are in short supply, is indexed by f. For the present example the constants were estimated (Bayliss 1987, modified by Caughley 1987) as:

$$r = -0.4 + 0.5[1 - \exp(-0.007V)]$$

The maximum rate of increase (i.e. when vegetation abundance is maximal) on a 3-monthly basis is $0.5 - 0.4 = 0.1$. Note that we have calculated the parameters for growth over 3 months, to remain consistent with the time frame for other parameters used in the model. On an annual basis r_{max} can be scaled up by simply multiplying by the four quarters in the year: $r_{max} = 0.4$. Hence, the population's maximum finite rate of increase over a year $\lambda = \exp(0.4) = 1.49$, a 49% increase per year.

12.5.4 Plant–kangaroo dynamics

So far we have taken a plant–herbivore system and dissected it into its component processes: plant growth, the herbivore functional response to changes in plant biomass, and the numerical response of the herbivore, in terms of its rate of increase, to the biomass of the plants.

The evaluation of these component influences upon a population's dynamics provides two bonuses. First, it furnishes a tight summary of the dynamic ecology of the system. Second, it furnishes that summary in terms of causal relationships rather than correlations. What follows is a short summary of the statistics of the Australian plant–kangaroo system described in detail by Caughley (1987).

Rainfall

	Mean (mm)	SD (mm)
December–February	62	59
March–May	57	47
June–August	59	34
September–November	61	44
Annual	239	107

These figures summarize 100 years of weather. There was no significant correlation of rainfall from one quarter to the next, nor between consecutive years.

Plant growth response

$$\Delta = -55.12 - 0.01535V - 0.00056V^2 + 2.5R$$

where Δ is the growth increment to ungrazed plant standing crop in kg/ha over 3 months, V is the plant standing crop in kg/ha at the beginning of those 3 months, and R is rainfall in mm over those 3 months. The 2.5 coefficient of R used here differs from Robertson's (1987) 3.946 for reasons given by Caughley (1987).

Feeding response

$$I = 86[1 - \exp(-0.029V)]$$

where I is intake of food in kg dry weight over 3 months, per red kangaroo, assuming a mean body weight of 35 kg and no shrubs in the pasture layer (Short 1987).

Numerical response

$$r = -0.4 + 0.5[1 - \exp(-0.007V)]$$

where r is the exponential rate of increase of red kangaroos on a 3-monthly basis.

The numerical response of the herbivore allows us to calculate the equilibrium level to which plant biomass will converge in a constant environment under the influence of an unrestrained population of herbivores, that is, the null isocline. It is the x-intercept of the regression of rate of increase of the herbivores against plant biomass or, put another way, the plant biomass at which rate of increase of the herbivore is zero (see Fig. 12.6). In the absence of seasonality, and of year-to-year variation in rainfall and temperature, this will be the equilibrium plant biomass imposed by grazing. The numerical response curve of this example was fitted as:

$$r = -d + a[1 - \exp(-fV)]$$

and the level of plant biomass V at which $r = 0$ is solved simply by setting r to zero and solving for V. Thus:

$$V = (1/f)\log_e[a/(a - d)]$$

which, when loaded with the values of the constants given here on a 3-monthly basis, yields:

$$V = (1/0.007)\log_e[0.5/(0.5 - 0.4)] = 230 \text{ kg/ha dry weight plant biomass}$$

This value is immensely important ecologically. It is the equilibrium level of plant biomass imposed by grazing in a constant environment. This is of some theoretical interest but of limited practical importance because environments are not constant. However, it is also the level of plant biomass above which the herbivore population will increase and below which it will decrease (the **critical threshold**), and that is true whether the environment is constant or variable and whether the density of herbivores is high or low.

Using similar logic, we can calculate the combination of kangaroo and plant densities at which consumption exactly matches regrowth by plants. This will occur

Fig. 12.7 Dynamics of kangaroos and plants over time, based on the Australian model discussed in the text.

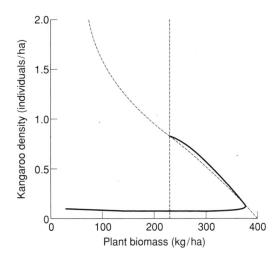

when $NI = \Delta$. We can rearrange terms to isolate kangaroo density on the left-hand side of the equation, $N = \Delta/I$. Both null isoclines for the kangaroo–plant system are plotted as broken lines in Fig. 12.7.

We can now reassemble the response functions of the system in their proper relationships to examine dynamics in the absence of stochastic variability in rainfall. We shall see what these in combination reveal about the system's dynamic behavior:

$$\frac{d}{dt}V(t) = \Delta[V(t), R(t)] - N(t)I[V(t)]$$

$$\frac{d}{dt}N(t) = N(t)r[V(t)]$$

Under a constant rainfall regime (in this case 60 mm per 3-month period), the system converges on the equilibrium, that is, the point of intersection of both null isoclines (Fig. 12.7). This shows that the equilibrium is stable, as one might have guessed, based on the negative slope of the plant null isocline. Convergence on the equilibrium is circuitous, involving a burst of plant growth, followed by plant decline as the kangaroo population stabilizes.

Rainfall can also be simulated as a sequence of random events from a normal distribution with a mean and standard deviation identical to the Australian data (Fig. 12.8), and the consequent changes in plant biomass and kangaroo numbers can be calculated accordingly.

Figure 12.9 demonstrates a typical time trend for kangaroos as generated by the equations describing the unpredictable rainfall (Fig. 12.8) and the responses to it of the plants and herbivores. The only external input other than starting conditions are the random values from the 3-monthly rainfall distributions whose observed means and standard deviations are given above. The kangaroo population trajectory is a mathematical consequence of that rainfall, as its effect feeds through to plant growth, herbivore population growth, and grazing pressure.

Fig. 12.8 A typical stochastic series of rainfall amounts per 3-month period drawn from a normal distribution with mean and variance equivalent to the Australian data. (After Caughley 1987.)

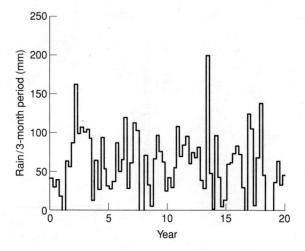

Fig. 12.9 A typical stochastic time series for kangaroos, using the model discussed in the text, for the rainfall sequence shown in Fig. 12.8.

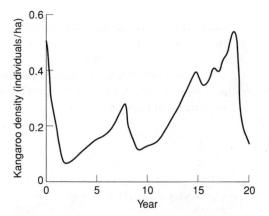

The rainfall of this region takes the form of high-amplitude, high-frequency fluctuations (Fig. 12.8). The herb layer, whether grazed or ungrazed, generates a similar trace of high-amplitude, high-frequency fluctuations as it reacts speedily to rainfall or the lack thereof. The fluctuations are paralleled by similar but more constrained fluctuations in the kangaroos' rate of increase as the population reacts dynamically to variations in food supply. The trend of kangaroo density differs from the rainfall regime, comprising fluctuations of high amplitude but low frequency. This result might have been predictable from first principles: present density is an integration of past rates of increase, not of present conditions. Initial conditions are not highly influential: the system remembers previous plant biomass for only 3 years but the memory of kangaroo density can linger for 10 years. As a consequence of the slow tracking of resources by kangaroos, there is a substantial time lag in response of kangaroos to changing climatic conditions. This lag imparts an irregular fluctuation over time, rather than constancy in abundance, despite the stability of the system under deterministic (constant climatic) conditions. Caughley (1987) has christened systems that show slow convergence on stochastically shifting equilibria as "centripetal."

12.6 Wolf–moose–woody plant dynamics in the boreal forest

12.6.1 *Models of the tri-trophic system*

Few natural systems have been studied in sufficient detail to supply all the necessary parameters that we observed in the Australian kangaroo and plant system. Fortunately, it is often possible to estimate plausible parameter values from allometric reasoning or historical data from a variety of sources, allowing us to make educated guesses about system dynamics in a generic sense (Yodzis and Innes 1988; Turchin 2003). We shall demonstrate this approach for moose, wolves, and woody plants in the boreal forests of North America. This is an important system to understand, because it occurs across much of the extensive boreal forest biome spanning North America. We use Turchin's (2003) parameter estimates for the interactive model.

First, we recognize from the outset that this is fundamentally a tri-trophic system, meaning that there are three trophic levels that interact in the food chain. The framework we shall use simply expands the consumer–resource model outlined at the beginning of the chapter to a third trophic level (P, for wolves) that feed on the second trophic level (N, for moose), that themselves feed on self-regulating plant resources (V). In all cases, we shall measure density in biomass (plants) or numerical abundance (for animals) per square kilometer. Mathematically, we can represent this interaction with the following system of equations:

$$\frac{\mathrm{d}}{\mathrm{d}t}V(t) = r_{max}\left[1 - \frac{V(t)}{K}\right] - N(t)\frac{aV(t)}{b + V(t)}$$

$$\frac{\mathrm{d}}{\mathrm{d}t}N(t) = eN(t)\left[\frac{aV(t)}{b + V(t)} - d\right] - P(t)\left[\frac{AN(t)}{B + N(t)}\right]$$

$$\frac{\mathrm{d}}{\mathrm{d}t}P(t) = EP(t)\left[\frac{AN(t)}{B + N(t)} - D\right]$$

where a is the maximum rate of plant consumption by a single moose, b is the plant biomass at which plant consumption is half of the maximum, d is the rate of plant consumption at which moose just sustain themselves, e is the efficiency of conversion of food intake into new moose, and A, B, D, and E represent the same set of parameters with respect to wolves.

We should note the similarity between the tri-trophic equations and the simpler consumer–resource model outlined at the beginning of the chapter. Resources have a self-regulating growth term, where the density-dependent term, $1 - V(t)/K$, reduces the growth rate proportionately with plant biomass. Plant consumption by moose is balanced against this positive contributor to resource abundance, with plant consumption expressed as the Michaelis–Menten form of the Type II functional response. Moose have a per capita growth function that depends on their intake of plants. Balanced against this is moose consumption by wolves. Finally, wolves have a per capita growth function that depends on their intake of moose, balanced against a constant per capita rate of mortality (presumably due to, for example, accidents, disease, and old age).

12.6.2 *Parameter estimation for the wolf–moose–woody plant system*

For a large part of the year moose browse on leaves and twigs of woody plants. Many species of plants contribute to the food supply of moose (Belovsky 1988). However, we know little about the web of ecological interactions within this plant guild, so we shall consider woody browse during winter (the period of the year when food is most often limiting to moose) as a single category. Edible biomass (measured in Mg/km²) is denoted V. Field data suggest that it is rare to observe higher browse availability than 100 g/m^2, which is equivalent to $K = 100$ Mg/km^2. Maximum moose density is thought to be 2 moose/km^2 (Messier 1994), and woody plant r'_{max} is estimated as 3.33 Mg/km^2/year (Turchin 2003). We see that the growth term for the edible plant biomass is maximized at low biomass, not at intermediate biomass, as it would be for a logistic growth function. The rationale for low edible biomass is that moose have access only to regrowing tissues, such as twigs and leaves, so that the rest of the plant functions as an ungrazeable reserve. Regrowth capacity should not be inversely affected by herbivory so long as it does not jeopardize plant survival. The indigestible component is the same kind of refuge demonstrated in Robertson's (1987) study of food plants fed upon by kangaroos in semi-arid Australian grasslands.

The maximum rate of plant consumption by moose was set at 2 Mg/individual/year, based on maximum values quoted in the feeding studies literature (Crête and Bédard 1975). Fitting various curves to Vivås and Sæther's (1987) studies of moose foraging in Norway suggests a foraging efficiency of $b = 40$ Mg/km^2. Moose can just meet their metabolic requirements at a level of intake of half the maximum, providing an estimate of $d = 1$ Mg/individual/km^2. Given a maximum exponential rate of increase of 0.2 for moose (Fryxell *et al.* 1988b) and values for all the other parameters, one can solve for e using the following relationship:

$$e = \frac{r_0}{\dfrac{aK}{b + K} - d}$$

yielding $e = 0.467$.

Rates of wolf consumption of moose are modeled as a Type II functional response, based on Messier's (1994) review of several moose–wolf studies throughout North America. Each of these studies provides one or more estimates of consumption rate by wolves at a given moose density. By combining all of the recorded data together in a single graph (Fig. 12.10), Messier was able to illustrate one of the most difficult kinds of ecological relationships, functional responses of large organisms under free-living conditions. Such patterns are essential to our understanding of consumer–resource interactions, yet are prohibitively costly to gather in a single study. Use of aggregate data is a very useful way to solve this problem.

Fig. 12.10 The consumption rate by wolves of moose in relation to moose density. (After Messier 1994.)

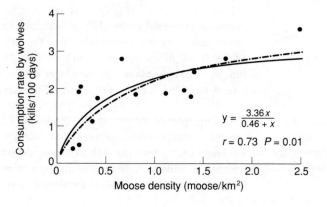

$$y = \frac{3.36x}{0.46 + x}$$

$r = 0.73$ $P = 0.01$

Fig. 12.11 Predicted
population dynamics
of moose (top) and
wolves (bottom) based
on Turchin's (2003)
tri-trophic model,
described in the text.

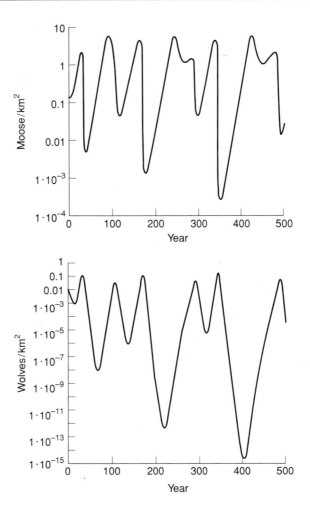

Scaling the wolf consumption rate to a yearly time frame yields estimates of $A = 12.3$ moose/wolf/year and $B = 0.47$ moose/km^2. According to Fuller and Keith (1980), each wolf needs to eat 0.06 kg of meat per day to meet maintenance requirements, whereas a population whose individuals eat 0.13 kg each day can grow at the maximal rate. This yields estimates of $D = 0.6$ and $E = 0.1$.

12.6.3 *Dynamics of the wolf–moose–woody plant system*

Combining these parameter values together, the outcome is a complex series of oscillations in moose and wolf abundance, which never quite repeat themselves (Fig. 12.11). This is a mild form of deterministic chaos, common in tri-trophic systems (Hastings and Powell 1991; McCann and Yodzis 1994). Even though the fluctuations are non-repetitive, the time between successive peaks tends to be several decades – a very protracted pattern of fluctuation.

The manner by which parameters for the wolf–moose–woody plant model were derived, using a set of observations gathered around the globe, makes it fairly unlikely that we can predict the dynamics of any given system. It does suggest, nonetheless, that this system should exhibit an inherent tendency towards protracted fluctuations that recur over a decade-long time scale. Moreover, the model suggests that these fluctuations will not necessarily converge on a stable limit cycle, as do consumer–resource models with only two trophic levels. Rather, we may expect to see inconsistency as each population progresses from peak to peak.

One obvious objection to this model is that it ignores the role of wolf territoriality. In most landscapes wolves form communal packs that partition the available habitat amongst themselves. Territorial strife among wolf packs can be intense, leading to substantial levels of mortality (Peterson *et al.* 1998). At least we should expect that the risk of this mortality should climb with wolf density, if only because of increasing frequency of encounters between members of different packs. One way to incorporate this effect is to make wolf mortality explicitly density dependent:

$$\frac{d}{dt}V(t) = r_{max}\left[1 - \frac{V(t)}{K}\right] - N(t)\frac{aV(t)}{b + V(t)}$$

$$\frac{d}{dt}N(t) = eN(t)\left[\frac{aV(t)}{b + V(t)} - d\right] - P(t)\left[\frac{AN(t)}{B + N(t)}\right]$$

$$\frac{d}{dt}P(t) = EP(t)\left[\frac{AN(t)}{B + N(t)} - D\right] - \frac{s_0 P(t)^2}{\gamma}$$

where the maximum density of wolves (recorded from field studies) $\gamma = 0.1$ and the maximum per capita rate of wolves $s_0 = 0.4$. This modification imposes an additional per capita mortality term that increases by s_0/γ with each unit increase in wolf density P.

Territorial effects of this sort often have a stabilizing influence. Such is the case with the wolf–moose–woody plant model: the addition of density-dependent mortality due to territorial strife changes the dynamics of the system from deterministic chaos to a stable limit cycle (Fig. 12.12). The level of strife is insufficient, however, to completely stabilize the system.

The best long-term data set available on both moose and wolves is from Isle Royale, a small island 40 km off the coast of Canada in Lake Superior that supports a mix of deciduous and coniferous vegetation species typical of the boreal forest on the mainland. Moose apparently invaded Isle Royale a century ago, whereas wolves arrived by ice in the 1940s. Estimated patterns of abundance on Isle Royale certainly suggest protracted fluctuations over time (McLaren and Peterson 1994; Peterson 1999; Post *et al.* 1999), with moose populations slowly fluctuating over time, with 25 years between successive peaks (Fig. 12.13).

It is difficult to conclusively tell from the Isle Royale time series data whether the system is cyclic or chaotic, because there are simply not enough data to evaluate even such a well-studied system. Such will nearly always be the case in slow-changing wildlife species. Nonetheless, the tri-trophic model seems to capture the fluctuating tendency of the Isle Royale system.

There are many other factors that could also contribute to the apparent instability of the Isle Royale populations. For example, complex changes over time in the age structure of moose could itself contribute to the propensity for fluctuations (Peterson and Vucetich 2003). Wolves are highly selective for specific age classes of prey, so changes in age distribution could translate into substantial changes in predation risk.

Fig. 12.12 Predicted
dynamics of the moose
(top) and wolf (bottom)
system with woody
plants when wolves
have additional density-
dependent mortality due
to territorial aggression,
as described in the text.

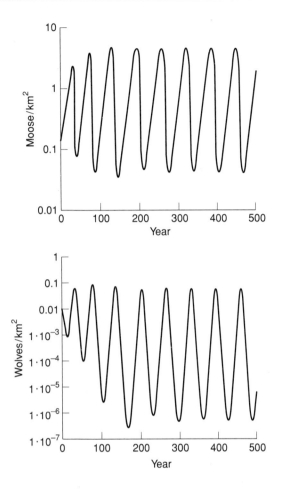

As we saw in Chapter 6 (see also Chapter 14), it can take many years for age distributions to stabilize in long-lived organisms. When age distributions are shaped by dynamic interactions with predators, this can be even more destabilizing. We also know that the wolf population on Isle Royale has much lower levels of genetic variability than do populations on the mainland (Wayne *et al.* 1991). This could influence wolf demographic parameters in unknown ways. Finally, there is evidence that complex interactions among climatic conditions, social grouping patterns of wolves, and predation risk of moose could contribute to instability. In years of deep snow, wolves form larger packs, which leads to increased rates of mortality on moose (Post *et al.* 1999). Nonetheless, the instability of this system seems to be intrinsic to the basic consumer–resource interactions (moose/vegetation and wolf/moose).

Truly long-term data for temperate zone carnivores (wolves and coyotes) and ungulates (moose or white-tailed deer, depending on location) are scarce. Data from the Hudson's Bay Company probably represent the lengthiest data set. These data suggest very slow oscillations in the abundance of wolves and coyotes during 1750–1900, with roughly two cycles per century (Turchin 2003). Although the Hudson's Bay data on deer skins are more fragmentary, they too suggest long-term cycles in abundance (Turchin 2003). Slow oscillations by white-tailed deer in Canada (Fryxell

Fig. 12.13 Population
dynamics of wolves (a)
and moose (b), as well
as annual growth of
balsam fir trees (c) on
Isle Royale. The solid
lines represent observed
values, the dotted lines
polynomial regressions.
The densities represent
total abundance of each
species recorded over
the entirety of Isle
Royale. (After Post
et al. 1999.)

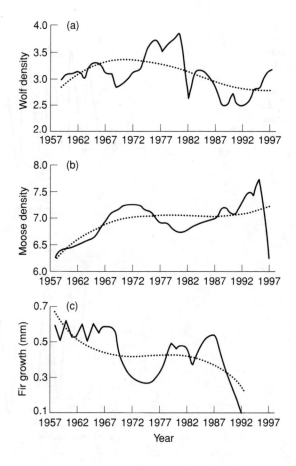

et al. 1988a) and moose in Finland (Lehtonen 1998) suggest that long-term oscillations are an important feature of some large mammal species.

12.7 Other population cycles

Long-term data for a number of other wildlife populations show pronounced cycles, first identified by Charles Elton (Elton 1924). Such cycles are sometimes regular, such as the 10-year cycle of snowshoe hares (Sinclair *et al.* 1993), and other times somewhat erratic, such as the 3–6-year cycle of voles (*Microtus agrestis* and *Clethrionomys rufocanus*) in northern Europe (Turchin and Hanski 1997). Such cycles can be explained in many ways. A short list of hypotheses include: unstable behavioral polymorphisms in cyclic populations (Chitty 1967; Krebs and Myers 1974); maternal effects transferred to offspring, imparting lagged density dependence (Inchausti and Ginzburg 1998); and coupled interactions between plants, herbivores, and/or carnivores (Hansson 1987; Turchin and Hanski 1997; Turchin and Ellner 2000; Turchin and Batzli 2001). We shall review the northern European vole and North American snowshoe hare populations and consider the logic underlying consumer–resource explanations for population cycles.

Some of the longest continuous studies of vole populations come from sites in Scandinavia, Finland, and Russia (Turchin 2003). These data point to a fascinating geographical pattern: populations at southern latitudes show little evidence of

Fig. 12.14 The latitudinal gradient in vole dynamics across northern Europe, with most northerly sites at the top of the figure. (After Turchin and Hanski 1997.)

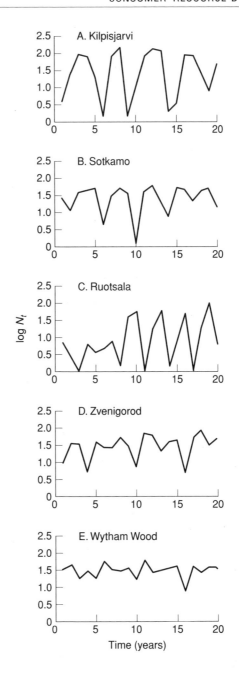

repetitive, cyclic dynamics, whereas populations from more northerly latitudes exhibit repetitive cycles or perhaps even chaotic dynamics over time (Fig. 12.14).

Many ecological variables change as one progresses from the Arctic Circle to more southerly latitudes, including temperature maxima and minima, precipitation, vegetation cover and composition, primary productivity, mammalian and avian community diversity, and human population density. Most important amongst these variables, however, is the transition from a suite of generalist predators (red foxes,

feral cats, badgers, and various owls, hawks, and other raptors) in the south to a narrow range of specialist predator species (primarily the least weasel, *Mustela nivalis*)
that predominate in the most northerly areas. The abundance of generalist predators
declines in northern latitudes because of the duration and depth of snow cover (Hansson
and Henttonen 1985).

Least weasels exhibit a Type II functional response to changes in vole density (Erlinge
1975). As we have already discussed, this pattern of foraging tends to destabilize prey
populations, because the per capita risk of mortality due to predators is inversely
related to prey density. However, generalist predators switch feeding preferences to
favor voles when they reach high density, but ignore them when they collapse to low
density (Erlinge *et al.* 1983; Korpimäki and Norrdahl 1991). As we show in Chapter 10, switching behavior can stabilize prey numbers, because the per capita risk
of mortality for prey due to predation increases with prey density, at least over some
prey densities. Because generalist predators can feed on a wide variety of other species,
they may not be dependent on vole numbers (Turchin and Hanski 1997). In the absence
of predators, vole population growth is self-regulating, due to density-dependent resource
limitation and territorial spacing among individual voles.

Turchin and Hanski (1997, 2001) linked specialist predation by weasels, generalist predation, and self-regulating population dynamics of voles with seasonal changes
in vole logistic growth. In keeping with the empirical data, their model predicted
better than alternative models complex cycles or chaos when generalist predators are
rare, but much more stable dynamics when generalist predators are common.

Data on the cyclical variation in abundance of snowshoe hares come from fur records
of the Hudson Bay Company in Canada (Fig. 8.3). These data show a regular
oscillation in numbers with a period of 10 years. Like the other examples we have
discussed, snowshoe hares interact not only with their food supplies but also with a
suite of carnivores that feed on them (Krebs *et al.* 2001b). Some of these carnivores,
especially the lynx, which is a specialist feeding on hares, also display a 10-year cycle
in abundance, slightly lagged behind that of the snowshoe hare. These characteristics suggest that the tri-trophic model might be a useful starting point in modeling
the dynamics of hare and lynx populations. King and Schaffer (2001) estimated parameters to model dynamics of the woody plant, hare, and lynx interaction. They found
that realistic parameter values generated cycles in hare and lynx abundance of 8–12
years, consistent with the historical data.

Unlike the previous examples we have discussed, however, there is an inherent
environmental cycle, the 11-year sunspot cycle, that apparently plays a crucial role
in generating the hare–lynx cycle (Sinclair *et al.* 1993). Snow depth is strongly influenced
by the sunspot cycle, as evidenced by ice cores taken from glaciers. Disentangling
the effect of the sunspot cycle from the endogenous rhythm of the tri-trophic
consumer–resource interaction presents a sizeable challenge.

King and Schaffer (2001) also used the tri-trophic model to explain the outcome
of a series of large-scale field experiments conducted in Kluane National Park,
Canada, during the 1980s and 1990s (Krebs *et al.* 1995, 2001b). The Kluane study
involved experimental manipulations of food availability, predation risk, and both of
these factors combined to tease apart bottom-up versus top-down trophic mechanisms.
The Kluane team found that each of the manipulations had considerable effect on
hare densities and hare demographic rates. Food addition doubled hare densities,
predator exclusion trebled hare densities, but both had an 11-fold effect on hare

densities relative to controls. The clear implication is that both bottom-up and top-down processes are important to the natural regulation of snowshoe hares. Despite these results, however, none of the treatments dismantled the hare cycle. This result may have arisen from the use of semi-permeable fencing in the experimental treatments, allowing hare populations within the exclosures to be driven by dynamics generated outside, via immigration.

The best interpretation of the existing information is that the snowshoe hare–lynx cycle is a complex tri-trophic interaction synchronized to some degree by the exogenous environmental rhythm of the sunspot cycle. These results suggest that coupled resource–consumer models can be a vital step in understanding complex patterns of population dynamics that can occur in natural ecosystems.

12.8 Summary

A resource is something an animal needs and whose consumption diminishes its availability to other consumers. Consumers and their resources often form a system in which the rate of increase of the resources is determined by the density of the animals eating them, and the rate of increase of the animals is determined by the density of the resources. Such a complex system can be studied only by breaking it down to its dynamic components, of which three dominate. First, there is the functional response of the animal, the rate of resource intake by a single consumer as a function of resource abundance. Second, there is the numerical response of the consumer, the rate at which its population increases as a function of the resource abundance. Finally, we require supplementary information on the growth rate of resources in relation to resource abundance. On the basis of these functional relationships, the full dynamic behavior of the system can be described. We illustrate this approach with two well-studied consumer–resource systems: kangaroos and their plants in Australia, and wolves, moose, and woody plants in North America. Interactive systems with these components can be deterministically stable (such as the Australian plant–kangaroo system) or unstable (such as the wolf, moose, and woody plant system). Deterministic instability is evident in the repetitive population fluctuations (stable limit cycles) or non-repetitive fluctuations (deterministic chaos). Even stable food chain models can show pronounced long-term fluctuations in response to stochastic environmental variability (centripetal systems). Two well-documented cyclical populations (voles in northern Europe and snowshoe hares in North America) have dynamics consistent with predictions of coupled consumer–resource models.

Part 2

Wildlife conservation and management

13 Counting animals

13.1 **Introduction**

The trick in obtaining a usable estimate of abundance is to choose the right method. What works in some circumstances is useless in others. Hence, we provide a broad range of methods and indicate the conditions under which each is most effective.

13.2 **Estimates**

Knowledge of the size or density of a population is often a vital prerequisite to managing it effectively. Is the population too small? Is it too large? Is the size changing and if so in what direction? To answer these questions we may have to count the animals, or we may obtain adequate information by way of an indirect indication of abundance. In any event we need to know when a census is necessary and how it might be done.

Although **census** is strictly the total enumeration of the animals in an area, we use the word in its less restrictive sense of an estimate of population size or density. That estimate may come from a total count, from a sampled count, or by way of an indirect method such as mark–recapture.

Closely related to the census is the **index**, a number that is not itself an estimate of population size or density but which has a proportional relationship to it. The number of whales seen per cruising hour is an index of whale density. It does not tell us the true density but it allows comparison of density between areas and between years. Indices provide measures of relative density and are used only in comparisons. They are particularly useful in tracking changes in rates of increase and decrease.

Almost all decisions on how a population might best be managed require information on density, on trend in density, or on both. There are many methods to choose from and these differ by orders of magnitude in their accuracy and expense. Hence before any censusing is attempted the wildlife manager should ask a number of questions.

- Do I need any indication of density and what question will that information answer?
- Is absolute density required or will an index of density suffice?
- Will a rough estimate answer the question or is an accurate estimate required?
- What is the most appropriate method biologically and statistically?
- How much will it cost?
- Do we have that kind of money?
- Would that money be better spent on answering another question?

This chapter outlines briefly the variety of available methods and their applications, providing references to where each is treated in detail.

13.3 **Total counts**

The idea of counting every animal in a population, or on a given area, has an attractive simplicity to it. It is the method used by farmers to keep track of the size of their flocks. No arithmetic beyond adding is called for and the results are easily inter-preted. That is why total counting was once very popular in wildlife management and why it is still the most popular method for censusing people.

Total counts have two serious drawbacks: they tend to be inaccurate and expen-sive. Nonetheless they have a place. The hippopotami (*Hippopotamus amphibius*) in a clear-water stretch of river can be counted with reasonable facility from a low-flying aircraft. The number of large mammals in a 1 km² fenced reserve can be determined to a reasonable level of accuracy by a drive count. It takes much organization and many volunteers, but it can be done. Every nesting bird can be counted in an adélie penguin (*Pygoscelis adeliae*) rookery, either from the ground or from an aerial photo-graph. That is an example of a "total count" providing an index of population size because more than half the birds will be at sea on any given occasion.

Total counting of large mammals over extended areas was common in North America up to 1950. Gill *et al.* (1983) described the system in Colorado:

> Biologists attempted to count total numbers of deer comprising the most important "herds" in the state. Crews of observers walked each drainage within winter range complexes and counted every deer they encountered. The sum of all counts over every drainage of a winter range was taken as the minimum population size of that herd (McCutchen 1938; Rasmussen and Doman 1943).

Total counting of large mammals from the air was a standard technique in Africa in the 1950s and early 1960s. Witness the total counts of large mammals on the 25,000 km² Serengeti–Mara plains (Talbot and Stewart 1964) and 20,000 km² Kruger National Park, South Africa:

> . . . trends in population totals, spatial distribution and social organization are obtained by means of surveys by fixed-wing aircraft. Due to the size of the Kruger National Park these (total count) surveys require three months to complete and are consequently undertaken only once annually (i.e. during the dry season from May to August (Joubert 1983)).

These massive exercises continued in Kruger until 1996 when they were abandoned due to cost. Similar methods are used to count pronghorn antelope in the USA (Gill *et al.* 1983). Total counts continue to be used on species that are highly clumped with wide spacing between clumps. For example, both African buffalo (*Syncerus caffer*) and African elephant (*Loxodonta africana*) live in widely dispersed large herds of several hundred animals in both Serengeti and Kruger National Parks, and total counting is still the best method of counting them. This is because the dispersion pattern of these species means that sample counts produce very high variances and hence wide confidence limits. A simulated transect sampling strategy for a known dispersion of buffalo showed that over 90% of the area had to be sampled before confidence limits were reduced to acceptable values (< 15% of the estimated total). Thus, total counting was more efficient because it was logistically easier than rigidly flown transects (Sinclair 1973). Similarly, the clumped distribution of pronghorn ante-lope (*Antilocapra americana*) in North Dakota produced such high variances from a variety of sampling strategies that Kraft *et al.* (1995) advised against using samples to estimate numbers.

13.4 Sampled counts: the logic

There are two important areas in which scientific thinking differs from everyday thinking: the selection of a random or unbiased sample and the choosing of an appropriate experimental control. Knowing how to sample, and knowing how to design an experiment that gives an unambiguous answer, are the two attributes distinguishing science from ideology. Sampling is the technique of drawing a subset of sampling units from the complete set and then making deductions about the whole from the part. It is used all the time in wildlife research and management, but often incorrectly.

The next section takes you through some of the mystery of what happens when we sample. It explores what actually happens when we sample a population in several different ways, thereby making the point that the true estimate is independent of whatever mathematical calculations are applied to the data.

13.4.1 *Precision and accuracy*

If a large number of repeated estimates of density has a mean that does not differ significantly from the true density then each estimate is said to be **accurate** or **unbiased**. Accuracy is a measure of **bias error**. If that set of estimates has little scatter, the estimates are described as **precise** or **repeatable**. Precision is a measure of **sampling error**. A system of estimation may provide very precise estimates that are not accurate, just as a system may provide accurate but imprecise estimates. Ideally both should be maximized, but often we must choose between one or other according to what question is being asked. For example, is density below a critical threshold of one animal per square kilometer? Here we need an accurate measure of density and may be willing to trade off some precision to get it. But if we had asked whether present density is lower than that of last year we would need two estimates each of high precision. Their accuracy would be irrelevant so long as their bias was constant. Most questions require precision more than accuracy. Precision is obtained by rigid standardization of survey methods, by sampling in the most efficient manner, and by taking a large sample.

13.4.2 *Bias errors*

Bias errors derive from some systematic distortion in the counting technique, the observer's ability to detect animals, or the behavior of the animals. Often, but not always, the bias produces an undercount. Thus biases can accrue from sampling schemes that do not properly sample all habitats, for example using roads that avoid hills or riverine areas; from observers missing animals on transects because there are too many animals, or because the observer is counting one group and so is distracted from seeing another, or simply due to fatigue; or from animals hidden in thickets, under trees or underwater.

The best way to measure bias error is to compare the census estimate with that from a known population. This, together with the use of a subpopulation of marked animals, mapping with multiple observers, line-transect sampling, and multiple counts on the same area, are reviewed in Pollock and Kendall (1987). Visibility corrections have been calculated by comparing fixed-wing aerial surveys of waterfowl with ground counts (the known or unbiased population) (Arnold 1994; Bromley *et al.* 1995; Prenzlow and Lovvorn 1996). A similar approach was used to estimate bias in counts of wood stork (*Mycteria americana*) nests in Florida (Rodgers *et al.* 1995) and great blue heron (*Ardea herodias*) nests in South Carolina (Dodd and Murphy 1995). Moose usually live in dense habitats where they are difficult to see. Rivest and Crepeau (1990) compared fixed-wing surveys of moose with the more accurate subsample surveys by helicopter to correct for visibility bias, an approach also used for counts of chicks in osprey (*Pandion haliaetus*) nests (Ewins and Miller 1995).

13.4.3 *Sampling frames*

Before an area is surveyed to estimate the number of animals on it, that area must be divided into **sampling units** that cover the whole area and are non-overlapping. The sampling units may comprise areas of land if we count deer, or trees if we count nests, or stretches of river if we count beavers or crocodiles. To allow us to sample from this **frame list** of sampling units, the list must be complete for the whole area. Hence the frame of units contains all the animals whose numbers we wish to estimate.

For purposes of explanation we use the first example: sampling units of land. The survey area may be divided up into units in any way the surveyor desires, into quadrats, transects, or irregular sections of land perhaps delimited by fences. The choice is a compromise between what is most efficient statistically and what is most efficient operationally.

13.4.4 *Sampling strategies*

Suppose that we wished to estimate the number of kangaroos or antelopes in a large area by counting animals on a sample of that area. Several strategies are open to us. We could sample quadrats or transects, we could select these sampling units systematically or randomly and, if the latter, we could ensure that each sampled unit occurred only once in the sample (sampling without replacement) or that the luck of the draw allowed units to be selected more than once (sampling with replacement). The efficiency of these systems will be demonstrated with the hypothetical data of Table 13.1, which may be thought of as the number of kangaroos standing on each square kilometer of an area totalling 144 km^2. In all cases one-third of the area will be surveyed. We can test the accuracy of the method by determining whether the mean of a set of repeated estimates is significantly different from the true total of 1737 kangaroos. The precision of a sampling system is indicated by the spread of those repeated and independent estimates, and that spread will be measured by the standard deviation of those estimates:

$$s = \sqrt{[(\sum x^2 - (\sum x)^2/N)/(N-1)]}$$

Table 13.1 A simulated dispersion of kangaroos on a 1 × 1 km grid of 144 cells. Marginal totals give numbers on 1 × 12 km transects oriented across the region and down it.

1	2	7	4	7	14	9	18	24	22	19	15	142
0	1	5	6	12	11	9	15	20	21	27	28	147
2	3	5	6	10	13	16	20	160	14	19	21	147
1	4	4	6	9	13	14	17	20	16	25	20	149
2	2	5	7	10	12	16	19	20	16	18	22	149
2	4	5	6	9	12	16	22	18	18	21	23	156
0	2	5	8	4	7	11	.13	17	16	21	30	134
1	0	4	9	8	10	11	16	14	20	17	17	127
0	4	2	7	8	11	11	11	12	19	22	21	128
0	2	5	8	8	12	16	20	24	25	23	25	168
1	0	4	9	8	8	8	17	17	14	18	22	126
2	5	7	6	12	12	13	15	20	21	20	23	156
12	29	58	82	105	135	150	203	222	222	250	269	1737

where x is an independent estimate of total numbers and N is the number of such repeated estimates.

We will first sample 1 km² quadrats randomly with replacement – **sampling with replacement** (SWR). The quadrats are numbered from 1 to 144 and a sample of 48 of these is drawn randomly. Quadrat numbers 27, 31, 50, and 53 were drawn twice and quadrat number 7 three times, but since these are independent draws they are included in the sample as many times as they are randomly chosen. The quadrat is replaced in the frame list after each draw, allowing it the chance of being drawn again. The number of kangaroos in this sample of quadrats totalled 523, and since we sampled only a third of the quadrats we multiply the total by 3 to give an estimate of animals in the study area: 1569.

Note that this answer is wrong in the sense that it differs from the true total known to be 1737 (i.e. it is not accurate). That disparity is called **sampling error**, and is quite distinct from **errors of measurement** resulting from failure to count all the animals on each sampled quadrat.

We now repeat the exercise by drawing a fresh sample of 48 units and get a sampled count of 493 kangaroos, which multiplies up to an estimate of 1479. The third and fourth surveys give estimates of 1836 and 1752. That exercise was repeated 1000 times with the help of a computer. The 1000 independent estimates had a mean of $\bar{x} = 1741$, very close to the true total of 1737. We can be confident, therefore, that this sampling system produces accurate (i.e. unbiased) estimates. The 1000 independent estimates had a standard deviation of $s = 153$, which tells us that there is a 95% chance that any one estimate will fall in the range $\bar{x} \pm 1.96s$ or 1741 ± 300, between 1441 and 2041. The standard deviation of a set of independent estimates is the measure of the efficacy of the sampling system and hence of the precision of any one of the independent estimates. It can be estimated from the quadrat counts of a single survey (see Section 13.5.1) and when estimated in this way it is called the standard error of the estimate. Hence the standard error of an estimate is a calculation of what the standard deviation of a set of independent estimates is likely to be.

With that background we can now compare the efficiency of several sampling systems.

13.4.5 *Sampling with or without replacement?*

When we use **sampling without replacement** (SWOR) a quadrat may be drawn no more than once, in contrast to the previous system which allowed, by the luck of the draw, a quadrat to be selected any number of times. We draw a unit, check whether that unit has been selected previously, and if so reject it and try again. Having drawn 48 distinct units we calculate density. The sampling is again repeated 1000 times, yielding 1000 independent estimates, each based on a draw of 48 units, of the number of animals we know to be 1737. Those 1000 estimates had a mean of 1743, and a standard deviation of 131, which is appreciably lower than the $s = 153$ accruing from sampling with replacement.

The gain in precision by sampling without replacement reflects the slightly greater information on density carried by the 48 distinct quadrats of each survey. Sampling without replacement is always more precise than sampling with replacement for the same sampling fraction, the relationship being:

$$s(\text{SWOR}) = s(\text{SWR}) \times \sqrt{(1 - f)}$$

where f is the sampling fraction, in this case 0.333. The s(SWR) from the 1000 repeated surveys was 153 and from this we could have estimated, without needing to run the simulation, that the precision of the analogous SWOR system would be about:

$$s = 153 \times \sqrt{0.666} = 125$$

Our empirical s(SWOR) is 131, which is much the same as the $s = 125$ predicted theoretically.

However, it is not as simple as that. The quadrats chosen more than once in a SWR sample are not surveyed more than once although they are included in the analysis more than once, and so the time taken for the survey is shorter. In the example only about 41 of the 48 units drawn in a SWR sample would be distinct units, the other seven being repeats. To compare the precision of a SWOR sample with that of a SWR sample entailing the same groundwork, we would have to draw by SWR about 58 units. Ten are repeats, "free" units that do not need to be surveyed a second time. Intuitively we would assume that the SWR sample of 48 distinct units and 10 repeats must give a more precise estimate than the SWOR sample with its 48 distinct units, none repeated. Not so. The smaller SWOR samples provide estimates more precise by a factor of $\sqrt{(1 - \frac{1}{2}f)}$. In all circumstances SWOR is more precise than SWR (Raj and Khamis 1958). Precision is increased by rejecting the repeats and cutting the sample size back to that of the analogous SWOR sample.

Why then, if sampling without replacement is always better, is sampling with replacement often used? First, when the sampling fraction is low, less than 15%, the precision of the two systems of sampling is similar. At $f = 0.1$ there is only a 5% difference in precision, reflecting the low likelihood of repeats at low sampling intensity. Most sampling intensity in wildlife management is of this order. Second, it is often convenient to sample with replacement when an area is traversed repeatedly by aerial-survey transects. There is not the same necessity to ensure that no transect crossed another or overlaps it. That is a useful flexibility for an aerial survey in a strong cross wind or for a ground survey in thick forest.

13.4.6 Transects or quadrats?

A frame of transects is a good or bad sampling system according to how it is oriented with respect to trends in density. The dispersion of Table 13.1 has a marked increase in density from left to right. The precision of the estimate of total numbers would be relatively high if the transects were oriented along this cline but low if oriented at right angles to it. That can be demonstrated empirically by sampling the column totals at one-third sampling intensity. Each column represents a transect and each survey comprises four transects randomly chosen. One thousand independent surveys produces a standard deviation of estimates of 512 for SWR and 427 for SWOR. If these transects had been oriented at right angles so that the rows rather than the columns formed the transects, the standard deviation of estimates of 1000 independent surveys would have been approximately 80 for SWR and 69 for SWOR. In this case precision is increased enormously by swinging the orientation of the transects through 90°.

Transects should go across the grain of the country rather than along it, they should cross a river rather than parallel it, and they should go up a slope rather than hug the contour. They should be oriented such that each transect samples as much as possible of the total variability of an area. In essence, we must ensure that the

Table 13.2 The effect of sampling system on the precision of an estimate. All systems sample one-third of an area of 144 km² containing the dispersion of kangaroos simulated in Table 13.1. Each sampling system is run 1000 times to provide 1000 independent estimates of the true total of 1737.

Sampling system	Mean estimate	Standard deviation of 1000 estimates
Large quadrats (n = 4)		
Random with replacement	1746	487
Random without replacement	1738	414
Small quadrat (n = 48)		
Random with replacement	1741	153
Random without replacement	1743	131
Transects parallel to the density cline (n = 4)		
Random with replacement	1732	80
Random without replacement	1734	69
Restricted random	1730	57
Systematic	1736	48

variation between transects is minimized and therefore that the precision of the estimate is maximized.

Much the same principle adjudicates between the use of quadrats as against transects. So long as the frame of transects is oriented appropriately, the resultant estimate will be more precise than that from a set of quadrats whose area sums to that of the transects. The more clumped is the distribution of the animals, the greater will be the gain in precision of transects over quadrats. A quadrat is likely to land in a patch of high density or a patch of low density whereas a transect is more likely to cut through areas of both. Table 13.2 shows that transects oriented along the cline in density of Table 13.1 provide estimates six times more precise than do quadrats of the same size and number.

13.4.7 *Random or non-random sampling?*

Sampling strategies grade from **strictly random** to **strictly systematic**. The region in between is described as **restricted random sampling**. One might decide, for example, to sample randomly but to reject a unit that abuts one previously drawn. Or one might break the area into zones (strata) and draw the same number of samples randomly from each zone. These two strategies depart from the requirement of strict random sampling whereby each sampling unit has the same probability of selection. The extreme is systematic sampling in which the choice of units is determined by the position of the first unit selected.

Systematic or restricted random sampling has several practical advantages over strict random sampling. First, it encourages or enforces sampling without replacement which, as we have seen, leads to a more precise estimate. Second, it reduces the disturbance of animals on a sampling unit caused by surveying an adjoining unit. That is particularly important in aerial survey where the noise of the aircraft can move animals off one transect onto another. Third, any deviation from strictly random sampling tends to increase the precision of the estimate because the sampled units together provide a more comprehensive coverage of total variability. Table 13.2 demonstrates this for our example. The standard deviation of 1000 independent surveys is lower for restricted random sampling than for random sampling without replacement, and lower still for systematic sampling.

Statisticians do not like non-random sampling because the precision of the estimate cannot be calculated from a single survey. The formulae given in Section 13.5.1 for calculating the standard error of an estimate are correct only when sampling units are drawn at random, and they will tend to overestimate the true standard error when restricted random or systematic sampling is used. But not always. If a systematically drawn set of sampling units tends to align with systematically spaced highs and lows of density, the standard error calculated on the assumption of random sampling will be too low and the estimate of density will be biased.

In practice this tends not to happen. It is entirely appropriate to sample systematically or by some variant of restricted random sampling and to approximate the standard error of the estimate by the equation for random sampling. One can be confident that the estimate is unlikely to be biased and that the true standard error is unlikely to exceed that calculated.

13.4.8 *How not to sample*

There are a number of traps that sampling can lure one into and which can result in a biased estimate or an erroneous standard error. Suppose one decided to sample quadrats but, for logistical reasons, laid them out in lines, the distance between lines being considerably greater than the distance between neighboring quadrats within lines. The standard error of the estimate of density cannot then be calculated by the usual formulae because the counts on those quadrats are not independent. Density is correlated between neighboring quadrats and this throws out the simple estimate of the standard error, which returns an erroneously low value. There are ways of dealing with the data from this design to yield an appropriate standard error (see Cochran (1977) for treatment of two-stage sampling and Norton-Griffiths (1973) for an example using the Serengeti wildebeest) but they are beyond the scope of this book. The simple remedy is to pool the data from all quadrats on each line, the line rather than the quadrat becoming the sampling unit. That procedure may appear to sacrifice information but it does not (Caughley 1977a).

Another common mistake is to throw random points onto a map and to declare them centers of the units to be sampled, the boundary of each being defined by the position of the point. In this case the requirement that sampling units cover the whole area and are non-overlapping is violated and the sampling design becomes a hybrid between sampling with replacement and sampling without replacement, leading to difficulties in calculating a standard error. There is nothing wrong with choosing units to be sampled by throwing random points on a map so long as the frame of units is marked on the map first. The random points define units to be selected. They do not determine where the boundaries of those units lie.

A third trap to watch for is a biased selection of units to be sampled. The most common source of this bias in wildlife management is the so called "road count" in which animals are counted from a vehicle on either side of a road or track. Roads are not random samples of topography. They tend to run along the grain of the country rather than across it, they go around swamps rather than through them, they tend to run along vegetational ecotones, and they create their own environmental conditions, some of which attract animals while others repel them.

13.5 Sampled counts: methods and arithmetic

Sampled counts of animals fall easily into two categories. There is first the method of counting on sampling units whose boundaries are fixed. We might for example walk lines and count deer on the area within 100 m each side of the line of march.

Or we might count all ducks on a sample of ponds, the shoreline of the pond providing a strict boundary to the sampling unit.

The alternative to fixed boundaries is unbounded sampling units (Buckland *et al.* 1993, 2001). Instead of restricting the counting to those animals within 100 m of a line of march, those outside the transect being ignored, we might count all the animals that we see. Since the observed density will fall away with distance from the observer, the raw counts are no longer an estimate of true density. They must therefore be corrected.

Of these two options (sampling units with boundaries and sampling units without boundaries) the first has immense advantages of simplicity and realism. If the transect width is appropriately chosen, what the observer sees is what the observer gets. The mathematics of such sampling are simple, elegant, and absolutely solid. In contrast, the accuracy of corrected density estimated from unbounded transects depends heavily upon which model is chosen for the analysis. There are many to choose from and they give markedly different answers for the same data. The advantage of unbounded transects is in all the sightings being used, none being discarded. Since the precision of an estimate is related tightly to the number of animals actually counted, any sampling scheme increasing the number of sightings also tends to increase the precision of the estimate. That is an advantage if the increased precision is obtained without the sacrifice of too much accuracy.

The choice of one or other system is often determined by density. If the species is rare then one might be tempted to use all the data one can get. If it is common one might be content to use the more dependable sampling units with fixed boundaries, knowing that fewer things can go wrong.

13.5.1 *Fixed boundaries to sampling units*

The appropriate analysis depends on whether the sampling units are of equal or unequal size, and how they are selected. Formulae were originally developed by Jolly 1969 (see also Norton-Griffiths 1978) based on Cochran (1977).

Notation

y = the number of animals on a given sampled unit
a = the area of a given sampled unit
A = the total area of the region being surveyed
n = the number of units sampled
D (or d) = the estimate of mean density
$SE(D)$ = the standard error of estimated mean density
Y = the estimate of total numbers in the region of size A
$SE(Y)$ = the standard error of the estimate of total numbers

The simple estimate (for equal-sized sampling units)

The simple estimate is used when sampling units are of constant size, as when the region being surveyed is a rectangle which can be subdivided into quadrats or transects. It will provide an unbiased, although imprecise, estimate, even when sampling units differ in size, but more appropriate designs are available for that case. We will explore this design at some length because most of the principles are shared with the other designs.

The region to be surveyed, of area A, is divided on a map or in one's head into an exhaustive set of non-overlapping sampling units, each of constant area a. Let us assume, for illustration, that the region is as given in Table 13.1, and that this region of $A = 144$ km^2 is to be sampled by $n = 4$ transects each of area $a = 12$ km^2. Sampling intensity is hence $na/A = 4 \times 12/144 = 0.333$.

In Table 13.1 the rows represent transects and the marginal totals the number of animals on each transect. Numbering the transects from 1 to 12 and selecting at random with replacement from this set we draw transect numbers 4, 8, 1, and 4. On surveying these transects we would obtain counts of

Transect:	1	4	4	8
Count:	142	149	149	127

Note that transect number 4 has been drawn twice, so in practice we survey only three transects although the count from transect number 4 enters the calculation twice.

Density is estimated as the sum of the transect counts $(142 + 149 + 149 + 127)$ divided by the sum of the transect areas $(12 + 12 + 12 + 12)$. Thus:

$$D = \Sigma y / \Sigma a = 567/48 = 11.81/\text{km}^2$$

The precision of that estimate is indexed by its standard error SE(D), which is itself an estimate of what the standard deviation of many independent estimates of density would be, each estimate derived from four transects drawn at random with replacement:

$$\text{SE}(D) = 1/a \times \sqrt{[(\Sigma y^2 - (\Sigma y)^2/n)/(n(n-1))]}$$

That is a slight approximation. To be exactly unbiased it should be multiplied by a further term $\sqrt{[1 - (\Sigma a)/A]}$, but that usually makes so little difference that it tends to be ignored.

The calculation tells us that this hypothetical distribution of estimates, each of them made in the same way as we made ours, with the same sampling frame and the same sampling intensity, only the draw of sampling units being different, would have a standard deviation in the vicinity of ± 0.43. In fact, that is likely to be an underestimate because it is based on only four sampling units, three degrees of freedom. With samples above 30 sampling units we can form 95% confidence limits of the estimate by multiplying by 1.96, but for smaller samples we must choose a multiplier from a Student's t-table corresponding to a two-tailed probability of 0.05 and the degrees of freedom (d.f.) of our sample. In the case of d.f. = 3, the multiplier is 3.182 and so the 95% confidence limits of our estimate of density are $\pm 3.18 \times 0.43 = \pm 1.37$.

The number of animals (Y) in the surveyed region can now be calculated as the number of square kilometers in that region (A) multiplied by the estimated mean number per square kilometer (D):

$$Y = AD = 144 \times 11.81 = 1701$$

It has a standard error of:

$$\text{SE}(Y) = \pm A \times \text{SE}(D) = \pm 144 \times 0.43 = \pm 62$$

Its 95% confidence limits are calculated as A multiplied by the 95% confidence limits of D:

$$\pm 144 \times 1.37 = \pm 197$$

We can check that against Table 13.2, which shows that the true total number (Y) is 1737 and so the estimate with 95% confidence of $Y = 1701 \pm 197$ is entirely acceptable.

If the sampling is without replacement the above formula for SE(D) yields an overestimate. The standard error for sampling without replacement is estimated by the formulation for the standard error with replacement multiplied by the square root of the proportion of the area not surveyed. This **finite population correction** or FPC is:

$$\text{FPC} = \sqrt{[1 - (\Sigma a)/A]}$$

The simple estimate may validly be used even when sampling units are of unequal size. The constant a is then replaced by the mean area of sampling units. The precision of the estimate will be lower (i.e. the standard error will be higher) than by the ratio method (see next subsection), but the estimate is unbiased and may be precise enough for many purposes.

The simple estimate, with minor modification, can be used when the total area A is unknown. One of us was forced to this exigency while surveying from the air a population of rusa deer (*Cervus timorensis*) in Papua New Guinea. The deer lived on a grassed plain, the area of which could not be gauged with any accuracy from the available map. The remedy was to measure the length of the plain by timing the aircraft along it at constant speed, and then to run transects from one side of the plain to the other at right angles to that measured baseline. The area of a sampling unit is entered as $a = 1$, even though they are of different and unknown areas. D then comes out as average numbers per transect rather than per unit area. Total numbers Y on the plain could then be estimated by replacing A by N, where N is the total number of transects that could have been fitted into the area. That is simply the length of the baseline divided by the width of a single transect. A similar approach was used for censusing wildebeest in the Serengeti (Norton-Griffiths 1973, 1978).

The ratio estimate (for unequal-sized sampling units)
This is the best method for a frame of sampling units of unequal size, as might be provided by a faunal reserve of irregular shape sampled by transects. Statistical texts warn that the estimate is biased when the number of units sampled is less than 30 or so, but the bias is usually so slight as to be of little practical importance. The number of units may be as low as two without generating a bias of more than a few percent.

The appropriate formulae are given in Table 13.3 and the notation at the beginning of Section 13.5.1. That for the standard error looks quite different from that of the simple estimate, but they are mathematical identities when the sampling units are of equal size. The ratio estimate is general, the simple estimate being a special case of it. Hence if these analyses are to be programmed into a calculator or computer, the ratio method is the only one needed.

Table 13.3 Estimates and their standard errors for animals counted on transects, quadrats, or sections. The models are described in the text.

Model	Density	Numbers
Simple		
Estimate	$D = \Sigma y/\Sigma a$	$Y = A \times D$
Standard error of estimate (SWR)	$SE(D)_1 = 1/a \times \sqrt{[(\Sigma y^2 - (\Sigma y)^2/n)/(n(n-1))]}$	$SE(Y) = A \times SE(D)_1$
Standard error of estimate (SWOR)	$SE(D)_2 = SE(D)_1 \times \sqrt{[1 - (\Sigma a)/A]}$	$SE(Y) = A \times SE(D)_2$
Ratio		
Estimate	$D = \Sigma y/\Sigma a$	$Y = A \times D$
Standard error of estimate (SWR)	$SE(D)_3 = n/\Sigma a \times \sqrt{[(1/n(n-1))(\Sigma y^2 + D^2\Sigma a^2 - 2D\Sigma ay)]}$	$SE(Y) = A \times SE(D)_3$
Standard error of estimate (SWOR)	$SE(D)_4 = SE(D)_3 \times \sqrt{[1 - (\Sigma a)/A]}$	$SE(Y) = A \times SE(D)_4$
PPS		
Estimate	$d = 1/n \times \Sigma(y/a)$	$Y = A \times d$
Standard error of estimate (SWR)	$SE(D) = \sqrt{[(\Sigma(y/a)^2 - (\Sigma(y/a))^2/n)/(n(n-1))]}$	$SE(Y) = A \times SE(d)$

SWR, sampling with replacement; SWOR, sampling without replacement. Notation is given in Section 13.5.1.

The probability-proportional-to-size (PPS) estimate

By the previous two methods all sampling units in the frame have an equal chance of being selected. By the PPS (probability-proportional-to-size) method the probability of selection is proportional to the size of the sampling unit. Suppose that the area to be surveyed is farmland. We might decide to declare the paddocks (or "pastures" or "fields," depending on which country you are in) as sampling units because the fences provide easily identified boundaries to those units.

If each sampling unit were assigned a number and the sample chosen by lot, we would use the ratio method of analysis. However, we might decide instead to choose the sample by throwing random points onto a map. Each strike selects a unit to be sampled, the probability of selection increasing with the size of the unit.

The PPS estimate has the advantages that it is entirely unbiased and that the arithmetic (Table 13.3) is simple. Its disadvantage is that it can be used only when sampling with replacement and so it is not as precise as the ratio method used without replacement. Hence this method should be restricted to surveys whose sampling intensity is less than 15%. The PPS estimate is a mathematical identity of the simple estimate and the ratio estimate when units of equal size are sampled with replacement.

13.5.2 Unbounded transects (line transects)

The observer walks a line of specified length and counts all animals seen, measuring one or more subsidiary variables at each sighting (angle between the animal and the line of march; radial distance, the distance between the animal and the observer at the moment of sighting; the right-angle distance between the animal and the transect). If we know the shape of the sightability curve relating the probability of seeing an animal on the one hand to its right-angle distance from the line on the other, and if an animal standing on the line will be seen with certainty, it is fairly easy to derive an estimate of density from the number seen and their radial or right-angle distances. We seek a distance from the line where the number of animals missed within that distance equals the number seen beyond it. True density is then the total seen divided by the product of twice that distance and the length of the line.

Therein lies the difficulty. That distance is determined by the shape of the sightability curve, which can seldom be judged from the data themselves. Consequently the shape of the curve must be assumed to some extent and the validity of the assumption determines the accuracy of the method.

We present only two of the many models available, mainly to give some idea of their diversity. The first is the Hayne (1949) estimate, which is derived from the assumption that the surveyed animals have a fixed flushing distance and will be detected only when the observer crosses that threshold. If k is the number of animals detected and r the radial distance from a detected animal to the observer:

$$D = (1/2L)\sum_k(1/r)$$

where L is the length of the line. Hence density is the sum of the reciprocals of the radial sighting distances divided by twice the length of the line.

It is implicit in Hayne's model that sin θ, the sine of the sighting angle, is uniformly distributed between 0 and 1, and that the theoretically expected mean sighting angle is 32.7°. Hence the reality of the model can be tested against the data. Eberhardt (1978) recommended tabulating the frequency of sin θ in 10 intervals of 0.1 (0.0–0.1, 0.1–0.2 . . . 0.9–1.0) and testing the uniformity of the frequencies by chi-square. He gave a worked example for a survey of the side-blotched lizard (*Uta stansburiana*). Robinette et al. (1974) and Burnham et al. (1980) suggested that most mean sighting distances tended to be around 40° or more, the latter authors being convinced that the Hayne estimate is used far too uncritically in wildlife management. Robinette et al. (1974) compared the accuracy of the Hayne estimate with that of eight other line-transect models, showing that when applied to inanimate objects or to elephants it tended to overestimate considerably. However, Pelletier and Krebs (1997) found that both the Hayne estimate and line-transect estimates provided relatively unbiased results when compared with a known population of ptarmigan (*Lagopus* species) in Yukon. Buckland et al. (1993) provide a starting point for reading further about line-transect methods.

Our second example is a non-parametric method developed by Eberhardt (1978) from work by Cox (1969). First, we choose arbitrarily a distance, Δ, perpendicular from the line. Eberhardt's estimate of density is:

$$D = (3k_1 - k_2)/4L\Delta$$

where k_1 and k_2 are the number of animals seen on either side of the line transect at distances that fall within the interval 0 to Δ and Δ to 2Δ, respectively. Eberhardt (1978) considered that the method is most useful as only a cross-check on the results of other methods because its estimate is likely to be imprecise. Precision is enhanced by choosing a large value of Δ but accuracy is enhanced by choosing a small Δ (Seber 1982).

Much of the present use of line transects in wildlife management stems from the belief that they are somehow more "scientific" than strip transects, just as there was once a belief that quadrats are statistically superior to transects. There are rare situations in which transect sampling will not work and where line-transect methodology might (e.g. in very thick cover). The unbounded line-transect method has advanced considerably with the use of the computer software DISTANCE (http://www.ruwpa.st-and.ac.uk/distance) developed by Buckland et al. (1993, 2001). Although the software is not easy to use, it is currently the most powerful tool for line censuses. In particular, it is most useful for rare observations, although it does require at least 30 records to be reliable. In addition, there must be time to make the necessary estimates of perpendicular distance from the line to the animal (or groups of animals). If there are insufficient observations of a particular species (or other category) in a station or habitat, then one can repeat the line survey until sufficient observations have been accumulated. The only proviso is that animals distribute themselves randomly with respect to the line and there is no spatial correlation between surveys. The method is particularly suitable for rare species such as carnivores and rare ungulates and birds. It is not suitable where there are large numbers of animals, for example ungulates on the Serengeti plains.

Note that none of these unbounded methods can be used in aerial survey. They are all anchored by the assumption that all animals on the line of march (equivalent to the inner strip marker of aerial survey) are tallied by the observer. That assumption does not hold for aerial survey because the ground under the inner strip marker is at a distance from the observer, because an animal under a tree on that line may be missed, and because an observer cannot watch all parts of the strip at once and may therefore miss animals in full view on that line. In addition, the speed of the aircraft makes the measurements of distances from the observer unfeasible.

The assumption that all animals on the line are counted can be relaxed if the probability of detecting animals on the line can be estimated. This is particularly important for marine mammals, where only a fraction of a group or pod are on the surface at any one time. The probability of detection on the line for harbor porpoises (*Phocoena phocoena*) was estimated to be only 0.292, which illustrates just how many remain unseen. Furthermore, this estimate was made by experienced observers; for inexperienced observers the sighting probability was only 0.079, that is, some 90% were missed. This shows the importance of training and experience (Laake *et al.* 1997).

The biologist must decide whether the statistical power of line transects justifies their practical application. Can the difficulty of measuring sighting distances and the unreliability of the resultant estimates be justified when an alternative with fewer problems is available? The line transect was originally introduced to circumvent the difficulty of counting all animals on a transect or quadrat. It cured that problem by replacing it with several new ones. Perhaps we should give some thought to ways of treating the original problem without introducing new ones. If animals are difficult to see on a transect of fixed width, why not walk two people abreast down the boundaries? If that does not work, put a third person between them. And so on.

13.5.3 *Stratification*

The precision of an estimate is determined by sampling intensity and by the variability of density among sampling units. Suppose there were two distinct habitats in the survey area and that, from our knowledge of the species, we could be sure that it would occur commonly in one and rarely in the other. If we surveyed those two subareas separately and estimated a separate total of animals for each, the combined estimate for the whole area would be appreciably more precise than if the area had been treated as an undifferentiated whole.

The process is called **stratification** and the subareas **strata**. By this strategy we divide an area of uneven density into two or more strata within which density is much more even. The strata are treated as if they were each a total area of survey, the results subsequently being combined. The estimate from each stratum will be called Y_h which has a standard error of SE(Y_h). Total numbers Y are estimated by $Y = \sum Y_h$. Its standard error is the square root of the sum of the variances of the contributing stratal estimates. The variance of an estimate is the square of its standard error. Here it is designated Var(est) to distinguish it from the variance of a sample designated s^2. Calculate $Var(Y_h) = [SE(Y_h)]^2$ for each stratum and then:

$$SE(Y) = \sqrt{\sum Var(Y_h)}$$

to give the standard error of the combined estimate of total numbers.

Optimum allocation of sampling effort

If our aim is to get the most precise estimate of Y as opposed to a precise estimate of each Y_h, sampling intensity should be allocated between strata according to the expected standard deviation of sampled unit counts in each stratum. That requires a pilot survey or at least approximate knowledge of distribution and density gained on a previous survey. Often we have nothing more than aerial photographs or a vegetation map to give us some idea of the distribution of habitat, and only a knowledge of the animal's ecology to guide us in predicting which habitats will hold many animals and which will hold few. This scant information in fact is sufficient to allow an allocation of sampling effort between strata that will not be too far off the optimum. The important point to understand is that for almost all populations the standard deviation of counts on sampling units rises linearly with density. From that can be derived the rule of thumb that the number of sampling units put into a stratum should be directly proportional to what Y_h is likely to be.

At first thought that is a daunting challenge – to guess each Y_h before we have estimated it – but it is easier if we break it down into components. First, guess the density in each stratum. It does not matter too much if this is wrong, even badly wrong, because all we need to get roughly right is the ratios of densities between strata. Second, multiply each guessed density by the mapped area of its stratum to give a guess at numbers in the stratum. Third, divide each by the total area to give the proportion of total sampling effort that should be allocated to each stratum. Table 13.4 shows the calculation for a degree block that can be divided into three strata from a vegetation map and to which a total of 10 hours of aerial survey has been allocated.

13.5.4 Comparing estimates

If the sampling units are drawn independently of each other, the estimates of density from two surveys may be compared. The surveys may be of two areas, or of the same area in two different years, or the same area surveyed in the same year by two teams

Table 13.4 Allocation of $E = 10$ hours of aerial survey among strata to maximize the precision of the estimate of animals in the total area.

Stratum (h)	Area (km^2) (A_h)	Guessed density (D_h)	Guessed numbers ($Y_h = A_h D_h$)	Proportion of total effort ($P_h = Y_h/\sum Y_h$)	Hours allocated ($E_h = P_h E$)
1	2,000	1	2,000	0.03	0.3
2	7,000	5	35,000	0.52	5.2
3	3,000	10	30,000	0.45	4.5
	12,000		67,000	1.00	10.0

or by different methods. A quick and dirty comparison is provided by the normal approximation, which is adequate if each survey covered more than 30 sampling units. The two estimates are significantly different when:

$$(\text{est}_1 - \text{est}_2)\sqrt{[\text{Var}(\text{est}_1) + \text{Var}(\text{est}_2)]} > 1.96$$

If sample sizes are too low, or if more than two surveys are being compared, the determination of significance should be made by one-factor analysis of variance. If the surveys are not independent, as when the same transects are run each year, a comparison may still be made by analysis of variance but with TRANSECTS now declared a factor in a two-factor analysis. Chapter 16 goes further into this and other uses of analysis of variance.

13.5.5 *Merging estimates*

If a comparison shows that two or more independent estimates of the same population are not significantly different, we may wish to merge them to provide an estimate more precise than the individual estimates. This is a procedure quite distinct from stratification where estimates from different populations are combined to give an overall estimate. Merging is restricted to the same population estimated more than once. We must make sure that environmental (e.g. different seasons) and biological (e.g. significant mortality or emigration) conditions do not differ between censuses. Merging is particularly powerful in obtaining a reduced confidence interval from a series of individual censuses each with very wide confidence intervals. If one obtains a single estimate with a wide confidence interval (say because too few samples were counted) then it will often pay to repeat the census as soon as possible and merge the two results.

There are two methods. First, there is a quick and dirty method, to be used only when the individual estimates were each made with about the same sampling intensity. The merged estimate $\hat{Y}$ can then be calculated as:

$$\hat{Y} = (Y_1 + Y_2 + Y_3 + \ldots + Y_N)/N$$

where there are N surveys. It has a variance of:

$$\text{Var}(\hat{Y}) = [\text{Var}(Y_1) + \text{Var}(Y_2) + \text{Var}(Y_3) + \ldots + \text{Var}(Y_N)]/N^2$$

Thus the merged estimate is simply the mean of the individual estimates, and its variance is the mean of the individual-estimate variances divided by their number. $\text{SE}(\hat{Y})$ is the square root of $\text{Var}(\hat{Y})$. From these the merged density estimate is $D = \hat{Y}/A$ which has a standard error of $\text{SE}(D) = \text{SE}(\hat{Y})/A$.

Second, a more appropriate method, particularly for surveys utilizing markedly different intensities of sampling, is provided by Cochran (1954), who also considers more complex merging. Here the contribution of an individual estimate to the merged estimate is weighted according to its precision. Letting $w = 1/\text{Var}(\hat{Y})$:

$$\hat{Y} = (w_1Y_1 + w_2Y_2 + w_3Y_3 + \ldots + w_NY_N)/(w_1 + w_2 + w_3 + \ldots + w_N)$$

with a variance of:

$$\text{Var}(\hat{Y}) = 1/(w_1 + w_2 + w_3 + \ldots + w_N)$$

13.6 Indirect estimates of population size

This section outlines some of the methods available for calculating the size of a population by techniques that do not necessarily depend on accurate counts of animals. The line-transect method could well come under this head but it is placed in "sampled counts" because it requires accurate counting of animals on the line.

13.6.1 Index-manipulation-index method

If we obtain two indices of population size, I_1 and I_2, the first before and the second after a known number of animals C was removed, the population's size can be estimated for the time of the first index by:

$$Y_1 = I_1 C/(I_1 - I_2)$$

The proportion removed is estimated as $p^* = (I_1 - I_2)/I_1$ and the proportion of those remaining as $q^* = 1 - p^*$. Following Eberhardt (1982), the variance of the estimate of Y can be approximated by:

$$\text{Var}(Y_1) \approx Y_1^2 (q^*/p^*)^2 (1/I_1 + 1/I_2)$$

from which $\text{SE}(Y_1) = \sqrt{\text{Var}(Y_1)}$. Eberhardt (1982) gives three examples from populations of feral horses. The data from his Cold Springs population were:

$$I_1 = 301$$

$$I_2 = 76$$

$$C = 357$$

$$p^* = 0.748$$

Thus, the population at the time of the first index is estimated as:

$$Y_1 = (301 \times 357)/(301 - 76) = 478$$

with a variance of that estimate of:

$$\text{Var}(Y_1) \approx 478^2 (0.252/0.748)^2 (1/301 + 1/76) = 428$$

from which $\text{SE}(Y_1) = \sqrt{428} = 21$.

The index-manipulation-index method assumes that the population is closed (no births, deaths, immigration, or emigration) between the estimation of the first and second indices. That assumption is approximated when the entire experiment is run over a short period.

13.6.2 Change-of-ratio method

If a population can be divided into two classes, say males and females or juveniles and adults, and one class is significantly reduced or increased by a known number of animals, the size of the population can be estimated from the change in ratio. Kelker (1940, 1944) introduced this method to estimate the size of deer populations manipulated by bucks-only hunting.

The two classes are designated x and y. Before the manipulation there was a proportion p_1 of x-individuals in the population, and p_2 after the manipulation which removed or added C_x x-individuals (additions are positive, removals negative) and C_y y-individuals: $C = C_x + C_y$. The size of the population before the manipulation may be estimated as:

$$Y_1 = (C_x - p_2C)/(p_2 - p_1)$$

As with the index-manipulation-index method, Kelker's method assumes that the population is closed. Hence the two surveys to estimate the class proportions must be run close together. Additionally, all removals or additions must be recorded and the two classes must be amenable equally to survey.

Cooper *et al.* (2003) have extended this approach using likelihood estimates of the ratios. When harvesting is highly skewed towards a single sex or age class, the change in these ratios provides information about the exploitation rate, and, when combined with absolute numbers removed, also provides information on absolute abundance.

13.6.3 Mark–recapture

Mark–recapture is a special case of the change-of-ratio method. A sample of the population is marked and released, a subsequent sample being taken to estimate the ratio of marked to unmarked animals in the population. From data of this kind we can estimate the size of the population, and with further elaboration (individual markings, multiple recapturing occasions) the rate of gain and loss.

The huge number of mark–recapture models available have been reviewed adequately by Blower *et al.* (1981) and in detail by Seber (1982) and Krebs (1999). Bowden and Kufeld (1995) present methods of estimating confidence limits for general mark–resighting calculations, using the example of Colorado moose (*Alces alces*). Here we outline the range of methods, provide an introduction to the most simple of these, emphasize their pitfalls, and mention some of the recent advances which might circumvent those pitfalls.

Petersen–Lincoln models

A sample of M animals are marked and released. A subsequent sample of n animals are captured of which m are found to be marked. If Y is the unknown size of the population then clearly:

$$M/Y = m/n$$

within the limits of sampling variation. With rearranging, that equation allows an estimate of populations size as:

$$Y = Mn/m$$

Intuitively obvious as that is, it is not quite right because of a statistical property of ratios that leads on average to a slight overestimation. The bias may be corrected (Bailey 1951, 1952) by:

$$Y = [M(n + 1)]/(m + 1)$$

which has a standard error of approximately:

$$SE(Y) = \sqrt{[(M^2(n + 1)(n - m))/((m + 1)^2(m + 2))]}$$

Fig. 13.1 Simulated replications of estimates of a population of 500 individuals by mark–recapture where 100 are marked and 50 captured. Note the positive skew of estimates and the fact that only a limited number of estimated values are possible.

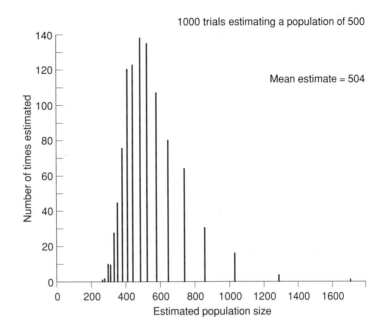

Fig. 13.1 Simulated replications of estimates of a population of 500 individuals by mark–recapture where 100 are marked and 50 captured. Note the positive skew of estimates and the fact that only a limited number of estimated values are possible.

These formulae are for "direct sampling," when the number of animals to be recaptured is not decided upon before recapturing. There are further variants for sampling with replacement and for inverse sampling (see Seber 1982).

Except for the unlikely case of half or more of the population being marked, the distribution of repeated independent estimates of population size is always strongly skewed to the right, a positive skew. (The direction of skew is the direction of the longest tail.) Figure 13.1 shows this effect from a computer simulation of 1000 estimates of a population of 500 animals containing 100 marked individuals. Each estimate is derived from a capturing of 50 animals. Apart from demonstrating the skew of estimates, the figure makes the point that only a limited number of estimated values is possible. With $Y = 500$ and $M = 100$ the probability of a given animal being marked is 0.2, and so the expected number of marked animals in a sample of 50 is 10. That would give a population estimate of $Y = 464$. If nine were recaptured the estimate would be $Y = 510$. No estimate between 464 and 510 is possible.

Since the estimates are skewed, the confidence limits of an estimate are also skewed and cannot easily be calculated from the standard error. Blower *et al.* (1981) recommended an approximating procedure. Let $a = m/n$. In a large sample the 95% confidence limits of a are approximately $\pm 1.96\sqrt{[a(1 - a)/n]}$. Since $Y = M/a$ the upper and lower 95% confidence limit of a can each be divided into M to give upper and lower 95% confidence limits of Y.

The Petersen estimate is the most simple of a family of estimation procedures. If animals are marked on more than one occasion and recaptured on more than one occasion, it is possible to estimate gains and losses from the population as well as its size. Seber (1982) describes most of the options.

The Petersen estimate depends on these assumptions:
1 all animals are equally catchable;
2 no animal is born or migrates into the population between marking and recapturing;

3 marked and unmarked animals die or leave the area at the same rate;
4 no marks are lost.

Assumption (2) is not needed when marked animals are recaptured on more than
one occasion, but the others are common to all elaborations of the Petersen estimate.
The least realistic is the assumption of equal catchability, which is routinely violated
by almost any population the wildlife manager is called upon to estimate (Eberhardt
1969). For this reason the Petersen estimate and its elaborations (Bailey's triple catch,
Schnabel's estimate, the Jolly–Seber estimate, and many others) are of limited utility
in wildlife management.

Frequency-of-capture models

Petersen models work only when all animals in the population are equally catchable.
Frequency-of-capture models are not constrained in that way but will work only if
the population is closed, if there are no losses from or gains to the population over
the interval of the experiment. That is easy enough to approximate by running the
exercise over a short period.

Animals are captured on a number of occasions, usually on successive nights, and
marked individually at the first capture. At the end of the experiment each individual caught at least once can be scored according to the number of times it was
captured. The data come in the form:

Number of times caught (i):	1	2	3	4	5	6	7	8	...	18
Number of animals (f_i):	43	16	8	6	0	2	1	0	...	0

which are from Edwards and Eberhardt (1967) who trapped a penned population of
wild cottontail rabbits for 18 days. Of these, 43 were caught once only, 16 twice,
eight three times, and so on. $\Sigma f_i = 76$ gives the number of rabbits caught at least
once and so the population must be at least that large. If we could estimate f_0, the
number of rabbits never caught, we would have an estimate of population size:

$$Y = f_0 + 76$$

Traditionally this has been attempted by fitting a zero-truncated statistical distribution (Poisson, geometric, negative binomial) to the data and thereby estimating the
unknown zero frequency. Eberhardt (1969) exemplifies this approach. More complex mark–recapture models use sophisticated analytical techniques to cope with
variation in the probability of capture due to time (seasonal trends, changes in weather),
variation among individual animals (site fidelity, sex differences, dominance relationships), prior trapping history (capture-shyness and capture-proneness), and various
combinations of these (Pollock 1974; Burnham and Overton 1978; Otis et al. 1978).
The fit of each model can be tested against the data and an objective decision made
as to which is the most appropriate model, often using information theory (see Chapter
15). The computations are too lengthy to be attempted by hand but several software
programs are freely available on the web: CAPTURE (White et al. 1982), SURGE
(Lebreton et al. 1992), and MARK (White and Burnham 1999).

Estimation of density

All previously reviewed mark–recapture methods yield a population size Y that
can be converted to a density D only when the area A relating to Y is known. In

most studies Y itself is meaningless because the "population" is not a population in the biological sense but the animals living on and drawn to a trap grid of arbitrary size.

Seber (1982) and Anderson *et al.* (1983) reviewed the methods currently used to estimate A as a prelude to determining density. Most rely on Dice's (1938) notion of a boundary strip around the trapping grid such that the effective trapping area A is the grid area plus the area of the boundary strip. Most of these methods are *ad hoc* and subject to numerous problems, or require large quantities of data to produce satisfactory estimates, or require supplementary trapping beyond the trapping grid.

Anderson *et al.* (1983) circumvented this problem with a method of mark–recapture that provides a direct estimate of density. The traps are laid out not in a grid but at equal intervals along the spokes of a wheel. Trap density therefore falls away progressively from the center of the web. The method pivots upon the assumption that the high density of traps at the center guarantees that all animals at the center will be captured. This is analogous to the assumption of line-transect methodology that all animals are tallied on the line itself. The data collected as "distance of first capture from the center of the web" are analysed almost exactly as if they were from a line transect (Buckland *et al.* 1993, 2001). This analysis can be run on the computer program DISTANCE (Laake *et al.* 1993).

13.6.4 *Incomplete counts*

The problem of estimating the size of a population from "total counts" known to be inaccurate has been approached from three directions. One family of methods requires a set of replicate estimates, the second requires two estimates, and the third provides an estimate known with confidence to be below true population size.

Many counts

Hanson's (1967) method assumes that all animals have the same probability of being seen but that this probability is less than one. Hence, whether a given animal is seen or not on a given survey is a draw from a binomial distribution. It follows from the mathematics of the binomial distribution that $Y = \bar{x}^2/(\bar{x} - s^2)$, where Y is the population size, $\bar{x}$ the mean of a set of (incomplete) counts, and s^2 the variance of those counts.

This method is not recommended because of the restriction that all animals have the same sightability. In practice sightability varies by individuals and between surveys. The variance of a set of replicate counts tends to be greater than their mean (a binomial variance is always less than the mean), indicating that the method is unworkable.

A modification of this method to circumvent that restriction was suggested by Caughley and Goddard (1972). It requires repeated counts made at two levels of survey efficiency (e.g. two sets of aerial surveys, one flown at 50 m and the other at 100 m altitude). However, Routledge (1981) showed by simulation that this method yields a very imprecise estimate unless the number of surveys is prohibitively large, and hence we do not recommend it.

The non-parametric **method of bounded counts** (Robson and Whitlock 1964) provides a population estimate from a set of replicate counts as twice the largest minus the second largest. Routledge (1982) dismissed this method also (as do we) because in most circumstances it greatly underestimates the true number.

Two counts

Caughley (1974) showed that if the counts of two observers of equivalent efficiency were divided into those animals (or groups of animals) seen by only one observer and those seen by both, the size of the population could be estimated. Henny *et al.* (1977) and Magnusson *et al.* (1978) extended the method to allow for the two observers being of disparate efficiency.

Essentially the method is a Petersen estimate, although animals are neither marked nor captured. Suppose that the entities being surveyed are stationary and that their individual positions can be mapped. Magnusson *et al.* (1978) surveyed crocodile nests and Henny *et al.* (1977) the nests of ospreys. If the area is surveyed independently twice, perhaps once from the ground and once from the air, the entities can be divided into four categories:

S_1 = the number seen on the first survey but missed on the second
S_2 = the number seen on the second survey but missed on the first
B = the number tallied by both surveys
M = the number missed on both

This is equivalent to a mark–recapture exercise. The first survey maps (marks) a set of entities, each of which may or may not be seen (recaptured) on the second survey. But unlike a true mark–recapture exercise the model is symmetrical and the first and second surveys are interchangeable.

If P_1 is the probability of an entity being seen on the first survey and P_2 the probability of its being seen on the second survey:

$$P_1 = B/(B + S_2)$$

$$P_2 = B/(B + S_1)$$

$$M = S_1 S_2/B$$

$$Y = [(B + S_1)(B + S_2)]/B$$

where Y is an estimate of the size of the population. The last equation may be corrected for statistical bias (Chapman 1951) to:

$$Y = [((B + S_1 + 1)(B + S_2 + 1))/(B + 1)] - 1$$

which has a variance given by Seber (1982) of:

$$\text{Var}(Y) = [S_1 S_2 (B + S_1 + 1)(B + S_2 + 1)]/[(B + 1)^2 (B + 2)]$$

Magnusson *et al.* (1978) reported that, although the method is based on the assumptions that the two surveys are uncolluded and that there is a constant probability of seeing an entity on a given survey (equal catchability), the second assumption is not critical. The population estimate is close enough even when the probability of being seen varies greatly between individuals.

Caughley and Grice (1982) extended the method to moving targets, dropping the requirement that the position of stationary entities must be mapped so that they could be identified as seen or not seen at the two surveys. Groups of emus (*Dromaius novaehollandiae*) were tallied simultaneously but independently by two observers seated in tandem on one side of an aircraft. Their counts of $S_1 = 7$, $S_2 = 3$, and $B = 10$ yielded $P_1 = 0.77$ and $P_2 = 0.59$, the population estimate being $Y = 22$ emu groups on the 843 km^2 of transects that they surveyed together, a density of 0.03 groups/km^2.

This method of simultaneous but uncolluded tallying carries two dangers, one technical, the other statistical. The two observers must not unconsciously cue each other to the presence of animals in their field of view and ideally should be screened from each other. Second, the chances of "marking" and "recapturing" an entity should be uncorrelated, but they are not because marking and recapturing occur at the same instant, the search images transmitted to each observer being nearly identical. Caughley and Grice (1982) showed by simulation that the effect of the close correlation was to underestimate density but that the underestimation became serious only when the mean of P_1 and P_2 was less than 0.5.

Known-to-be-alive

Most estimates of population size require that the manager makes a leap of faith. There is seldom any certainty that the population fits the assumptions of the model, nor whether the estimate is wildly inaccurate, nor whether the confidence limits have much to do with reality. The more complex is the model the greater is the uncertainty. Many ecologists, particularly those working on small mammals, have decided that the work needed to achieve an unbiased estimate is not worth the effort. They would prefer an estimate which, although perhaps inaccurate, is inaccurate in a predictable direction and which does not depend on a set of assumptions of dubious reliability. Hence the **known-to-be-alive** estimate, the number of animals that the researcher knows with certainty to be in the study area. These estimates for small mammal populations are usually made by trapping an area at high intensity over a short period. Each animal is marked at first capture, the estimated population size being simply the number of first captures. Such estimates are acknowledged as underestimates but they have the advantage of yielding a real number, not an abstract concept, to work with.

Known-to-be-alive estimates are often the most appropriate in wildlife management. There are several problems of conservation and of harvesting for which an overestimate of density may lead to inappropriate management action. An underestimate, on the other hand, may simply produce inefficient but entirely safe management. The penalty for a poor estimate is often distributed asymmetrically around the true population size. It is not good to overestimate the number of individuals of an endangered species. It is not safe to apply a harvesting quota, known to be safe for the population size you estimated, to a population that is much smaller than you thought. Where the undesirable consequences of an overestimate are considerably greater than those accruing from an underestimate, the known-to-be-alive number is often the most appropriate estimate to work with.

13.7 Indices

An index of density is some attribute that changes in a predictable manner with changes in density. It may be the density of bird nests, or the density of tracks of brown bears, or the number of minke whales (*Balaenoptera acutorostrata*) seen per cruising hour. A common index is the pellet or fecal dropping count. This is often used in studies of deer. It was used for the endangered marsh rabbits (*Sylvilagus palustris*) in Florida, where pellet counts were closely correlated with radiotelemetry estimates (Forys and Humphrey 1997). Active burrow entrances were used for ground squirrel populations (van Horne *et al.* 1997), and call counts for mourning dove (*Zenaida macroura*) densities (Sauer *et al.* 1994). The North American Breeding Bird Survey is a standardized method in which some 2000 routes are sampled in June each year. The number of singing birds of each species is scored (Droege and Sauer 1989).

These indices would reveal something about the density of birds, mammals, or whales. Without knowing anything about the proportional relationship between the index and the abundance of the animal we could be confident that if the index halved or doubled it would reflect roughly a halving or doubling of animal density. Formally, that holds only when the relationship between index and density is a straight line that passes through the point of zero index and zero density.

Indices of density, if comparable, are useful for comparing the density of two populations or for tracking changes in the density of one population from year to year. Often a comparison is all we need. The relevant question may be not how large is the population but whether it has declined or increased under a particular regime of management. In such circumstances the accuracy of an index is irrelevant; precision is paramount.

Let us compare an aerial survey designed to yield an estimate of absolute density with one designed to yield an index of density, as was conducted for pronghorn antelope in Colorado (Pojar *et al.* 1995). The first maximizes accuracy, the second precision. The "accurate" survey would probably inspect small quadrats by circling at a low but varying height above the ground. That is a good way to see animals but it is a technique difficult to standardize between pilots. The "precise" survey would sample transects from a fixed height above ground at a constant speed. Since there is no requirement that all the animals be counted on the sampled units, only a fixed proportion being sought, the survey variables are set according to how easily they may be standardized. Ground speed is higher than for an "accurate" survey to allow the pilot to maintain constant ground speed safely even with a strong tail wind. Height above ground is set higher so that the inevitable variations in height will be proportionally less than at low level. Plus or minus 10 m around a height of 30 m results in large variations in search image. The same variation around 90 m has little effect. We might choose a transect width of 50 m per observer for an accurate survey but 200 m for a precise survey. The precision of the estimate is approximately proportional to the square root of the number of animals actually tallied (Eberhardt 1978) and so, although proportionally fewer will be seen on a 200 m strip, we choose the wider strip to increase the absolute number that we see.

Consistency and rigid standardization of techniques are crucial when estimating an index. A good observer is not one who gets a high tally but one who has a consistent level of concentration and who produces results of high repeatability.

All the rules of sampling and of analysis hold as well for indices as for absolute counts of animals. Remember, however, that indices are useful only in comparisons and, therefore, the quantity to be estimated is the difference between two indices. The variance of an estimate of difference is the sum of the variances of the two estimated indices. As a rule of thumb we should measure the two indices with a precision such that each standard error is less than a third of the difference we anticipate. Hence an index must often be estimated much more precisely than is a one-off estimate of population size or density.

Errors in indices can be estimated by comparing results with a known population similar to the way we estimate bias errors in counts (see Section 13.4.2) (Eberhardt and Simmons 1987). For example, the number of sightings of fallow deer (*Dama dama*) in France along a transect (the index) was calibrated against a known population. The sighting index was found to be an effective standardized method to detect trends in the population (Vincent *et al.* 1996).

13.7.1 *Known-to-be-alive used as an index*

Although known-to-be-alive is sometimes used as a one-off estimate of population size, it is more often used to track trends in population size. The operating rules governing these two uses are quite different. In the first exercise we seek the most accurate estimate we can get. In the second we seek consistency of method among several estimates such that their bias is held constant. In the first case we put in as much work as possible. In the second we put in precisely the same amount of sampling on each surveying or capturing occasion. Otherwise the trend in the estimates may reflect no more than variation in capturing effort.

A variant of this aberration of effort, very common in ecological research, is to boost the number known to be alive (because they were caught) on a given occasion by the number of individuals not caught on that occasion but which must have been there because they were caught on both previous and subsequent occasions. Although the accuracy of the estimate of absolute numbers is thereby enhanced, the consistency of the string of estimates is thereby lowered. Estimates for the earlier occasions are inflated relative to those of later occasions, the rate of increase being underestimated if density is rising and the rate of decrease being overestimated if density is falling.

13.8 **Summary**

Animal numbers can be estimated by total counts, sampled counts, mark–recapture, or various indirect methods. In each case the usefulness of the method is determined by how closely its underlying assumptions are matched by the realities of what the animals do and how difficult they are to see, trap, or detect. The range of methods provided should allow wildlife managers to choose one that will be adequate in any given circumstance.

14 Age and stage structure

14.1 Age-specific population models

All wildlife populations have individuals of different ages. The vital rates (i.e. birth rates and probabilities of survival and mortality) often vary with age. Hence, a population composed of old individuals might well exhibit a different potential for growth than does a younger population. Assessing these kinds of processes requires an age-specific model of population growth (Caswell 2001).

The standard technique is to use a Leslie matrix model, named in honor of the British ecologist who pioneered this approach (Leslie 1945, 1948). This involves multiplying age-specific population densities by a transition matrix (A). The top row in A reflects the probability of survival (p) from the previous age class multiplied by fecundity (m) at age x. The subdiagonal reflects the age-specific survival probabilities, p_0, p_1, etc.:

$$A = \begin{pmatrix} p_0 \cdot m_0 & p_1 \cdot m_1 & p_2 \cdot m_2 & p_3 \cdot m_3 \\ p_0 & 0.0 & 0.0 & 0.0 \\ 0.0 & p_1 & 0.0 & 0.0 \\ 0.0 & 0.0 & p_2 & p_3 \end{pmatrix}$$

Provided that we have an initial age distribution, we can apply the Leslie matrix to estimate the abundance of individuals in each age group in subsequent years. This is done by multiplying the matrix A by the age vector n:

$$n = \begin{pmatrix} 9.3 \\ 5.8 \\ 3.5 \\ 1.7 \end{pmatrix}$$

which gives the initial densities of each group, the youngest at the top.

A reminder in matrix algebra may be helpful here. Provided that an age vector has the same number of rows as the matrix has rows and columns, we can multiply them in the following manner. The first subscript refers to the row and the second subscript refers to the column:

$$n_{0,t+1} = A_{0,0} \cdot n_{0,t} + A_{0,1} \cdot n_{1,t} + A_{0,2} \cdot n_{2,t} + A_{0,3} \cdot n_{3,t}$$

$$n_{1,t+1} = A_{1,0} \cdot n_{0,t} + A_{1,1} \cdot n_{1,t} + A_{1,2} \cdot n_{2,t} + A_{1,3} \cdot n_{3,t}$$

and so forth for the other age groups 2 and 3. This would obviously be rather cumbersome to calculate by hand for very long. Fortunately, there is a simple way to automate the procedure, using matrix operations:

$$\mathbf{n}^{<t+1>} = \mathbf{A} \cdot \mathbf{n}^{<t>}$$

The "<>" notation used here refers to the age-specific abundance in year t. Hence, $\mathbf{n}^{<2>}$ stands for the abundances $n_{0,2}$, $n_{1,2}$, $n_{2,2}$, and $n_{3,2}$ present in year 2.

To add some flesh to these theoretical bones, we can supply some arbitrary values for the age-specific rates of survival (p) and fecundity (m):

$$\mathbf{p} = \begin{pmatrix} 0.5 \\ 0.9 \\ 0.5 \\ 0.1 \end{pmatrix} \qquad \mathbf{m} = \begin{pmatrix} 0.5 \\ 2.2 \\ 2.9 \\ 3.7 \end{pmatrix}$$

This leads to the age-specific totals shown in the matrix called $\mathbf{n}$. The first column of the matrix is the initial density of individuals in each age group, with newborns on the top row and old individuals on the bottom row. Each column thereafter represents densities in successively later years:

$$\mathbf{n} = \begin{pmatrix} 9.3 & 19.51 & 22.36 & 32.01 & 43.75 & 58.94 & 80.97 & 110.01 & 150.05 \\ 5.8 & 4.65 & 9.76 & 11.18 & 16.01 & 21.88 & 29.47 & 40.49 & 55.01 \\ 3.5 & 5.22 & 4.19 & 8.78 & 10.06 & 14.41 & 19.69 & 26.52 & 36.44 \\ 1.7 & 1.92 & 2.8 & 2.37 & 4.63 & 5.49 & 7.75 & 10.62 & 14.32 \end{pmatrix}$$

Model predictions of total population density, obtained by summing the densities of all age groups present at any point in time ($N_t = \Sigma n_{x,t}$), are shown in Fig. 14.1.

After an initial period of adjustment, the population settles into a pattern of geometric (i.e. exponential) population growth, whose finite rate of increase depends on the integrated combination of age-specific parameters. As the population settles into a geometrical growth pattern, the proportions of each age class in the population ($W_{x,t}$)

Fig. 14.1 Population growth over time predicted by the Leslie matrix model with constant survival and reproduction discussed in the text.

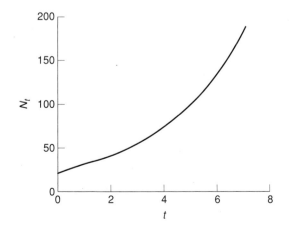

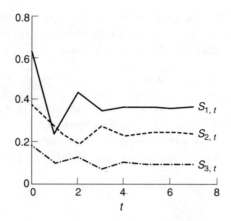

converge on a stable age distribution, as illustrated in Fig. 14.2. Each age-specific proportion $W_{x,t}$ is given by:

$$W_{x,t} = \frac{n_{x,t}}{N_t}$$

So, over time, we can be sure of two things: (i) the population will eventually grow geometrically; and (ii) once this happens, the proportions of individuals in each age group will also become constant. The findings of geometric increase in N and stable age distribution imply that the following mathematical statements are equivalent: $N_{t+1} = \lambda * N_t$ and $\mathbf{n}^{<t+1>} = \mathbf{A} * \mathbf{n}^{<t>}$, where N and λ are **scalars** (i.e. single, countable numbers) and $\mathbf{A}$ and $\mathbf{n}$ are **matrices** or **vectors**. In other words, a simple model of geometric increase ($N_{t+1} = \lambda N_t$) yields the same results as the Leslie matrix model ($\mathbf{n}^{<t+1>} = \mathbf{A} \cdot \mathbf{n}^{<t>}$). This means that we can estimate λ (the finite annual rate of increase) from the transition matrix $\mathbf{A}$, by something called the **dominant (largest) eigenvalue** of the transition matrix. The largest of the eigenvalues (1.36) is the finite rate of population increase (λ) once the population has reached a stable age distribution:

$$\text{eigenvals}(\mathbf{A}) = \begin{pmatrix} 1.36 \\ -0.49 + 0.38i \\ -0.49 - 0.38i \\ -0.03 \end{pmatrix}$$

Hence, after the initial period of uncertainty, the total population would increase by 36% per year (because $\lambda = 1.36$). This period of uncertainty is generally two or three generations, where generation is defined as the typical time that elapses between a mother's birth and that of her daughters (discussed in more detail in Chapter 6).

Just as there is a simple means of estimating the eventual rate of population growth (dominant eigenvalue), there is an equally simple way to predict the eventual proportion of individuals in each age group. We calculate the so-called "right eigenvector" corresponding to the "dominant eigenvalue" of the transition matrix $\mathbf{A}$:

$$\mathbf{W}_{raw} = \text{eigenvec}(\mathbf{A}, 1.36) \qquad \mathbf{W}_{raw} = \begin{pmatrix} 0.91 \\ 0.33 \\ 0.22 \\ 0.09 \end{pmatrix}$$

It is easier to interpret these values if we transform them into proportions:

$$\mathbf{W} = \frac{\mathbf{W}_{raw}}{\sum_x \mathbf{W}_{raw_x}} \qquad \mathbf{W} = \begin{pmatrix} 0.59 \\ 0.22 \\ 0.14 \\ 0.06 \end{pmatrix}$$

In other words, once the rate of growth has stabilized, newborns will comprise 59%, 1-year-olds will comprise 22%, 2-year-olds will comprise 14%, and older individuals will comprise 6% of the population.

The discussion of right eigenvectors and eigenvalues can be unnerving for many biologists, even for hardened professionals, but do not worry too much. Although the terms sound mysterious, the meaning of eigenvalue and eigenvector is actually fairly simple. Once a population has converged on its stable age distribution, thereafter every year the total population (N) increases by a multiplicative factor λ (the dominant eigenvalue), meaning that each age group in the population also increases year-to-year by the same factor. So, an eigenvector is just a string of numbers (the stable age distribution) that produces exactly the same outcome when multiplied by the constant λ as by the transition matrix $\mathbf{A}$. Fortunately, we can use this string of numbers in a very practical way, because it tells us the relative proportion of individuals we can expect eventually to see over time in each age category.

14.2 Stage-specific models

In many organisms it makes more sense to think about different demographic stages or size classes, rather than specific age classes. This can also be a convenient means of approximating the dynamics of long-lived organisms, by lumping age groups into stages, because often we do not have information on exact ages. Such an approach is known as a Lefkovitch stage-class model (Lefkovitch 1965; Caswell 2001). This involves multiplying stage-specific population densities by a transition matrix ($\mathbf{A}$). The top row in $\mathbf{A}$ reflects the probability of survival for stage class i multiplied by its fecundity (f_i). The diagonal reflects the probability of surviving and remaining within stage i (p_i); the subdiagonal represents the probability of surviving and growing into the next stage (g_i):

$$\mathbf{A} = \begin{pmatrix} p_0 & f_1 & f_2 & f_3 & f_4 & f_5 & f_6 \\ g_0 & p_1 & 0 & 0 & 0 & 0 & 0 \\ 0 & g_1 & p_2 & 0 & 0 & 0 & 0 \\ 0 & 0 & g_2 & p_3 & 0 & 0 & 0 \\ 0 & 0 & 0 & g_3 & p_4 & 0 & 0 \\ 0 & 0 & 0 & 0 & g_4 & p_5 & 0 \\ 0 & 0 & 0 & 0 & 0 & g_5 & p_6 \end{pmatrix}$$

Box 14.1 The
Lefkovitch stage-class
model for the
loggerhead sea turtle
(Crouse *et al.* 1987).

There are seven stages in the model. The youngest stage (0) represents eggs and hatchlings. The next two stages (1 and 2) represent small and large juveniles. Subadults are represented by the next stage (3). All individuals beyond stage 3 are capable of breeding. Stage 4 represents novice breeders. The last two stages correspond to young (5) and older (6) adults. Crouse *et al.* (1987) estimated the proportion of individuals growing into the next class, by assuming a stable age distribution, perhaps a risky assumption. On this basis, they derived the following Lefkovitch matrix model:

$$f = \begin{pmatrix} 0 \\ 0 \\ 0 \\ 0 \\ 127 \\ 4 \\ 80 \end{pmatrix} \quad p = \begin{pmatrix} 0.0 \\ 0.737 \\ 0.661 \\ 0.691 \\ 0.0 \\ 0.0 \\ 0.809 \end{pmatrix} \quad g = \begin{pmatrix} 0.675 \\ 0.049 \\ 0.015 \\ 0.052 \\ 0.809 \\ 0.809 \\ 0 \end{pmatrix}$$

$$A = \begin{pmatrix} p_0 & f_1 & f_2 & f_3 & f_4 & f_5 & f_6 \\ g_0 & p_1 & 0 & 0 & 0 & 0 & 0 \\ 0 & g_1 & p_2 & 0 & 0 & 0 & 0 \\ 0 & 0 & g_2 & p_3 & 0 & 0 & 0 \\ 0 & 0 & 0 & g_3 & p_4 & 0 & 0 \\ 0 & 0 & 0 & 0 & g_4 & p_5 & 0 \\ 0 & 0 & 0 & 0 & 0 & g_5 & p_6 \end{pmatrix}$$

When we multiply this Lefkovitch matrix by a vector of stage groups, then the use of the two sets of diagonals allows some individuals in each stage to mature into the next stage, while keeping others in the same stage as before. Just as we saw for simple age-structured models, the stage-class model predicts geometric increase (or decrease, depending on the magnitude of λ) after the age structure has equilibrated.

A good example of the application of this kind of model involves the loggerhead sea turtle (*Caretta carreta*), a marine species that lays its eggs in sandy beaches of the southeastern USA. Demographic parameters are difficult to estimate for a long-lived species like the loggerhead, which roams widely across the Atlantic Ocean. Hence, accurate age-specific data are unavailable, as are age-specific estimates of fecundity. Crude data are available, however, on the relative survival rates and fecundities of different stages: eggs, hatchlings, and mature individuals. There is additional predictable variation in fecundity stemming from body size. Crouse *et al.* (1987) developed a Lefkovitch stage-class model (Box 14.1) to evaluate which of seven life stages would be most responsive to conservation efforts. This is a considerable simplification of the 54 age classes that would be needed for a full Leslie matrix model.

14.3 Sensitivity and elasticity of matrix models

The largest eigenvalue of the loggerhead sea turtle transition matrix equals 0.95, implying that the population cannot sustain itself (a value of 1.0 is required for sustainability, i.e. a stationary population). Age- or stage-specific models also offer useful insights into possible remedies to counteract such population declines. By modifying vital rates, one can interpret the effectiveness of possible conservation actions

Box 14.2 Deriving the elasticity of matrix models (Caswell 1978).

For small changes in λ as a function of small changes in any element of the transition matrix ($\mathbf{A}_{ij}$), elasticity is defined as:

$$\frac{d}{d\log_e \mathbf{A}_{ij}}\log_e \lambda = \frac{\mathbf{A}_{ij}}{\lambda}\cdot\frac{d}{d\mathbf{A}_{ij}}\lambda$$

and

$$\frac{\mathbf{A}_{ij}}{\lambda}\cdot\left(\frac{d}{d\mathbf{A}_{ij}}\lambda\right) = \frac{\mathbf{A}_{ij}}{\lambda}\cdot\left(\frac{\mathbf{V}_i\cdot\mathbf{W}_j}{\mathbf{V}\cdot\mathbf{W}}\right)$$

So elasticity equals:

$$\frac{\mathbf{A}_{ij}}{\lambda}\cdot\left(\frac{\mathbf{V}_i\cdot\mathbf{W}_j}{\mathbf{V}\cdot\mathbf{W}}\right)$$

where $\mathbf{V}\times\mathbf{W}$ in the denominator refers to the scalar or dot product obtained by multiplying together the column vector representing the stable stage distribution ($\mathbf{W}$) and the row vector of reproductive values ($\mathbf{V}$).

that might be taken. In the case of loggerheads, for example, it might be possible to protect nesting sites on beaches or alternatively devote larger effort to improving survival while animals are out at sea. How can we evaluate these options?

There is an easy (but not very exact) way to do this. We substitute different values for each parameter in the transition matrix and see which one has the biggest effect. You should do so for yourself, to determine which parameter is most conducive to improving the growth rate λ. A more elegant answer can be obtained by using calculus to determine how sensitive λ is to a proportionate change in each parameter ($\mathbf{A}_{ij}$) in the transition matrix, where i refers to row and j refers to column. This is termed by ecologists the **elasticity** of λ, and the derivation is presented in Box 14.2 (based on Caswell 1978; De Kroon *et al.* 1986):

$$\frac{\mathbf{A}_{ij}}{\lambda}\cdot\left(\frac{\mathbf{V}_i\cdot\mathbf{W}_j}{\mathbf{V}\cdot\mathbf{W}}\right)$$

where $\mathbf{W}$ is the vector corresponding to the stable age distribution and $\mathbf{V}$ is the vector of reproductive values (defined below).

We have already introduced two key concepts in our discussion of age-structured (Leslie matrix) models: the dominant eigenvalue and right eigenvectors. These useful tools also apply to stage-structured (Lefkovitch matrix) models. There is another useful string of numbers, needed to calculate elasticity, called the **vector of reproductive values**. It is the left eigenvector (as opposed to the stable age distribution, which is the right eigenvector) that corresponds to the dominant eigenvalue λ. Left eigenvector just means that the order of the multiplication shown above is reversed. The left eigenvector is a string of numbers that satisfies the following equality:

$$\mathbf{V}^{\mathrm{T}}\cdot\mathbf{A} = \lambda\cdot\mathbf{V}^{\mathrm{T}}$$

Box 14.3 Calculation
of elasticities of the
Lefkovitch matrix for
the loggerhead sea turtle
(based on data from
Crouse *et al.* 1987).

The stable age distribution for the loggerhead turtle can be readily calculated from the right eigenvector of the transition matrix **A**, where 0.947 is the dominant eigenvalue of **A**:

$$\mathbf{W}_{raw} = \text{eigenvec}(\mathbf{A},\ 0.947)$$

$$\mathbf{W}_{raw} = \begin{pmatrix} 0.29 \\ 0.94 \\ 0.16 \\ 9.48 \times 10^{-3} \\ 5.21 \times 10^{-4} \\ 4.45 \times 10^{-4} \\ 2.62 \times 10^{-3} \end{pmatrix}$$

In order to make these values easier to interpret, we convert them to the proportion in each age group, by dividing each element by the sum of all of the elements:

$$\mathbf{W} = \frac{\mathbf{W}_{raw}}{\sum_i \mathbf{W}_{raw_j}}$$

This yields the following stable age distribution:

$$\mathbf{W} = \begin{pmatrix} 0.21 \\ 0.67 \\ 0.11 \\ 6.73 \times 10^{-3} \\ 3.69 \times 10^{-4} \\ 3.16 \times 10^{-4} \\ 1.85 \times 10^{-3} \end{pmatrix}$$

In other words, after the loggerhead turtle model has proceeded for a number of years, we would expect to see 21% of the population being composed of eggs or hatchlings, 67% juveniles, and the rest (12%) subadults and adults.

In the case of the loggerhead sea turtle, the vector **V** depicting stage-specific reproductive value is as follows:

$$\mathbf{V} = \begin{pmatrix} 1 \\ 1.4 \\ 6 \\ 116 \\ 569 \\ 507 \\ 588 \end{pmatrix}$$

We can doublecheck to verify that this is indeed the left eigenvector by performing the two different multiplications alluded to in the text:

$$\mathbf{V}^T = (1 \quad 1.4 \quad 6 \quad 116 \quad 569 \quad 507 \quad 588)$$
$$\mathbf{V}^T \cdot \mathbf{A} = (0.95 \quad 1.33 \quad 5.71 \quad 109.74 \quad 537.16 \quad 479.69 \quad 555.69)$$
$$\lambda \cdot \mathbf{V}^T = (0.95 \quad 1.33 \quad 5.68 \quad 109.85 \quad 538.84 \quad 480.13 \quad 556.84)$$

With these values we can estimate elasticities for every parameter in the transition matrix:

$$\mathbf{S}_{ij} = \frac{\mathbf{A}_{ij}}{\lambda} \cdot \left(\frac{\mathbf{V}_i \cdot \mathbf{W}_j}{\mathbf{V} \cdot \mathbf{W}} \right)$$

$$\mathbf{S} = \begin{pmatrix} 0 & 0 & 0 & 0 & 0.01 & 3.27 \times 10^{-4} & 0.04 & 1 \\ 0.05 & 0.18 & 0 & 0 & 0 & 0 & 0 & 0.37 \\ 0 & 0.05 & 0.12 & 0 & 0 & 0 & 0 & 0.25 \\ 0 & 0 & 0.05 & 0.14 & 0 & 0 & 0 & 0.09 \\ 0 & 0 & 0 & 0.05 & 0 & 0 & 0 & 0 \\ 0 & 0 & 0 & 0 & 0.04 & 0 & 0 & 0 \\ 0 & 0 & 0 & 0 & 0 & 0.04 & 0.23 & 0 \end{pmatrix}$$

where the T means that we are transposing the vector, so that it is now a row, rather than a column, vector. Stage-specific reproductive values for the loggerhead sea turtle are calculated in Box 14.3.

If we look carefully through the elasticities for the loggerhead turtle given in Box 14.3 we see that there is a great deal of variation. Changing some parameters obviously has more important consequences than changing others. The largest elasticities correspond to the probabilities of survival and remaining within the adult and juvenile stage classes. If one has a finite amount of money, energy, and time to devote to conserving this species, it would be most effectively spent on improving adult or juvenile survival. Interestingly, most conservation effort before the paper of Crouse *et al.* (1987) had been devoted to enhancing breeding success on the beaches. Although such efforts were no doubt useful, the elasticity calculations suggest that this activity may not be as useful as concentrating on improvement of survival at sea. Turtle excluder devices (TEDs) can greatly reduce mortality at sea by turtles which get caught accidentally in fish and shrimp trawls. The elasticity calculations suggest that application of TEDs should be the most efficient means of improving long-term viability of the population. Let us assume that such devices improve survival of old adults from 81% to 95%. Would this be sufficient to ensure long-term viability?

The demographic modified matrix obtained by changing adult survival to 95% leads to a finite rate of increase $\lambda = 1.01$, which is indeed just sufficient to allow sustainability. Hence, strenuous enforcement of TEDs would be sufficient to allow population recovery. Given the slender margin, however, other conservation practices are also called for, such as improved breeding success and enhanced hatchling survival to the juvenile stage. In practice, enhanced usage of TEDs has led to dramatic improvement in loggerhead turtle survival, a genuine conservation success story!

14.4 Short-term changes in structured populations

Although the Leslie matrix and the geometric models make similar predictions in the long term, they definitely make different predictions in the short term, before the age distribution has had a chance to stabilize. This may be particularly important in conservation and management of many wildlife species that tend to be long lived. For example, a recent study of Soay sheep on the island of Hirta demonstrated that age structure can be crucial to understanding dynamics over time (Coulson *et al.* 2001a). As we showed in Chapter 8, Soay sheep tend to have a strong threshold response to changes in population density: as density rises, survival drops precipitously. Different age groups vary in their degree of density dependence and sensitivity to weather conditions (Catchpole *et al.* 2000; Coulson *et al.* 2001a). As a result, a population dominated by young animals would have quite different population dynamics in the short term from one with a more equitable distribution of age groups (Fig. 14.3).

Many wildlife populations have age distributions that are far from stable (Owen-Smith 1990; Coulson *et al.* 2001a, 2003; Lande *et al.* 2002). Under these circumstances, it would be useful to have a reliable means of quantifying which demographic parameters have the greatest short-term impact on population growth. A promising approach has recently been outlined by Fox and Gurevitch (2000), based on use of the full set of eigenvalues and eigenvectors, rather than just the dominant set. Application of this new approach in an endangered cactus species (*Coryphanta robbinsorum*) demonstrated that the key demographic parameters to improve population growth in the short term differed considerably from those identified by standard elasticity assessment (Fox and Gurevitch 2000).

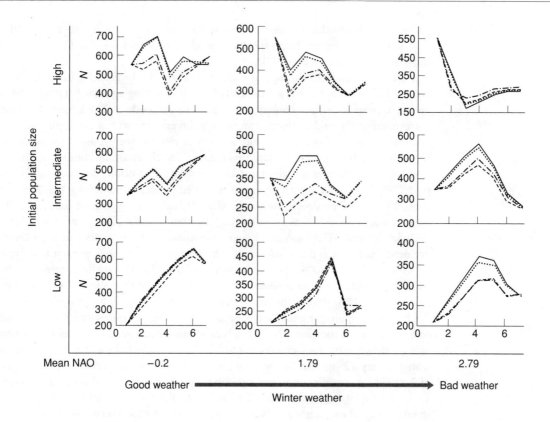

Fig. 14.3 Predicted short-term population dynamics of Soay sheep on the island of Hirta depend on the initial age structure, weather, and population density (Coulson *et al.* 2001a). Each individual graph displays the results of four simulations with identical weather conditions and identical starting population size. In each graph, each line represents the outcome of one simulation with different initial age and sex structure chosen from the data (see table below). The *y*-axis shows different starting population sizes and the *x*-axis different mean values for the simulated weather distributions. NAO stands for the North Atlantic Oscillation, a standard aggregate measure of weather conditions in the northern hemisphere (Stenseth *et al.* 2002).

Age/sex structure (%)

	Female lambs	Female yearlings	Female adults	Older females	Male lambs	Male adults	Older males
————	19	5	48	7	13	7	1
··············	17	8	41	8	18	8	0
·—·—·	23	10	24	11	19	13	0
------	22	13	17	13	16	19	0

14.5 **Summary**

The dynamic behavior of a population – whether it increases, decreases, or remains constant – is determined by its age- or stage-specific mortality and fecundity rates interacting with the underlying distribution of ages or stages in the population. The long-term rate of population change can be calculated through use of a transition matrix composed of these vital rates. Determination of the elasticity of the rate of population change due to slight modifications of the vital rates can be a useful means of evaluating alternative conservation and management options, as illustrated by the case study with loggerhead turtles.

15 Model evaluation and adaptive management

15.1 **Introduction**

In ecology, a **model** is a hypothesis that is usually expressed mathematically. A mathematical description allows a more precise definition of the hypothesis than does a verbal description, and this precision is particularly important for complex, non-linear processes. We can often construct more than one model to describe a process, and these alternatives are equivalent to alternative hypotheses. In this chapter, we explore the methods for choosing between such alternative models or hypotheses.

In Chapter 16, we will introduce the concept of statistical inference, which uses standardized criteria for decision-making to help ensure that decisions are not swayed by personal opinion or pressure brought to bear by politicians or the public. Despite its widespread use, however, statistical inference is not the only, nor even necessarily the best, way to choose among a wide variety of alternative hypotheses, whether these arise in the quest for "pure" or more "applied" knowledge (Johnson 1999; Anderson *et al.* 2000; Guthery *et al.* 2001; Johnson and Omland 2004). Statistical tests are effective at ruling out null hypotheses. The trouble is, the null hypothesis is sometimes an explanation that we need not seriously entertain, so rejecting it is not helpful for increasing our understanding of observations. For example, the null hypothesis in many wildlife habitat studies is that animals have no habitat preferences. We would be astounded if this ever proved true, so what progress do we make in rejecting it?

There are far fewer examples of hypothesis testing that are directed at evaluating a suite of alternative models or hypotheses that vary subtly from one another. It is hard enough to gather enough data to discriminate between random versus "significant" patterns of association, let alone tease apart subtle variants. More importantly, however, classic statistical methods are often impossible to use when alternative models are not special cases of more general models. This situation is particularly common in the kind of non-linear models that we find in ecology. Such "non-nested" models, in the jargon of professional statisticians, present special problems for finding a suitable statistical test.

Statistical inference is also plagued by "statistical" versus "biological" significance. You will recall that in statistical hypothesis testing, a P value of less than 0.05 is taken to mean that there is a remote probability (1 in 20) that an observed pattern could have been produced by chance alone. This probability is quite sensitive, however, to the amount of data that go into the assessment. Endangered species are often plagued by a crucial lack of data. This can preclude sufficient sample sizes and replicated treatments needed for significance testing, leaving us with no reliable option for making management decisions or informed scientific judgments, if we rely on standard statistical approaches. Even when we have a large sample of data that

253

yields a statistically significant result (that is, yields a *P* value less than 0.05), this may be of trivial biological significance. Hence, slavish adherence to statistical significance alone can distract one from the real issues at hand. It is often much more important to decide which factor has the strongest effect on the pattern or variable of interest.

Recognizing these limitations, ecologists have developed an alternative branch of statistics that allows them to evaluate a series of alternative models of the same phenomena (Hilborn and Mangel 1997; Burnham and Anderson 1998; Anderson *et al.* 2000; Johnson and Omland 2004). The philosophical spirit of this approach, called **model evaluation**, is not so much to discriminate significant from insignificant factors, but rather to decide which of many competing explanations is most consistent with the facts at hand, so that one can make an informed judgment about the best course of management action.

15.2 Fitting models to data and estimation of parameters

While model fitting can be readily applied to experimental data (e.g. Hobbs *et al.* 2003), it is applied most frequently to evaluation of observational data that are routinely gathered by many wildlife agencies, such as mark–recapture studies (Jorgenson *et al.* 1997), spatial distribution data (Fryxell *et al.* 2004), or annual censuses of abundance (Hebblewhite *et al.* 2002; Taper and Gogan 2002). Early in the exercise of model fitting one should ensure that no major changes have occurred over time in the way that data have been gathered, or if they have, that some method exists directly to compare data gathered at different points in time. Discrepancies in the way data are gathered are more frequent than one would like. Biologists are constantly tempted to modify observational techniques, either to improve the ease of observation or to improve the repeatability of observations. Such changes in methodology can be a good idea, but they can make it difficult to compare data from different eras. Methodological changes are often neglected when someone (e.g. a consultant) decides to analyze the cumulative historical data. For this reason, any analysis team ought to include at least one person familiar with the original data; even better if that person has gathered the data themselves.

As an example of a long-term data set we shall consider the census data on migratory wildebeest in Serengeti National Park illustrated in Fig. 8.7. Population estimates in this system date back to the early 1960s, when the Serengeti Research Institute was first established. It was recognized early that aerial counting was perhaps the best way to monitor the broad expanses of savanna grasslands and broadleaf woodlands that comprise Serengeti National Park. Counting methods were established early, with little deviation over the years despite different observers and technological changes in aircraft and navigation equipment. Serengeti wildlife ecologists used a stratified aerial count design, with photographs taken at known altitude used to count individuals in the center of wildebeest aggregations and visual counts made in areas with lower numbers of animals.

Results of these censuses over the past 40 years show a rapid increase in wildebeest abundance over the first 20 years, with subsequent leveling off and erratic fluctuation around roughly 1,250,000 individuals. There are numerous ways one could mathematically depict this general pattern, however, and they all have different management implications with respect to long-term viability of the wildebeest population. Which model is most consistent with the data? We'll use a formal model evaluation, based on information criteria, to find out.

Table 15.1 The instantaneous rate of increase calculated for the Serengeti wildebeest population.

Year	N (thousands)	Rate of increase (r)
1958	190.000	0.108835146
1961	263.262	0.130874700
1963	356.124	0.104751424
1965	439.124	0.047919596
1967	483.292	0.090021783
1971	629.777	0.109589002
1972	773.014	0.124420247
1977	1440.000	−0.142352726
1978	1248.934	0.03443494
1980	1337.979	−0.050802822
1982	1208.711	0.050754299
1984	1337.849	−0.077244421
1986	1146.340	0.012747402
1991	1221.783	−0.095578878
1994	917.204	0.070080128
1998	1165.908	0.110475091
1999	1302.096	−0.044661447

There are a great many ways that we can interpret underlying causes of this population growth. Ecological interactions among wildebeest, other large mammals, and the rest of the environment could influence the patterns of wildebeest population growth and natural regulation. We can begin by determining how fast the population is growing. A useful way to do this is to convert the census data into estimates of the exponential rate of increase (r) between sequential population estimates. We recall from Chapter 6 that the growth rate for a population can be expressed either as the finite growth rate ($\lambda = N_{t+1}/N_t$) or as its exponential equivalent ($r = \log_e(N_{t+1}/N_t)$). The exponential growth rate is especially convenient when population censuses or estimates are timed irregularly, rather than occurring every year, because it can be readily translated into shorter or longer time intervals by simple multiplication or division operations. We calculate the natural log of the ratio of subsequent to initial population abundance for each time interval and divide this ratio by the number of years between successive population estimates (τ):

$$r_t = \log_e(N_{t+\tau}/N_t)/\tau$$

We encourage you to check the procedure by calculating the first two or three estimates of r by hand. Why do we need to divide by τ? In many cases, we will not have annual data to work from. In these cases, we can handle the irregular timing between censuses by dividing by the number of years between them, τ. The result of these calculations for the Serengeti wildebeest is shown in Table 15.1. Once the values of r have been calculated, we can readily translate back into values of λ by exponentiation: $\lambda = e^r$.

The next step is to fit a mathematical relationship to the multiple estimates of r recorded over time. The accepted convention in such matters is to find a mathematical model whose values best fit the existing data, where "best fit" means to minimize the sum of squared deviations between the model estimates and the observed data. The recorded estimates for Serengeti wildebeest certainly seem to decline with

Fig. 15.1 Predicted
(line) and observed
(circles) exponential
rates of increase shown
by Serengeti wildebeest
in relation to population
density.

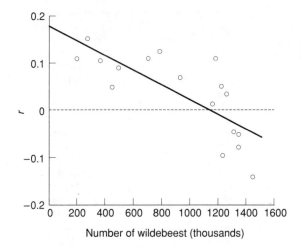

Fig. 15.1 Predicted
(line) and observed
(circles) exponential
rates of increase shown
by Serengeti wildebeest
in relation to population
density.

increasing wildebeest abundance, so one obvious model candidate would be a linear decline in r with N. In mathematical terms, this means we postulate the linear model $r(i) = a + b*N(i)$, where a is the intercept of the straight line and b is the slope of the relationship between N and r in any particular census i. These values can be estimated from any linear regression package. In this case, the intercept = 0.18 and the slope = −0.00016.

We can plot the observed value of r for each time interval versus population abundance at the beginning of the interval and then overlay these values with those predicted by the linear model whose parameters we have just estimated (Fig. 15.1). The linear model seems to do a reasonable job of predicting the rate of population growth shown by Serengeti wildebeest. The consistent tendency for deviation below the regression line at either very low or very high wildebeest densities, coupled with deviation above the regression line at intermediate densities, suggests that a curve might fit these data even better. We will address this possibility shortly.

15.3 Measuring the likelihood of models in light of the observed data

We now consider how we choose between several possible models that could represent our observations, in this case the trend in wildebeest numbers. We have illustrated one model of trend, the straight line, and we observed that a curve may be a better model. However, there could be several types of curve. Therefore, we need to decide how "likely" each of several alternative mathematical models might be, based on their ability to explain the census data obtained for the Serengeti wildebeest. First, we describe the distribution of residual variability around the postulated relationships (i.e. the models). Before we go any further, however, we are going to convert our equation for the linear relationship between the exponential growth rate (r_i) and population density (N_i) estimated at the beginning of time interval i into a form more familiar to students of wildlife biology, based on ecologically relevant parameters. The intercept a is usually called r_{max} by ecologists (see Chapter 6). It is the maximum exponential rate of increase of which the population is capable, applying under optimal growth conditions of low population density and high food availability. The slope b refers to $-r_{max}/K$, where K is known as the ecological carrying capacity, that is, the maximum number or density of animals that can be sustained over the long term in a particular place (Chapter 6). After making these conversions, $r_{max} = 0.18$ and $K = 1142$. In the case of the Serengeti data, there are 17 estimates of r_i (a

Fig. 15.2 Frequency distribution of residuals around the Ricker logistic regression of r versus N (histogram) and a normal probability curve with the same mean and variance, plotted against these data.

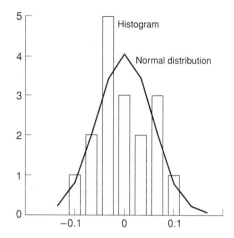

sample size that we will call n). For each of these n estimates of r, we can calculate the deviation, $resid_i$, around the regression line:

$$resid_i = r_i - r_{max}\left(1 - \frac{N_i}{K}\right)$$

The standard deviation of these residuals ($\sigma = 0.055$) can then be calculated by squaring each residual, summing together the squared residuals, and dividing by the total number of replicates.

In our example (Fig. 15.2), the residuals seem to be reasonably well depicted by a normal distribution, although the sample size is too small to be totally sure. On the presumption that the residuals do follow a normal distribution with mean of zero and $\sigma = 0.055$, we assess how well each model explains the existing data through use of the concept of likelihood (Λ). For example, we could use this approach to evaluate the likelihood that the carrying capacity of Serengeti wildebeest is any particular value:

$$\Lambda = \prod_{i=0}^{n-1} \frac{1}{\sigma\sqrt{2\pi}} \exp\left[\frac{-\left[r_i - r_{max}\left(1 - \frac{N_i}{K}\right) \right]^2}{2\sigma^2} \right]$$

Likelihood is proportional to the probability that a given model is correct, given a particular set of data. Likelihood is calculated from the probability function defining the residual variability that affects each estimate (in this case the normal distribution) and the value predicted by the model. Hence, we can use the equation defining the normal distribution, which we obtain from any statistics textbook, with an expected value for each observation, derived from the Ricker logistic model (see Section 15.4.1).

Because likelihoods are often very small or very large numbers, it is customary to evaluate their negative natural log-transformed values (termed the **negative log-likelihood**). This transforms the function from a "dome" to a "valley" shape, in which the most likely parameter is the value that is at the bottom of the valley (Fig. 15.3):

Fig. 15.3 Negative log-likelihood for a range of values of K in the Ricker logistic model applied to the wildebeest population in a Serengeti ecosystem.

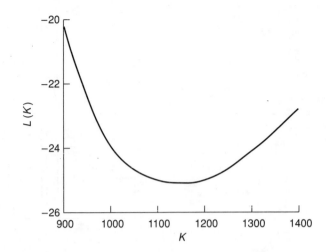

$$L(K) = n\left[\log(\sigma) + \frac{1}{2}\log(2\pi)\right] + \sum_{i=0}^{n-1} \frac{\left[r_i - r_{max}\left(1 - \frac{N_i}{K}\right)\right]^2}{2\sigma^2}$$

The profile curve shown in Fig. 15.3 is useful in demonstrating how the negative log-likelihood would change with different parameter values. We see that the value of K that minimizes the negative log-likelihood is the value identified using least-squares regression. The minimum log-likelihood is the maximum likelihood, so we refer to the "best-fit" value of K as a maximum likelihood estimate. Just as we can use likelihood to evaluate the most probable parameter values, in analogous fashion we can use likelihood to assess the plausibility of alternative models to explain the observed data.

15.4 Evaluating the likelihood of alternative models using AIC

We now want to decide which of numerous mathematical expressions best represent the data we have at hand. A famous physicist named Ludwig Boltzmann provided the theoretical basis for making such decisions. This concept was reformalized 75 years later by Kullback and Liebler, using the concept of "information." Information can be loosely defined as the "distance" between the true causal relationship and an approximating model. Interestingly, we do not really need to know the "true" relationship in order to evaluate the predictive ability of the alternative model. That is a good thing, because in ecology we will never know the real relationship or even the real parameters for any of the candidate explanatory models. Nonetheless, we can still evaluate the plausibility of each of the alternative models, by using the likelihood (Λ) and the number of parameters (p) in the estimating function.

    The most general approach that can be applied to non-nested models is Akaike's information criterion (AIC) or one of its variants (Burnham and Anderson 1998). AIC is derived from Kullback–Liebler information (Akaike 1973), when applied to experimental or field data whose parameters must be estimated and for whom the form of mathematical relationships is inexact. AIC values become smaller with increasing likelihood, but larger with each increase in the number of parameters. One would ordinarily expect that the likelihood of models should go up when a larger number of parameters are available to model the data. On the other hand, the AIC approach applies a penalty to each parameter p in the model. This enables us to identify the most parsimonious model (i.e. the model that explains the data reasonably well, using a moderate number of parameters). For further details, the interested reader should consult the comprehensive treatise on AIC by Burnham and Anderson (1998). Perhaps the easiest way to introduce the procedure here is to apply it to our data for the Serengeti wildebeest.

15.4.1 *Ricker logistic model*

The Ricker logistic model that we have fitted to the Serengeti wildebeest data (we will call it model 1) has three parameters, because we estimated r_{max}, K, and the standard deviation of the residuals around the linear regression line.

$$AIC_1 = -2\log_e(\Lambda_1) + 2\,p_1\left(\frac{n}{n - p_1 - 1}\right)$$

$$AIC_1 = -42.346$$

where p_1 is the number of parameters estimated from the data and n is data sample size. This AIC value only has meaning relative to those arising from other plausible models. We now consider three such candidates in the following sections.

15.4.2 *Theta logistic model*

The theta logistic model is appropriate for populations that have a threshold curvilinear relationship between r and N. According to this model, growth rates change little with density when N is modest, but the density-dependent response becomes much steeper as N becomes large. In other words, the intensity of density-dependent processes becomes disproportionately severe at high population densities. If there is a threshold effect, the theta logistic might be a more plausible model. In the theta logistic model, exponential growth rates are the following non-linear function of N, with the precise pattern dictated by three parameters: r_{max}, K, and θ:

$$F(N, r_{max}, K, \theta) = \left[r_{max}\left[1 - \left(\frac{N}{K}\right)^{\theta}\right]\right]$$

Most statistical programs have a routine to estimate the parameters for non-linear relationships such as this (including MATHCAD supplied with this book). In this case, the parameters are estimated as follows:

$$r_{max} = 0.105$$

$$K = 1.241 \times 10^3$$

$$\theta = 5.946$$

With these values we can plot the theta logistic curve against the data (Fig. 15.4). Then, we calculate the residual variance or mean-squared error (MSE):

Fig. 15.4 Predicted (line) and observed (circles) exponential rates of increase shown by Serengeti wildebeest in relation to population density, based on the theta logistic model.

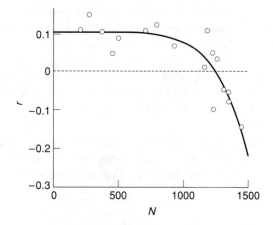

$$\text{MSE} = \sum_{i=0}^{n-1} \frac{\left[r_i - \left[r_{\max}\left[1 - \left(\frac{N_i}{K}\right)^{\theta}\right]\right]\right]^2}{n}$$

From this we obtain the residual standard deviation:

$$\sigma = \sqrt{\text{MSE}}$$

$$\sigma = 0.04$$

The likelihood calculation for the theta logistic model is calculated in a similar manner as we did for the Ricker model, except that we modify the expected value and the residual variance:

$$\Lambda_2 = \prod_{i=0}^{n-1} \frac{1}{\sigma\sqrt{2\pi}} \exp\left[\frac{-\left[r_i - \left[r_{\max}\left[1 - \left(\frac{N_i}{K}\right)^{\theta}\right]\right]\right]^2}{2\sigma^2}\right]$$

$$\Lambda_2 = 1.681 \times 10^{13}$$

$$p_2 = 4$$

Note that we now have four parameters ($r_{\max}$, K, θ, and the standard deviation of the residuals around the linear regression line), necessary for the more complex, nonlinear model:

$$\text{AIC}_2 = -2\log_e(\Lambda_2) + 2p_2\left(\frac{n}{n - p_2 - 1}\right)$$

$$\text{AIC}_2 = -49.573$$

Fig. 15.5 Predicted (line) and observed (circles) exponential rates of increase shown by Serengeti wildebeest in relation to population density, based on the Gompertz model.

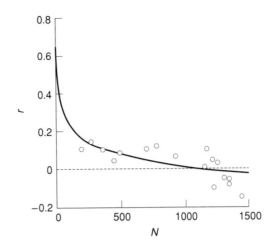

15.4.3 Gompertz logistic model

The Gompertz model is based on a logarithmic curvilinear relationship between r and N, which is steepest at small values of N, becoming progressively less steep at larger values of N (Fig. 15.5). It has been applied to many invertebrate and small mammal populations, but is less common in species of larger wildlife. As a first step, we calculate the natural logarithm of N:

$$X = \log_e(N)$$

We then proceed as before to estimate the slope (β) and intercept (α) of r versus X:

$$\alpha = 0.657$$

$$\beta = -0.093$$

Residual variance or mean-squared error is calculated as follows:

$$\mathrm{MSE} = \sum_{i=0}^{n-1} \frac{[r_i - (\alpha + \beta \log_e(N_i))]^2}{n}$$

$$\sigma = \sqrt{\mathrm{MSE}}$$

$$\sigma = 0.062$$

The likelihood calculation for the Gompertz model is:

$$\Lambda_3 = \prod_{i=0}^{n-1} \frac{1}{\sigma\sqrt{2\pi}} \exp\left[\frac{-[r_i - (\alpha + \beta \log_e(N_i))]^2}{2\sigma^2} \right]$$

$$\Lambda_3 = 1.015 \times 10^{10}$$

Fig. 15.6 Predicted (line) and observed (circles) exponential rates of increase shown by Serengeti wildebeest in relation to population density, based on the geometric growth model.

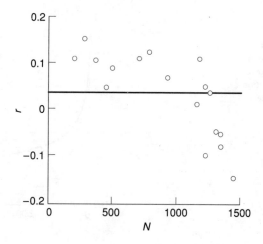

There are three parameters, because we estimated α, β, and the residual variation in r:

$$p_3 = 3$$

$$\text{AIC}_3 = -2\log_e(\Lambda_3) + 2p_3\left(\frac{n}{n - p_3 - 1}\right)$$

$$\text{AIC}_3 = -38.236$$

15.4.4 *Geometric growth model*

As our final model, we consider an old standby, the geometric growth model (Chapter 6). This model assumes that there is no change in growth rates as the population increases over time (Fig. 15.6):

$$r_{\max} = \text{mean}(r)$$

The residual variance or mean-squared error is:

$$\text{MSE} = \sum_{i=0}^{n-1} \frac{(r_i - r_{\max})^2}{n}$$

$$\sigma = \sqrt{\text{MSE}}$$

$$\sigma = 0.085$$

The likelihood calculation for the geometric growth model is:

$$\Lambda_4 = \prod_{i=0}^{n-1} \frac{1}{\sigma\sqrt{2\pi}} \exp\left[\frac{-(r_i - r_{\max})^2}{2\sigma^2}\right]$$

$$\Lambda_4 = 5.24 \times 10^7$$

There are two parameters, because we estimated the mean and residual variation in r:

$$p_4 = 2$$

$$\text{AIC}_4 = -2\log_e(\Lambda_4) + 2p_4\left(\frac{n}{n - p_4 - 1}\right)$$

$$\text{AIC}_4 = -30.692$$

15.4.5 *Evaluation of the models*

The model with the lowest AIC score is taken as the best. Smaller scores imply a better fit to the data. We can evaluate the merit of each model relative to the best model, using the difference between their AIC scores (ΔAIC_i = AIC score of model i minus the AIC score of the best model). If $\Delta\text{AIC} < 4$, then the difference in explanatory power is considered trivial by statistical experts so the models provide similar explanation (Burnham and Anderson 1998).

On the basis of this simple evaluation, one would conclude that the threshold theta logistic model is the best (most parsimonious) descriptor of the observed changes in Serengeti wildebeest abundance recorded over the past 40 years, followed by the Ricker logistic model ($\Delta\text{AIC} = 7.2$). The Gompertz logistic model ($\Delta\text{AIC} = 11.4$) and geometric growth model ($\Delta\text{AIC} = 18.9$) are even less consistent with the observed data.

We can more formally assess the likelihood of each of the competing models by calculating their Akaike weights. The Akaike weight for a given model i is calculated by dividing the likelihood of that model ($\exp[-\Delta\text{AIC}_i/2]$) by the summed likelihoods of all the competing models ($\Sigma\exp[-\Delta\text{AIC}_j/2]$):

$$w_i = \frac{\exp(-\Delta\text{AIC}_i)}{\sum_{j=1}^{4}\exp(-\Delta\text{AIC}_j)}$$

The Akaike weight for the theta logistic model is 0.99, whereas the Akaike weights of all the other models are less than 0.01. This implies that there is a 99% probability that the theta logistic model is the best (i.e. the most parsimonious) model in the set considered. This is not to say that the theta logistic model is correct – only that it offers the most parsimonious (efficient) prediction of the data, balancing the need for accurate prediction with a tolerably small number of parameters. Akaike weights offer a practical means of assessing how seriously to take alternative predictions derived from each of the models.

The simplest biological interpretation of this pattern is that wildebeest require some resource, such as food of suitable quality, whose availability is strongly related to wildebeest abundance. This interpretation is further strengthened by the observation that wildebeest survival rates measured locally are positively related to the amount of green grass available per individual animal, which in turn varies with monthly rainfall (Mduma *et al.* 1999).

To simplify our example, we evaluated only four alternative models. One can readily imagine other plausible models that might be even more useful in predicting changes in wildebeest abundance over time, such as age-dependent models or models incorporating interactions with predators. These new models should also be ranked in terms of their AIC scores. A wise researcher or wildlife manager would consult all of the models whose ΔAIC values were within a difference of 2 of the leading model, because all such models are consistent with the evidence. The robustness of future population projections can then be re-evaluated as further evidence is accrued. This learning procedure is especially useful if comparisons can be made under conditions in which plausible models make different predictions.

15.5 Adaptive management

Nowhere is model evaluation more important than in the emerging concept of adaptive management (Walters 1986; Walters and Holling 1990). No ecosystem is completely understood. As a consequence, we can never predict with certainty how any ecosystem will respond to human intervention, such as harvesting (Chapter 19), or conservation programs (Chapter 18). What this means, of course, is that any management policy that we choose to adopt should be viewed as an experiment whose outcome is uncertain. Good wildlife managers have always recognized this, at least subconsciously, and accordingly gather data to monitor the status of species with which they are charged. Where "adaptive" management departs from simply "good" management is in formalizing a mechanism by which management policies can be improved over time, by reducing at least some of the uncertainties.

One way to go about this is to make the best use possible of historical data to judge which model out of many possible models is most useful. We have already demonstrated such an analysis to explain the demographic patterns of Serengeti wildebeest. In sifting through the possible candidate explanatory models, information theory (likelihood, AIC, and the like) provides a powerful set of tools for data analysis. This process is known as **passive adaptive management**, because it relies on natural variability to expose the underlying relationships. Without wide variation in wildebeest abundance, for example, we would have had great difficulty discriminating among the three subtle, but dynamically different, classes of density-dependent model that might apply.

The variation in numbers of wildebeest occurred through a fortuitous accident, the unintentional elimination of rinderpest. This event immediately suggests a radically different approach: manipulative experimentation with the express intention of clarifying our understanding of the system. Such intentional experimentation at the ecosystem level is known as **active adaptive management**. In principle, all of the attributes of good experimental design (Chapter 16) should be incorporated: use of controls versus experimental treatments, replication, and a factorial design to identify possible interactions among processes. In practice, however, it is difficult to implement an ideal experimental design at the enormous spatial scale at which wildlife management policy is typically conducted. Consequently, these difficulties argue against using conventional inferential statistics to evaluate the outcome of active adaptive management, such as the ANOVA designs described in Chapter 16. Rather, information theory and Bayesian methods of analysis offer a more practical toolkit, based on the objective of finding the best explanatory model among many possible models, rather than definitively rejecting a null model. Active adaptive management policies trade off the short-term goal to maximize some output, such as a harvest, against the long-term goal of gaining greater understanding of important ecological, physical, or social processes. In that sense, active adaptive management is rather like industrial research and development: reinvesting current revenue to enhance future profits.

Table 15.2 Hypothetical matrix of net benefit from two alternative stocking models, showing that the best harvest of mallard ducks is obtained from current stock values, versus higher stock values.

Policy options	Alternative models (hypotheses)	
	Current stock is optimal	Higher stock is optimal
Maintain status quo	1.0	1.0
Increase stock	0.5	1.2

After Walters and Holling (1990).

One unfortunate aspect of this terminology is that "active" seems a great deal more attractive than "passive" adaptive management. It is not always clear that this is so. Experimentation implies consciously ignoring the current "best" model to explore alternatives. This can only come at a cost, perhaps a great cost, if the alternative models prove to be inferior. For example, Anderson (1975) suggested that higher harvests of mallard ducks could be sustained if the breeding stock is allowed to climb considerably above the current levels. Choosing to explore this option would require considerable reduction in harvest quotas for hunting enthusiasts and resultant loss of revenue to agencies, tourist operators, and equipment suppliers. If Anderson's hypothesis is incorrect, the cost of learning would be a reduction in harvests (and profits). Small wonder that this may not be an attractive option for everyone concerned! One way to evaluate the wisdom of embarking on an active adaptive experiment is to evaluate the costs and benefits of sticking with the untested status quo hypothesis versus adopting an experiment to test an alternative hypothesis. This can be symbolized by a decision matrix (Table 15.2).

For example, let us say that the current stocking rate of mallard ducks really is most productive. If we increase the stock to a less productive level, this would lead to a reduction in average harvest from 1.0 to 0.5 units. On the other hand, if mallards really are more productive at a higher stock size, then increasing the stock may increase yields by a given amount, say 20%. If each model is equally plausible, based on our current data, then we can calculate the expected payoff by simply averaging the possible outcomes across each row.

In this example, the expected payoff is 1.0 for the status quo option (= [1.0 + 1.0]/2), whereas the experimental option has an expected payoff of 0.85 (= [0.5 + 1.2]/2). This suggests that the experiment is too costly and/or too unprofitable to warrant testing. In contrast, if the payoff in the lower right-hand cell of the matrix was 1.75, then the expected payoff would be 1.125 and the experiment would be justifiable. The point is that minor improvements in yield, accompanied by major costs, may not justify experimentation.

Even when the decision to adopt an experimental procedure is justifiable, many individuals may not value slight improvements in management efficiency if it interferes with their personal recreational values. That situation may explain why moose or trout populations close to population centers in North America are probably well below the level of maximal productivity. Alternatively, some resource users would not want to take any risk of losing income, simply because there are no other economic options. Hence, the payoff matrix will often vary among different special interest groups.

While passive adaptive management might be attractive to risk-averse decision-makers, it also has its perils. For one thing, it can lead to lost opportunity for all

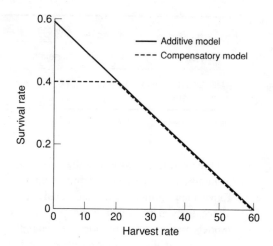

Fig. 15.7 Schematic representation of the compensatory (broken line) and additive (solid line) models for survival in waterfowl in relation to harvest levels.

resource users, should several models continue to make similar predictions. For another, passive management makes it difficult to discriminate between good management and good luck. High harvests could accrue by chance during a series of good years, despite application of a wrong model.

As an example of the potential utility of active adaptive management, let us consider waterfowl harvesting, specifically mallard ducks, in more detail. Harvest quotas for a variety of ducks are determined in part using a sophisticated system of stratified aerial surveys criss-crossing the extensive area of breeding habitat on the North American prairies (Nichols *et al.* 1995). Density levels and pond availability are used to predict stochastic variability in duck recruitment rates, and these recruitment rates are interpreted as a harvestable surplus (Anderson 1975). Much of the stochastic variability in demographic parameters stems from variation in rainfall on the prairies. Wet weather generates large numbers of small ponds and pothole lakes on the prairies, which in turn generates increased success in offspring recruitment. Banding records, obtained from the recovery of identification bands (in the hunting season), allow an estimation of mortality rates. This information on offspring recruitment and mortality is then used in quantitative population models to predict safe harvesting levels year-to-year. The remarkable consistency in duck numbers over time attests to the robustness of this program (Nichols *et al.* 1995).

There are indications, nonetheless, that the harvest allocation for some waterfowl species could be considerably improved. A key uncertainty in the harvest evaluation procedure is whether mortality is compensatory or not (Anderson 1975; Williams *et al.* 1996). In this context, perfect compensation means that increased duck mortality due to harvest has no effect on overall duck mortality, at least over some range of harvest rates, because survival in the wild adjusts perfectly to losses imposed by man (Fig. 15.7). The alternative hypothesis is that there is no compensation, hunting mortality is in addition to natural mortality, and so total mortality is linearly related to harvest rates (Fig. 15.7). Current data are inadequate to discriminate between these two hypotheses, yet they have critical implications with respect to both the risk of over-harvesting, particularly in poor years, and the optimal harvest policy (Anderson 1975). Simulation models have been used to show that by far the most efficient way to decide which of these alternative models is correct is through active adaptive management (Nichols *et al.* 1995; Williams *et al.* 1996). Indeed, this may be the only realistic way to reduce the uncertainty in biological processes, at least within our lifetime.

Such an active adaptive management procedure has been implemented, despite the inherent difficulty in coordinating agencies and resource users in a variety of jurisdictions (Nichols *et al.* 1995). If this kind of coordinated model evaluation can be conducted for waterfowl, it can be used even more readily for less mobile species. The key may be that there has been a long-standing tradition in waterfowl management to apply biomathematical models to the production of recruitment and harvest management. Such models have been applied rarely to wildlife species, for which harvesting policy is often developed in a more haphazard fashion. The adaptive approach demonstrates a more productive option.

15.6 **Summary**

Statistical hypothesis testing is not always the best way to make informed decisions about causal factors associated with wildlife population dynamics, because of preoccupation with rejection of null hypotheses rather than evaluation of the merits of a suite of more plausible models. We outline an alternative approach to inference that is based on information theory, allowing one to decide which model or suite of models offers the best explanation for existing patterns of data. Such an approach complements the practical need to make the best management decisions possible on the basis of incomplete scientific information. A cornerstone of all model evaluation procedures is some measure of goodness-of-fit between models and data. Such model evaluation is an essential component of adaptive management regimes, where alternative explanatory models are vigorously pursued using historical data or experimental perturbation. We show how adaptive management can be used to improve management of harvesting in migratory waterfowl populations in North America.

16 Experimental management

16.1 **Introduction**

In practical terms, there are two different modes of wildlife management: that in which management decisions flow from personal experience and received wisdom, and that in which they are based upon data and analysis. For want of better names we will call these the "traditional" and "experimental" modes of wildlife management, respectively. The value of the traditional mode should not be underestimated. Its dominant characteristic is conservatism, a lack of interest in trying out new ideas. That is sometimes a strength rather than a weakness because most new ideas turn out to be wrong (Caughley 1985). However, some new ideas are useful and these are best identified by the experimental approach. In this chapter we explain how a technical judgment can be evaluated, by posing it as a question (hypothesis).

One way to evaluate a hypothesis is through a standard statistical test of inference. We provide guidelines for designing effective experiments and outline some standard statistical methods of analysis of such experiments. We describe the use of replication to sample the natural range of variability and the use of controls to render the conclusions unambiguous. However, standard statistical tests are sometimes inadequate for identifying which of many possible hypotheses provide the "best" explanation for the observations at hand, and for deciding which course of management would be most effective. In order to pursue these issues, we need a different approach, namely model evaluation and inference. In Chapter 15, we outlined some of the principles of model evaluation and showed how this method can provide a powerful tool in the resource-manager's arsenal.

16.2 Differentiating success from failure

Wildlife management is not like civil engineering. The theory and practice of civil engineering is placed on public display every time a bridge is built. No expertise is needed to interpret that test. If the bridge remains operational for the length of its design life, the engineers got it right. If it collapses they got it wrong, and we look forward to hearing the details of how and why they got it wrong at the subsequent court case.

Wildlife management differs from civil engineering in a number of respects. First, the managers are not erecting something new but acting as custodians of something already there. They are not responsible for the initial conditions but these often constrain their options.

Second, in civil engineering the question is usually obvious. In ecology the appropriate question is seldom obvious. Choosing the appropriate question is the most difficult task, much more difficult than answering that question. Good design does not correct an inadequate grasp of the problem.

Third, criteria for success and failure are seldom tight and often are not available to the public. Compare these two statements:

1 The provision of nest boxes for wood ducks will increase the size of the population.

2 The provision of nest boxes for waterfowl will benefit their overall ecology.

The first is a hypothesis testable against a predicted outcome. The second is of the type that covers everything and probably cannot be disproved. What is an "overall ecology?" How is it measured? What species should it be measured on? Wildlife management objectives that are framed in an unverifiable form (example 2) are not very useful whereas those in the form of testable hypotheses (example 1) allow us to learn more about the system.

Fourth, even when there are such criteria to judge success, the wildlife manager is seldom in complete control of the situation and hence cannot be held fully and personally responsible for the outcome. A failure is usually referable to the acts of many people, often interacting with changes in the environment (sometimes referred to as acts of God).

Fifth, the wildlife and its habitat usually forms a robust ecological system. Within rather wide limits that system will absorb the most inappropriate or irrelevant of management and still look good.

Because the criteria for success are often fuzzy in wildlife management, the outcomes of different management systems are sometimes difficult to rank. For example, when managing deer populations do we shoot only bucks, shoot only does, shoot 70% bucks and 30% does, shoot 30% bucks and 70% does, shoot neither? All these schemes have been tried and all have been reported as highly successful. Highly successful with what end in view? How highly successful? Perhaps we ask those questions less often than we should.

Agriculture made a major advance because R.A. Fisher invented the "analysis of variance" and because a few agriculturalists recognized that here was a technique that could differentiate the effect of different management treatments. More importantly they believed that differentiation was necessary. Wildlife management can learn from the history of agriculture by incorporating more statistical design in management programs.

16.3 Technical judgments can be tested

In contrast to the **value judgment** discussed above, the wisdom of a **technical judgment** can be evaluated according to strict criteria. If a manager decides that supplementary feeding will increase the density of quail then that can be tested and the decision rated right or wrong. If a manager decides that elephants must be culled because, if not, they will eliminate *Acacia tortilis* trees from the area, that decision is right or wrong and it can be demonstrated as right or wrong by an appropriate experiment. Note that the decision on whether the local survival of the acacia justifies the proposed reduction of elephants is a value judgment and hence not testable.

So there are value judgments and technical judgments and these must not be confused one with the other. Technical judgments can be tested and should be tested. By this means we learn from our failures as well as from our successes. A recurring theme of this book is that wildlife management advances only when the efficacy of a management treatment is tested. For that to happen the technical decision as to the appropriate treatment must be stated in a form that predicts a verifiable outcome.

16.3.1 *Hypotheses*

Research questions are usually phrased in positive form such as, for example, does the mean body weight of black bears (*Ursus americanus*) change as we move from the equator towards the pole? That question is most easily tested statistically if we frame it in negative form, the so-called **null hypothesis** denoted H_0: mean body weight of black bears does not change with latitude. If this hypothesis is falsified by data showing that mean weights are not the same at all sampled latitudes, we reject the null hypothesis in favor of an **alternative hypothesis** H_a. Whereas a question can generate only one null hypothesis, there may be a number of competing alternative hypotheses. In the bear example the alternative to no change of weight with latitude may be an increase with latitude, a decrease with latitude, a peak in the middle latitudes, or a trough in the middle latitudes.

The procedures by which we test hypotheses make up the realm of statistical analysis. They come quite late in the research sequence, which proceeds in the following manner:

1 pose a research question (usually our best guess or prediction as to what is going on);
2 convert that to a null hypothesis;
3 collect the data that will test the null hypothesis;
4 run the appropriate statistical test;
5 accept or reject the null hypothesis in the light of that testing;
6 convert the statistical conclusion to a biological conclusion.

Most statistical tests estimate the probability that a null hypothesis is false. A probability of say 10% is often interpreted loosely as meaning that there is only a 10% chance of the null hypothesis being true. That is not quite right. Suppose our null hypothesis states that there is no difference in bill length between the females of two populations of a particular species. We draw a sample from each, perform the appropriate statistical test for a difference, and find that the test statistic has a probability of (say) 10% for the sample sizes that we used. That 10% is the estimated probability of drawing two samples as different or even more different in average bill length as those that we drew *if the populations from which they were drawn did not differ in that estimated attribute.*

If there really is no difference between the two populations in average bill length, then the probability returned by the statistical test will be in the region of 50%. This implies that the chance of drawing more disparate samples than those we actually drew is the same as the chance of drawing less disparate samples than those we actually drew. A probability greater than 50% means that the two samples are more similar than we would expect from random sampling of identical populations. If that probability approaches 95% or so, we should investigate whether the sampling procedure was biased.

Statistical tests deal in probabilities, not certainties. There is always a chance that we are wrong. Such errors come in two forms, the **Type 1 error** (also known as an alpha error) in which the null hypothesis is rejected even though true, and the **Type 2 error** (beta error) when the null hypothesis is accepted even though false. Following Zar (1996), the relationship between the two kinds of error can be shown as a matrix:

	If H_0 is true	If H_0 is false
If H_0 is rejected	Type 1 error	No error
If H_0 is accepted	No error	Type 2 error

Obviously we are not keen to make either kind of error. The probability of committing a Type 1 error is simply the specified significance level. The probability of committing a Type 2 error is not immediately specifiable except that we can say that it is related inversely to the significance level for rejecting the null hypothesis. The two kinds of error cannot be minimized simultaneously, except by increasing the sample size. Hence, we need a compromise level of significance that will provide an acceptably small chance of rejecting a factual null hypothesis, but which is not so small as to generate too large a chance of committing a Type 2 error. Experience has indicated that a 5% chance of rejecting the null hypothesis when it is true provides reasonable insurance against both kinds of error. We therefore conventionally specify the 5% probability as our significance level, although that level is essentially arbitrary and little more than a gentlemen's agreement. What is not arbitrary is that the hypothesis to be tested and the level of significance at which the hypothesis is rejected must be decided upon *before* the data are examined and preferably before they are collected. Otherwise the whole logic of testing is violated.

Our standard statistical tests concentrate on minimizing Type 1 errors. The extent to which they minimize Type 2 errors is called **power**. Depending on context, avoidance of Type 2 errors may be more important than ensuring the warranted rejection of the null hypothesis.

16.3.2 *Asymmetry of risk*

Converting a statistical result back into a biological conclusion is not at all straightforward. The classical null hypothesis method is at its best when testing whether a treatment has an effect, the treatment representing a cost and the response a benefit. An example might be supplementary feeding to increase the clutch size of a game bird. Here the feeding costs money and time, and we will use it operationally only if an adequate response is clearly demonstrated. First, the null hypothesis must be rejected, and then the difference in response between experimental control and treatment must be evaluated to determine whether the cost of the treatment is justified by the size of the response in fecundity. If the null hypothesis (no effect of treatment) is not rejected we are simply back where we started and no harm is done. Type 1 and Type 2 errors are both possible, both are inconvenient, but neither is catastrophic. A Type 1 error leads to unnecessary expense until the mistake is identified; a Type 2 error results in a small sacrifice in the potential fecundity of the game bird population.

Null hypothesis testing is less effective and efficient when the treatment itself is a benefit and the lack of treatment is itself a cost. Suppose a marine fish stock appears to be declining although there is considerable year-to-year variation in the index of abundance used: catch per unit effort. Further, there are good reasons to suspect that the fishing itself is heavy enough to precipitate a decline. The null hypothesis is that fishing has no effect on population size. In this case the failure to reject the "no-effect" null hypothesis is not sufficient reason to operate on the assumption that the fishing is having no effect. At the very least one would first want to know something about the power of the test. In this case the cost of making a Type 2 mistake greatly outweighs the benefit of getting it right. The effect of continuing to fish when one should have stopped could be disastrous and irreversible, whereas unnecessary cessation of fishing results only in a temporary cost until fishing resumes. This is an **asymmetry of risk**. It is particularly prevalent in work on endangered species where an error can result in extinction. Asymmetry of risk demands conservative interpretation of statistical results.

16.4 **The nature of the evidence**

16.4.1 *Experimental evidence*

A management treatment may be successful or it may be a failure. If the first, the manager needs to know whether the success flowed from the treatment itself or whether it would have happened anyway. Otherwise an expensive and unnecessary management scheme might run indefinitely. Alternatively, the management may not achieve its stated aim, in which case the manager must first establish that fact without doubt and then find out why. Was the failure caused by some extraneous factor that formed no part of the treatment? Was the entire management treatment inappropriate or only a part of it? Would a higher intensity of the treatment have been successful whereas the same treatment at a lower intensity was not?

To find out, the management must be run as an experiment. There are rules to designing an experiment that are there for one very important reason: if they are broken the questions the experiment is designed to answer cannot be answered unambiguously.

Suppose a manager wished to increase the density of quail in an area by supplementing their supply of food with wheat. How that is done determines whether anything will be learned from the exercise. There is a graded series of approaches, ranging from useless in that they yield no verification of the worth of the treatment, through suggestive in that their results allow a cautious choice between alternative interpretations, to definitive where the results can be interpreted without error:

1 Grain is scattered once a month but density is not monitored. The manager assumes that since the treatment *should* increase the density of quail that it *will* increase their density.

This is no test because the outcome of the treatment is assumed rather than observed.

2 The manager measures density on two occasions separated by 1 year, the first before supplementary feeding is instituted. If density were higher on the second occasion the manager might assume that the rise resulted from the feeding.

This is the classic fallacy of "before" being taken as a control on "after." Interpretation of the result rests on an assumption that the density would have remained stable without supplementary feeding, and there is no guarantee of that. It may, for example, have been increasing steadily for several years in response to a progressive and general increase of cover.

3 The manager designates two areas, one on which the birds are fed (the treatment) and the other on which they are not (the experimental control). The density of quail is measured before and after supplementary feeding is instituted. If the proportionate increase in density on the treatment area is greater than that on the control area, the difference is ascribed to the effect of feeding.

This design is a radical improvement but still yields ambiguous results: the difference in rate of increase may reflect a difference between the two sites rather than between the two treatments. We say that the effect of site and the effect of treatment are **confounded**. Perhaps the soil of one site was heavy and that of the second light, the vegetation on the two sites thereby reacting differently to heavy winter rainfall and the quail reacting to that difference in plant growth.

Flawed as it is, this design is often the only one available, particularly if the treated area is a national park. In such cases the control must be chosen with great care to ensure that it is in all important respects similar to the treated area, and the response variable should be monitored on each area for some time before the treatment is instituted to establish that it behaves similarly in the two areas. Another way around this problem is to reverse the two treatments and see if the same result is obtained.

4 The effect of such local and extraneous influences on the results of the experiment is countered by replication. Suppose six sites are designated, three of which are treated by feeding and the other three left as controls. The category of the site is determined randomly. Before and after measurements of density are made at the same time of the year in all six areas. The biological question: does supplementary feeding affect density? is translated into a form reflecting the experimental design: is the difference in quail density between treatments (feeding versus not feeding) greater than the difference between sites (replicates) within treatments?

This is an appropriate experimental design in that the outcome provides an unambiguous test of the hypothesis. Its efficiency and precision could be increased in various ways but its logic is right.

The form of an experimental design is dictated by logic rather than by the special requirements of the arithmetic subsequently performed on the data. This is an immensely important point. If the manager has no intention of applying powerful methods of analysis to the data, that in no way sanctions shortcuts in the basic experimental design.

Another common fallacy is the belief that although a logically designed experiment is necessary for publication in a scientific journal the manager need not bother with all that rigmarole if the only aim is to find out what is going on. The manager might then simply run an "empirical test" like the second or third example given above without realizing that the measurements do not reveal what is going on.

16.4.2 *Why replicate the experiment?*

Suppose we wish to determine whether grazing by deer affects the density of a species of grass. The experimental treatment is grazing by a fixed density of deer and the experimental control is an absence of such grazing. We cannot simply apply the two treatments each to a single area because no two areas are precisely the same. We would not know whether the measured difference in plant density was attributable to the difference in treatment or whether it reflected some intrinsic difference between the two areas. There will always be a measurable difference between areas in the density of any species whatever one does, or does not do, to those areas.

We can postulate that a difference between treated areas is caused by the disparate treatments applied to them only when the difference between the treatments is appreciably greater than the difference within treatments. To determine the scale of variation within the "population" of treatments we must look at a sample of areas that have received the same treatment. The minimum size of a sample is two. Thus, we must designate at least two areas as grazing treatments and two as controls. A sample of three is better.

The density of a plant species is usually measured within small quadrats scattered over a treatment area. Fifty might be measured in each. Those 50 quadrats are not replicates. They are subsamples of a single treatment and their invalid use as "replicates" is called **pseudo-replication**. Sampling within a treatment is not treatment replication. Data from such subsamples could be fed into an analysis of variance, which would then provide what might appear to be a rigorous test of the hypothesis, but that is an illusion. The arithmetic procedures have been fulfilled but the logic is not satisfied. The result is actually a test of whether the combination of the treatment and the intrinsic characteristics of a single area differ from another treatment combined with the intrinsic characteristics of another area. We say that area and treatment are **confounded**. Their individual effects cannot be disentangled. No strong test of the effect of the treatments themselves is possible unless those treatments are replicated.

Replicates are not meant to be similar. They are meant to sample the natural range of variability. Consequently, one does not look for six similar sites to provide three treatments and three controls. One picks six sites at random. A common excuse offered for a lack of replication in management experiments (and even in research experiments) is that sites similar enough to act as replicates could not be located. Such an excuse is not valid and points to a lack of understanding of the nature of evidence.

These principles carry over to all other forms of comparison. We cannot conclude from two specimens that parrots of a given species have a higher hemoglobin count near the tops of mountains than at lower altitudes. We get no further forward by taking a number of blood samples (subsampling) from the two individuals. Instead, we must test the blood of several parrots from each zone, look at the variation within each group of parrots, and then calculate whether the average difference between groups is greater than the difference within groups. Hence, we must replicate. The arithmetic of such a comparison can be extracted from any book on statistical methods. That is the easy part. The difficult part is getting the logic right.

16.5 Experimental and survey design

Experimental design has its own vocabulary. The thing that we monitor, in this case the density or rate of increase of the quail, is the **response variable**. That which affects the response variable, in this case WHEAT, is a **factor**. In our imaginary experiment the factor we examined had two **levels**: no supplementary feeding of wheat and some supplementary feeding of wheat (Fig. 16.1). Equally, its levels could have been set at 0, 30, 70, and 250 kg of grain per hectare per month as in Fig. 16.2. The levels of a factor need not be numbers as in that example. The levels of factor HABITAT, for example, might be pine, oak, and grassland. The levels of factor ORDER might be first, second, third, and fourth. The levels of factor SPECIES might be mule deer, white-tailed deer, and elk.

Suppose we wished to examine the effect of two management treatments simultaneously. Instead of looking at the effect of just wheat on density of quail we might wish also to examine the effect of supplying extra salt. There are now two factors: WHEAT and SALT. The questions now become:

1 Does WHEAT affect density?
2 Does SALT affect density?
3 Is the effect of WHEAT on density influenced by the level of SALT, and vice versa?

In statistical language the last question deals with the interaction between the two factors, whether their individual effects on density are additive (i.e. independent of each other) or whether the effect of a level of one factor changes according to which level of the other factor is combined with it. Section 16.6.3 considers interactions in greater detail.

Figure 16.3 gives an appropriate experimental design for such a two-factor experiment. Its main features are that each level of the first factor is combined with each level of the second, that there are therefore $2 \times 4 = 8$ cells or treatments, that each treatment is replicated, and that the number of replicates per treatment is the same for all treatments.

16.5.1 Controls

A control is that level of a factor subjected to zero treatment. That is not to say that it is necessarily left undisturbed. Everything done to the other levels must also be done to the control, other than for the manipulation that is formally the focus of the treatment. If vehicles are driven over the quail plots to distribute the grain they must also be driven over the control areas.

Controls must obviously be appropriate, and often a good deal of thought is needed to ensure that they are. We have previously dealt with the mistake of declaring "before treatment" a control on "after treatment" (see Section 16.4) but there are more subtle traps to keep in mind. If the treatment is an insecticide dissolved in a solvent, then the control plots must be sprayed with the solvent minus the insecticide. If treated birds are banded then the control birds must be banded. If animals are removed from the field to the laboratory for treatment and then released, the control animals must also be subjected to that disturbance. And so on.

16.5.2 *Sample size*

There is no general answer to the question "How many replicates are necessary?" other than the trite "at least two per treatment." It depends upon the number of treatments to be compared, the average variance among replicates within treatments, and the magnitude of the differences one expects or is attempting to establish. These may be estimated from a pilot experiment or from a previous experiment in the same area.

As a general rule, however, the fewer the treatments the more replicates needed per treatment, but there is little to be gained from increasing replication beyond 30 degrees of freedom for the residual. Suppose the experiment had three factors with i levels in the first, j in the second, and k in the third. There are thus ijk treatments and $ijk(n - 1)$ degrees of freedom in the residual, where n is the number of replicates per treatment.

16.5.3 *Standard experimental designs*

Most questions on the effect of this or that management treatment have a similar logical structure, even though they deal with different animals in different conditions. The most common questions lead to standard experimental designs.

One factor, two levels

Figure 16.1 represents the simplest design that will provide an answer that can be trusted. It evaluates the operational null hypothesis that supplementary feeding with

Fig. 16.1 Minimum one-factor experimental design.

Factor: WHEAT (2 levels)
Response variable: Density or rate of increase of quail

Design logic

WHEAT	No WHEAT
Rep Rep	Rep Rep

Design layout

Site 1 WHEAT added	Site 2 No WHEAT
Site 3 No WHEAT	Site 4 WHEAT added

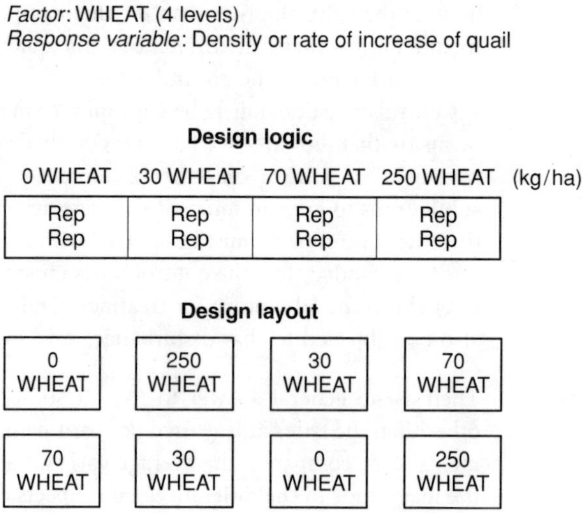

Fig. 16.2 One-factor
experimental design
where the factor has
more than two levels.

Factor: WHEAT (4 levels)
Response variable: Density or rate of increase of quail

Design logic

0 WHEAT   30 WHEAT   70 WHEAT   250 WHEAT   (kg/ha)

Rep Rep	Rep Rep	Rep Rep	Rep Rep

Design layout

0 WHEAT	250 WHEAT	30 WHEAT	70 WHEAT
70 WHEAT	30 WHEAT	0 WHEAT	250 WHEAT

wheat has no effect on the density (or rate of increase) of quail. What is tested, how-ever, is the statistical null hypothesis that the difference between treatments is not significantly greater than the difference between replicates within treatments. If the experiment rejects that null hypothesis we accept as highly likely the alternative hypo-thesis that supplementary feeding affects the dynamics of quail populations living in conditions similar to those of the populations being studied.

This design tests the effect of only one factor (WHEAT) and evaluates it at only two levels (no wheat and some wheat). Note that the diagram of design logic calls for two replicates at each level. The diagram of design layout shows that the treatments are interspersed: thus we do not have the zero treatments (i.e. controls) bunched together in one region and the wheat-added treatments in a second region.

One factor, several levels
This design (Fig. 16.2) is similar to the last, with the difference that the effect of supplementary feeding with wheat is evaluated at four levels: 0, 30, 70, and 250 kg/ha of wheat distributed each month. It allows an answer to two questions: first, whether supplementary feeding has any effect at all upon the density of the quail; and second, whether that effect varies according to the level of supplementary feeding. An answer to the second question allows a cost–benefit analysis on the optimum level of supplementary feeding. Treatment replication and interspersion of treatments is maintained.

Two factors, two or several levels per factor
In this design (Fig. 16.3) the effect of supplementary feeding on quail density is evaluated in tandem with an evaluation of a second factor, the provision of rock salt.

Although the two factors could have been evaluated by two separate experiments there are large advantages in combining them within the same experiment. It pro-vides an answer to a question that might prove to be of considerable importance: do the two factors interact?

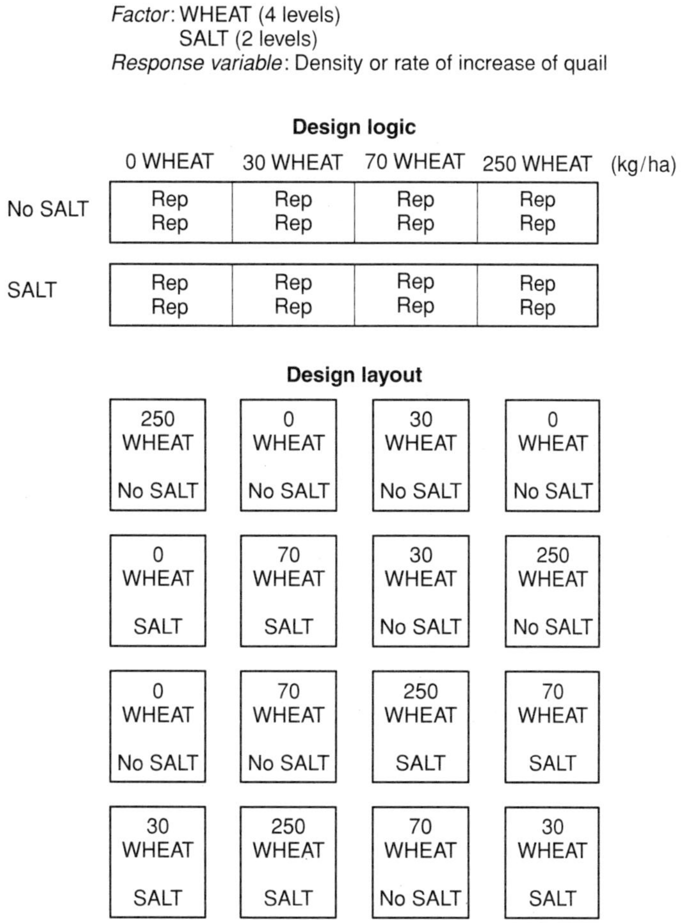

Fig. 16.3 Two-factor experimental design. Do the two factors act independently of each other or do they interact? If the latter, do they reinforce each other or oppose each other?

Factor: WHEAT (4 levels)
　　　　SALT (2 levels)
Response variable: Density or rate of increase of quail

Design logic

	0 WHEAT	30 WHEAT	70 WHEAT	250 WHEAT	(kg/ha)
No SALT	Rep Rep	Rep Rep	Rep Rep	Rep Rep	
SALT	Rep Rep	Rep Rep	Rep Rep	Rep Rep	

Design layout

250 WHEAT No SALT	0 WHEAT No SALT	30 WHEAT No SALT	0 WHEAT No SALT
0 WHEAT SALT	70 WHEAT SALT	30 WHEAT No SALT	250 WHEAT No SALT
0 WHEAT No SALT	70 WHEAT No SALT	250 WHEAT SALT	70 WHEAT SALT
30 WHEAT SALT	250 WHEAT SALT	70 WHEAT No SALT	30 WHEAT SALT

Hypothetically, additional salt in the diet of quail might affect their physiology and hence their dynamics, particularly in sodium depleted areas, and the same may be true of supplementary feeding with wheat. However, suppose that supplementary feeding has an effect only when there is adequate salt in the diet. In such circumstances two separate experiments would produce the fallacious conclusion that, whereas salt has an effect, wheat has none. The interactive relationship between the two factors would have been missed and the resultant management would have been inappropriate. One looks for an interaction by calculating whether the effect of the two factors in combination is greater or less than the addition of the two effects when the factors are evaluated separately. That is achieved by ensuring that each level of the first factor is run in combination with each level of the second. The factors are said to be mutually **orthogonal** (at right angles to each other).

The design logic (Fig. 16.3) is seen as a simple extension of the logic of the one-factor design and the design layout continues to adhere to replication and interspersion of treatments. Since there are now eight treatments, each with two replicates, the interspersion of treatments can best be achieved by laying them out either in a systematic manner as with a Latin square or, as in the example, assigning their positions on the ground by random numbers.

Fig. 16.4 Design logic
for a two-factor
experiment on the effect
of sheep and of rabbits
on the biomass of
pasture.

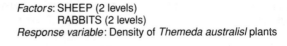

Factors: SHEEP (2 levels)
           RABBITS (2 levels)
Response variable: Density of *Themeda australisl* plants

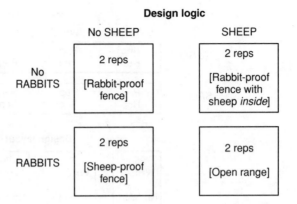

Design logic

16.5.4 *Weak-inference designs*

Very often a field experiment breaches one or more rules of experimental design and so no longer answers unambiguously the question being posed. Such an occurrence has two causes: an unfortunate mistake or a necessary choice.

Mistakes

Very often there may be no logistical or technical justification for using an inappropriate design. Such a flaw is simply a mistake. One of the most common in ecological and wildlife research is pseudo-replication (= subsampling), used under the misapprehension that it constitutes treatment replication (Hurlbert 1984). In this case site and treatment are confounded (see Section 16.4.1).

A second common mistake is the unbalanced design. Figure 16.4 illustrates an experiment to evaluate the effect of grazing by sheep and rabbits on the density of a species of grass. There are two factors (SHEEP and RABBITS), each with two levels (presence and absence). "Presence" for sheep is taken as the standard stocking rate, and that for rabbits the prevailing density. Variation of rabbit density across the area is taken care of by the replication.

The four treatments may be symbolized by a code in which 1 indicates presence and 0 indicates absence. Most of the practical details of setting up such a trial are simple. A rabbit-proof fence around a quadrat excludes both sheep and rabbits (treatment R0 S0). A sheep-proof fence excludes sheep but allows rabbits in (R1 S0). A quadrat to measure the effect of sheep and rabbits together is simply an unfenced square marked by four pegs (R1 S1). Thus, three of the four treatments are easily arranged. They can be set up and then temporarily forgotten, the experimenter returning after several months or even years to harvest the data.

The final treatment (R0 S1) cannot be managed in this way. No one has yet invented a fence that acts as a barrier to rabbits while allowing sheep free access to the quadrat. Hence R0 S1 must be handled differently. It requires a rabbit-proof fence around the quadrat to exclude rabbits (as for R0 S0) but with sheep at standard stocking density *within* the enclosure. That treatment cannot be set up and then left untended. Sheep need water and husbandry. Hence, that treatment is often left out of the experiment. There results a set of data in which the individual effects of sheep on vegetation cannot be disentangled from the effects of rabbits, the total justification of the experiment in the first place.

We have seen many such incomplete experiments set up, often at some expense. They provide estimates of the effect on the vegetation of rabbits alone and of sheep and rabbits together, but not of sheep alone. The effect of sheep alone cannot be obtained indirectly by subtraction because that works only where the two effects are additive (i.e. no interaction). But a significant interaction can quite safely be assumed because each blade of grass eaten by a sheep is no longer available to a rabbit, and vice versa.

Necessary compromises

There are a number of problems that involve passive observation of a pattern or process not under the researcher's manipulative control. In these circumstances a tight experimental design is sometimes not possible, or alternatively the problem may not be open to classical scientific method. In many fields, for example astronomy, geology, and economics, such problems are the rule rather than the exception. A common example from ecology is the environmental impact assessment (EIA). As Eberhardt and Thomas (1991) put it: "the basic problem in impact studies is that evaluation of the environmental impact of a single installation of, say, a nuclear power station on a river, cannot very well be formulated in the context of the classical agricultural experimental design, since there is only one 'treatment' – the particular power-generating station." In fact the problem is even more intractable: EIA studies do not test hypotheses. However, EIAs are still necessary. That they generate only weak inference is no good argument against doing them.

Weak inference results also from a second class of problems: where tight experimental design is theoretically possible but not practical. In such circumstances we may have an unbalanced design, or poor interspersion of treatments, or insufficient replication or even no replication. Again the results are not useless but they must be treated for what they are: possibilities, which may be confirmed by further research. This brings us into the realm of meta-analysis where replication of complete studies is the answer (Johnson 2002).

Weak inference is seldom harmful and can be very useful so long as its unreliability is recognized. Weak inference mistaken for strong inference can be ruinously dangerous.

16.6 Some standard analyses

There are several possible analyses available for any given experimental or survey design. Sometimes they give much the same answer and sometimes different answers. The former reflect only that there is more than one way of doing things; the latter reflect differing assumptions underlying the analyses. Hence, it is important to know what a particular analysis can and cannot do lest one chooses the wrong one. For example, chi-square tests are used only on frequencies (i.e. counts that come as whole numbers); analysis of variance (called also A of V or ANOVA) can deal both with frequencies and with continuous measurements. The Student's t-test is a special case of ANOVA and shares its underlying assumptions.

We will use ANOVA to introduce a broad class of analyses appropriate for the majority of experimental and survey designs. Any statistical textbook will take this discussion further and present additional analytical options.

16.6.1 *One-factor*
ANOVA

The one-factor ANOVA tests the hypothesis that the response variable does not vary with the level of the factor. The alternative hypothesis is that the response variable differs according to the level of the factor, either generally increasing or decreasing with its level, or going up then down or the reverse, or varying in an unsystematic manner.

Our example (Box 16.1) comprises counts of kangaroos on randomly placed east–west transects, each 90 km in length, on a single degree block in southwest Queensland, Australia. The question of particular concern is whether there is an order effect in days of survey. Did the kangaroos become increasingly disturbed by the aircraft and therefore seek cover whenever one was heard? Or did they become progressively habituated to the noise such that more were seen each day as the survey

Box 16.1 Red kangaroos counted on the Cunnamulla degree block (10,870 km^2) in August 1986. Each replicate is the number of kangaroos counted on a transect measuring 0.4 km by 90 km.

Day 1	Day 2	Day 3
96	71	28
38	45	43
80	45	29
35	67	36
50	31	37
55	28	59
38	84	lost
64	70	lost

$n_1 = 8$ $n_2 = 8$ $n_3 = 6$
$T_1 = 456$ $T_2 = 441$ $T_3 = 232$
$\bar{x}_1 = 57.0$ $\bar{x}_2 = 55.1$ $\bar{x}_3 = 38.7$
$k = $ number of classes $= 3$
$N = $ number of samples $= n_1 + n_2 + n_3 = 22$
$\sum X_{ij} = 96 + 38 + \ldots + 37 + 59 = 1129$
$\sum X_{ij}^2 = 96^2 + 38^2 + \ldots + 37^2 + 59^2 = 66{,}251$
$\sum(T_i^2/n_i) = 456^2/8 + 441^2/8 + 232^2/6 = 59{,}273$

Main effects sum of squares (SS):
$\sum(T_i^2/n_i) - (\sum X_{ij})^2/N = 59{,}273 - 1129^2/22 = 1335$

Residual SS:
$\sum X_{ij}^2 - \sum(T_i^2/n_i) = 66{,}251 - 59{,}273 = 6978$

Total SS:
$\sum X_{ij}^2 - (\sum X_{ij})^2/N = 66{,}251 - 1129^2/22 = 8313$

ANOVA

Source	SS	d.f.	MS	F
Main effect	1335	$k - 1 = 2$	667.5	$\dfrac{667.5}{367} = 1.8$
Residual	6978	$N - k = 19$	367	
Total	8313	$N - 1 = 21$		

$F = 1.8$ with 2 d.f. in the numerator and 19 in the denominator. The probability is 0.19, too high to argue for rejection of the null hypothesis that observable density does not differ by day of survey.

progressed? The null hypothesis is that the average seen per transect per day is independent of the day order.

Note that factor DAY contains three levels, the first day, the second day, and the third day. The last contains only six replicates in contrast to the eight of the first two days. It will make the point that the arithmetic of one-factor ANOVA does not require that the design is balanced (i.e. the number of replicates is the same for all levels). The analysis can be run without balance although the result must be interpreted more cautiously. Balance should always be sought, if not necessarily always attained.

The analysis of Box 16.1 leads to an F ratio (named for R.A. Fisher who invented analysis of variance) testing the null hypothesis. Appendix 1 gives its critical values. The probability of 20% is too high to call the null hypothesis into serious question. That value is the probability of drawing by chance three daily samples as disparate or more disparate than those we did draw, when there is no difference in density or sightability between days. We would require a probability value of around 10% before we became suspicious of the null hypothesis, and one below 5% before we rejected the null hypothesis in favor of some alternative explanation.

16.6.2 Two-factor ANOVA

A two-factor ANOVA tests simultaneously for an effect of two separate factors on a response variable and for an interaction between them. Even though the arithmetic is simply a generalization of the one-factor case, the two-factor ANOVA differs in kind from the one-factor because of the interaction term. There are also a number of other differences, but we will get to them after we have considered an example.

Data for a two-factor ANOVA are laid out as a two-dimensional matrix with the rows representing the levels of one factor and the columns the levels of the other. These are interchangeable. Each cell of the matrix contains the replicate readings of the response variable, whatever it is. Table 16.1 outlines symbolically and formally the calculation of the sums of squares and degrees of freedom for the four components into which the total sum of squares is split: the effect on the response variable of the factor represented by the rows, the effect of the factor represented by the columns, the effect of the interaction between them (of which more soon), and the remaining or residual sum of squares which represents the average intrinsic variation within each treatment cell and which therefore is not ascribable to either the factors or their interaction.

Box 16.2 provides a set of data amenable to a two-factor ANOVA. As with the one-factor example they are real data from an aerial survey whose purpose was to establish whether the counts obtained on a given day were influenced by the disturbance or habituation imparted by the survey flying of previous days. However, two species were counted this time, the red kangaroo and the eastern gray kangaroo, and since they might well react in differing ways to the sound of a low-flying aircraft their counts

Table 16.1 Calculations of sums of squares for two-factor ANOVA.

ROW effect	$(1/nc)\sum T_i^2 - (1/nrc)T^2$	d.f. $= r - 1$
COLUMN effect	$(1/nr)\sum T_j^2 - (1/nrc)T^2$	d.f. $= c - 1$
ROW $\times$ COLUMN effect	$(1/n)\sum T_{ij}^2 - (1/nc)\sum T_i^2 - (1/nr)\sum T_j^2 + (1/nrc)T^2$	d.f. $= (r - 1)(c - 1)$
Residual	$\sum X_{ijk}^2 - (1/n)\sum T_{ij}^2$	d.f. $= rc(n - 1)$
Total	$\sum X_{ijk}^2 - (1/nrc)T^2$	d.f. $= rcn - 1$

T_{ij}, total of replicates in the cell at the ith row and jth column; T_i, total of replicates in the ith row; T_j, total of replicates in the jth column; T, grand total; r, number of rows; c, number of columns; n, number of replicates per cell.

Box 16.2 Red kangaroos and gray kangaroos counted on the Cunnamulla degree block ($10,870$ km^2) in June 1987. Each replicate is the number of kangaroos counted on a transect measuring 0.4 km by 90 km.

	Day 1	Day 2	Day 3	T_i
Red kangaroos	45	19	18	
	17	51	44	
	8	8	61	
	28	11	35	
	26	72	65	
	48	34	76	
	53	67	52	
	62	27	30	
$T_{ij} =$	287	289	381	957
$\bar{x} =$	35.9	36.1	47.6	
Gray kangaroos	66	27	27	
	52	47	66	
	34	13	75	
	8	16	104	
	35	101	109	
	36	150	170	
	42	116	51	
	65	66	14	
$T_{ij} =$	338	536	616	1490
$\bar{x} =$	42.5	67.0	77.0	
$T_j =$	625	825	997	$T = 2447$

r = number of rows = 2
c = number of columns = 4
n = number of replicates per cell = 8

$(1/nc)\sum T_i^2 = (1/24)(957^2 + 1490^2)$ $= 130,665$
$(1/nr)\sum T_j^2 = (1/16)(625^2 + 825^2 + 997^2)$ $= 129,079$
$(1/n)\sum T_{ij}^2 = (1/8)(287^2 + 289^2 + \ldots + 616^2)$ $= 136,506$
$(1/nrc)T^2 = (1/48)(5,957,809)$ $= 124,746$
$\sum X_{ijk}^2 = 45^2 + 17^2 + \ldots + 14^2$ $= 184,081$

Sum of squares

ROW (SPECIES)	$130,665 - 124,746$	$= 5,919$
COLUMN (DAYS)	$129,079 - 124,746$	$= 4,333$
ROW × COLUMN	$136,506 - 130,665 - 129,079 + 124,746$	$= 1,508$
Residual	$184,081 - 136,506$	$= 47,575$
Total	$184,081 - 124,746$	$= 59,335$

ANOVA

Source	SS	d.f.	MS	F
ROW (SPECIES)	5,919	$r - 1 = 1$	5919	5.22
COLUMN (DAYS)	4,333	$c - 1 = 2$	2166	1.91
SPECIES × DAYS	1,508	$(r - 1)(c - 1) = 2$	754	0.67
Residual	47,575	$rc(n - 1) = 42$	1133	
Total	59,335	$rcn - 1 = 47$		

are kept separate for purposes of analysis. Red kangaroos and eastern gray kangaroos are now the two levels of the factor SPECIES.

In the ANOVA at the bottom of the box, the sum of squares of each source of variation is divided by the respective degrees of freedom to form a mean square. (Mean square is just another name for variance.) The three sources of variance of interest, those of the two factors and their interaction, are divided (in this case) by the residual mean square to form the F ratios that are our test statistics. That for SPECIES is 5.22 and we check that for significance by looking up the F table of Appendix 1. It will show that an F with 1 degree of freedom in the numerator, and 42 (say 40) in the denominator, will have to exceed 4.08 if the magic 5% or lower probability is to be attained. We therefore conclude that the disparity in observed numbers between reds and grays, 957 as against 1490, is more than a quirk of sampling, that gray kangaroos really were more numerous than reds on the Cunnamulla block at the time of survey.

In like manner we test for a day effect. The trend in day totals – 625, 825, and 997 kangaroos – suggests that the animals are becoming habituated to aircraft noise and hence progressively more visible, day-by-day. The F tables show however that, with degrees of freedom 2 and 42, a one-tail probability of 5% or better would require $F = 3.23$. Ours reached only 1.91, equivalent to a probability of 16%, and so we are not tempted to replace our null hypothesis (no day effect) with the alternative explanation suggested by a superficial look at the data.

The F ratio for interaction was less than 1, indicating that the mean square for interaction is less than the residual mean square. It cannot, therefore, be significant and we do not even bother to look up the probability associated with it.

16.6.3 *What is an interaction?*

In the last section we tested an interaction, and found it non-significant, without really exploring what question we were answering. A non-significant interaction implies that the effect of one factor on the response variable is independent of any effect that may be exerted on it by the other factor, that the two factors are each operating alone. The effect of the two factors acting together is exactly the addition of the effects of the two factors each acting in the absence of the other.

If an analysis produces a significant interaction, the relationship should be examined by graphing the response variable against the levels of the first factor. Figure 16.5 shows the kind of trends most commonly encountered. A significant interaction is telling you that no statement can be made as to the effect on the response variable of a particular level of the first factor unless we know the prevailing level of the second factor. The graph will make that clear.

It is entirely possible for an ANOVA to reveal no main effect of the first factor, no main effect of the second, but a massive interaction between them. A graphing of the response variable will reveal a crossing over pattern as in the last graph of Fig. 16.5.

16.6.4 *Heterogeneity of variance*

The main assumption underlying ANOVA is that the variance of the response variable is constant across treatments. The means may differ (and that is in fact what we are testing to discover) but the variances remain the same. A violation of this assumption can seriously bias the test. Consequently, we need to test for heterogeneity of variance and, if we find it, either transform the data to render the variances homogeneous or use an alternative method such as analysis of deviance that does not employ the assumption of homogeneity.

Fig. 16.5 Common
forms of interaction in
two-factor ANOVAS. ns,
Not significant; sig,
significant.

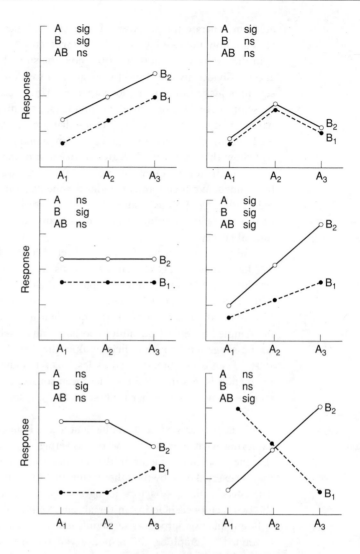

The most common test for heterogeneity of variance is Bartlett's. It can be found in almost all statistical texts. Recent work has shown, however, that it is too sensitive. Analysis of variance is an immensely robust test that performs well even when the assumptions of the analysis are not met in full. It copes well with minor heterogeneity of variance and with deviations from normality. About the only thing that throws it out badly is bimodality of the response variable. A better test is Cochran's C, whose test statistic is simply the largest variance in a cell divided by the sum of all cell variances. For the two-factor ANOVA given in Box 16.2 the largest cell variance is returned by the replicate counts of gray kangaroos on day 2. It is $s^2 = 2566$. The sum of all six variances is $\Sigma s^2 = 6796.6$, and so Cochran's $C = 2566/6796.6 = 0.378$. Looking up a table of the critical values of Cochran's C (Appendix 2) reveals that the test statistic would have to exceed $C = 0.398$ (d.f. = 7 per variance and there are six variances) to represent a significant departure from homogeneity of variance. We can thus choose to analyze without transformation.

In many biological cases the variance of the response variable rises with the mean. That is particularly true of counts of animals that tend to fit a negative binomial distribution. A transformation of the counts to logarithms after adding 0.1 (to knock out the zeros) will usually stabilize the variances. If the animals are solitary rather than gregarious their counts are more likely to fit a Poisson distribution, which is characterized by the mean and the variance being identical. Transformation to square roots after adding 3/8 to each will homogenize the variances. The variances of almost all body measurements regress on their means and the data need log transformation. Use an arcsine transformation if data come as percentages.

16.6.5 *Are the factors fixed or random?*

Before any data are collected, let alone analyzed, we must decide precisely what questions are to be asked of them. Take the example of comparing counts of kangaroos on successive days. We were asking whether the act of flying over the study area on one day influenced the counts obtained the next day. The influence could conceivably be negative (disturbance forced the kangaroos into cover in front of the surveying aircraft) or positive (the kangaroos became progressively habituated to aircraft noise).

Now take another question: do viewing conditions differ between days of survey? That question would be answered by counts obtained by days sampled at random. We would want those days to be spaced rather than consecutive as they were in answering the question about the effect of day order. Otherwise the answers to the two questions would be confounded and we would not know which was being answered by a significant day effect. In the question concerning an order effect of days, the factor DAY is said to be fixed. No arbitrary selection of any three days will do. The days have to follow each other without gaps between them.

Whether a factor is declared fixed or random determines both the question being asked and the denominator of the F ratio that answers it. Table 16.2 shows the appropriate choice of denominators. For two of the three-factor models there is no explicit test for the significance of some of the factors. There are various messy approximations available (see Zar 1996, appendix) but it is far better to rephrase the question to one logically answerable from consideration of the data.

Let us generalize the difference between fixed and random factors. A fixed factor is one whose levels cover exhaustively the range of interest. MONTHS therefore usually constitute a fixed factor because they index seasons. YEARS may be fixed or random depending on context. REGIONS may be fixed if the levels of that factor are the only ones of interest. If they are simply a random sample of regions, and any other selection of regions would serve as well, the factor REGIONS is random.

Note that questions change according to whether the factor is fixed or random. Suppose that the response variable is growth rate of a species of pine and we wish to test for a difference among three soil types (factor SOILS) covering the entire range of soil types of interest. The factor SOILS is thus fixed. If the question concerns the best region in which to plant a plantation of that species, and there are four and only four regions that are possible candidates, the choice of regions is fixed and the appropriate denominator of the F testing a difference of growth rate among the soil types of those regions is the residual mean square. However, if we ask the more general question of whether grow rate differs among soil types across regions in general, any random selection of a set of regions will suffice and the denominator of the F testing SOILS is the mean square of the interaction between SOILS and REGIONS (Table 16.2).

Table 16.2 The mean square providing the denominator for an F-ratio testing significance of a factor or interaction (i.e. source of variation).

Model	Source	Denominator of F
One-factor ANOVA		
A fixed	A	MS_e
A random	A	MS_e
Two-factor ANOVA		
A fixed, B fixed	A	MS_e
	B	MS_e
	AB	MS_e
A fixed, B random	A	MS_{AB}
	B	MS_e
	AB	MS_e
A random, B random	A	MS_{AB}
	B	MS_{AB}
	AB	MS_e
Three-factor ANOVA		
A fixed, B fixed, C fixed	A	MS_e
	B	MS_e
	C	MS_e
	AB	MS_e
	AC	MS_e
	BC	MS_e
	ABC	MS_e
A fixed, B fixed, C random	A	MS_{AC}
	B	MS_{BC}
	C	MS_{ABC}
	AB	MS_e
	AC	MS_e
	BC	MS_e
	ABC	MS_e
A fixed, B random, C random	A	XXX
	B	MS_{BC}
	C	MS_{BC}
	AB	MS_{ABC}
	AC	MS_{ABC}
	BC	MS_e
	ABC	MS_e
A random, B random, C random	A	XXX
	B	XXX
	C	XXX
	AB	MS_{ABC}
	AC	MS_{ABC}
	BC	MS_{ABC}
	ABC	MS_e

XXX, no explicit test is possible; MS_e, mean square of the residual.

16.6.6 *Three-factor ANOVA*

The three-factor ANOVA follows the general lines of the two-factor except that seven rather than three questions are now being addressed simultaneously. Table 16.3 gives the sums of squares and degrees of freedom.

Box 16.3 gives an example, again kangaroos counted by aerial survey, but with factor YEARS added. YEARS is fixed because we wish to test specifically whether density

Table 16.3 Calculation of sums of squares and degrees of freedom for three-factor ANOVA.

ROW effect	$(1/ncl)\sum T_i^2 - (1/nrcl)T^2$	d.f. $= r - 1$
COLUMN effect	$(1/nrl)\sum T_j^2 - (1/nrcl)T^2$	d.f. $= c - 1$
LAYER effect	$(1/nrc)\sum T_k^2 - (1/nrcl)T^2$	d.f. $= l - 1$
RC interaction	$(1/nl)\sum T_{ij}^2 - (1/ncl)\sum T_i^2 - (1/nrl)\sum T_j^2 + (1/nrcl)T^2$	d.f. $= (r-1)(c-1)$
CL interaction	$(1/nr)\sum T_{jk}^2 - (1/nrl)\sum T_j^2 - (1/nrc)\sum T_k^2 + (1/nrcl)T^2$	d.f. $= (c-1)(l-1)$
RL interaction	$(1/nc)\sum T_{ik}^2 - (1/ncl)\sum T_i^2 - (1/nrc)\sum T_k^2 + (1/nrcl)T^2$	d.f. $= (r-1)(l-1)$
RCL interaction	$(1/n)\sum T_{ijk}^2 - (1/nl)\sum T_{ij}^2 - (1/nr)\sum T_{jk}^2 - (1/nc)\sum T_{ik}^2$ $+ (1/ncl)\sum T_i^2 + (1/nrl)\sum T_j^2 + (1/nrc)\sum T_k^2 - (1/nrcl)T^2$	d.f. $= (r-1)(c-1)(l-1)$
Residual	$\sum X_{ijkm}^2 - (1/n)\sum T_{ijk}^2$	d.f. $= (n-1)rcl$
Total	$\sum X_{ijkm}^2 - (1/nrcl)T^2$	d.f. $= nrcl - 1$

changed between 1987 and 1988. We do not wish to test the more general question of whether kangaroo populations remain stable with time. The ANOVA shows that of the three main effects, the three first-order interactions, and the one second-order interaction, only the main effect of kangaroo species is significant. We conclude therefore that the two species certainly differed in density but that there was insufficient evidence to identify a day effect or a change in density between years. Neither did any factor appear to interact with any other.

16.7 Summary

Testing of wildlife management treatments requires rigorous definition of the expected outcome of the treatments. Once a verifiable outcome is posed as a hypothesis, the data to test it can be collected by following the logic of experimental design. Insufficient replication of treatments to sample the range of natural variability is a common shortcut, but it nullifies the point of the exercise.

The principles illustrated in this chapter can be summarized as the basic rules of experimental design. There are exceptions to several of them, but until the manager or researcher learns how and in what circumstances they may safely be varied, these should be followed in full.

1 To determine whether a factor affects the response variable under study, more than one level of that factor should be examined. The levels may be zero (control) and some non-zero amount, or they may be two or more categories (e.g. habitat types) or non-zero quantities (e.g. altitudinal bands).

2 "Before" is a poor control for "after," because subsequent trends can be caused by other influences unrelated to the treatment under study.

3 Treatments must be replicated, not subsampled. (See Hurlbert (1984) for an excellent exposition of the pitfalls of "pseudo-replication" in ecological research.)

4 The number of replications per treatment (including the control treatment) should be as close as possible to equal across treatments.

5 Treatments must be interspersed in time and space. Do not run the replications of treatment A and then the replications of treatment B. Mix up the order. Do not site the replications of treatment A in the north of the study region and the replications of treatment B in the south. Mix them up.

6 If the influence of more than one factor is of interest, each level of each factor should be examined in combination with each level of every other factor (factorial design).

7 If an extraneous influence (site in the quail example) is likely to be correlated with one of the designated factors, either it should be declared a factor in its own right and the design modified accordingly or its range should be covered at random by

Box 16.3 Red kangaroos and gray kangaroos counted on the Cunnamulla degree block (10,870 km^2) in 1987 and 1988. Each replicate is the number of kangaroos counted on a transect measuring 0.4 km by 90 km.

	June 1987			June 1988		
	Day 1	Day 2	Day 3	Day 1	Day 2	Day 3
Red kangaroos	45	19	18	72	31	28
	17	51	44	32	29	9
	8	8	61	94	34	48
	28	11	35	27	47	38
	26	72	65	66	21	91
	48	34	76	55	49	138
	53	67	52	102	29	67
	62	27	30	41	67	106
$T_{ijk} =$	287	289	381	489	307	525
$\bar{x} =$	35.9	36.1	47.6	61.1	38.4	65.6
Gray kangaroos	66	27	27	116	65	10
	52	47	66	81	57	22
	34	13	75	63	41	43
	8	16	104	25	33	63
	35	101	109	75	120	74
	36	150	170	77	28	101
	42	116	51	59	62	76
	65	66	14	59	17	82
$T_{ijk} =$	338	536	616	555	423	471
$\bar{x} =$	42.3	67.0	77.0	69.4	52.9	58.9

ANOVA

Source	SS	d.f.	MS	F	$F_{0.05}$*
ROW (SPECIES)	4,551.3	1	4551.3	4.7	4.0
COLUMN (DAYS)	3,227.3	2	1613.6	1.6	3.1
LAYER (YEARS)	1,086.8	1	1086.8	1.1	4.0
RC interaction	1,018.1	2	509.0	0.5	3.1
CL interaction	4,681.6	2	2340.8	2.3	3.1
RL interaction	1,708.6	1	1708.6	1.7	4.0
RCL interaction	1,444.7	2	722.3	0.7	3.1
Residual	86,359.4	84	1028.1		
Total	104,077.7	95			

*$F_{0.05}$ is the critical level that must be exceeded by the observed F to qualify for a probability equal to or less than 5%.

the replication. Thus its influence is factorized out in the first option or randomized across treatments in the second.

8 All of these rules may be broken, but that degrades the design to one yielding neither strong inference nor an unambiguous conclusion. Such results are still useful so long as their dubious nature is fully appreciated and declared. Environmental impact assessments are just such examples.

17 Conservation in theory

17.1 Introduction

In this chapter we deal with theory that has been developed to account for why and how populations become extinct. Most of that theory deals with extinction as a consequence of low numbers, the various difficulties that a population can get into when it is too small. A second class of extinction processes – those caused by a permanent and deleterious change in the population's environment – is less well served by theory, but promising new approaches are under development.

17.2 Demographic problems contributing to risk of extinction

Demography deals with the probability of individuals living or dying and, if they live, the probability that they will reproduce. Those individual probabilities, accumulated over all individuals in the population, determine what the population as a whole will do next, whether it will increase, decrease, or remain at the same size. Three effects can influence the population outcome underlain by those individual probabilities: individual variation, short-term environmental variation, and environmental change. These will be examined in turn, particularly in the context of the likelihood of the population going extinct.

17.2.1 Effect of individual variation

A population's rate of increase is determined by the age-specific fecundity rates interacting with the age-specific mortality rates, but its value is predictable only when the population has a stable age distribution. If it does not have a stable age distribution, or if numbers are low, the actual rate of increase may vary markedly in either direction from that predicted by the life table and fecundity table (see Chapter 6). This effect is called **demographic stochasticity**.

Take the hypothetical case of a population of 1000 large mammals whose intrinsic rate of increase is $r_m = 0.28$. A female can produce no more than one offspring per year. The population is at a low density of $D = 0.01/\text{km}^2$ so there will be little competition for resources and consequently rate of increase will be close to r_m. On average the probability of an individual surviving 1 year is $p = 0.9$, and the probability that a female will produce an offspring over a year is $b = 0.95$. The beginning of the year is defined as immediately after the birth pulse, at which time the population contains 500 males and 500 females. By the end of the year the population will have been reduced by natural mortality to about 900 (i.e. 1000×0.9) and these animals produce about 428 offspring (450×0.95) at the next birth pulse. The population therefore starts the next year with about 1328 individuals ($900 + 428$), having registered a net increase over the year at about the rate $r = \log_e(1328/1000) = 0.28$, or 32%. The actual outcome will be very close to those figures because the differences in demographic behavior between individuals tend to cancel out.

Table 17.1 Probabilities for the population outcome over a year of a population comprising two individuals, one of each sex. The chance of an individual surviving the year is $p = 0.9$ and the chance of the female producing an offspring at the end of that year is $m = 0.95$.

N_{t+1}	What happened	Probability of outcome		r
		Symbolic	Numerical	
0	Both die	$(1 - p)^2$	0.01	$-\infty$
1	One dies	$2p(1 - p)$	0.18	-0.69
2	Both live, no offspring	$p^2(1 - m)$	0.0405	0.0
3	Both live, one offspring	p^2m	0.7695	0.41
			1.0000	

Table 17.2 Deviation from expected rate of increase resulting from stochastic variation. The influence of one individual on variance in r is taken as $\mathrm{Var}(r)_1 = 0.5$.

N	Expected r	Var(r)	SE(r)	95% confidence limits of r
10	0.28	0.05	0.224	± 0.500
50	0.28	0.01	0.100	± 0.202
100	0.28	0.005	0.071	± 0.139
500	0.28	0.001	0.032	± 0.063
1000	0.28	0.0005	0.022	± 0.043

Now consider a subset of this population restricted to a reserve of 200 km². The density of 0.01/km² translates to a population size of two individuals. These two obviously cannot increase by 32% to 2.64 individuals as the large-population estimate would imply. They can only increase to 3, or remain at 2, or decline to 1 or even to 0. Table 17.1 gives the probabilities of those outcomes.

Table 17.1 shows that the most likely outcome is 3 animals and a rate of increase of $r = 0.41$. But even though the population is "trying" to increase, the actual rate of increase may by chance vary between a low of minus infinity to a high of $r = 0.41$. Hence the demographic behavior of a small population is determined by the luck and misfortune of individuals. It is a lottery. That of a large population is ruled by the law of averages. We say that the outcome for a small population is stochastic and for a large population deterministic.

The extent to which actual r is likely to vary from its deterministic value in a constant environment is measured by $\mathrm{Var}(r) = \mathrm{Var}(r)_1/N$, where $\mathrm{Var}(r)_1$ is the component of variance in r attributable to the demographic behavior of an average individual. For a population with a relatively low r_m, as in our hypothetical example, $\mathrm{Var}(r)_1$ will be in the region of 0.5. We adopt that value for purposes of illustration. $\mathrm{Var}(r)$ declines progressively as population size N rises. Variance of r at any population size N can be estimated for this "population" as $\mathrm{Var}(r) = 0.5/N$.

Table 17.2 shows that at a population size of $N = 50$ the effects of small numbers and hence necessarily unstable age distribution can result in a rate of increase varying (at 95% confidence) between $r = 0.48$ (i.e. $0.28 + 0.202$) and $r = 0.08$ (i.e. $0.28 - 0.202$). At $N = 10$ the possible outcomes vary between a high rate of increase and a steep decline. In this example the deterministic rate of increase becomes a safe guide to the actual rate of increase only after the population has attained a size of several hundred.

Although the details are special the message is general: populations containing fewer than about 30 individuals can quite easily be walked to extinction by the random demographic variation between individuals, even when those individuals are in the peak of health and the environment is entirely favorable.

17.2.2 The effect of environmental variation

We have seen that the demographic behavior of a population in a constant environment is broadly predictable when it contains several hundred individuals. The larger the population the tighter the correspondency between actual rate of increase and the expected deterministic rate of increase. However, individual variation is by no means the most important source of variation in r. Year-to-year variation in environmental conditions has a more profound effect and, unlike the effect of individual variation, does not decline with increasing population size. It is called **environmental stochasticity**.

The most important source of environmental variation is yearly fluctuation in weather. Weather has a direct effect on the demography of plants, invertebrates, and cold-blooded vertebrates. Their rates of growth are often a direct function of temperature as measured in degree-days. Wildlife is largely buffered against the direct effect of temperature and humidity, the influence being indirect through food supply.

We denote as $Var(r)_e$ the variance in r caused by a fluctuating environment. It can be measured as the actual year-to-year variance in r exhibited by a population whose size is large enough to swamp the effect of variance in r due to individual variation. We recommend that such a population should contain at least 5000 individuals. Even so, $Var(r)_e$ will be overestimated because its measurement contains a further component of variance introduced by the sampling variation generated in estimating the year-to-year rates of increase.

The major influence of environmental variation on the probability of extinction is its interaction with the effect of individual variation. Thus it becomes progressively more important with decreasing population size, even though its average effect on r is independent of population size.

17.3 Genetic problems contributing to risk of extinction

In the next few sections we examine some of the ways in which genetic malfunction may contribute to the extinction of a population. But first we provide a brief introduction to population genetics for those who have not studied it previously. Those who have can skip to Section 17.4.

17.3.1 Heterozygosity

A chromosome may be thought of as a long string of segments, called loci, each locus containing a gene in paired form. The two elements of that pair, one contributed by the individual's mother and the other by its father, are called alleles and they can be the same or different. The chromosomes of vertebrates and vascular plants contain around 100,000 loci.

Suppose the gene pool of a population contains only two alleles for locus A. These will be referred to as A_1 and A_2. Any individual in that population will thus have one of three combinations of alleles at that locus: A_1A_1 or A_1A_2 or A_2A_2. If the first or third combination obtains, the individual is **homozygous** at that locus, if the second **heterozygous**. The proportions of the three combinations in the population as a whole are called **genotypic frequencies**. Which will be the most common depends on the frequencies (proportions) of the two alleles in the population as a whole. Suppose the frequency of the A_1 allele is $p = 0.1$ and therefore that A_2 is $q = 0.9$ (because the sum of a complete set of proportions must equal 1), then the frequencies of the three genotypes will be:

Genotypes	A_1A_1	A_1A_2	A_2A_2
Genotypic frequencies	p^2	$2pq$	q^2
=	0.01	0.18	0.81

Note that the genotypic frequencies (proportions) also sum to 1.

That relationship between allelic frequencies and genotypic frequencies is the Hardy–Weinberg equilibrium law. Formally it holds only when the population is large, its individuals mate at random, and there is no migration, mutation, or selection. In practice, however, it is highly robust to deviations from these assumptions and can be accepted as a close approximation to the actual relationship between allelic frequency and genotypic frequency for two alleles at a single locus. The Hardy–Weinberg equilibrium holds equally for more than two alleles at a locus so long as it is calculated in terms of one allele against all the others. Alleles of the two types A_1 and not-A_1 (not-A_1 = $A_2 + A_3 + A_4$, etc.) also take Hardy–Weinberg equilibrium proportions.

The number and frequency of different alleles at a locus can be determined fairly easily by electrophoresis or DNA sequencing. If p_{ij} is the frequency of allele i at locus j in the population as a whole, the proportion of individuals heterozygous at that locus may be estimated as:

$$h_j = 1 - \sum_i p_{ij}^2$$

providing that the number of individuals n_j examined for locus j is greater than 30. If fewer, h_j is underestimated but can be corrected by multiplying by $2n_j/(2n_j - 1)$. Thus the more alleles at a locus the higher the value of h_j, and the less diverse the frequencies of the alleles the higher is h_j.

Mean heterozygosity is estimated as:

$$H = (1/L)\sum_j h_j$$

where L is the number of loci examined. H varies considerably between species for reasons that are not understood. As estimated by one-dimensional electrophoresis of loci controlling production of proteins, H ranges between 0.00 and 0.26 for mammals with an average at about 0.04. Heterozygosity estimated in the same way yields $H = 0.036$ for both white-tailed and mule deer (Gavin and May 1988) and $H = 0.029$ for leopards. Figure 17.1 shows the frequency distribution of H for 169 species of mammals. Note first that the distribution is shaped like a reverse J: most species have a low H but a few break out of that pattern to return a high H. Second, a substantial proportion (10.6%) of mammalian species are homozygous at all loci examined by electrophoresis.

Genetic variability can be reported also as the proportion of polymorphic loci in the population (i.e. the proportion of loci for which there is more than one allele within the population as a whole). This is not the same as H above. If all but one individual in the population were homozygous at locus A that locus is nonetheless scored as polymorphic. The proportion of polymorphic loci within the population is therefore higher than the average heterozygosity (the proportion of heterozygous loci in an average individual), usually about three times higher. Further, it is an

Fig. 17.1 Frequency distribution of mean heterozygosity H for 169 mammalian species. (Data from Nevo *et al.* 1984.)

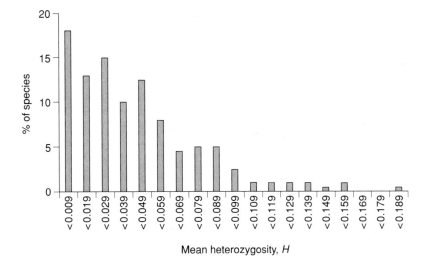

unstable statistic tending to increase as sample size increases. We recommend against its use.

The level of heterozygosity within a population can be calculated by DNA sequencing or gel electrophoresis. Here we consider only the latter. If the function of a given gene is to direct the synthesis of a protein, an individual with heterozygous alleles at that locus will produce two versions of that protein, but only one if homozygous. Identifying the individual as one or the other is a simple technical exercise. A sample of its blood plasma is placed on a slab of gel and subjected to a weak electrical current which causes the plasma proteins to migrate through the gel at a speed specific to each protein. The protein of interest, say transferrin, can be differentiated from the other proteins by staining with Nitroso-R salt (1-nitroso-2-naphthol-3,6-disulfonic acid) and the different versions of that protein, corresponding to different alleles, can be identified by their banding pattern on the gel. By this means we can readily identify which alleles are present at the locus controlling transferrin synthesis and hence whether the individual is heterozygous at that locus. The same applies for other loci and other proteins. A precise value for H can be obtained by examining about 30 loci for each of about 30 individuals, and so H can be monitored fairly easily over time to detect a slump in overall heterozygosity.

17.3.2 *Additive genetic variance*

So far we have dealt only with single genes in isolation, those that either produce an effect or do not. Electrophoretic analysis depends on these: a particular protein is either produced or not according to whether a particular allele is present at the locus of interest. This is the realm of Mendelian genetics, which deals with qualitative characters.

Adaptive evolution, however, is based mainly on the selection of quantitative characters such as foot length or timing of conception. These vary in a continuous fashion. Such a character is influenced by many genes. The variation in a quantitative trait between individuals has two sources, environmental and genetic. The genetic variance can be partitioned into three components: that resulting from the effect of alleles within and between loci (additive genetic variance), that resulting from dominance effects, and that resulting from interaction between loci. The additive genetic variance is by far the most important and for present purposes can be discussed as if it constituted the total genetic variance.

The intricacies of genetic variance are beyond the scope of this book but all the necessary points can be made by using heterozygosity H as an index of additive genetic variance. Those factors that influence additive genetic variance also influence heterozygosity and they do so by similar amounts and in similar ways. The mathematics describing the dynamics of the two are much the same.

At this stage we make an important point on a subject that is widely misunderstood: the nature of "**genetic diversity**." It is usually conceptualized by conservationists as the number of distinct alleles in the population. The loss of one of those alleles is seen as a reduction in genetic diversity and therefore a bad thing. That notion of genetic diversity is trivial and entirely inappropriate to conservation management. Rather, the important measure is **genetic variance**, which can be conceptualized as a parameter closely akin to heterozygosity H which is the proportion of loci that are heterozygous in an average individual. Consequently the amount of heterozygosity carried within a sample comprising a couple of dozen individuals will closely approximate the amount of heterozygosity within the population from which that sample was drawn. By extension, the amount of genetic variance carried by a relatively small sample of the population will closely approximate the magnitude of the genetic variance of that population as a whole. It is the genetic variance of the population that we seek to conserve, not the "genetic variation" or "genetic diversity" represented by the total number of distinct alleles within the population.

That misunderstanding carries over to populations established by translocation. It is often argued that since these were usually started by a nucleus of only a few individuals, and since those individuals must carry only a small fraction of the genetic variability of the population from which they were drawn, the subsequent robust health of the population that developed by the liberation indicates that a population needs little genetic variability. Those populations tend either to increase rapidly or to go extinct (often apparently by simple demographic stochasticity) within the first few years following translocation. If the former, they are shortly free of any further loss of genetic diversity. The condition that gets a population into the greatest trouble is a history of small population size that persists for many generations. That seldom occurs for a wildlife population founded by translocation.

17.3.3 *Drift and mutation*

In the absence of immigration and mutation, the number of different alleles at a locus in the population as a whole can either remain constant or decrease. It cannot increase. In practice it will always decrease because alleles will be lost under the influence of non-random mating and unequal reproductive success between individuals. Heterozygosity thus decreases also. Its rate of decline is a function of population size N, the proportion of heterozygous loci in the population as a whole being reduced by the fraction $1/(2N)$ per generation. Over one generation H changes according to:

$$H_1 = H_0[1 - 1/(2N)]$$

and over t generations:

$$H_t = H_0[1 - 1/(2N)]^t$$

which may alternatively be written:

$$H_t = H_0 e^{-t/2N}$$

After $t = 2N$ generations the population's heterozygosity will have dropped to 0.37 (i.e. e^{-1}) of its initial value at time $t = 0$. This holds as well for a single locus as it does across all loci. The loss of additive genetic variance is exactly analogous and conforms to the same equations. The process is called **random genetic drift**.

The rate of mutation at a single locus is about 10^{-6} per gamete per generation. However, most evolution is in terms of phenotypic characters that change in very small steps. These quantitative characters are controlled by many genes. Mutation within such a gene complex is much more frequent than at a single locus, closer to 10^{-3} or even 10^{-2} per gamete per generation.

Heterozygosity (and hence additive genetic variance) will change over one generation by an amount ΔH according to:

$$\Delta H = -H/2N + m$$

where m is the input of heterozygosity by mutation. Its equilibrium is solved by setting ΔH to zero:

$$H = 2Nm$$

which informs us that for any population size N there will be an equilibrium between mutational input of additive genetic variance and loss of it by drift. What varies, however, is the value of H at equilibrium. It will be higher when N is large and lower when N is small. The population size N must remain constant for many generations for such an equilibrium to establish.

17.3.4 *Selection*

Selection generally comes in two forms: directional and stabilizing (or normalizing). **Directional selection**, that which moves a trait in one direction, is the stuff of evolution. **Stabilizing selection**, on the other hand, eliminates extremes of a trait and holds the trait to its optimum for the current environment. There is a third form, **disruptive selection**, where individuals with intermediate traits are selected against whereas individuals with extreme forms for the traits are favored, but it is much rarer than the other forms of selection.

At any particular time most selection will be of the stabilizing type. Any tendency for the breeding season to expand, for example, will be attacked continuously by stabilizing selection. An offspring born during a seasonally inappropriate time of the year has little chance of survival and so that mistake, if it has a genetic basis, will be swiftly corrected. The relaxation of that selection in captivity is presumably the reason why the breeding season of captive populations tends to expand after several generations.

Stabilizing selection is essential to maintaining the fitness of a wild population. Ironically, it is one of the strongest forces reducing additive genetic variance and heterozygosity. We must therefore avoid an obsession with maximizing variance to the exclusion of maintaining fitness by reducing the number of deleterious alleles.

17.3.5 *Inbreeding depression*

Suppose the A_1 is a recessive, its effect being masked by A_2 when the two occur together at the locus. If additionally A_1 is slightly deleterious when its effect is expressed, it can have damaging effects on fecundity and survival of the population if its frequency increases by the statistical luck of random pairing of alleles in each generation. The gene pools of most populations contain many of these sublethal recessives, about enough to kill an individual three times over if by chance they all occurred on its chromosomes in homozygous form and were therefore all expressed in its phenotype. Thus a decline in heterozygosity tends to lead to a decline in fitness.

Genetic malfunction may follow as a consequence of small population size. The following sequence may be triggered if a population becomes too small:

1 The frequency of mating between close relatives rises and random genetic drift increases,

2 which leads to reduced heterozygosity in the offspring,

3 which exposes the effect of semi-lethal recessive alleles,

4 which reduces fecundity and increases mortality,

5 which causes the population to become smaller yet, and that trend may continue until extinction. The population must be held at low numbers for several generations before that process is initiated. A short bout of low population size has little effect on heterozygosity.

Loss of fitness during inbreeding can be traced largely to the process of fixation (i.e. reduction of alleles at a locus to one type) of deleterious recessive alleles. In mammals, mortality is 33% higher for the offspring of parent–offspring or full sibling matings than for the offspring of unrelated parents (Ralls *et al.* 1988). Hybrid vigor is largely the reverse, the masking of the effect of those recessives; but it might also contain a component of heterosis where the heterozygote is fitter than either homozygote.

Inbreeding does not automatically lead to inbreeding depression. It is seldom reported for populations larger than a couple of dozen individuals. Nor does low heterozygosity necessarily lead to inbreeding depression. Note that the average individual of most wild populations is heterozygous at less than 10% of loci. A population that has survived a bout of inbreeding may come out of it with enhanced fitness because inbreeding exposes deleterious recessives and allows them to be selected out of the population. That is precisely the method used by animal breeders to remove deleterious alleles. Homozygosity causes an immediate problem only when the allele is deleterious. Nonetheless inbreeding often does produce inbreeding depression. That possibility must always be kept in mind if a population is small. Section 17.6 addresses the question: how small is too small?

17.3.6 *How much genetic variation is needed?*

The cheetah (*Acinonyx jubatus*) has a low level of heterozygosity (O'Brien *et al.* 1983). Stephen O'Brien and his colleagues (O'Brien *et al.* 1985b, 1986) and Cohn (1986) studied the genetic variance of captive populations of that species originating from southern Africa. A standard electrophoretic analysis of 52 loci ($n = 55$ individuals) discovered a heterozygosity value of $H = 0.00$ as compared with $H = 0.063$ for people and 0.037 for lions. A more refined "two-dimensional" electrophoretic analysis, separating the proteins first by electrical charge (as above) and then by molecular weight, uncovered rather more variability and yielded $H = 0.013$ for cheetahs as against 0.024 for people analyzed by the same method. A further sample from East Africa returned $H = 0.014$ (O'Brien *et al.* 1987).

Is the cheetah in peril? It is possible that as a direct consequence of the low heterozygosity the cheetah produces sperm of low viability, its rate of juvenile mortality is abnormally high, and it is particularly susceptible to disease. All these claims have been made but no causal relationship has been established between these putative defects and the peculiarities of the genotype. Alternatively the cheetah may be in no danger of demographic collapse despite its low genetic variability. In support of that is its widespread distribution, which was even wider in the recent past, particularly in Asia. Contraction of range over the last 1000 years has been no greater than that of the lion, another widespread species but with a standard level of heterozygosity. For both species the contraction of range seems to be a result of excessive human predation rather than of a diminished genetic fitness. As far as we know there is no evidence from the wild suggesting that the cheetah is faced by a level of risk beyond those hazards imparted by a rising human population (Caughley 1994).

The cheetah clearly has low genetic variance but well within the range exhibited by mammalian species (see Fig. 17.1). The suggestion that it is in demographic peril as a consequence of that modest genetic variance earns no support from what is known of its ecology. There are two messages transmitted by this example. The first is special: we need more disciplined information on the cheetah in the field to determine whether its diminished genetic variance is associated with demographic malfunction. The second is general: by genetic theory currently followed in conservation biology the cheetah should be in demographic trouble, but there is no convincing evidence for that and considerable but circumstantial evidence to the contrary. A plausible alternative hypothesis is that present genetic theory overestimates the amount of genetic variation needed to sustain an adequate level of individual fitness. One should not jump to the conclusion that a species is in danger simply because it has a low H. There is too much evidence to the contrary.

We cannot yet lay down general rules as to the minimum genetic variance required for adequate demographic fitness. Nor can we define a minimum viable population size (genetic). We need much more research on the incidence of inbreeding depression in the field, on the population size and the period over which that size must be maintained before inbreeding depression becomes a problem, and on the relationship between heterozygosity and fitness.

17.4 Effective population size (genetic)

It can happen that the size of a population appears large enough to avoid genetic malfunction but that the population is acting genetically as if it were much smaller. The proportion of genetic variability lost by random genetic drift may be higher than the computed theoretical $1/(2N)$ per generation because that formulation is correct only for an "ideal population." In this sense "ideal" means that family size is distributed as a Poisson, sex ratio is 50 : 50, generations do not overlap, mating is strictly at random, and rate of increase is zero. This introduces the notion of **effective population size** in the genetic sense, the size of an ideal population that loses genetic variance at the same rate as the real population. The population's effective size (genetic) will be less than its census size except in special and unusual circumstances.

Perhaps the greatest source of disparity between census size and effective size is the difference between individuals in the number of offspring they contribute to the next generation. In the ideal population their contribution has a Poisson distribution, the fundamental property of which is that the variance equals the mean. Should the variance of offspring production between individuals exceed the mean number of offspring produced per individual, the effective population size will be smaller than the census size. In the unlikely event of variance being less than the mean the effective population size is greater than the census size and the population is coping better genetically than one might naively have expected. The effective population size N_e corrected for this demographic character can be calculated as:

$$N_e = (NF - 1)/[F + (s^2/F) - 1]$$

where F is the mean lifetime production of offspring per individual and s^2 is the variance of production between individuals. It indicates that when mean and variance are equal, N_e approximates to N. Since males and females sometimes differ in mean and variance of offspring production, this equation is often solved for each sex separately and the sex-specific N_e values summed.

Genetic drift is minimized when sex ratio is 50 : 50. Effective population size (genetic) in terms of sex ratio is given by:

$$N_e = 4/[(1/N_{em}) + (1/N_{ef})]$$

where N_{em} and N_{ef} are the effective numbers of males and females as corrected above for variation in production of offspring. The series below shows the relationship numerically:

Sex ratio	50 : 50	60 : 40	70 : 30	80 : 20	90 : 10
N_e	N	$0.96N$	$0.84N$	$0.64N$	$0.36N$

Further corrections can be made for other sources of disparity between real and ideal populations. These considerations are often important in *Drosophila* research and in managing the very small populations in zoos, but they have little utility in conservation. Rather than attempting to estimate N_e for a threatened population you should simply assume as a rule of thumb that $N_e = 0.4N$, and that the censused population is losing genetic variability at a rate appropriate to an ideal population less than half its size.

17.5 Effective population size (demographic)

The **effective population size (demographic)** is the size of a population with an even sex ratio and a stable age distribution that has the same net change in numbers over a year as the population of interest.

17.5.1 *Effect of sex ratio*

If a species is polygamous, and most species of wildlife are, a disparate sex ratio may have a large effect upon net change in numbers over a year and hence effective population size (demographic) at the beginning of that year. Net change in numbers over a year can be calculated as:

$$\Delta N = NpbP_f - N(1 - p)$$

where P_f is the proportion of females in the population, p is the probability of surviving the year averaged over individuals of all ages within a stable age distribution, and b is the number of live births produced per female at the birth pulse terminating the year. It indicates that net change in numbers in a population of any given size is a linear function of the proportion of females in the population. The regression of ΔN on P_f has a slope of Npb and an intercept (i.e. the value of ΔN when $P_f = 0$) of $-N(1 - p)$.

The hypothetical example in Section 17.2.1 had a population size of $N = 1000$, a probability of survival per individual per year of $p = 0.9$, and a fecundity rate of $b = 0.95$ live births per female per year. Those values fed into the above equation yield ΔN as 541 when $P_f = 0.75$, as 328 when $P_f = 0.5$, and as 114 when $P_f = 0.25$.

The relationship can be rearranged to estimate N_e, the effective population size (demographic), as:

$$N_e = N(pbP_f + p - 1)/(0.5pb + p - 1)$$

For the above example, N_e is solved as 1653 when $P_f = 0.75$, as 1000 when $P_f = 0.5$, and as 347 when $P_f = 0.25$. Thus a disparate sex ratio may have a significant effect on a population's ability to increase from low numbers, enhancing that ability when females predominate and depressing it when males dominate. Ungulate populations that have crashed because they have eaten out their food, or because a drought has cut their food from under them, often end the population slide with a preponderance of females. They are thus in better shape demographically to recover from the decline than if parity of sex ratio had been retained. Note here an important point: the sex ratio minimizing genetic drift (50 : 50) is not that maximizing rate of increase (disparity of females). Hence the appropriate "effective population size" depends on context. The genetic version is appropriate for a small, capped population in a zoo. The demographic version is often more appropriate in the wild where the aim is usually to stimulate the growth of an endangered species.

17.5.2 *Effect of age distribution*

Similarly, an equation for effective population size (demographic) can be written to correct for the effect of an unstable age distribution. It requires knowledge of the age distribution, population size, age-specific fecundity, and age-specific mortality (see Sections 6.3 and 6.4 for calculating these). However, these estimates are difficult to obtain in practice, particularly for a small population. When faced by the urgent task of diagnosing the cause of the decline of an endangered species, instead of wasting valuable research time on estimating its age-distribution version of effective population size (demographic) you should simply understand the principle and appreciate that, because of instability of age distribution, the population's rate of change may be higher or lower than would be expected from a simple tallying of numbers. Suppose the population experienced a severe drought in the previous year that killed off the vulnerable young and old animals. The age distribution will now be loaded with animals of prime breeding age, and the birth rate, measured as offspring per individual, will therefore be much higher than usual. In consequence, rate of increase will be higher than usual. Alternatively the population may have experienced a mild winter such that the age distribution is loaded with young animals below breeding age. Birth rate at the next birth pulse is then lower than usual and so is rate of increase.

17.6 **How small is too small?**

Two values for minimum viable population size (genetic) are commonly quoted: 500 and 50. The difference between them reflects the differing assumptions upon which they are based. The 500 figure (Franklin 1980) is the effective population size at which the heritability of a quantitative character stabilizes at 0.5 on average (i.e. 50% of quantitative phenotypic variation is inherited and 50% is environmentally induced). The figure of 0.5 is the heritability coefficient for bristle number in *Drosophila*, and quantitative characters in farm animals often have a heritability coefficient of that general magnitude. The assumption is that such a level of heritability reflects a genetically healthy population. The genetic variance needed to enforce such a heritability coefficient has an equilibrium (where loss by genetic drift is balanced by gain from mutation) of $N_e = 500$ in the absence of selection, and that is taken as a safe lower limit for population size.

The estimate of 50 comes from the observation of animal breeders that a loss of genetic variance of 1% per generation causes no genetic problems. Since that rate is $1/(2N_e)$ we can write $0.01 = 1/(2N_e)$, rearrange it to $N_e = 1/(2 \times 0.01)$, and solve it as $N_e = 50$.

The arbitrary nature of both estimates of minimum viable population size will be readily apparent. Neither should be accepted as anything more than general speculation. No single number estimates have been offered for minimum viable population size (demographic). It is clearly a function of $Var(r)_1$ interacting with $Var(r)_e$ (see Sections 17.2.1 and 17.2.2) and so will vary among species and among populations within species. However, for most populations it is likely to be higher than the corresponding minimum viable population size (genetic).

17.7 Population viability analysis

Population viability analysis (PVA) is a procedure by which we estimate the probability of persisting (or its converse, extinction) over a specified time interval (Boyce 1992). Depending on the biological facts known for the population in question, PVAs can be based on exponential, density-dependent, interactive predator–prey, metapopulation, or even age- or stage-dependent models (Boyce 1992; Morris and Doak 2002). Regardless of structural details, all PVAs use estimated variation in demographic parameters to add noise to simulated populations. By repeating such Monte Carlo models many times, one can assess the probability of falling below an arbitrary critical value (termed a **quasi-extinction threshold**). Why is this critical level not set at zero? First, many otherwise useful models do not have reliable behavior as density approaches zero (for example, in the logistic model a population always increases near zero). Second, we know that demographic stochasticity would often doom any population that spent too long at "too low" a density. Third, we might feel that crisis management is called for below this arbitrary threshold.

17.7.1 *PVA based on the exponential growth model*

Some populations are so small that they are unlikely to experience any major changes in net recruitment due to increasing density. If so, we would expect the population to grow according to an exponential population model, such as:

$$N_{t+1} = e^r N_t$$

To make this more realistic we shall look at a real example (Fig. 17.2), the Yellowstone population of grizzlies, which is the largest remnant population left in the continental USA.

For the period 1959–82, the Yellowstone grizzly population hovered around 35–40 female bears (Eberhardt *et al.* 1986). Such small population levels are often considered dangerously low, due to the risk of demographic stochasticity or Allee effects. This has raised conservation concerns for the long-term viability of grizzlies

Fig. 17.2 Population dynamics over time of female grizzly bears in Yellowstone National Park during 1959–82. (After Eberhardt *et al.* 1986.)

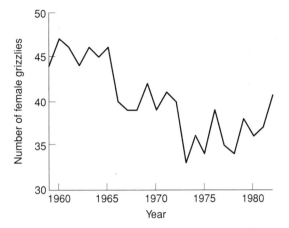

in Yellowstone. The average exponential rate of population change (r) recorded over this period was −0.00086, with a standard deviation of 0.08. Since r was negative there was cause for concern because if it was negative for long enough there would be extinction.

One can use these demographic data, simple as they are, to set up a simple stochastic simulation of the population dynamics of Yellowstone grizzlies, following the Monte Carlo approach outlined in Chapter 8:

$$N_{t+1} = N_t e^{\mu + \varepsilon_t}$$

where μ is the mean exponential growth rate recorded in the past (−0.00086) and ε_t is the magnitude of environmental variation simulated for year t, drawn from a normal distribution with a mean of zero and a residual standard deviation equivalent to that recorded in the past ($\sigma = 0.08$).

We assume that a value of 10 bears is the lower critical threshold. By setting the lowest value on the y-axis to 10, we can readily monitor when our simulated bear population falls below the critical threshold (Fig. 17.3). Repetition of this process

Fig. 17.3 Simulated dynamics of a grizzly bear population with mean and variance in population growth rate similar to that of the Yellowstone population during 1959–82 and starting with the population size recorded in 1982. Note that the simulated population has just reached the critical "quasi-extinction" threshold of 10 animals in year 97.

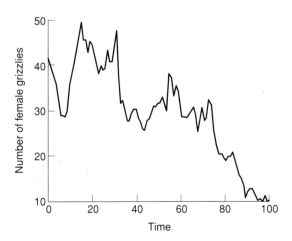

Fig. 17.4 Results of
100 replicate Monte
Carlo simulations of
the Yellowstone grizzly
population, based
on an exponential
growth model with
$\mu = -0.00086$ and
$\sigma = 0.08$. The lower
critical density is
arbitrarily set at 10
individuals. Between
5% and 10% of the
replicates tend to reach
the critical threshold
within a century.

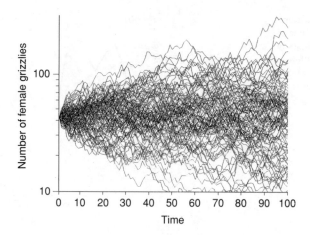

many times suggests roughly an 8–10% risk of extinction within the next century,
or one extinction event every 10,000–20,000 years (Fig. 17.4).

17.7.2 *PVA based on the diffusion model*

There is a more mathematically elegant way to estimate extinction risk in populations with exponential population growth, using what is known as a "diffusion model"
(Lande 1993). We follow the discussion of diffusion models in Morris and Doak (2002).
A diffusion model can be visualized as if a vast number of beads are released at a
single point near the bottom of an infinitely deep river. Over time, the cloud of beads
would spread, due to individual beads bouncing up or down due to turbulence as
they flow downstream, just as the population trajectories for the geometric growth
model tend to spread over time (Fig. 17.4). A probability density formula known
as the inverse Gaussian distribution is used to predict the analogous distribution of
population densities over time, with residual variability in the exponential growth
rate of the population generating the turbulence. This equation generates a bell-shaped
distribution of population densities at each point in time, with the degree of spread
of the bell-shaped curve growing wider with every year. The initial position at which
the beads are released corresponds to the starting population density N_c. This is
important, because more beads will strike the bottom when they are released low
in the water column, just as the probability of extinction is highest for populations
starting at initially low numbers. The quasi-extinction threshold is denoted N_x, μ is
the average exponential rate of increase (which equals the arithmetic mean of $\log_e \lambda$),
and σ^2 is the variance of $\log_e \lambda$. The probability of extinction in any given year t and
the cumulative probability of extinction are calculated according to the equations shown
in Box 17.1.

We will illustrate the application of the diffusion equation using population
censuses of females during 1959–82 from the Yellowstone grizzly bear population
to estimate mean r and the standard deviation in r. As before, we will assume a lower
critical threshold $N_c = 10$ animals and use the 1982 census total as the initial density $N_x = 41$.

The predicted probability of extinction increases over time, because it takes several
poor years in sequence for a population of 41 grizzlies to crash to below 10 individuals (Fig. 17.5). The probability declines at very long times, simply because the
losers have already disappeared whereas any winners have likely grown well out of

Box 17.1 Equations for calculating the probability of extinction in any given year t and the cumulative probability of extinction.

This diffusion model predicts the probability of extinction in any given year, $P(t)$, as well as the cumulative probability for all years before and including t, $G(t)$. N_c is initial density and N_x is the quasi-extinction threshold.

$$P(t) = \frac{d}{\sqrt{2\pi\sigma^2 t^3}} \exp\left[\frac{-(d + \mu t)^2}{2\sigma^2 t}\right]$$

where

$$d = \log_e(N_c) - \log_e(N_x)$$

By integrating the time-specific equation from minus infinity to t, we get the following cumulative function:

$$G(t) = \Phi\left(\frac{-d - \mu t}{\sqrt{\sigma^2 t}}\right) + \exp\left(\frac{-2\mu d}{\sigma^2}\right)\Phi\left(\frac{-d + \mu t}{\sqrt{\sigma^2 t}}\right)$$

where

$$\Phi(z) = \left(\frac{1}{\sqrt{2\pi}}\right)\int_{-\infty}^{z} \exp(-y^2)\, dy$$

Fig. 17.5 The predicted probability of extinction in any future year, based on the diffusion equation approximation for an exponentially growing grizzly bear population with demographic parameters identical to the Yellowstone population studied during 1959–82.

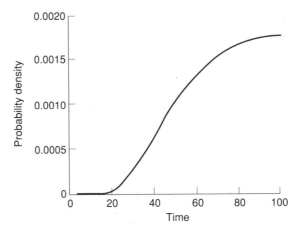

the danger zone. The peak and spread of this curve depends, obviously, on the mean rate of growth, as well as the variance.

The diffusion model predicts that the cumulative risk of extinction for Yellowstone grizzlies should tend to increase over time, initially at an accelerating rate, then later at a diminishing rate (Fig. 17.6). For a population whose mean exponential growth rate $\mu < 0$, as this one is, eventual extinction is certain, provided one waits long enough. The question is: how long might this process take? By year 100, the diffusion model estimates a 9% probability of extinction (Fig. 17.6), just as we found from our earlier simulations.

In reality the Yellowstone grizzly population managed to recover to much higher numbers in the 1990s (Fig. 17.7). Hence the risk of imminent extinction in the early

Fig. 17.6 The cumulative probability of extinction over a given span of time for a grizzly bear population with demographic parameters identical to the Yellowstone population studied during 1959–82. $\sigma = 0.08$.

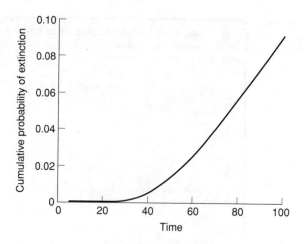

Fig. 17.7 Population dynamics of female Yellowstone grizzlies from 1959 to 1995. (Based on the combined data of Eberhardt *et al.* 1986 and Haroldson 1999.)

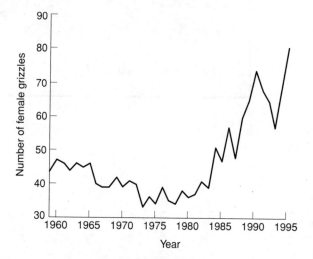

1980s proved to be a false alarm, at least in this case, as one might expect in the majority of cases (> 90%). Nonetheless, the PVA approach gives us a formal starting point in evaluating extinction risk.

17.7.3 *Strengths and weaknesses of PVA*

Population viability analysis models are attractive because they supply hard numbers for the kind of uncertain (stochastic) processes that threaten small populations. This has led to widespread proliferation of PVA models, as discussed by several in-depth reviews (Boyce 1992; Morris and Doak 2002). However, we should be cautious about the reliability of projected extinction risks, for a number of reasons. First, we rarely have precise, reliable estimates of growth rates in any population, let alone those that are under threat. Minor errors in our estimates of demographic parameters (because the number of years of data is too short, for example) multiply geometrically over time, leading to inflated (or deflated) estimates of extinction risk (Ludwig 1999; Fieberg and Ellner 2000). Projection of extinction risk beyond a short time ahead (10–20% of the length of the time series) provides high uncertainty that the extinction risk estimates have little value (Fieberg and Ellner 2000). In other words, 40 years of data on grizzly bears might allow us to make reliable estimates of extinction over the next 5 or 10 years, but certainly not over the next 100 years.

This problem is compounded when we have no idea about the reliability of population estimates in any given year. This is obviously less of an issue in the rare cases where every individual is recognizable. Most populations are known only from samples taken from a small fraction of the inhabited range, leading to considerable uncertainty in true abundance. Observation errors play a key role, because they convey a false impression about the true magnitude of environmental and demographic stochasticity, as well as biased estimates (usually downward) about the strength of density dependence. As a consequence, even well-studied populations may yield biased predictions of extinction risk (Ludwig 1999).

Data for well-studied populations illustrate that catastrophic climatic events play an important role in causing population collapse. Such catastrophes tend to be difficult to predict using Monte Carlo simulations, particularly when long-term census data are lacking (Coulson *et al.* 2001b). Most importantly, application of PVAs is founded on an underlying faith that conditions (e.g. climate, habitat availability, and human interference) will hold far into the future (Coulson *et al.* 2001b). For example, the Yellowstone grizzly bear data for 1959–82 provide a substantially higher risk of extinction than that of the later demographic data, suggesting a change in environmental conditions.

On the basis of these quantitative weaknesses, some critics have claimed that population viability analysis is virtually meaningless (Ludwig 1999; Coulson *et al.* 2001b). Moreover, preoccupation with the stochastic dynamics of small populations ignores the ecological, physical, and anthropomorphic causes of population decline (Caughley 1994; Harcourt 1995; Walsh *et al.* 2003). Other conservation biologists argue that, while not infallible, PVA might still be quite useful as a means of comparing relative extinction risk among populations or in various subpopulations of a single species, or of assessing the relative risks associated with alternative management actions (Lindenmayer and Possingham 1996; Brook *et al.* 2000; Morris and Doak 2002).

To resolve these different views, Brook *et al.* (2000) gathered demographic data from 21 well-studied populations. They used the first half of the data for each time series to parameterize PVA models, and then used the resulting PVA models to predict the outcome of the second half of each data set. They concluded that the risk of decline closely matched predictions and that there was no significant bias in predictions. They also found few major differences in the quality of predictions of any of the most common models that are commercially available to decision-makers. Coulson *et al.* (2001b) countered that the 21 data sets considered by Brook *et al.* were unrepresentative of endangered species most likely to be candidates for PVA. Rare organisms are, by their very nature, poorly understood. Nonetheless PVA is a widely accepted tool for risk assessment by both field biologists and decision-makers.

17.8 Extinction caused by environmental change

Despite the preoccupation of most population viability analyses with stochastic extinction processes, the most common cause of extinction is a critical change in the organism's environment. This is distinct from year-to-year fluctuation due to either demographic or environmental stochasticity. We identify the new environment as the driving variable responsible for the population's decline and the population may be driven to extinction by its action. A population seldom "dwindles to extinction." It is pushed. If you can identify the agent imparting the pressure and neutralize that pressure you can save the population.

The three most common causes of driven extinctions, roughly in order of importance, are (i) contraction and modification of habitat; (ii) unsustainable harvesting by humans; and (iii) introduction of a novel pathogen, predator, or competitor into the environment (Hilton-Taylor 2000). Case studies are considered in some detail in Chapter 18. Before we consider the historical record, however, we briefly consider how theoretical models can provide useful insight into each of these processes.

17.8.1 *Extinction threat due to introduction of exotic predators or competitors*

The first major way in which humans wreak havoc on threatened species is through modification of trophic relationships within a pre-existing community. Often this is via introduction of a competitor and/or predator for which an endemic species is poorly prepared. This is particularly common for endemic species on islands that have evolved for considerable periods without risk of predation. Such species are poorly equipped to cope with a novel predator – the typical traits for wariness, stealthy lifestyle, and inconspicuous coloration have not provided any selective advantage and may have disappeared. This sets the scene for a brief, but sadly inevitable, slide into oblivion once a novel predator has arrived.

The existing body of predator–prey theory is sometimes perfectly suitable for understanding such processes. If a predator is particularly efficient (high rates of capture even at low prey densities, high efficiency of conversion of prey into new predators) then prey densities would be expected to plunge suddenly to dangerously low levels, for which there is a high probability of extinction due to demographic or environmental stochasticity. A particularly graphic example is the brown tree snake (*Boiga irregularis*), introduced onto the island of Guam in the 1950s (Savidge 1987). In the course of two decades, this generalist predator has spread rapidly across the island. This range expansion coincided with the rapid decline and (in some cases) disappearance of 11 native species of forest birds on the island.

Special circumstances sometimes apply, however, that are far from obvious. For example, introduction of rabbits onto a number of islands in the South Pacific has apparently triggered collapse of endemic bird species which were previously well able to withstand predation. A case in point is loss of an endemic parakeet and a banded rail species from Macquarie Island (Taylor 1979). This has been attributed to hyperpredation (see Chapter 21), a form of apparent competition by which an exotic prey that is capable of withstanding predation subsidizes population growth by a resident predator (Smith and Quin 1996). Endemic prey then decline because they are more vulnerable to predation than the exotic species.

This scenario of asymmetric apparent competition (Holt 1977) induced via subsidies to a common predator population was developed theoretically by Courchamp *et al.* (2000b). They represented the hyperpredation process with the following system of equations:

$$\frac{\mathrm{d}}{\mathrm{d}t}B(t) = r_{\max_B}B(t)\left[1 - \frac{B(t)}{K_B}\right] - C(t)\mu_B B(t)\frac{\alpha B(t)}{R(t) + \alpha B(t)}$$

$$\frac{\mathrm{d}}{\mathrm{d}t}R(t) = r_{\max_R}R(t)\left[1 - \frac{R(t)}{K_R}\right] - C(t)\mu_R R(t)\frac{R(t)}{R(t) + \alpha B(t)}$$

$$\frac{\mathrm{d}}{\mathrm{d}t}C(t) = \frac{[\lambda_B\mu_B\alpha B(t)^2 + \lambda_R\mu_R R(t)^2]C(t)}{\alpha B(t) + R(t)} - \nu C(t)$$

Fig. 17.8 Simulation of hyperpredation, leading to collapse of an endemic prey species when an exotic alternative prey species is translocated into the system. The following parameter values were used: $\alpha = 3$, $r_{max_B} = 0.1$, $r_{max_R} = 2.0$, $K_B = 1000$, $K_R = 5000$, $\mu_B = 0.1$, $\mu_R = 0.1$, $\lambda_B = 0.01$, $\lambda_R = 0.01$, and $v = 0.5$. (After Courchamp et al. 2000b.)

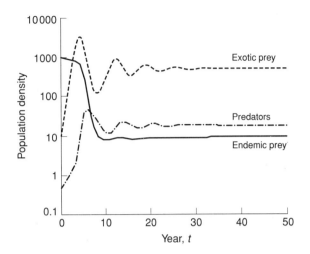

where B represents the density of the endemic prey species, R represents the density of the exotic prey species, and C represents the density of the predator that both prey have in common. You should note the similarity in structure to other consumer–resource models that we discussed in Chapter 12. Both prey species have logistic patterns of growth in the absence of predation, with maximum rate of increase dictated by r_{max} and carrying capacity dictated by K. The predator is assumed to have a Type I (linear) functional response to changes in prey density, with an attack rate of μ on each prey type. The predator is assumed to have a fixed preference for the endemic species, the magnitude of which is proportional to α (defined as the frequency of captures of preferred endemic prey to less preferred exotic prey when both are equally abundant). Nonetheless, the actual fraction of each prey in the predator's diet changes proportionately with shifts in their relative abundance, according to the ratio $\alpha B/(R + \alpha B)$. The presumption used in the model is that exotic prey individuals are much less easy to catch than endemic prey (α is much greater than 1). Any prey individuals that are successfully attacked are converted into new predators with an efficiency of λ, but predators also have a constant per capita rate of mortality v.

Depending on the magnitude of the key parameters, these equations can lead to several different outcomes of conservation interest: (i) extinction of endemic prey, but perpetuation of exotics and predators; (ii) extinction of exotics, but perpetuation of endemic prey and predators; (iii) extinction of predators, but perpetuation of both prey; and (iv) coexistence of all three species. A common outcome is depicted in Fig. 17.8: introduction of the exotic leads to high predator density, collapse of endemic prey to dangerously low levels at which demographic or environmental stochasticity threaten extinction, and substantial numbers of exotics. This scenario tends to play out when the exotic species has much higher carrying capacity and intrinsic growth rate than the endemic species (both of which are often true of introduced pest species like rabbits) and the predator tends to have a stronger probability of encountering endemic prey than exotic prey (which is also often true when the endemic has little or no prior experience with predation).

A superb example of this kind of process is seen on the Channel Islands off the coast of California. Several islands have had accidental translocations of an exotic herbivore, the feral pig (*Sus scrofa*). In response to exploding pig populations,

Fig. 17.9 (a) The number of golden eagle sightings on the Channel Islands, California, increased as the feral pig population increased. (b) The abundance of the Channel Island fox (●), as indicated by trapping, declined due to depredation by eagles, and this allowed the increase of the endemic skunks (□). (Data from Roemer *et al.* 2002.)

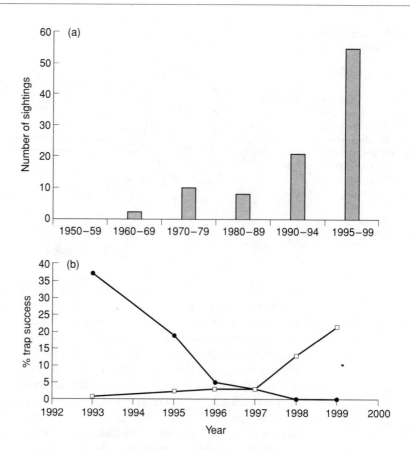

golden eagles (*Aquila chrysaetos*) have recently immigrated into the Channel Islands from the mainland and begun to breed successfully. Rising eagle abundance has led to the rapid decline of an endemic prey species, the island fox (*Urocyon littoralis*) (Fig. 17.9). Foxes have completely disappeared from two of the islands and have experienced a 10-fold decline in abundance on the largest island (Santa Cruz) following colonization by eagles in the early 1990s (Roemer *et al.* 2002; Courchamp *et al.* 2003). The decline in fox abundance has led to a parallel increase in a competitor, the island spotted skunk (*Spilogale gracilis amphiala*). Although there has been a successful effort to translocate some eagles away from the islands, several pairs still remain on the island, thwarting fox recovery. This presents a conservation conundrum: do we harm or remove a protected species (the golden eagle) in order to save an endangered species (the island fox)? An obvious countermeasure is to cull feral pigs. The model predicts that, without timely reduction in eagle abundance, pig eradication could inadvertently lead to heightened predation pressure on foxes, perhaps even doom them rather than helping them (Courchamp *et al.* 2003). This is a clear demonstration of the utility of trophic models as a means of evaluating alternative conservation actions.

17.8.2 *Extinction threat due to unsustainable harvesting*

For most big game species, harvesting does not pose a conservation threat. Indeed, such species usually become entrenched as "game" because their life-history attributes (high reproductive capacity, broad geographical distribution, ability to tolerate interference by hunting humans) make them relatively robust. Modern exceptions to this are species whose male ornaments (horn, tusks, antlers, or other body parts) make them particularly attractive to humans, regardless of the cost and energy required to kill them. There are several obvious examples: black rhinos, elephants, and big cats (lions, leopards, and tigers). When the profit from a rhino horn can exceed a rural African's expected income for a decade, it is not surprising that overharvesting occurs.

In many cases, though, the ornaments of interest appreciate in value as animals get older. When successful breeding depends on having adequate numbers of mature males, it may make a good deal of conservation sense only to harvest the oldest males, who have already bred, rather than harvest indiscriminately. This principle is illustrated in lions (Whitman *et al.* 2004). Male lions are attractive as trophies to many tourist hunters, particularly old males with a full mane. Lions ordinarily live, feed, and breed in a stable social group called a pride. A typical pride in East Africa comprises a group of six or so breeding females, often sisters, and a coalition of two or three males. Male coalitions come and go, whereas females usually remain within a given pride for their entire life. Given the short pride tenure (2–3 years) that any male can expect to hold, rapid breeding is essential to their reproductive success. Because of this, incoming coalitions often kill all the cubs surviving from their predecessors. This brings all the mothers rapidly into estrous, allowing the new batch of males a chance to sire offspring. However, if males turn over too quickly, infanticide outstrips successful reproduction and the population declines.

Whitman *et al.* (2004) developed a detailed, spatially explicit model of individual lions, each of whom lived, bred, and died within 5–10 computer prides, based on long-term studies conducted in Serengeti National Park and Ngorongoro Crater (Packer *et al.* 2005). They used this model to consider the impact of harvesting of males by sport hunting. At typical quotas for the East African savanna, their model suggests that indiscriminate harvesting of all mature males with full manes (those 4 years and older) is unsustainable. The reason is that removal of males by hunters before they would ordinarily lose their position at the head of a pride causes new males to come in and so there is too much infanticide for the lion population to cope with. On the other hand, the model suggests that hunters *could* harvest all the males they might want that are 8 years old or older. These males have already bred, by and large, so they are expendable. Since they are also the most attractive as trophies it is a win–win situation, so long as hunters can tell how old each male is before shooting him. Whitman *et al.* (2004) showed that the amount of black on a lion's nose provides a reliable indicator of the age of that individual. This simple harvesting strategy, combined with a reliable clue to age, might prove vital in conserving African lions in the long term.

This kind of enlightened harvest policy might prove useful for other trophy species as well. For example, harvesting of horns from male saiga antelopes (*Saiga tatarica*) is an important source of income for people living around the Caspian Sea. Recent data, however, suggest that so many male saiga antelopes are now being removed that breeding by females is compromised (Milner-Gulland *et al.* 2003). This is an example of how harvesting can induce an Allee effect (Courchamp *et al.* 2000a; Petersen and Levitan 2001), whereby population growth rates begin to decline once a population falls below a critical lower threshold.

17.8.3 *Extinction threat due to habitat loss*

The most serious challenge currently facing most threatened bird and mammal species is habitat contraction and loss. It is rare that habitat loss would pose a serious risk for a generalist species, capable of living in a wide range of places. It is specialists for whom habitat loss is most crucial – those species whose survival depends critically on particular, usually rare, habitats for survival and successful reproduction. Habitats supply numerous attributes: food, protective cover from predators, denning sites, shelter from inclement weather, and access to mates. This means that habitat needs are probably unique for every species. Nonetheless, there are models that predict the effects of dwindling supply, size, and spatial distribution of habitat patches in a metapopulation framework (see Chapter 6).

One approach uses incidence functions to characterize probabilities of extinction and recolonization for specific patches (Hanski 1994, 1998). Because extinction is often negatively related to population size and small patches tend to hold small populations at the best of times, extinction is usually modeled as a negative function of patch size. Colonization rates tend to be low when patches are widely spaced, so distance among patches is often a critical variable in incidence functions. Data on the sequence of local extinction and recolonization events allow one to estimate incidence functions across a matrix of possible sites. These functions can then be solved, either using matrix techniques or via simulation, to evaluate the long-term probabilities of persistence (Hanski 1998). Data for the Glanville fritillary (*Melitaea cinxia*) show that this approach has high predictive capability, at least in well-studied species (Hanski 1998; Lindenmayer *et al.* 2003).

An alternative approach is the software package ALEX (Possingham and Davies 1995). This software provides a flexible structure for modeling successional change and other temporal variation within patches, as well as accommodating a variety of methods to model movements among patches, age structure, and demography within and among patches. ALEX has been successful in predicting some key metapopulation attributes of a wide range of vertebrate species in fragmented forest patches in southeastern Australia (Lindenmayer *et al.* 2003).

Some biologists have suggested that a particularly useful way to predict the effects of habitat loss may be to explicitly link such processes with behavioral models of patch use and resource selection (Goss-Custard and Sutherland 1997). By understanding the factors guiding resource use, one can predict how changes in habitat availability and population density might translate into alterations in birth and death rates across the population (Stillman *et al.* 2000). This kind of approach has been most thoroughly developed for shorebirds (Sutherland and Anderson 1993; Sutherland and Allport 1994; Goss-Custard and Sutherland 1997; Percival *et al.* 1998). For example, humans compete with oystercatchers via commercial harvesting of the large bivalves that oystercatchers prefer. Using behaviorally based models, Stillman *et al.* (2001) predicted the demographic impact of bivalve fisheries on oystercatcher populations.

In principle, this kind of behaviorally based approach could be applied to other wildlife species. For example, behavioral modeling of patch selection by Thomson's gazelles has been used to predict patterns of movement across the Serengeti plains and demonstrate the importance of unrestricted access to large areas of rangeland for the long-term viability of these grazers (Fryxell *et al.* 2004, 2005).

17.9 Summary

A population may, by chance, be forced to extinction by year-to-year variation in weather or other environmental factors. When the population is small it may exhibit a random walk to extinction because its dynamics at low numbers are determined by the unpredictable fortunes of individual members. Genetic drift and inbreeding depression may also operate at low numbers to reduce fitness and thereby lower numbers even further. Their effect increases with declining population size, modified further by the sex and age structure of the population. The probability of population extinction over a specified time span can sometimes be estimated using population viability analysis (PVA). There are many different ways to develop PVAs, some of which derive from diffusion approximation to stochastic population growth. Extinction due to demographic or environmental stochasticity is less common than habitat modification or the introduction, usually by people, of a new element into the environment. This is commonly a new predator, competitor, or pathogen. Sometimes the new factor is simply an unsustainable level of harvesting by humans. The population is then driven to extinction rather than dropping out by chance. These processes can be incorporated in a variety of modeling frameworks that usefully augment and extend the PVA approach.

18 Conservation in practice

18.1 **Introduction**

In Chapter 17 we examined the ways in which demographics and genetics contribute, at least potentially, to the risk that a population will go extinct. The extinction of a species does not differ in kind. The species goes extinct because the last population of that species goes extinct. Here we review actual extinctions or near extinctions to show what are the commonest causes of extinction in practice. We then describe how to detect such problems and how to treat a population in danger.

18.2 **How populations go extinct**

Extinctions may be divided into two categories, driven extinctions and stochastic extinctions.

1 *Driven extinction*: whereby a population's environment changes to its detriment and rate of increase falls below zero. The population declines. Perhaps this lowering of density frees up resources to some extent, or lowers the rate of predation, but this is not sufficient to counteract the force of the driving variable, and the population finally goes extinct. Included in this category are extinctions caused by environmental fluctuation and extinctions caused by catastrophes. The latter are viewed here as simply large environmental fluctuations.

2 *Stochastic extinction*: whereby a population fails to solve the "small population problem." The effect of chance events, which would be trivial when numbers are high, can have important and sometimes terminal consequences when numbers are low.

(a) *Extinction by demographic malfunction*: whereby a population goes extinct by accident (chance) because it is so small that its dynamics are determined critically by the fortunes of individuals rather than by the law of averages. In those circumstances a population is quite capable, by chance configuration of age distribution or sex ratio, of registering a steep decline to extinction over a couple of years even though its schedules of mortality and fecundity (see Sections 6.3 and 6.4) would result in an increase if the age distribution were stable.

(b) *Extinction by genetic malfunction*: whereby a population at low numbers for several generations loses heterozygosity to the extent that recessive semi-lethals are exposed, average fitness therefore drops, and the population declines even further and ultimately to extinction. The loss of an allele from the genotype is an event resulting from the lottery of random mating. Although each individual loss is unpredictable, the average rate of loss, as a function of population size, can be predicted fairly accurately.

These mechanisms do not exclude each other entirely but they are sufficiently distinct that we treat them separately. Although the relative contribution of these mechanisms is unknown, enough anecdotal information is available to suggest that the driven extinction is by far the most prevalent. Extinction by demographic malfunction is probably the second most important but it usually requires that the population is driven to low numbers before demographic stochasticity can operate. Many examples have been documented, particularly for introductions. Extinction by genetic malfunction appears to come a distant third. Genetic problems have a low priority in saving a natural population from extinction. They are more relevant to managing a population in captivity or one whose size is so small and its future so bleak that it should be in captivity.

We will now look at a few examples of species that have become extinct, or came close to extinction, to give us a feel for the range of possibilities.

18.2.1 *Effect of habitat change and fragmentation*

Many extinctions appear to have been caused by habitat changes (Griffith *et al.* 1989; Brooks *et al.* 2002), but the precise mechanism of population decline is usually difficult to determine retrospectively. One form of habitat loss is through fragmentation of continuous habitat into patches. Over time these patches become smaller and the gaps between them become larger. The ratio of edge to interior habitat of the patches becomes larger (Temple 1986). We have seen this clearly in the fragmentation of the eastern hardwood forests of North America since settlement in the 1600s, and in the eucalypt woodlands of Australia in the last century (Saunders *et al.* 1993). This process occurred in the 1300s in New Zealand with the arrival of Maori (Worthy and Holdaway 2002) and much earlier in Europe during medieval times. Fragmentation is seen most commonly in the transformation of forest or woodland into farmland, but also in the change from native grassland into agriculture. The hostility of the matrix is important too. Thus, a matrix of young regenerating forest or even exotic plantation is less hostile for animals in old growth forest patches than a matrix of agriculture. Human residential development is yet more hostile (Friesen *et al.* 1995). A further aspect we need to consider is the type of forest involved. In northern boreal forests of Europe and Canada, containing widespread, migratory bird species, there is much less effect of fragmentation on species richness and ability to colonize than in tropical forests with their highly restricted distributions of birds (Haila 1986).

Fragmentation of habitat has a number of consequences:

1 Species that require interior forest habitats (many bird species), away from the edge, experience reduced habitat and hence population reductions (Saunders *et al.* 1993). In a long-term experiment where forest fragments of different sizes were constructed in the Amazon forest, the ecosystem showed aspects of degradation within the patches (Laurance *et al.* 2002). Many bird species avoided even small clearings less than 100 m across. Edges were avoided and the type of matrix affected movement. In both England and the eastern USA, extinctions of bird species occurred once a critical percentage of the original habitat was destroyed (McLellan *et al.* 1986).

2 Species that need to disperse through intact habitat (many reptiles, amphibians, ground-dwelling insects) are prevented from doing so and their populations are reduced to isolated pockets with potential demographic and genetic consequences. In fragmented parts of the northern boreal forests of North America, the foraging movements of the three-toed woodpecker (*Picoides tridactylus*) are highly constrained because this species strongly avoids open areas (Imbeau and Desroches 2002). Dispersal of crested tits (*Parus cristatus*) in Belgium was restricted in pine forest fragments relative to continuous forest, and this probably reduced their ability to settle in preferred habitats (Lens and Dhondt 1994). Northern spotted owls (*Strix occidentalis*) also

suffer higher mortality when dispersing across unsuitable habitat between patches (Temple 1991).

3 The greater length of habitat edge allows incursions of predators from outside the patch, increasing the predation rate on interior forest species. We discuss such a case for birds in the eastern hardwoods of North America in Section 10.7 (Fig. 10.10b) (Wilcove 1985; Temple and Carey 1988). Nesting success of ovenbirds (*Seiurus auro-capillus*), red-eyed vireos (*Vireo olivaceus*), and wood thrushes (*Hylocichla mustelina*) in these deciduous woodlands was reduced both by higher nest predation and by increased brood parasitism from the brown-headed cowbird (*Moluthrus ater*) (Donovan *et al.* 1995). In general, fragmentation results in the synergistic inter-action of several deleterious factors, particularly habitat decay, reduced dispersal of animal populations, and increased risk of predation (Hobbs 2001; Laurance and Cochrane 2001). However, species respond differently to fragmentation of habitat. Species that do not move far (insects, reptiles, some forest birds) are more restricted than are highly mobile taxa (many birds, mammals, long-lived species, generalist preda-tors) (Debinski and Holt 2000).

There follow examples where extinctions or steep declines were associated with a change in habitat and where that change probably caused extinction.

The Gull Island vole

The Gull Island vole (*Microtus nesophilus*) was discovered and described in 1889. It was restricted to the 7 ha Gull Island off Long Island, New York. Fort Michie was built there in 1897, its construction requiring that most of the island (and thus the vole's habitat) be coated with concrete. The species has not been seen since.

The hispid hare

The hispid hare (*Caprolagus hispidus*) once ranged along the southern Himalayan foothills from Nepal to Assam but is now restricted to a handful of wildlife sanctu-aries and forest reserves in Assam, Bengal, and Nepal. This short-limbed rabbit-like hare depends on tall dense grass formed as a successional stage maintained either by monsoon flooding or by periodic burning (Bell *et al.* 1990). The hare's decline reflects fragmentation of suitable habitat by agricultural encroachment. Most of the surviv-ing populations are now isolated in small pockets of suitable habitats in reserves. Much of the natural grassland has been lost to agriculture, forestry, and flood con-trol and irrigation schemes. What remains is modified, even within the reserves, by unseasonal burning and grass cutting for thatching material.

Like many endangered species, especially those that are small or inconspicuous, neither density nor rate of decline has been measured. The current status was deter-mined from searches of the few remaining pockets of tall grassland. There is some evidence that contraction of the species into pockets of favorable habitat renders indi-vidual hares more vulnerable to predation.

Wallabies and kangaroos

Some of the more dramatic examples of driven extinctions involve the ecology of a significant segment of the fauna being disrupted by large-scale and abrupt habitat changes. In Australia there has been a substantial depletion of the Macropodoidea (about 50 species of kangaroos, wallabies, and rat-kangaroos) following European settlement (Calaby and Grigg 1989).

The changes in the macropod fauna included the extinction of six species and the decline of 23 species, of which two died out completely on the mainland but still occur in Tasmania. The rates of decline are often difficult to estimate because the year in which the decline began is seldom known and indices of population size are seldom available. Land clearing and extensive sheep farming were in full swing in Australia by 1840 and declines in macropods were evident by the late 1800s. Many of the declines followed sweeping changes to habitat as land was cleared for agriculture or grazing by sheep modified the vegetation. The smaller macropods (< 5 kg) were the most affected, only one species (*Macropus greyi*) of the larger macropods going extinct. Calaby and Grigg (1989) emphasized the difficulty of determining the cause of declines retrospectively but considered that the evidence strongly suggested that the declines of nine species could be referred to the effects of land clearing, nine to modification of vegetation by sheep, five to introduced predators (foxes and cats), and seven to unknown causes.

The sheep grazing and woodland clearing that led to the decline or extinction of at least 17 species concomitantly benefited five species of the larger macropods, which increased in numbers. A further four large macropods and four of the 11 species of the smaller rock-wallaby (*Petrogale*) have changed little in numbers from the time of the European settlement to the present day.

18.2.2 The effect of loss of food

The Mundanthurai sanctuary in southern India was classified as a tiger reserve in 1988. Tigers (*Panthera tigris*) live in dense vegetation with access to water, but are restricted to core areas of protected reserves and avoid areas frequented by humans. Since 1988 cattle have been removed and fires controlled so that a dense vegetation of exotic *Lantana camara* and other species unpalatable to wild ungulates has grown up. Consequently the large ungulate species that comprise the food of tiger, and that require grassland, have declined. Both tigers and leopards (*P. pardus*) have declined with their food supply in the reserve (Ramakrishnan *et al.* 1999).

18.2.3 The effect of introduced predators

The Lord Howe woodhen

The Lord Howe woodhen (*Tricholimnas sylvestris*) is a rail about the size of a chicken. It lives on the 25 km^2 Lord Howe Island in the southwest Pacific, 700 km off the coast of Australia. Lord Howe was one of the few Pacific islands, and the only high one, that was not discovered by Polynesians, Melanesians, or Micronesians before European contact, and which therefore suffered none of the man-induced extinctions common on Pacific islands over the first millennium AD. Humans set foot on it for the first time in 1788, at which time it hosted 13 species of land birds of which nine became extinct over the next 70 years or so.

The story of the Lord Howe woodhen is related by Hutton (1991, 1998). The island was visited regularly for food (which no doubt included woodhen) and water by sailing ships in the late eighteenth and early nineteenth centuries, and finally settled permanently in 1834. Pigs were introduced before 1839, dogs and cats before 1845, domestic goats before 1851, and the black rat (*Rattus rattus*) in 1918 from a shipwreck. By 1853 the woodhen's range was restricted to the mountainous parts of the island, and by 1920 its range had apparently contracted to the summit plateau (25 ha) of Mt Gower, a 825 m (2700 ft) mountain almost surrounded by near vertical cliffs rising out of the sea. The summit plateau is a dreary place: dripping moss forest and perpetual cloud, a rather different place from the coastal flats that used to be the bird's habitat. This was obviously a species on its last legs, but that problem was not recognized until 1969 after which the population was studied intensively. The population was stable at between eight and 10 breeding pairs, although in one year it went down to six pairs. Apparently no more than 10 territories could be fitted into the 25 ha of space and so we can be confident that the population on Mt Gower, and probably that of the entire species, did not exceed that size over the 60 years between 1920 and 1980.

The obvious candidate for the contraction of population size and range was the black rat, which has been implicated in the extinction of several species of birds on islands. In this case, however, the rat does not appear to be implicated. The wood-hen kills rats, and in any event rats are more common on the summit of Mt Gower than on any other part of the island. The culprit instead appeared to be the feral pig, which will kill and eat incubating birds and will destroy the nest and eggs. Pigs cannot accomplish the minor feat of mountaineering needed to reach the summit of Mt Gower. The pigs were shot out in the 1970s and the cats by 1980 with the consent of the islanders, who now ban domestic cats.

In 1980 a captive breeding center was established on the island at sea level and seeded with three pairs from Mt Gower. Thirteen chicks were reared in the first season of captivity, 19 in the second, and 34 in the third. The birds were released and the captive breeding terminated at the end of 1983. The population reached its maximum at about 180 birds, 50–60 breeding pairs, and that number seems to saturate all suitable habitat on the island, mainly endemic palm forest. A by-product of the pig and cat control is the expansion of breeding colonies of petrels, shearwaters, and terns.

The Stephen Island wren

The Stephen Island wren (*Xenicus lyalli*), the only known completely flightless passerine, was discovered in 1894. It lived on a 150 ha island in Cook Strait which separates the North Island and South Island of New Zealand. Subfossil remains indicate that it was previously widespread on both main islands but became extinct there several centuries before European settlement, part of the extinction event that followed the colonization of New Zealand by the Polynesians about AD 900. The causal agent of its extinction on the mainland was probably the Polynesian rat (*Rattus exulans*) introduced by the Polynesians.

The wren was extinguished by a single domestic cat, the pet of the lighthouse keeper, Mr Lyall. He was the only European to see the species alive and then on but two occasions, both in the evening. He said it ran like a mouse and did not fly, a fact confirmed subsequently from the structure of the primary feathers. The first one he saw was dead, having been brought in by the cat. Subsequently the cat delivered a further 21, 12 of which eventually found their way to museums. Then it brought in no more. The species went extinct in the same year that it was discovered (Galbreath 1989).

18.2.4 *The side effects of pest control*

The effects of pest control often exceed the original intentions of the control exercise.

The black-footed ferret

The sinuously elegant black-footed ferret (*Mustela nigripes*) provides an example of a species paying the price for the control of another. This account of its narrow escape from extinction is taken from Seal *et al.* (1989), Cohn (1991), and Biggins *et al.* (1999).

The black-footed ferret once ranged across most of the central plains of North America from southern Canada to Texas. Its lifestyle is closely linked to that of the prairie dog (*Cynomys leucurus*), a squirrel-like rodent that lived in huge colonies on the plains. The ferret feeds mainly on prairie dogs but can feed also on mice, ground squirrels, and rabbits. However, 90% of its diet comprises prairie dogs. The ferret lives in the warrens or burrow systems of the prairie dogs and hence that species provided the ferret with both its habitat and a large proportion of its food supply. Around the turn of the century there was a concerted effort to eradicate the prairie dog, which was viewed as vermin by ranchers. It was seen as competing with sheep and cattle for grass and its burrow systems made riding a horse most unsafe. The prairie dogs were poisoned, trapped, and shot in their millions by farmers and by government pest controllers. As the prairie dogs went, so did the ferrets. By the middle of the century they were judged to be extinct, but in 1964 a small population was discovered in South Dakota. That colony died out in 1973. In 1981 a colony was discovered in Wyoming. Careful censusing produced an estimate of 129 individuals in 1984, but by the middle of 1985 the population had declined to 58 animals and within a few months was down to 31. Canine distemper was diagnosed in this population and it might well have been the cause of that decline (see Section 11.11.3).

With the population obviously threatened there was an attempt to capture the remaining animals to add to an already established captive breeding colony. Five were caught in 1985, 12 in 1986, and one in 1987. By February 1987 the last known wild black-footed ferret was in captivity. Captive breeding was successful, and 49 and 37 ferrets were released in 1991 and 1992 into their former range in Wyoming (Biggins *et al.* 1999).

18.2.5 *The effect of poorly regulated commercial hunting*

The type example of serious declines caused by hunting is provided by the history of commercial whaling. It demonstrates the effect of the discount rate (see Section 19.8) upon the commercial decision determining whether a sustained yield is taken or whether the stock is driven to commercial extinction.

Market economics will act to conserve a commercially harvested species only when that species has an intrinsic rate of increase r_m (see Section 6.2.1) considerably in excess of the commercial discount rate, the interest a bank charges on a loan to a valued customer. Hence, when a species is harvested commercially, the yield must be regulated by an organization whose existence and funding is independent of the economics of the industry that it regulates; otherwise it will necessarily endorse the quite rational economic decisions of the industry, which may well be to drive a stock to very low numbers and then to switch to another stock.

The muskoxen in mainland Canada

Unregulated commercial hunting reduced the muskoxen (*Ovibos moschatus*) on the arctic mainland of Canada to about 500 animals by 1917. In that year the species was protected by the Canadian government. The size of the historic populations will never be known but, ironically, documentation of the purchase of the muskox hides from native hunters by the trading companies was detailed. Barr (1991) collated the records and estimated that a minimum of 21,000 muskoxen were taken between 1860 and 1916. Their hides were shipped to Europe as sleigh and carriage robes, replacing bison robes after that species had been reduced almost to extinction.

Commercial hunting appears to be the overriding cause of the virtual extinction of the muskoxen on the Canadian mainland. Legislative protection successfully reversed the trend. Muskoxen now number about 15,000 on the mainland and have reoccupied almost all of their historic range (Reynolds 1998). The conservative hunting quotas introduced in the 1970s did not stop that recovery.

18.2.6 *The effect of unregulated recreational hunting*

Recreational hunting is intrinsically safer than commercial hunting because sport hunters operate on an implicit discount rate of zero. Sport hunting hence has an enviable record of conserving hunted stocks. Instances of gross overexploitation are rare but not unknown. The overhunting of the "forty-mile" caribou herd in Yukon has already been described in Section 10.7.1.

The Arabian oryx

The Arabian oryx (*Oryx leucoryx*) is a spectacular antelope whose demise in the wild and its subsequent re-establishment from captive stock is related by Stanley Price (1989) and Gordon (1991). Its original distribution appears to have included most of the Arabian Peninsula, but by the end of the nineteenth century the remaining Arabian oryx were divided into two populations. A northern group lived in and around the sand desert of north Saudi Arabia known as the Great Nafud and a southern group occupied the Rub' al Khali (the Empty Quarter) of southern Arabia.

The northern population became extinct about 1950. The range of the southern population declined from about 400,000 km² in 1930, to 250,000 km² in 1950, to 10,000 km² in 1970. Within the next couple of years the population was reduced to six animals in a single herd. They were shot out on October 18, 1972.

Recreational hunting caused this extinction. The countries of the Arabian Peninsula are essentially sea frontages, the inland boundaries being little more than lines on a map. There is little control over activities in the hinterlands. Oil company employees and their followers used company trucks for hunting trips and seemed to have been at least partly responsible for the decline. Then there were the large motorized hunting expeditions originating mainly from Saudi Arabia. These were self-contained convoys that included fuel and water tankers. The vehicles and support facilities allowed large areas to be swept each day with efficient removal of the wildlife. Their main quarry was bustards and hares secured by hawking, but gazelles and Arabian oryx were also chased. One such party crossed into the Aden Protectorate (now the People's Democratic Republic of Yemen) in 1961 and killed 48 Arabian oryx, about half the population of that region. In the 1960s large parties from Qatar would each year capture Arabian oryx with nets in the hinterland of Oman, trucking them the 900 km back to replenish the captive herd of Shaikh Kasim bin Hamid.

18.2.7 *The effect of competition with introduced species*

Hawaiian birds

More bird species have been introduced to the Hawaiian islands than to any other comparable land area. Of 162 species introduced, 45 are fully established and 25 have secured at least a foothold (Scott *et al.* 1986). These exotic species have been suggested as one of the causes of the decline and extinction of the native birds.

Mountainspring and Scott (1985) estimated the geographic association within pairs of the more common small- to medium-sized insectivorous forest passerines. After statistically removing the effect of habitat they showed that a higher proportion of exotic/native pairs of forest birds were negatively associated than were pairs of indigenous birds. They suggested that these results reflected competition, mainly for food.

The Japanese white-eye (*Zosterops japonicus*) became the most abundant land bird in Hawaiian islands after being introduced to Oahu in 1929 and to the island of Hawaii in 1937. It feeds on a wide variety of foods and is fairly catholic in its choice of habitats. It shares the range of at least three native species with similar food habits. Although causality cannot be demonstrated conclusively, particularly in retrospect, there is strong inference that the Japanese white-eye was implicated in the decline of the Hawaii creeper (*Oreomystis mana*) in the 1940s.

18.2.8 *The effect of environmental contaminants*

Rachel Carson's (1962) classic book *Silent Spring* raised the alarm about the effects on birds of DDT (dichlorodiphenyltrichloroethane) and other organochlorines. In particular, these chemicals caused eggshells to become abnormally thin and fragile (Ratcliffe 1970; Cooke 1973). Because these chemicals accumulated in the food chain it was the species at the top of the food chain, the raptors, which suffered the effects most. Nesting success declined precipitously and so raptor populations collapsed (Hickey and Anderson 1968; Cade *et al.* 1971). The chemical industry initially denied these unwelcome side effects but by the 1970s the evidence was overwhelming and DDT was banned in most countries. As a result, we have seen a rebound in the populations of several species of raptors, such as peregrine falcon (*Falco peregrinus*) in North America and common buzzard (*Buteo buteo*) in Europe.

The issue of contaminants remains with new pesticides and herbicides. For example, the so-called second generation of anticoagulants, such as brodifacoum, are both highly toxic to birds and mammals and persistent, so that they increase the risk of secondary poisoning of non-target species (Eason *et al.* 2002). Monofluoracetate (1080) is commonly used in baits to kill mammal pests in Australia and New Zealand, but there are impacts on non-target birds and mammals (Spurr 1994, 2000).

We give two examples.

The Indian vultures

In Pakistan and India cattle and water buffalo (*Bubalus bubalus*) are treated with a non-steroidal anti-inflammatory drug, diclofenac, to counter the effects of trauma and disease. When vultures feed on carcasses of these domestic animals they die from the toxic effects of diclofenac, to which they appear particularly sensitive. Three species are affected, the white-backed vulture (*Gyps bengalensis*), the long-billed vulture (*G. indicus*), and the slender-billed vulture (*G. tenuirostris*). The evidence suggests renal failure in the birds.

Between 1990 and 2003 vulture populations have declined from tens of thousands to a level where captive breeding is now required. At least 95% of the populations have died in 10 years. Research is now required to identify alternative drugs that are safe for vultures but remain effective for livestock (Green *et al.* 2004).

The California condor

The story of the California condor (*Gymnogyps californianus*) reveals a little about the realities of conservation: the gaps between theory and practice and the overwhelming need to determine, not assume, the causes of the decline.

The California condor was probably abundant and widely distributed in southern North America during the Pleistocene. Later it figured in the ceremonies and myths of the prehistoric and historic Indians and caught the eye (and trigger finger) of the early European explorers. California condors ranged from the Columbia River south into New Mexico in the 1800s but by 1940 their range had contracted to a small area north of Los Angeles. Koford's (1953) estimate, based on sightings, of only 60 individuals surviving by the early 1950s was probably low. Annual surveys by simultaneous observations of known concentrations were begun in 1965 but abandoned in 1981 because they were judged to be subject to unacceptable error. Photographic identifications were then used to generate a total count of 19–21 birds in 1983 (Snyder and Johnson 1985). The decline continued until, in 1985, the last eight wild individuals were caught and added to the captive flock.

The causes of the initial decline were probably shooting and loss of habitat, but the supporting evidence is anecdotal. Low productivity caused by an insufficiency of food was suggested as a cause of the decline during the 1960s. Road-killed deer were cached at feeding stations in 1971–73 to alleviate the perceived shortage of food (Wilbur et al. 1974). That program was run for an insufficient time to determine whether supplementary feeding was associated with increased productivity.

The connection between toxic organochlorines and eggshell thinning in birds was established in the late 1960s, but the resulting flurry of studies focused on bird-eating and fish-eating birds because avian scavengers were assumed to be less at risk. The possibility of a causal association between environmental toxins and the later decline of the condor was recognized in the mid-1970s, but determination of the specific role of organochlorines in that decline was delayed (Kiff 1989). Eggshell samples from California condors had been collected in the late 1960s but for various reasons, including mishaps to the samples, analyses were delayed until the mid-1970s. The negative correlation between eggshell thickness and DDE (dichlorodiphenyldichloroethane) levels was significant: the shells were thinner and their structure different from shells collected before 1944.

It was known that condor eggs often broke but the cause was open to debate. Even the monitoring activities themselves were suspected as being the cause. The evidence for organochlorines was circumstantial but it led Kiff (1989) to conclude that "DDE contamination probably had a very serious impact on the breeding success of the remnant population in the 1960s, leading to a subsequent decline in the number of individuals added to the pool of breeding adults in the 1970s." In 1972 DDT was banned in the USA. The few eggs measured after 1975 had thicker eggshells and this led to guarded optimism. In March 1986, however, an egg laid by the last female to attempt breeding in the wild was found broken. Its thin shell was suspiciously reminiscent of the "DDE thin-eggshell syndrome." In the meantime, analysis of tissue from wild condors found dead in the early 1980s revealed that three of the five had died from lead poisoning, probably from ingesting bullet fragments in carrion. Other condors had elevated lead levels in their blood (Wiemeyer et al. 1988). Recognition of the deleterious effects of yet another toxin in the condor's food supply led to provision of "clean" carcasses just before the last condors were taken into captivity.

18.2.9 *The effect of introduced diseases*

Extinctions caused by disease are particularly difficult to identify in retrospect. Moreover, on theoretical grounds disease is unlikely to be a common agent of extinction. In their review of pathogens and parasites as invaders, Dobson and May (1986b) noted the improbability of a parasite or pathogen driving its host to extinction unless it had access to alternative hosts.

Hawaiian birds

Avian malaria and avian pox have been suggested as contributing to the decline of the Hawaiian birds (Warner 1968). Migratory waterfowl may have provided a reservoir for avian malaria on the Hawaiian islands, and the continuous reintroduction by migration may have maintained a high level of infection in the face of a decline in host numbers (Dobson and May 1986b). Alternatively, avian malaria may have been carried by introduced birds such the common myna (*Acridotheres tristis*), which may themselves have maintained the disease at a high level because they are not greatly affected by it.

Originally there were no mosquitos on Hawaii capable of spreading malaria. The accidental introduction of mosquitos in 1826 and their rapid spread throughout the islands coincided with the decline of many species of birds. Six of 11 endemic passerines died out by 1901 on Oahu before their habitats had been modified (Warner 1968). Experiments showed that the Hawaiian passerines, especially the honeycreepers, are highly susceptible to malaria, much more so than are the introduced species (Warner 1968).

Avian malaria is a factor in restricting the present distribution of native birds on Hawaii, lending credence to the suggestion that it is implicated in the extinction of other species. Scott *et al.* (1986) noted that elevations above 1500 m that were free of mosquitos hosted the highest densities of native birds, especially of the rarer passerines.

18.2.10 *The effect of multiple causes*

The heath hen

The history of the heath hen is related by Bent (1932). We use here the summary and interpretation of that history presented by Simberloff (1988).

> Probably the best-studied extinction is that of the heath hen (*Tympanuchus cupido cupido*). This bird was originally common in sandy scrub-oak plains throughout much of the northeastern United States, but hunting and habitat destruction had eliminated it everywhere but Martha's Vineyard by 1870. By 1908 there were 50 individuals, for whom a 1600 acre refuge was established. Habitat was improved and by 1915 the population was estimated to be 2000. However, a gale-driven fire in 1916 killed many birds and destroyed habitat. The next winter was unusually harsh and was punctuated by a flight of goshawks; the population fell to 150, mostly males. In addition to the sex ratio imbalance, there was soon evidence of inbreeding depression: declining sexual vigor. In 1920 a disease of poultry killed many birds. By 1927 there were 13 heath hens (11 males); the last one died in 1932. It is apparent that, even though hunting and habitat destruction were minimized, certainly by 1908 and perhaps even earlier, the species was doomed. Catastrophes, inbreeding depression and/or social dysfunction, demographic stochasticity, and environmental stochasticity all played a role in the final demise.

18.3 **How to prevent extinction**

The previous sections summarized 12 examples of extinction or steep decline. The decline of the heath hen and the Hawaiian birds may be attributable to several factors but research on those species has not adequately revealed the causes of the declines. The extinction of several species of wallaby seems very likely to reflect habitat modification. The extinction of the Stephen Island wren and the decline of the Lord

Howe woodhen were unambiguously caused by an introduced predator. The extinction of the Arabian oryx in the wild (although subsequently re-established from captive stock) and the near extinction of the muskoxen were also caused by predation, this time by people. The black-footed ferret went extinct in the wild because the source of its habitat and most of its food supply – the prairie dog – was greatly reduced in density by control operations.

These examples implicate only a few potential causes of decline. Probably the most important is modification or destruction of habitat. The local extinction of several mammals from the sheep rangelands of Australia appears to have been caused by habitat changes induced by sheep introduced in the mid-1800s. Twelve of the original 38 species of marsupials and six of the original 45 species of eutherians (endemic rodents and bats) no longer live in that region (Robertson *et al.* 1987).

The first step in averting extinction is to recognize the problem. Many species have slid unnoticed to the brink of extinction before their virtual absence was noticed. The smaller mammals and birds, and the frogs and reptiles, are more likely to be overlooked than are the large ungulates and carnivores.

The second step is to discover how the population got into its present mess.
- Is the cause of decline a single factor or a combination of factors?
- Are those factors still operating?
- If so can they be nullified?

The cause of a decline is established by application of the researcher's tools of trade: the listing of possible causes and then the sequential elimination of those individually or in groups according to whether their predicted effects are observed in fact. This is the standard toolkit of hypothesis production and testing.

It is essential that the logic of the exercise is mapped out before the task is begun. The listing of potential causes is followed by a formulation of predictions and then a test of those predictions. The efficiency of the exercise is critically dependent on the order in which the hypotheses are tested. Get that wrong and a 3-month job may become a 3-year project. In the meantime the population may have slid closer to the threshold of extinction, so time is important.

Box 18.1 gives a specimen protocol for determining the cause of a population's decline. The example comes from the decline of caribou on Banks Island in the Canadian arctic. The first aerial surveys of the island in 1972 revealed an estimated population of 11,000 caribou. Subsequent surveys in the 1980s traced a dwindling population that numbered barely 900 caribou by 1991. Since then the population has stabilized, being 1195 in 2001 (N. Larter, pers. comm.). The muskoxen during the same time increased from 3000 to 46,000, leading to fears that there were too many muskoxen for the good of the caribou (Gunn *et al.* 1991). The population continued to increase to 64,600 by 1994, and then slowed to about 69,000 by 2001 (N. Larter, pers. comm.). Particularly severe winters restricted foraging for the caribou and caused dieoffs, at least in 1972–73 and 1976–77. The frequency of severe winters with deep snow and freezing rain increased during the 1970s and 1980s. Caribou and muskoxen differ in lifestyles and responses to winter weather.

An example of how difficult it can be to get the logic of diagnosis right is provided by research and treatment of the endangered Puerto Rican parrot (*Amazona vittata*). This strikingly attired bird has been the focus of some 50 years of intensive conservation efforts, including some 20,000 hours of observations of ecology and behavior (Snyder *et al.* 1987). The parrot may have numbered more than 1 million historically

Box 18.1 Hypotheses to be tested to discover the cause of the decline of caribou on Banks Island, Northwest Territories of Canada.

Hypotheses to account for the decline:
Either:
   A Food shortage
   B Increased predation

If (A) then mechanisms may be
   A1 Increase in weather events such as freezing rain that affect availability of food
   A2 Competition for food with muskoxen which are increasing
   A3 Caribou themselves reducing the supply of food

If (B) then mechanisms may be
   B1 Wolf predation
   B2 Human predation

The food shortage hypotheses (A) may be tested against the predation hypotheses (B) by checking body condition. Hypotheses A predict poor body condition and low fecundity during a population decline; hypotheses B predict good condition and high fecundity during a decline.
   If this test identifies the A hypotheses as the more likely, then A1 is separated from A2 and A3 by its predicting a positive rate of increase in some years. A2 and A3 predict negative rates of increase in all years.
   A2 (competition with another species) is separated from A3 (competition between caribou) by checking for concomitant decline of caribou where muskoxen are not present in the same climatic zone.

but by the early nineteenth century its distribution had contracted severely with the clearing of much of the forest of Puerto Rico. By the 1930s it was estimated as 2000 and by the mid-1950s, when the first intensive studies started, its numbers had collapsed to 200. Only 24 parrots were left in 1968 when rescue efforts were resumed. Despite a high-profile effort, including a captive breeding program started in 1968, little progress can be reported. The number of parrots in the wild population numbered only 21–23 before the 1992 breeding season (Collar *et al.* 1992) and was still only around 30 birds in 2004 (J. Wunderle, pers. comm.). The cause of the decline has not yet been identified unambiguously.

18.4 Rescue and recovery of near extinctions

Once a decline in a species is recognized and the causes are determined, the problem can be treated. The species accounts in the preceding sections give some idea of the range of management actions available to rescue a species from the risk of extinction. Sometimes, all it takes is a legislative change such as a ban on hunting (as with the Canadian muskoxen). More usually, active management (such as predator control and captive breeding for the Lord Howe Island woodhen) is necessary. The management actions needed to reverse the fortunes of a declining species are seldom more than conventional management techniques unless a species is in desperate straits. Then a whole new set of techniques may be called under the heading of *ex situ*. *Ex situ* techniques preserve and amplify a population of an endangered species outside its natural habitat. Thereafter it can be reintroduced. The Lord Howe Island woodhen and the Arabian oryx are examples of such reintroductions.

Reintroducing a species to the area from which it died out should not be attempted without some understanding of why the species went extinct there in the first place. Stanley Price (1989) describes the reintroduction of the Arabian oryx with captive stock, and details the considerations that should precede a reintroduction. Similar procedures were considered for the release of Przewalski's horse (*Equus caballus*) in Mongolia (Ryder 1993), and the addax (*Addax nasomaculatus*) in Niger (Dixon *et al.* 1991). In any event the liberated nucleus should be large enough to avoid demographic malfunction. Twelve individuals are an absolute minimum for an introduction. Twenty are relatively safe. Cade and Jones (1993) detail the successful captive breeding and reintroduction of the Mauritius kestrel (*Falco punctatus*). In the 1970s it was down to two breeding pairs, but by the 1990s some 235 birds had been reintroduced and established in new habitats.

When the cause of a local extinction is unknown, and when we therefore do not know whether the factor causing the extinction is still operating, a trial liberation should precede any serious attempt to repopulate the area. The 20 or more individuals forming the probe are instrumented where possible (e.g. with radiocollars) and monitored carefully to determine whether they survive and multiply or, if not, the cause of their decline. If the latter, the factor operating against the species can be identified and countermeasures can then be formulated. It is worth noting that a closely related species may be used as a probe when it is too risky to use individuals of the endangered species. For example, a successful probe release of Andean condors (*Vultur gryphus*) cleared the way for the release of two California condors from a captive breeding population in 1992 (Collar *et al.* 1992).

Short *et al.* (1992) showed the importance of probing for reintroductions of several wallaby species. Of 10 liberations into areas where the species had once been present but had died out, all failed. Of 16 liberations into areas where the species had not previously occurred, about half were successful. Apparently the factors that had caused the original extinctions of the first category were still operating. The authors suggested that exotic predators were probably the dominant factor causing the original extinctions and militating against successful reintroduction.

18.5 Conservation in national parks and reserves

National parks and reserves are pre-eminently important as instruments of conservation. In these areas alone the conservation of species supposedly takes precedence over all other uses of the land. Debate over whether protected areas, such as national parks, or community conservation areas are best for conservation is probably unnecessary because both have their advantages and disadvantages, as outlined in Box 18.2.

18.5.1 What are national parks and reserves for?

On one level that question is trite and it leads to the equally trite answer that parks and reserves are to conserve nature. When the question is refined to "what are the precise objectives of *this* park," the answer must be more concrete. However, even the general question is not as trite as it might seem. It is instructive to follow the history of ideas about the function of reserves, of which national parks can serve as the type example. Here we summarize those changing perceptions as outlined by Shepherd and Caughley (1987).

The national park idea has two quite separate philosophical springs whose streams did not converge until about 1950. The first is American, exemplified by the US Act of 1872 proclaiming Yellowstone as the world's first national park. The intent was to preserve scenery rather than animals or plants. Public hunting and fishing were at first entirely acceptable.

The second spring is "British colonial," with the Crown asserting ownership over game animals and setting aside large tracts of land for their preservation. The great national parks of Africa grew out of these game reserves, some physically and the others philosophically. Wildlife was the primary concern and scenery came second

Box 18.2 The advantages and disadvantages of protected areas such as national parks compared with community conservation areas.

Advantages of protected areas

1 Will protect fragile habitats (swamps, tundra, islands, endangered species). For example, the only breeding grounds of the whooping crane (*Grus americana*) occur within the Wood Buffalo National Park, Canada, and the only known location of the Madagascan tomato frog (*Dyscophus antongilii*) is in a single pond in the north of Madagascar.

2 Will protect large species that cannot coexist with humans, for example large carnivores and herbivores.

3 Can act as ecological baselines or benchmarks to monitor human disturbance outside (Arcese and Sinclair 1997; Sinclair 1998).

Disadvantages of protected areas

1 They do not represent all ecosystems or communities, often being selected for other reasons.

2 They are often too small to maintain viable populations, particularly of species that are adapted to live in large groups or that migrate across international borders (e.g. migrating caribou, bison, saiga antelope (*Saiga tatarica*), shorebirds (Charadriidae)).

3 Can alienate local indigenous peoples excluded by central governments.

Advantages of community conservation areas

1 Can represent species not included in protected areas, for example non-charismatic species (lower animals, microbes, fungi).

2 Can co-opt support of local peoples if benefits accrue to them.

Disadvantages of community conservation areas

1 Tend to protect only species of direct benefit to humans, and ignore the rest, which is the vast majority.

2 Excludes species that are detrimental to humans.

3 Tend to discount the future due to (i) increasing human population demands on the ecosystem and (ii) accelerating economic expectations from the system even with stationary human populations. These result in species loss and ecosystem decline.

---

if at all. The first was Kruger National Park established in 1926 on a game reserve proclaimed in 1898. Serengeti, in Tanzania, was gazetted in 1947 following from a reserve established in 1927. Kenya's first was established in 1946 on the Nairobi common.

All national parks established for 40 years or more have had their objectives and their management modified several times. The more influential fashions in park theory, listed here roughly in order of appearance over the last 100 years, are not mutually exclusive. They tend to be added to rather than replacing the previous ones.

1 The most important objective is to conserve scenery and "nice" animals. The aim translated into restricting roads and railways and attempting to exterminate the carnivores. Banff National Park, Canada, has such a history.

2 The most important objective is the conservation of soil and plants. This aim was a direct consequence of the rise of the discipline of range management in the USA during the 1930s. Its axiom was (and still is) that there is a "proper" plant composition and density. Enough herbivores were to be shot each year to hold the pressure of grazing and browsing at the "correct" level. An ecosystem could not manage itself. If left to its own devices it would do the wrong thing (Macnab 1985).

3 The most important objective is the conservation of the physical and biological state of the park at some arbitrary date. In the USA, South Africa, and Australia that date marked the arrival of the first European to stand on the land. This is the source of much of the controversy in Yellowstone National Park.

4 The fashion shifted to the conservation of representative examples of plant and animal associations. The wording is from Bell's (1981) definition of the function of national parks in Malawi, but the objective underlies the management of many national parks in many other countries.

5 The most important objective is the conservation of "biological diversity" (or biodiversity). This catch phrase had two meanings. It was sometimes used in the sense of "species diversity" (MacArthur 1957, 1960) whereby the information-theory statistic of Shannon and Wiener could be used to estimate the probability that the next animal you saw would differ at the species level from the last. The statistic is maximized for a given number of species when all have the same density. Within park management the idea translated as "the more species the better." The second meaning dealt with associations rather than species: the more diverse a set of plant associations the better the national park. For example, Porter (1977) defined the objectives of the Hluhluwe Game Reserve in Natal as "To maintain, modify and/or improve (where necessary) the habitat diversity presently found in the area and thus ensure the perpetuation and natural existence of all species of fauna and flora indigenous to the proclaimed area."

6 The most important objective is the conservation of "genetic variability." The phrase can be defined tightly and usefully (e.g. Frankel and Soulé 1981), but within the theory and practice of park management it lacked focus. It was tossed around with little or no attempt to define or understand what it means, whether the variability sought was in heterozygosity, in allelic frequency, or in phenotypic polymorphism. In practice it again translated into "the more species the better."

7 The most recent objective differs in kind from the six previous objectives. Frankel and Soulé (1981) express it thus: "the purpose of a nature reserve [in which category they include national parks] is to maintain, hopefully in perpetuity, a highly complex set of ecological, genetic, behavioral, evolutionary and physical processes and the coevolved, compatible populations which participate in these processes." Don Despain (quoted by Schullery 1984) puts it more plainly: "The resource is wildness."

18.5.2 *Processes or states?*

The first six objectives listed above identify biological states as the things to be conserved. The seventh identifies biological processes as the appropriate target of conservation. At first glance Frankel's and Soulé's purpose of a nature reserve appears also to require the maintenance of states because it refers to the conservation of populations. However, populations are not states in the sense that plant associations are states. A plant association has a species composition. Its component populations must have a ratio of densities one to the other that remains within defined limits. If those limits are breached the plant association has changed into another kind of plant association. A population, however, is not defined by ratios. The ratio of numbers in one age class relative to those in another, or the ratio of males to females, has no bearing on its status as a population.

The management of a national park will be determined by whether the aim is to conserve biological and physical states by suppressing processes or whether it is to preserve processes without worrying too much about the resultant states. There are three options:

1 If the aim is to conserve specified animal and plant associations that may be modified or eliminated by wildfire, grazing, or predation, then intervene to reduce the intensity of wildfire, grazing, or predation.

2 If the aim is to give full rein to the processes of the system and to accept the resultant, often transient, states that those processes produce, then do not intervene.

3 A combination of both: if the aim is to allow the processes of the system to proceed unhindered unless they produce "unacceptable" states, then intervene only when unacceptable outcomes appear likely.

18.5.3 *Effects of area*

Within any group of islands (e.g. the Antilles, Indonesia, Micronesia) big islands tend to contain more species than do small islands. Size as such is not the only influence on the number of species – distance to the mainland, for example, plays a part – but area alone provides a close prediction. The relationship between the number of species and the size of the area within which they were surveyed is known as a species–area curve.

Algebraically it takes the form:

$$S = CA^z$$

in which S is the number of species of a given taxon (e.g. lizards, forest birds, vascular plants), A is the area, C is the expected number of species on one unit of area (usually 1 km^2) and z indexes the slope of the curve relating the number of species to the number of square kilometers.

Table 18.1 shows the relationship between species number and land area for Tasmania and the islands between it and the Australian mainland (Hope 1972). These were all linked to each other and to the Australian mainland up to about 10,000 years ago, the subsequent fragmentation reflecting rise of sea level at the end of the Pleistocene. The number of marsupial herbivores that they carry therefore reflects differential extinction without reciprocal immigration over the last 10 millennia. The estimated $z = 0.18$ is low for islands, being closer to that expected for areas within continents, and it probably reflects the recent continental nature of those islands. Box 18.3 shows how C and z are calculated from these data.

The expected number of species on one unit of area, C, varies according to latitude, elevation, ecological zone, taxonomic group, and the units in which A is measured. In contrast z tends to be quite stable. For most taxa and groups of islands

Table 18.1 Relationship between the number of species of herbivorous marsupials and area of land on Tasmania and the islands between it and the Australian mainland.* The "expected" number is calculated as $S = 1.70A^{0.18}$ (see Box 18.3 for calculation).

Island	Area (km^2)	Observed species	Expected species
Tasmania	67,900	10	12.6
Flinders	1,330	7	6.3
King	1,100	6	6.0
Cape Barren	445	6	5.1
Clarke	115	4	4.0
Deal	20	5	2.9
Badger	10	2	2.6
Prime Seal	9	2	2.5
Erith-Dove	8	3	2.5
Vansittart	8	2	2.5
West Sister	6	2	2.3

*Number of species as at AD 1800. Only islands larger than 5 km^2 are included.
Data from Hope (1972).

Box 18.3 Estimating the
constants of a
species–area curve.

A species–area curve takes the form $S = CA^z$, where:

S = number of species
A = area, in this case always expressed as km^2
C = expected number of species on an area of 1 km^2
z = slope of the curve relating species number to area

Taking the data of Table 18.1 as our example, first convert area and species number to logarithms. Any base will do but we will use logs to the base e. Label log area as x and log species number as y:

x	y
11.126	2.3026
7.193	1.9459
7.003	1.7918
6.098	1.7918
4.745	1.3863
2.996	1.6094
2.303	0.6931
2.197	0.6931
2.079	1.0986
2.079	0.6931
1.792	0.6931

We now calculate these:

$$n = 11$$
$$\text{mean } x = 4.510 \qquad\qquad \text{mean } y = 1.336$$
$$\Sigma x = 49.61 \qquad\qquad \Sigma y = 14.70$$
$$\Sigma x^2 = 315.2 \qquad\qquad \Sigma xy = 82.58$$
$$(\Sigma x)^2/n = 223.7 \qquad\qquad (\Sigma x)(\Sigma y)/n = 66.30$$
$$SS_x = \Sigma x^2 - (\Sigma x)^2/n = 91.5 \qquad SS_{xy} = \Sigma xy - (\Sigma x)(\Sigma x)/n = 16.28$$

The constants of the species–area curve are now solved:

$$z = SS_{xy}/SS_x$$
$$= 16.28/91.5$$
$$= 0.18$$
$$C = \text{antilog}(\Sigma y/n - z\Sigma x/n)$$
$$= \exp(1.336 - 0.18 \times 4.51)$$
$$= 1.70$$
$$\text{Thus } S = 1.7A^{0.18}$$

it lies between 0.2 and 0.4. At the midpoint, 0.3, an increase or decrease of area by a factor of 10 results in a doubling or halving respectively of the number of species (by virtue of $10^{0.3} = 2$). Thus, when $A = 1$, $S = C$ irrespective of the value of z; and when $A = 10$ and $z = 0.3$, $S = 2C$.

The relationship is the same if we count the number of species on nested areas of progressively larger size on a continent. Here the value of z tends to be lower, usually around 0.15. It implies that a reduction of area by a factor of 10 reduces the number of species by a factor of only 1.4 ($10^{0.15} = 1.41$). The difference between that exponent of 0.3 for islands and 0.15 for continents probably reflects the easier dispersal between contiguous areas of land against between islands. These relationships are particularly important for determining optimum sizes of reserves.

18.5.4 Is one big national park better than two small national parks?

Suppose we have the money necessary to acquire 100 km² of land for conversion into national parks. If the aim were to conserve the maximum number of species for a long time, should we go for one park of 100 km² or two each of 50 km²? Obviously a number of factors would influence our choice, but let us assume that the overriding aim is to maximize the number of species of mammals within the single large reserve or the alternative two smaller reserves. Let us assume that 1 km² will on average contain 20 species in this region (i.e. $C = 20$). Further, we know that $z = 0.15$ for mammals in this region. Thus, a national park of 100 km² would contain about 40 species of mammals ($S = CA^z = 20 \times 100^{0.15} = 40$) whereas a park of 50 km² would hold about 36 mammals ($S = 20 \times 50^{0.15} = 36$). Whether we favor one park of 100 km² or two of 50 km² each comes down to how many species are held in common by the two smaller parks. That will depend on the extent to which they differ in habitat and on the distance between them.

The efficacy with which a reserve system conserves species and communities thus depends on the size of the reserves and, more importantly, on where they are – their dispersion relative to the distribution patterns of species. Margules et al. (1982) warn against using data-free geometric design strategies (big is better than small, three is better than two, linked is better than unlinked, grouped is better than linear).

18.5.5 Effects of corridors

Corridors between reserves provide the benefit of increasing the size of populations and thereby decreasing the chance of demographic malfunction. However, the overall benefit of corridors is not at all clear cut and must be decided upon case by case. Lindenmayer (1994) lists the factors that might influence the use of corridors:
1 The biology, ecology, and life history of the species.
2 Habitat suitability, including the degree of original vegetation integrity, length, and width.
3 Location of corridors in the landscape.
4 The type of disturbance in the matrix surrounding fragments and corridors.
5 Suitability of the matrix habitat.
There is a conceptual problem with corridors. By definition these are strips of habitat that are too small for the species of interest to live in permanently (e.g. too close to the edge of forest for interior forest birds, or too narrow to support a territory). Such strips may be suitable for wide-ranging species, such as rodents, that would benefit from the cover provided by forest or shrubs to allow safer movement relative to movement over fields. However, such species would probably traverse these open habitats if corridors did not exist. In contrast, sedentary species such as interior forest birds (the New Zealand kokako (Callaeas cinerea) is a good example of a highly territorial bird that flies poorly and moves little through dense forest) are unlikely to venture into corridors because they are unsuitable habitat. Thus, species that would benefit most from corridors, the reluctant travelers, are the ones least likely to use them, and vice versa. Corridors could also act as sinks, trapping animals in them but preventing successful breeding (Saunders and Hobbs 1991).

A good example of the effect of corridors is illustrated in Fig. 18.1. Narrow strips of eucalypt woodland act as corridors, or more accurately as stepping stones, to

Fig. 18.1 The number of species of birds in isolated patches of eucalypt habitat in Western Australia is related to the distance to roadside strips that act as corridors or stepping stones. (Data from Fortin and Arnold 1997.)

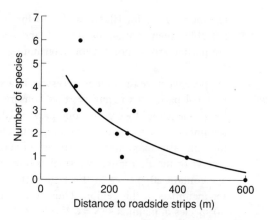

connect isolated fragments of original eucalypt woodland in Western Australia. These corridors result in higher species richness of birds, so that the closer the corridors to a patch the higher the species number (Fortin and Arnold 1997). Kangaroos (*Macropus robustus*) also use corridors to move between remnant patches of eucalypt woodland in Western Australia (Arnold *et al.* 1993).

Golden lion tamarins (*Leontopithecus r. rosalia*) need corridors between patches of tropical forest habitat, all that remains within a sea of sugarcane fields in Brazil. Rodents, such as chipmunks and voles, can use fencerows and hedgerows as corridors (La Polla and Barrett 1993; Bennett *et al.* 1994). Hill (1995) showed how some poorly dispersing insects (butterflies, dung beetles) use corridors but others do not.

The efficacy of corridors needs to be assessed on a case-by-case basis. Thus, Simberloff and Cox (1987) used the Seychelles islands of the Indian Ocean to make the point that corridors are not always beneficial. The Seychelles contained 14 endemic land birds when Europeans arrived in 1770. Land clearing, fires, and the introduction of rats and cats devastated the archipelago over the subsequent two centuries but resulted in the extinction of only two of those species. Losses were limited partly because no corridors (isthmuses) linked the islands. Introduced predators and fires were unable to reach all the islands.

The potential advantages and disadvantages of conservation corridors as summarized by Noss (1987) and Saunders and Hobbs (1991) are presented in Box 18.4.

18.5.6 *Effects of initial conditions*

Parks are chosen for a number of reasons: great scenery, many species, a cherished plant association, or a set of interesting landforms. Sometimes the area chooses itself, being deemed good for little else.

Most national parks established since 1960 (the majority) have been chosen with some care. They are designed to conserve the plant and animal communities and/or their associated ecological processes in a particular climatic zone. Having decided upon the zone the next step is to choose an area within that zone which samples or epitomizes that zone. The decision is determined first by what land is available for conversion to a park. It is then determined by whether a piece of available land is large enough, or can be made large enough by accretion of adjacent areas, to serve as a national park. Finally a choice is made between the various areas of land that meet the above criteria.

Box 18.4 Potential advantages and disadvantages of conservation corridors. (From Noss 1987.)

Potential advantages of corridors
1 Increased immigration rate to a reserve, which could:
   (a) increase or maintain species richness (as predicted by island biogeography theory);
   (b) increase population sizes of particular species and decrease probability of extinction (provide a "rescue effect") or permit re-establishment of extinct local populations;
   (c) prevent inbreeding depression and maintain genetic variation within populations.
2 Provide increased foraging area for wide-ranging species.
3 Provide predator-escape cover for movements between patches.
4 Provide a mix of habitats and successional stages accessible to species that require a variety of habitats for different activities or stages of their life cycles.
5 Provide alternative refugia from large disturbances (a "fire escape").
6 Provide "greenbelts" to limit urban sprawl, abate pollution, provide recreational opportunities, and enhance scenery and land values.

Potential disadvantages of corridors
1 Increased immigration rate to a reserve which could:
   (a) facilitate the spread of endemic diseases, insect pests, exotic species, weeds, and other undesirable species into reserves and across the landscape;
   (b) decrease the level of genetic variation among populations or subpopulations, or disrupt local adaptations and co-adapted gene complexes ("outbreeding depression").
2 Facilitate spread of fire and other abiotic disturbances ("contagious catastrophes").
3 Increase exposure of wildlife to hunters, poachers, and other predators.
4 Riparian strips, often recommended as corridor sites, might not enhance dispersal or survival of upland species.
5 Cost, and conflict with conventional land preservation strategy to preserve endangered species habitat (when inherent quality of corridor habitat is low).

At this stage the choice of land is determined mainly by which area contains the greatest internal diversity of habitats and also by which contains the largest number of species. The two tend to be correlated. These two criteria of choice have an effect upon extinction rates within the reserve. They ensure that the park will have an **over-diversity** of species and habitats. If an area contains a diversity of habitats it will on average contain little of each. A species dependent on a single habitat will therefore be represented on average by a small population within such a reserve.

If the area contains a diversity of species, several of those species will be near the edge of their ranges and so will be living outside their environmental optima. Such species will be at low density within the reserve. If the reserve becomes an ecological island the number of species it contains will be much higher than that predicted by the species–area curve and it can therefore be expected to lose species. Those are effects of sampling a region by an area containing most of the characteristics of the whole region.

Another consequence of choosing a diversity of habitats is that the main habitat of interest (e.g. rainforest, savanna, taiga) tends to be sampled near the edge of its distribution to allow other habitats to interdigitate with it. If the climate changes then the habitat of interest is likely to be lost.

To summarize, to select an area suitable for conservation, one should:
1 choose an area containing a moderate rather than a high number of species;
2 include only a moderate number of habitats within the same reserve;
3 position a reserve as close as possible to the center of distribution of the habitat of greatest interest.

18.5.7 *Culling in parks and reserves*

Whether or not the densities of mammals should be controlled artificially in a national park is a matter of some contention (Chase 1987), as illustrated by the papers in the *Wildlife Society Bulletin* (number 3, 1998) that discuss culling in general and the debate on Yellowstone National Park in particular. White *et al.* (1998) present the case for culling, particularly the elk, while Singer *et al.* (1998a,b), Boyce (1998), Frank (1998), and Detling (1998) present other viewpoints. Our own prejudices are to avoid culling in parks and reserves except in rare, special, and well-defined circumstances.

18.6 Community conservation outside national parks and reserves

The principles of conservation discussed above with reference to parks and reserves hold also for conservation outside those reserves. There are, however, a few important differences. In general, protected areas cover no more than about 10% of the terrestrial global surface, which means from our species–area equation (see Section 18.5.3) that only about 50% of the world's species are included. Thus, at least half of our terrestrial biota must be conserved in human-dominated systems. Box 18.2 outlines the pros and cons of community conservation approaches.

Some species or associations of species occur only rarely in reserves because parks and reserves do not capture a representative sample of the biota. In Australia, for example, few reserves contain forest types that grow on sites of high fertility. Most such sites were incorporated into state forests or alienated from common ownership before the reserve system was established. The koala (*Phascolarctos cinereus*) is dependent on such sites and so almost all attempts to conserve koalas must be made outside the reserve network where the manager does not have the same control over land use practices.

Legislation is the main means by which conservation is advanced outside reserves. Various practices, such as the killing of nominated species, are banned. Less commonly there are controls over land clearing, thereby protecting the habitat of species that dwell in forest and woodland. Activities on land owned by the people as a whole, even though that land is not designated as a conservation reserve, may be subject to environmental impact assessment (EIA). Laws governing conservation outside reserves should take legal precedence over forestry and mining law.

18.7 International conservation

Conservation is the responsibility of sovereign nations unless the issue is subject to international treaty (polar bears, ivory trade, migratory birds) or unless the problem occurs on the high seas (whales and pelagic fish stocks), on essentially unclaimed land (Antarctica) or on land under disputed sovereignty (parts of the high Arctic).

18.7.1 *IUCN Red Data Books*

The International Union of Nature and Natural Resources (IUCN) issues "Red Data Books" listing threatened species. Four categories are recognized, their exact wording varying according to the taxon. What follows is generalized.

Extinct (Ex)
Species not definitely located in the wild during the last 50 years.

Endangered (E)
Taxa in danger of extinction and whose survival is unlikely if the causal factors continue operating.

Included are taxa whose numbers have been reduced to a critical level or whose habitats have been so drastically reduced that they are deemed to be in immediate danger of extinction. Also included are taxa that are possibly already extinct but have definitely been seen in the wild in the past 50 years.

Vulnerable (V)

Taxa believed likely to move into the "endangered" category in the near future if the causal factors continue to operate.

Included are taxa of which most or all populations are decreasing because of over-exploitation, extensive destruction of habitat, or other environmental disturbance; taxa with populations that have been seriously depleted and whose ultimate security has not yet been assured; and taxa with populations which are still abundant but under threat from severe adverse factors throughout their range.

Rare (R)

Taxa with small world populations that are not at present "endangered" or "vulnerable," but are at risk.

These taxa are usually localized within restricted geographical areas or habitats or are thinly scattered over a more extensive range.

Indeterminate (I)

Taxa known to be "endangered," "vulnerable," or "rare" but where there is not enough information to say which of the three categories is appropriate.

Of these categories the "endangered" and "vulnerable" are the most important and there is widespread agreement on what the terms mean. "Rare" is not a particularly useful category of extinction risk and probably should not be used as such. If rarity itself is the cause of the risk, in the sense that the population size is at a level low enough to place it in danger of demographic or genetic malfunction, then it should be placed in one of the categories of threat.

The information from which the Red Data Books are produced is extracted largely by the Species Survival Commission (SSC) of IUCN, which is a network of the world's most qualified specialists in species conservation that serve on a voluntary basis. The various groups and their membership are listed in the SSC Membership Directory published by IUCN.

18.7.2 *The role of CITES*

CITES is the acronym for "Convention on International Trade in Endangered Species of Wild Fauna and Flora." The convention regulates trade in species of wildlife that are perceived to be at risk from commercial exploitation. There are 99 countries that are party to the convention.

The teeth of the convention are contained in its appendices listing the species covered by CITES. Article II of the convention decrees that:

1 Appendix I shall include all species threatened with extinction which are or may be affected by trade. Trade in specimens of these species must be subject to particularly strict regulation in order not to endanger further their survival and must only be authorized in exceptional circumstances.

2 Appendix II shall include:

(a) all species which although not necessarily now threatened with extinction may become so unless trade in specimens of such species is subject to strict regulation in order to avoid utilization incompatible with their survival; and

Table 18.2 Number
of species covered by
Appendices I and II
of the Convention on
International Trade in
Endangered Species of
Wild Fauna and Flora
(CITES) as at 2004.
This can be updated at
http://www.cites.org/
eng/disc/species.shtml.
Roughly 5000 species
of animals and 28,000
species of plants are
protected by CITES
against overexploitation
through international
trade.

	Appendix I (endangered)	Appendix II (threatened)
Mammals	228 spp. + 21 sspp. + 13 pops	369 spp. + 34 sspp. + 14 pops
Birds	146 spp. + 19 sspp. + 2 pops	1401 spp. + 8 sspp. + 1 pop.
Reptiles	67 spp. + 3 sspp. + 4 pops	508 spp. + 3 sspp. + 4 pops
Amphibians	16 spp.	90 spp.
Fish	9 spp.	68 spp.
Invertebrates	63 spp. + 5 sspp.	2030 spp. + 1 sspp.
Plants	298 spp. + 4 sspp.	28,074 spp. + 3 sspp. + 6 pops
Total	827 spp. + 52 sspp. + 19 pops	32,540 spp + 49 sspp. + 25 pops

spp., species; sspp., subspecies; pops, populations.

(b) other species which must be subject to regulation in order that trade in specimens of certain species referred to in the above sub-paragraph may be brought under effective control.

3 Appendix III shall include all species, which any Party identifies as being subject to regulation within its jurisdiction for the purposes of preventing or restricting exploitation, and as needing the cooperation of other parties in the control of trade.

4 The Parties shall not allow trade in specimens of species included in Appendices I, II, and III, except in accordance with the provisions of the present Convention.

Table 18.2 gives the number of species covered by Appendices I and II of CITES as of 2004.

18.8 Summary

Extinctions can be driven by a permanent change to a species' environment (e.g. a new predator, disease or competitor, or modification of its habitat) or can result from stochastic events. Driven extinctions are the most common. Stochastic extinctions are the chance fate of small populations: factors that would be swamped in a large population can have serious consequences for the individuals of a small population. The critical step in averting extinction is to follow the logical pathway of hypothesis testing to diagnose the cause of the decline. A species can seldom be rescued until the factors driving the decline are identified and removed. Rescue and recovery operations are standard wildlife management practices (e.g. regulation of harvest, predator control) but sometimes more elaborate steps such as captive breeding and translocations are called for. Reserves or national parks and community conservation all play a key role in the nurturing and recovery of endangered species.

19 Wildlife harvesting

19.1 Introduction

In this chapter we consider how to estimate an appropriate offtake for a wildlife population. It differs according to whether the population is increasing or whether it is stable, and whether or not the environment fluctuates from year to year. Wildlife is harvested for many different purposes. Sport hunting usually takes a sample of the population during a restricted season and often with a restriction placed on the sex and age of the harvest. Harvesting for sport is a complex activity whose product is as much a quality of experience as it is meat or trophies. On the other hand the purpose of commercial hunting or hunting for food is simply to harvest a product such as meat and skins.

Both recreational and commercial wildlife harvesting are controversial but it is not the purpose of this chapter to delve into that controversy. Whether or not one considers it is appropriate and ethical to harvest a population of a given species depends more on one's view of life than on what may be happening to the population. There is an ethical aspect, however, that is fundamental to wildlife harvesting: the operation, be it for recreation or profit, must result in a sustainable offtake, a yield that can be taken year after year without jeopardizing future yields.

In all but special circumstances the strategy of sustainable harvesting is simple: it is to harvest the population at the same rate as it can increase. Hence, a population increasing at 20% per year can be harvested sustainably at around 20% per year. That proportion of the population can be taken year after year, with the result that the population is held to an induced rate of increase of zero. The use of "rate of increase" as the appropriate harvesting rate is uninfluenced by whether the population is actively spreading, whether it is subject to predation or not, and whether sources of mortality are additive or compensatory.

The details of sustained-yield harvesting differ according to whether a harvested population's key resources are renewable or not, how the harvested population uses those resources, and various interactions between the resources and the harvested population. Most importantly, the dynamics of harvested populations depend on the regulatory strategy (legal limits) used to set harvest levels. We start with a consideration of the most simple harvest strategy: application of a constant harvest quota from year to year.

19.2 Fixed quota harvesting strategy

Most unharvested populations have a rate of increase which, when averaged over several years, is close to zero. Hence the sustained yield for such a population is also zero. Before such a population can be harvested for a sustained yield it must be stimulated to increase.

A population can most easily be stimulated into a burst of growth by increasing the level of a limiting resource or, much more rarely, reducing the level of predation to which it is subject. The key resource may be nest sites or cover, but in most cases it is food. For example, red kangaroos increase if the pasture biomass exceeds about 200 kg/ha dry weight and decline if it is below that level (Bayliss 1987). The easiest way to increase the average amount of food available to an individual animal is to reduce the number of animals competing with it for that food. The average standing crop of food then rises and the amount available to an individual animal is thereby increased. As a direct consequence the fecundity of individuals is enhanced, and mortality, particularly juvenile mortality, is reduced. The population enters a regime of increase as it climbs back towards its unharvested density.

19.2.1 Harvesting trades off yield against density

The trade-off between yield and density is the most important thing to know about sustained-yield harvesting. In general, the further the density is reduced the higher is the yield as a percentage of population size. The maximum rate of sustainable yield is the population's intrinsic rate of increase, but that rate of population growth is obtained only when food (or whatever other resource is limiting) is at a maximum, which in turn usually occurs only when the population is at minimum density.

Whereas sustained rate of yield (absolute yield divided by population size) tends to increase as density is reduced, the same is not true of the absolute sustained yield. If the population is reduced just a little, the induced rate of increase will be small and the sustained yield will be a small proportion of a relatively large population. The absolute yield will be modest. If the population is drastically reduced, the induced rate of increase will be large and the sustained yield will be a large proportion of what is now a relatively small population. Again the absolute yield will be modest. The highest yield is taken from a density at which the induced rate of increase multiplied by the density is at a maximum. It tends to be at intermediate density levels. For example, suppose that a population grows in a manner well described by the Ricker logistic model (see Chapter 6), with projected population size next year, $f(N)$, predicted from current population density in the following way:

$$f(N) = N \exp\left[r_{max}\left(1 - \frac{N}{K}\right)\right]$$

Net recruitment, $R(N)$, will be defined as the difference between population density in year t and population density the previous year, *in the absence of harvesting*:

$$R(N) = f(N) - N$$

Note that both positive and negative net recruitment values are possible, with negative values occurring when $N > f(N)$ and positive values occurring when $N < f(N)$. Fixed harvest policies are predicated on sustained use of the net recruitment, treating it as a surplus that can be safely harvested without harming resource sustainability in the long term. For example, imagine that we, as population managers, set the harvest at 20 units (Fig. 19.1).

Fig. 19.1 Net recruitment in the absence of harvest in relation to population density, plotted relative to an arbitrary constant level of harvest. At a given harvest quota, there are stable (circle) and unstable (×) equilibria. At population densities above ×, the population would tend to converge on the stable equilibrium. Perturbation of the population below ×, on the other hand, would lead to eventual extinction.

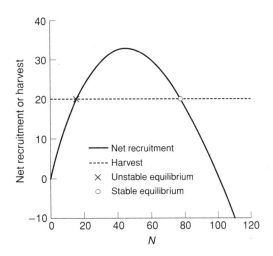

19.2.2 The two levels of sustained yield in constant quota systems

For a population conforming to the relationship between sustained yield and population size or density given in Fig. 19.1, a sustained yield of a given size may be taken from either of two densities, at the points at which the horizontal harvest line intersects the hump-shaped net recruitment curve. They comprise what is known as a **sustained-yield pair**. The member of the pair taken from the lower density is to be avoided because its harvesting requires more effort than is required to harvest the same yield from the higher density. Even more importantly, any reduction of population density below the level of the sustained-yield pair would inevitably lead to overharvesting relative to growth capacity of the population. When a constant number of animals is taken each year from a previously unharvested population, provided that there is no stochastic variation in weather or other factors that influence net recruitment, the population will decline and stabilize at the upper density for which that harvest is a sustained yield. Should that number exceed the maximum sustained yield, the population would inevitably decline to extinction.

19.2.3 The maximum sustained yield

The harvest that intersects the peak of the hump-shaped net recruitment curve is known as the **maximum sustained yield** (MSY). Harvesting a population at the MSY should never be contemplated. It imparts instability to the population's dynamics. The MSY can be taken only from the unique MSY density. If the population density has, for environmental reasons (such as drought or crusted snow), dropped below that value then the MSY represents an overharvest and the population's density is reduced further. Continued harvesting of the MSY will make the problem worse and even lead to extinction. This can be visualized more readily by considering a concrete example.

19.2.4 Harvesting in a stochastic environment: trapping of martens in Ontario

The American marten (*Martes americana*) is a forest carnivore species in the family Mustelidae (the weasel group) and is an important economic resource for many commercial trappers. It has a widespread distribution across the more northerly reaches of the USA and most of Canada. The marten lives in large tracts of coniferous and mixed forests, feeding on a wide variety of small mammals and berries, augmented occasionally by birds' eggs, insects, and other assorted invertebrate prey. Adult

Fig. 19.2 Year-to-year variation in the exponential annual rate of increase (r) of martens in the Bracebridge District near Algonquin Provincial Park, Ontario, in relation to variation in deermouse abundance (measured in captures per 1000 trap-nights). (After Fryxell *et al.* 1999, 2001.)

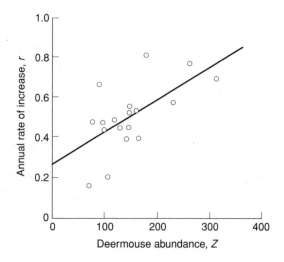

Fig. 19.3 Year-to-year variation in the exponential annual rate of increase (r) of martens in the Bracebridge District near Algonquin Provincial Park, Ontario, in relation to variation in marten abundance (minimum number alive). (After Fryxell *et al.* 1999, 2001.)

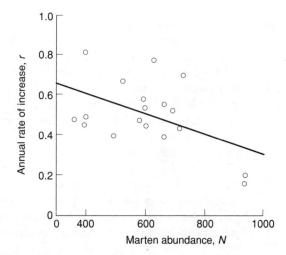

martens typically defend territories, with those of males being roughly twice the size of the female territories that they overlap. Young martens are born in a small litter in late spring, stay in a den with the mother until late summer or early fall, and then disperse over large distances seeking out their own breeding territory.

Like many other carnivores (see Chapter 10), marten abundance tends to track year-to-year changes in their major prey species (snowshoe hares, red-backed voles, red squirrels, and deermice in Ontario). In the area around Algonquin Provincial Park, for example, the exponential rate of increase by martens (r) varies four-fold with similar scale of variation in their major prey species (Fig. 19.2).

Like many territorial species, however, there is an additional density-dependent effect of marten numbers on their own annual rate of change, beyond the variation imposed by the food supply (Fig. 19.3). One explanation for this density dependence is that dispersing juveniles cannot obtain suitable sites for breeding territories when marten abundance is high, effectively cutting them off from contributing to further population growth. A second possibility is that the frequency of deadly aggression

Fig. 19.4 Marten net recruitment in the Bracebridge District of Ontario in relation to marten abundance (Fryxell *et al.* 2001). Three curves are shown, corresponding to average prey abundance over the 45 years of monitoring (Fryxell *et al.* 1999), as well as the highest and lowest levels of prey abundance on record. Net recruitment in any given year is expected to fall in the area bracketed by the broken lines. (After Haydon and Fryxell 2004.)

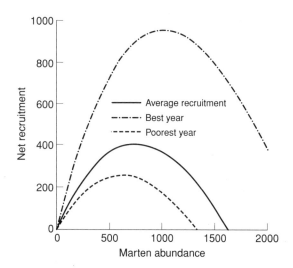

among territory-holders might escalate with increasing marten abundance. Such a mix of direct density dependence (interference competition, see Chapter 8) and indirect density dependence through variation in food supplies (scramble competition) has strong stabilizing properties (Fryxell *et al.* 1999).

As we discuss in Chapter 15, Akaike's information criterion is a useful means of evaluating alternative models of population growth. The model with both density and prey dependence fits the estimates of r substantially better ($AIC_c = -20.5$) than either a density-dependent model ($AIC_c = -11.3$) or a prey-dependent model ($AIC_c = -14.9$), so we shall use this as our best deterministic model forecasting the next population:

$$f(N, Z) = N \exp(a + bN + cZ)$$

where the intrinsic growth base rate ($a = 0.57$) is diminished by the density-dependent coefficient ($b = -0.0005$) multiplied by marten abundance (N), but augmented by the prey-dependent coefficient ($c = 0.0016$) multiplied by prey abundance (Z). This relationship (Fig. 19.4) predicts that the net recruitment should be a dome-shaped function of marten abundance, with the height of the dome dictated by prey abundance in a given year.

Of course, there is still some additional unexplained variation in r (in this case, with a standard deviation of approximately 0.10) that is due to variability in environmental conditions, such as weather, represented by the variable (ε), added to the other predictor variables in a model forecasting stochastic population growth:

$$g(N, Z, \varepsilon) = N \exp(a + bN + cZ + \varepsilon)$$

We are now in a suitable position to model different harvesting strategies that could be applied to this population. We start with a constant quota system, with annual quotas set at the expected MSY (423 martens per year, based on average prey abundance and $\varepsilon = 0$). Year-to-year variation in marten abundance can be modeled by the following set of equations:

Fig. 19.5 Predicted variation in marten abundance under a constant quota harvest policy, with annual marten harvests set at the maximum sustainable yield ($H = 423$ martens per year), starting from a population size of 400 animals in 1973.

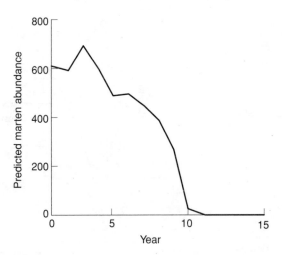

Fig. 19.6 Predicted variation in marten harvests under a constant quota harvest policy, with annual marten harvests set at the maximum sustainable yield ($H = 423$ martens per year), starting from a population size of 400 animals in 1973.

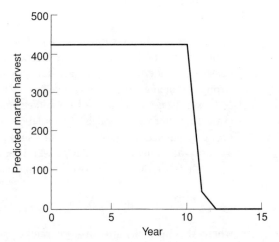

$$N_{t+1} = \begin{vmatrix} 0 & \text{if} & g(N_t, Z_t, \varepsilon_t) \le \text{MSY} \\ g(N_t, Z_t, \varepsilon_t) - \text{MSY} & \text{otherwise} \end{vmatrix}$$

Note that we have set a condition to ensure that the population never declines below zero. We use a random number generator drawn from a normal distribution with mean $= 0$ and standard deviation (σ) $= 0.10$, appropriate to the observed data, to generate population dynamics, starting from the average carrying capacity ($N = 1600$) recorded in the Ontario study. We use the observed fluctuations in prey recorded during 1973–87 to generate environmental variation in food supply for the martens. The results (Figs 19.5 and 19.6) demonstrate that the marten population cannot sustain a constant quota at the MSY for any appreciable length of time.

The important lesson we obtain from this example is this: given that all harvested wildlife species live in stochastic environments in which weather conditions and food supplies are expected to vary widely, wildlife managers should keep the harvesting rate well below MSY. A margin of error of about 25% below the estimated maximum sustained yield is appropriate; more where year-to-year variation in weather is above average.

19.3 Fixed proportion harvesting strategy

If a constant proportion of animals is taken each year from a previously unharvested population, the population will decline and stabilize, depending on the rate of harvesting, at any level between unharvested density and the threshold of extinction (Fig. 19.7). So long as the harvest rate (h) does not exceed the maximum intrinsic rate of population growth (r_{max}), harvesting a fixed proportion of the population should settle the population upon a stable population density, generating a sustained yield, even in the presence of stochastic variation in environmental conditions. Provided a wildlife or fisheries manager had perfect information on abundance in any given year before the harvest, a proportionate strategy would be guaranteed to allow sustainable harvests forever. The trouble is, wildlife or fisheries managers rarely have up-to-date information on current abundance, and they must set harvest levels (through licensing or allocation of quotas) long before annual recruitment is known. Indeed, the most common situation is that managers know little more about the population than the information gained from the harvest success during the previous year. Such information would permit, at best, a forecast of the current recruitment, based on an index of population abundance at the end of the previous harvest. In other words, population managers must guess the current level of abundance in setting an annual quota whose magnitude is intended to be a constant fraction of the population. The uncertainty thus introduced considerably increases the risk of unintentional overharvesting.

19.3.1 Fixed proportion harvesting of martens in a stochastic environment

As an example of applying a fixed proportion harvest strategy with uncertainty in current population levels, let us consider once again stochastic population growth by martens. As before, we presume that marten population growth tracks changes in abundance of their prey, as well as being influenced by weather conditions or other stochastic environmental features. We presume that the wildlife manager has knowledge of past population abundance, based on the previous year's harvest, which when combined with an assessment of prey abundance allows a forecast of the current

Fig. 19.7 Net recruitment in the absence of harvest in relation to population density, plotted relative to a constant proportionate harvest. The intersection of the net recruitment curve and the harvest line identifies the stable equilibrium, at which offtake equals the growth increment to the population.

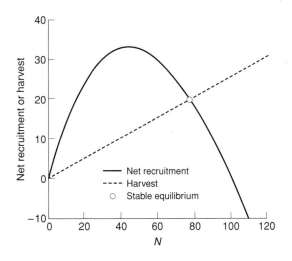

population size to which we apply an adjustable quota approximating a fixed proportion of the population. To enable direct comparison with the earlier fixed quota model, we choose a harvest proportion (h = 36%) whose yield under equilibrium conditions would be identical to the MSY (423 martens per year). Incidentally, this is quite close to the 34% harvest applied to the real marten population during 1973–91 (Fryxell *et al.* 2001). Population managers used the proportion of juvenile animals in the previous year's harvest as an index of growth potential for the population, based on the assumption (correct in this case) that reproductive success is inversely related to population abundance. Hence, monitoring age structure allowed them to maintain approximately constant harvest intensity, despite large swings in marten abundance over time. The equation of population change is calculated in the following manner:

$$N_{t+1} = \begin{vmatrix} 0 & \text{if} & g(N_t, Z_t, \varepsilon_t) \le hf(N_t, Z_t) \\ g(N_t, Z_t, \varepsilon_t) - hf(N_t, Z_t) & \text{otherwise} \end{vmatrix}$$

We have once again set a condition to ensure that the population never declines below zero. The simulation results (Figs 19.8 and 19.9) demonstrate that a fixed proportion harvest strategy is much more sustainable than a fixed quota strategy, at least in this case.

Note that harvests vary over time to a greater extent using a fixed proportion harvest strategy (Fig. 19.9) than they did under the fixed quota system (Fig. 19.6). This difference in variation arises because harvests are adjusted to absorb the impact of stochastic environmental variation, so quotas drop in years of poor per capita recruitment whereas harvest goes up in years of above-average per capita recruitment. This variation in harvest provides a stabilizing influence to fixed proportion harvest systems. However, a proportionate harvest strategy can still produce overharvesting when there are several years with unusually low recruitment, because managers do not know with certainty how many individuals have been recruited to the population.

In the not-so-distant past, fixed quota systems tended to be the norm in harvested fish and wildlife populations. Proportionate harvesting policies have become predominant in more recent years, mainly because of the improvement in resource conservation (Rosenberg *et al.* 1993).

Fig. 19.8 Predicted variation in marten abundance under a fixed proportion harvest policy, with annual marten harvests set at the proportion (h = 0.36) approximating the maximum sustainable yield (H = 423 martens per year), starting from a population size of 400 animals in 1973.

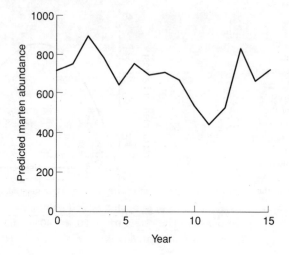

Fig. 19.9 Predicted variation in marten harvests under a fixed proportion harvest policy, with annual marten harvests set at the proportion ($h = 0.36$) approximating the maximum sustainable yield ($H = 423$ martens per year) under average environmental conditions, starting from a population size of 400 animals in 1973.

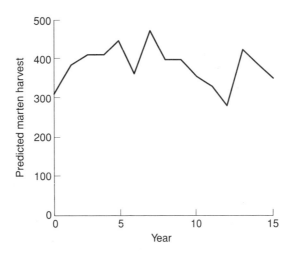

19.3.2 Constant effort harvesting strategy

A harvest can be controlled either by placing a quota on offtake or by controlling harvesting effort. The latter can be regulated by setting a hunting season or by limiting the number of people harvesting the population. The essence of controlling effort is that there is no direct attempt to control the number of animals harvested. An important outcome from controlling effort is that a constant *proportion* of the population is being harvested. Suppose, for example, that each day harvesters effectively sample 2% of the area inhabited by a harvested species. We can mathematically depict this by saying that the catchability coefficient (q) = 0.02. If hunters remove all animals that are encountered in the area that they sample, then the yield (H) would increase with both effort (E) and population density (N) in the following manner:

$$H(N) = qEN$$

Note that fixed effort policies have a built-in mechanism to reduce exploitation levels should resource density decline to dangerously low levels, because harvest levels also drop automatically with declines in resource abundance. Note also that a conservative effort level might yield a similar equilibrium harvest as more extreme effort, which could encourage a more moderate policy (the equilibrium harvest is the point at which the yield function intersects the net recruitment curve). Both of these characteristics tend to have positive moderating influences on population dynamics, as we shall demonstrate later in the chapter.

The control of harvest by quotas has an intuitive appeal because there is a direct relationship between the prescription and the result. In contrast, with the harvest regulated by control of harvesting effort an intermediate step has been inserted between prescription and outcome. Administrators tend to favor regulation by quotas because the size of the yield is directly under their control.

In fact, the disadvantages of regulating effort are more conceptual than real. Regulation of effort is usually a safer and more efficient means of managing harvested populations than is regulation by quota. Harvesting a constant number of animals each year is inefficient when the population is subject to large, environmentally induced, swings in density. The quota must be set low enough to be safe at the lowest anticipated density, or alternatively the size of the population must be censused each year before the harvesting season, the quota being adjusted according to the estimate. In addition, regulation by quota is unsafe when the quota is near the MSY. As mentioned in the previous section, the density at equilibrium with that yield is unstable such that a small environmental perturbation may trigger a population slide towards extinction.

If yield is controlled indirectly by limiting harvesting effort (e.g. by limiting the number of hunters), but with no further restriction on yield per unit of effort, those dangerous sources of instability we mention above are eliminated. A fixed effort system will, within limits, harvest the same proportion of the population at high and low density. Yield tracks density, the system automatically producing a higher yield when animals are abundant and a lower yield when they are scarce. A regulatory mechanism is built into the harvesting system itself and it is thus fairly safe so long as the appropriate harvesting effort has been calculated correctly. That is not difficult because fine-tuning of the appropriate effort does not destabilize the system in the way that fine-tuning a quota can. Also, because of the built-in regulation there is not the same need for frequent monitoring.

19.4 Fixed escapement harvesting strategy

The traditional means of setting harvests have been fixed quotas, fixed proportion, and fixed effort policies. In recent years, conservation biologists have argued that the truly safest option is to practice what is called fixed escapement harvesting. The premise of fixed escapement policies is this: rather than trying to maintain high levels of harvests in the face of stochastic variation in resource levels, managers instead choose to place conservation needs ahead of that of resource users. The general procedure is to harvest only "excess" animals above a target threshold, termed the **escapement**. After recruitment takes place the excess is removed by harvesters. This guarantees that recruitment never falls below the threshold, even if it means that no animals are harvested in some years (Lande *et al.* 1994, 1997). For example, imagine that escapement is set at 75 individuals, for a population with $r_{max} = 1$ and $K = 100$. Harvests are set according to the following formula, where $f(N)$ is the net recruitment function in the absence of stochastic variation in the environment (Fig. 19.10):

$$H(N) = \begin{vmatrix} 0 & \text{if} & f(N) \leq \text{escape} \\ f(N) - \text{escape} & \text{otherwise} \end{vmatrix}$$

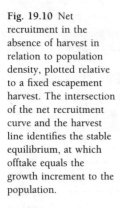

Fig. 19.10 Net recruitment in the absence of harvest in relation to population density, plotted relative to a fixed escapement harvest. The intersection of the net recruitment curve and the harvest line identifies the stable equilibrium, at which offtake equals the growth increment to the population.

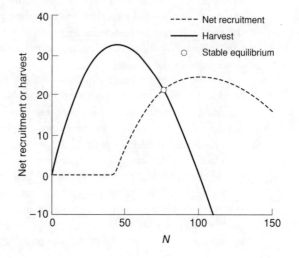

This escapement policy is obviously safe, at least in principle. It only becomes unsafe when we are uncertain of the abundance or recruitment. Hence, wildlife or fisheries managers would usually have to forecast abundance at the beginning of the harvest period from past levels of abundance, based on past harvest statistics themselves. To demonstrate how one might model a fixed escapement strategy, we shall again use the stochastic recruitment model for martens. The fixed escapement ($v = 723$) was chosen to approximate the marten abundance at which the MSY occurs under average conditions of prey abundance and weather. The equation of change in marten abundance (N) in the next time interval is defined below:

$$N_{t+1} = \begin{vmatrix} 0 & \text{if} & g(N_t, Z_t, \varepsilon_t) < f(N_t, Z_t) - v \\ g(N_t, Z_t, \varepsilon_t) - [f(N_t, Z_t) - v] & \text{otherwise} \end{vmatrix}$$

Although marten abundance varies considerably less (Fig. 19.11) than was the case for fixed quota (Figs 19.5 and 19.6) or fixed proportion (Figs 19.8 and 19.9) policies, harvest varies even more according to this policy (Fig. 19.12).

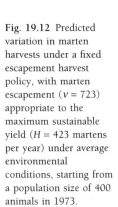
Fig. 19.11 Predicted variation in marten abundance under a fixed escapement harvest policy, with an escapement ($v = 723$) appropriate to the maximum sustainable yield ($H = 423$ martens per year), starting from a population size of 400 animals in 1973.

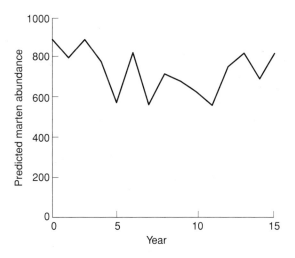

Fig. 19.12 Predicted variation in marten harvests under a fixed escapement harvest policy, with marten escapement ($v = 723$) appropriate to the maximum sustainable yield ($H = 423$ martens per year) under average environmental conditions, starting from a population size of 400 animals in 1973.

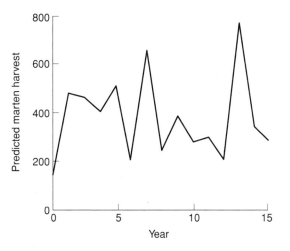

Although fixed escapement policies have the strongest conservation potential, the high variability in harvest levels can make fixed escapement less attractive to managers and resource users than other options. However, in a world that increasingly places multiple values on wildlife populations, including touristic, ecological, and ethical values well beyond their immediate recreational or commercial hunting value, fixed escapement policies seem likely to increase in attractiveness in the future.

19.5 Harvesting in practice: recreational

Most harvesting of wildlife for recreational hunting has been managed largely by trial and error. This approach works well when populations and their habitats are good at looking after themselves, when intrinsic rates of increase are high, when rate of increase and density are related by tight density-dependent negative feedback, and when the population size is kept above the level synonymous with the maximum sustained yield (N_{MSY}). These conditions describe what generally happens in traditional wildlife management. The populations and their habitats have been resilient because it is only such animals that evolve into game species.

In practice, one cannot have both high density and high yield, except while density is temporarily being reduced. It cannot stay there for long because it will track rapidly towards its equilibrium. Most managers seek to maximize offtake and so would like to raise yield and lower density. This is the cause of the frequent clashes between hunters and managers over whether females and young should be harvested. In their review of harvest management for game birds, Robertson and Rosenberg (1988) explain that "In America the potential harvest is assessed on an annual basis and the activity of sportsmen controlled by bag and season limits. These restrictions rarely aim to achieve MSY, partly due to the reluctance of sportsmen and managers to reduce breeding populations, a situation often wrongly referred to as over-shooting."

The results of this misunderstanding mostly kept offtake well below the maximum sustained yield and game populations were seldom threatened by the harvesting. However, the majority of sustained-yield estimates popular during the last few decades were overestimates and would have resulted in extinction of populations had they been applied rigorously.

The trick with managing a population for sustained yield is to play it safe. We estimate MSY on what information is available to us (usually the trend of population indices under a known constant offtake or constant effort), we refine that estimate of MSY as often as we can or at least as often as our monitoring system allows, but we keep the harvest well below the MSY. We make certain that our estimate of population size remains well above the estimate of N_{MSY}. Remember that in the early stages of managing a population for sustained yield our estimate of both the current population size N and the N_{MSY} may be wildly inaccurate. Unless we have done this sort of thing experimentally several times before we may not appreciate how inaccurate our estimates are likely to be: we must allow ourselves a wide margin of error. Remember that the standard error of an estimate tells us nothing about the accuracy of that estimate. Our monitoring of population size will let us know in plenty of time when we need to ease off harvesting effort.

19.6 Harvesting in practice: commercial

There is no difference in principle between harvesting for commercial benefit and harvesting for recreational benefit. Both are based on the sustainable yield concept suitably cushioned by a margin of error. However, in practice there are a number of pitfalls to the management of sustained commercial offtake.

Game harvesting comes in two forms: game ranching (or game farming) and game cropping. The difference is in degree rather than kind, but essentially game ranching seeks to bring the animals under human control, as in the farming of domesticants, whereas game cropping is the harvesting of wild populations. Game ranching is a spectrum of activities that overlap conventional animal husbandry at one end (e.g. the reindeer industry in Finland and Russia) and game cropping at the other end. It is beyond the scope of this book. The interested reader might try papers in Beasom and Roberson (1985) and Hudson *et al.* (1989) for a general overview of game ranching, and the volume edited by Bothma (1989) for a thorough treatment of game ranching in southern Africa. Because game cropping can be an important conservation threat, however, we consider it in greater detail below.

19.7 Age- or sex-biased harvesting

In large mammal species, harvesting is often directed at males rather than females or focused on older rather than younger age groups. This is often imposed through tag or license restrictions on hunters. The intent of such a policy is to guarantee the perpetuation of the breeding segment of the population. Age- and sex-structured harvest models of Norwegian moose, for example, suggest that the optimal policy would be to concentrate harvests on calves and old males, while rarely removing females from the population (Sæther *et al.* 2001). This makes intuitive sense, if the population is well mixed and all harvested individuals are equally valuable.

Male-biased harvesting is not always the best policy. In Scottish red deer, for example, the proportion of males born tends to decline with population density, and the mortality of older animals increases relative to females (Clutton-Brock *et al.* 2002). As a result, red deer populations have naturally skewed sex ratios, even without sex-biased hunting. But males bring in far more income to landowners than do females, so highly skewed sex ratios are economically disadvantageous. The remedy is increased culls of females to keep female density less than 60% of the ecological carrying capacity (Clutton-Brock *et al.* 2002). In other species, extreme male-biased culls could lead to some females being unable to conceive (Ginsberg and Milner-Gulland 1993; Milner-Gulland 1997), obviously outweighing any slight advantage accruing from age-biased harvesting.

In order to develop sex-biased harvest models, however, detailed information is required on age- and sex-specific survival rates, reproductive rates, and how these demographic parameters are affected by changes in population density. The last point is crucial: without reliable information on density-dependent parameters, even though detailed life tables are available, *a priori* identification of optimal harvesting levels is impossible. The requisite demographic information is available for few wildlife species, so male-biased harvests should be viewed with an appropriate degree of caution.

19.8 Bioeconomics

In addition to the biological complexities inherent to renewable resource systems, there are additional complicating factors when wildlife species are harvested commercially. This is because market dynamics, limitations on harvest controls, and potential conflict between short-term versus long-term goals also influence the levels of harvest effort. This has important effects on the economic equilibria for effort levels and the risk of resource overexploitation. For example, imagine that a population has a carrying capacity $K = 100$ individuals, maximum intrinsic growth rate $r_{max} = 1$, catchability coefficient $q = 0.02$, and effort $E = 20$ units. Annual net recruitment $(R(N))$

Fig. 19.13 Revenue and
costs under constant
environmental
conditions as a function
of effort. The
intersection of the
revenue curve and the
cost line identifies the
economic equilibrium,
at which cost equals the
potential gain.

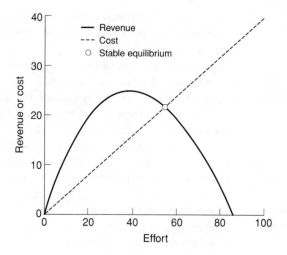

as a function of abundance is dictated by the Ricker logistic equation ($f(N)$) and proportionate harvest ($H(N)$) by the fixed effort equation (see Fig. 19.5):

$$R(N) = f(N) - N$$

$$H(N) = qEN$$

At equilibrium, $H(N) = R(N)$ and $N = H(N)/qE$. By substitution and doing some algebraic rearrangement (we encourage you to try this yourself), we can obtain the following solution for the harvest at equilibrium as a function of effort $H_{eq}(E)$:

$$H_{eq}(E) = qEK\left[1 - \frac{\log_e(1 + qE)}{r}\right]$$

This implies that there is an equilibrium harvest level for each effort that might be
exerted by resource users. We are now going to use this information to calculate the
most profitable level of effort to invest. We assume that the revenue scales with the
equilibrium harvest (with price per unit catch $p = 0.75$) and that harvesting costs
(C) escalate linearly with effort (with cost per unit effort $c = 0.40$) (Fig. 19.13):

$$costs(E) = cE$$

$$revenue(E) = pH_{eq}(E)$$

Presumably resource users want to maximize profit (which we will denote Π), which
is the difference between revenue and costs:

$$\Pi(E) = pH_{eq}(E) - cE$$

When economists discuss the cost of a particular activity, they are usually referring
to the opportunity cost. This is the difference between a given economic activity and

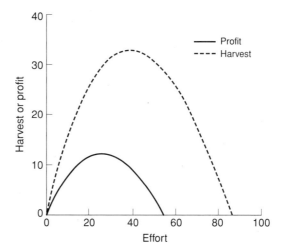

Fig. 19.14 Profit and equilibrium harvest under constant environmental conditions as a function of effort.

alternative ways of earning income. So the costs and benefits of resource use are measured relative to other forms of economic investment.

Note that the level of effort that maximizes profits (Fig. 19.14) is below the level of effort that maximizes sustainable harvest (Clark 1976, 1990). This offers yet another argument against maximizing the sustainable yield: it is less profitable than a lower rate of harvesting. That is probably a good thing for conservation: it implies that a monopolistic resource user ought to choose a relatively safe exploitation effort in order to serve its own selfish needs. For this to occur, however, the resource user needs control over decisions regarding resource use. In other words, they need to own the property rights to the resource. There would be little point in sustaining the resource unless these property rights continue long into the future.

A problem arises when access to the resource industry is unregulated, a situation known as **open access**. This is common in many game-cropping and small-scale fisheries operations around the globe. Incoming resource users may not necessarily care about depressing the profit margin, so long as they still find resource exploitation more profitable than other ways of economic investment. The threshold at which exploitation becomes unattractive in this case would be around 50–60 units of effort, rather than the optimal 25 or so units of effort. We will term this the **economic equilibrium**. Higher effort implies a lower resource abundance than that which optimizes profit. Thus, open access (at least in game cropping and fisheries) can logically lead to depressed harvest levels, depressed profits, and depressed resource abundance (Clark 1976, 1990). Everybody loses.

So far, however, we have not discussed how prices figure into the precarious balance. Before we jump into the special circumstances that influence renewable natural resources, we need a brief reminder about the so-called law of supply and demand. In economic parlance, both supply and demand are assumed to respond to the price of commodities. When the price of a commodity is high, entrepreneurs are encouraged to produce more of it, whereas there is much less incentive when prices are low (Fig. 19.15). On the other hand, high prices discourage consumer demand, whereas low prices encourage it (Fig. 19.15). These contrasting responses on the part of consumers and producers lead to an intersection between supply and demand curves. In a free, open competitive market system, pricing should adjust over time until the

Fig. 19.15 Idealized
relationship between
supply and demand in
an open, competitive
market.

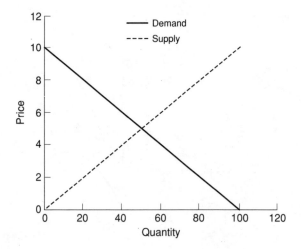

economic equilibrium is reached at which supply matches demand. This kind of open market is highly idealized: in practice there may be collusion between suppliers, artificial price supports imposed by external agencies (such as governments), or other barriers that mitigate against a free market. It may be fairly reasonable, however, as a depiction of economic interactions in a rural economy, such as that supplying game meat in equatorial Africa or southeast Asia (Clayton *et al.* 1997; Fa and Peres 2001).

Now we are going to consider how sustainable resource levels are liable to respond to variation in prices (Clark 1990; Milner-Gulland and Mace 1998). We will use some of the relationships already derived, and rearrange terms to calculate a new relationship based on price (p) and equilibrium harvest (H_{eq}), the latter the relevant measure of production in a renewable resource system. Recall the following:

$$pqEN - cE = 0$$

$$H_{eq} = N \exp\left[r_{max}\left(1 - \frac{N}{K}\right)\right] - N$$

By dividing both sides of the top equation by effort and then rearranging terms we get $N = c/(pq)$. One then substitutes $c/(pq)$ wherever N appears in the second equation, leaving the following algebraic relationship between equilibrium harvest level and resource price:

$$H_{eq} = \left(\frac{c}{pq}\right)\exp\left[r_{max}\left[1 - \frac{\left(\frac{c}{pq}\right)}{K}\right]\right] - \left(\frac{c}{pq}\right)$$

We plot this relationship with harvest levels on the horizontal axis and price on the vertical axis (Fig. 19.16), so that the similarity to the classic supply–demand curve is apparent (Fig. 19.15). To complete the analysis, however, we need to add a demand line.

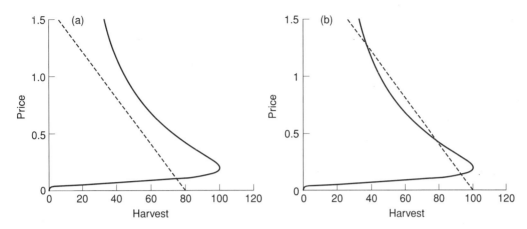

Fig. 19.16 Supply (solid) and demand (broken) curves for a renewable natural resource at (a) low and (b) high levels of demand.

If demand is modest relative to renewable resource supplies, then the classic supply–demand relationship occurs. There is a single intersection between the supply and demand curves, defining the economic equilibrium (Fig. 19.16a). Harvests should build over time and resource prices should fall until the supply meets demand. Due to the severe non-linearities in the supply curve, however, this ideal situation can be rapidly transformed into a more troubling scenario. Imagine, for example, that economic development leads to slight overall increase in consumer demand (Fig. 19.16b). This results in a situation in which there are three points of intersection: at high, moderate, and low prices, and low, moderate, and high harvest levels. Now the system is poised to flip between dramatically different economic equilibria with slight changes in resource levels or economic performance. The high price/low harvest combination is particularly worrisome, because it arises because of a severely overharvested resource, teetering dangerously near extinction. It is nonetheless profitable enough to justify the huge effort required to eek out a meager harvest. Further increase in demand leads to a singular economic equilibrium, this time in the risky zone.

This scenario is particularly likely when the demand curve descends quite sharply. An economist would characterize such a demand curve as being "inelastic," meaning that it requires quite substantial change in price to suitably alter consumer behavior (by definition, inelastic means that a 1% change in price achieves less than a 1% change in demand). For some wildlife products, such as ivory, there is evidence that consumer demand is determined solely by income, regardless of price (Milner-Gulland 1993). Should demand be highly elastic (much flatter in a plot of price versus quantity demanded), multiple economic equilibria become far less likely. Food products, because there is greater room for substitution, are often highly elastic, which should offer some comfort. On the other hand, small changes in rural incomes can lead to substantial changes in diets, with bush meat and other wildlife food products rapidly becoming highly prized.

Ivory-harvesting presents a fascinating case study (Milner-Gulland and Beddington 1993). Male elephants have substantially larger tusks than females, so any rational harvesting scheme would be biased towards males (Milner-Gulland and Mace 1998).

Rather than culling males at random, however, the optimal economic decision would be to collect ivory from individuals that have died naturally or harvest only senescent males (Basson *et al.* 1991). Elephant tusks increase in size exponentially with age and large tusks are worth more, gram for gram, than small tusks, because large tusks are much more valuable to carvers. As a result, the optimal economic solution should reinforce conservation needs.

19.9 Game cropping and the discount rate

There are two quite distinct phases to a cropping operation: first, the population must be reduced below its unharvested density (capital reduction), and then it must be harvested at precisely the rate it seeks to bounce back (sustained-yield harvesting). Biologists tend not to think too much about the capital-reduction phase because they look forward to the prospect of a yield sustainable into the indefinite future.

If you were offered $1000 now as against $1000 in 10 years' time you would take the money now. However, if you were offered $400 now as against $1000 in 10 years' time, the decision is no longer clear cut. Against money in the hand you are offered a guarantee of sure future benefit, but the monetary value of that future benefit is unclear. How much is $1000 in 10 years actually worth? A simple answer is that it is worth a present sum which, when prudently invested, yields $1000 ten years hence. If we assume that capital expands at about 10% per year, then $1000 in 10 years is worth $385 now, or even less if the currency is inflating. With this knowledge the answer to $400 now or $1000 in 10 years' time is clear. Take the $400 now; it is worth more. By the same reasoning a game animal harvested in 10 years' time is worth nothing like an animal harvested now. All future earnings must be discounted by the time it takes to get the money, and the economics of the harvesting operation can be dictated by the ratio of present to future earnings.

Discounting can be represented fairly simply, by rearranging the terms of the exponential growth model. Instead of growing exponentially, however, the present value of future harvests (PV) declines exponentially at the discount rate δ:

$$PV(t) = \text{profit } e^{-\delta t}$$

What this means, of course, is that future profits are not valued as highly as current profits. The higher the discount rate, the less the future is valued. A concrete example may prove illuminating. A mature Bolivian mahogany tree currently brings in around $396, for the lumber it can supply (Gullison 1998), yet it takes roughly a century for such a tree to mature. At the 17% inflation rate in Bolivia in the early 1990s ($\delta = 0.17$), the present value of a mahogany tree worth $396 a century from now is around 1 penny, almost 10 times more expensive than required for replanting. Small wonder that replanting is a low priority for most logging firms!

The present value of all future harvests can be calculated by integration:

$$PV = \int_0^\infty e^{-\delta t}\, \Pi\, dt$$

If the per unit price $p = 0.75$, cost per unit effort = 0.5, $r_{max} = 1$, and $K = 100$, then for a discount rate of 5% the total present value of all future harvests would be $246 at the optimal effort level ($E = 25$). Change in the discount rate to 10% (still very realistic in most economic models) halves the value of all future harvests. Severe discounting can therefore change the attractiveness of sustainable resource utilization for commercial interests.

Biologically, the rational scheme for harvesting is to reduce the population to a density allowing a suitable sustained yield and then take any excess population recruitment that accrues thereafter, via either harvesting a constant proportion of the population or maintaining a constant escapement at this target. But this biologically sustainable strategy does not necessarily maximize economic gain. Colin Clark's (1976, 1990) superb treatise on the economics of harvesting natural resources shows unambiguously that the best biological strategy and the best economic strategy coincide only when a population's maximum rate of increase greatly exceeds the discount rate. Rabbits and herrings go into that category. When maximum rate of population increase is lower than the discounting rate, however, the real money is made more by capital reduction than by sustained yield. It may even be economically clear-sighted to make a total trade-off, taking all revenue by capital reduction and sacrificing all future sustained yield. This strategy maximizes net revenue, discounted to present value, when the population's maximum rate of increase is below about 5% per year. This can be the economic justification for the extinction of a slow-growing population (and maybe even a species). Blue whales and redwood trees are obvious examples of such slow-growing organisms, as would be many threatened or endangered species. In such cases, economic incentives clearly cannot be relied upon to serve the greater good.

Discounted valuation of future profits is more likely for a privately owned renewable resource. This should be less likely for a publicly owned resource, because the public's discount rate should be much lower. Indeed, one could argue that environmental stewardship argues against any discounting at all, on purely ethical grounds. Publicly owned resources can sometimes take on the character of privately owned resources, however, when the people managing the resource and the people harvesting the resource imagine that they, and not the people as a whole, own the resource.

Any scheme to harvest a publicly owned renewable resource necessarily involves three parties: the owners of the resource (the people), the harvester of the resource (usually a private company), and the manager of the resource (a government agency) that regulates the harvesting. According to constitutional theory, the people in the agency are supposed to act for the owners of the resource, but often they become locked into a symbiotic relationship with the harvesters as if those two groups were themselves joint owners of the resource. Technical advice on sustainable yield offered by the agency's own research branch is commonly ignored by its policy and planning branch when it conflicts with the short-term requirements of the industry. Thus, the ecological aberrations that necessarily follow from the economic implications of the discount rate, and against which the people in the managing agency are employed to guard, often dominate the harvesting operation. Forestry and fisheries provide numerous examples, and commercial wildlife harvesting is not necessarily immune.

19.10 Summary

The way in which a safe sustained yield is estimated depends on the population's growth pattern that, in turn, is determined by the relationship between the population and its resources. The yield can be estimated in terms of either a numerical offtake or an appropriate harvesting proportion. The consequences of any harvesting policy should always be evaluated in the light of stochastic variation in population growth. In a stochastic world, harvesting a fixed proportion of the population is much less likely to lead to population collapse than is fixed quota harvesting. Perhaps the safest option of all is to strive to maintain populations above a critical threshold (known as a fixed escapement). Although from an ecological viewpoint recreational harvesting and commercial harvesting do not differ in principle, in practice commercial harvesting sometimes has greater potential to exceed sustainable levels of harvests. This is particularly likely when resources are a common property, for which multiple users compete. Under this circumstance, the economically rational behavior will be for resource users to continue to enter the industry until resources are reduced to dangerously low levels. The "discount rate" of economic analysis can also encourage overharvesting by imparting a greater value to present yields than to future yields, particularly when the maximum annual rate of resource growth is less than the discount rate. Risk due to discounted valuation once again tends to be most pronounced when resources are common, rather than private property. Particularly stringent conservation measures are called for under these circumstances.

20 Wildlife control

20.1 Introduction

We show that a control operation is similar to a sustained-yield exercise but is conceptually more complex. The objective must be defined precisely, not in terms of the number of pest animals removed but according to the benefit derived therefrom. Methods include mortality control, fertility control, and various indirect manipulations. A detailed analysis of this topic can be found in two textbooks by Hone (1994, 2004).

20.2 Definitions

"Control" has three meanings in wildlife research and management. The first two deal with manipulating animal numbers, the third with experimentation. "Control" is used first in the sense of a management action designed to restore an errant system to its previously stable state by reducing animal numbers. We speak of controlling an outbreak of mice in a grain store or wheat-growing district. The action is temporary.

The second use of "control" has to do with moving a system away from its stable state to another that is more desirable. The animals are reduced in density and the new density enforced by continuous control operations. The word is here used in a somewhat different sense than its use in engineering. There, a "control" (e.g. a governor on an engine) stops an intrinsically unstable system from shaking itself apart. It is a regulator. That connotation is inappropriate to wildlife management (although it has been so employed on occasion) because, except in special circumstances, the original state is more stable than that created by the control operation.

"Control" is used in a third sense within the parlance of experimental design. As Chapter 15 explains at length, an experimental control is the absence of an experimental treatment. That meaning of the word is usually obvious from the context except when the experiment tests the efficacy of a control program (i.e. "control" in one or other of the first two senses). The control operation is then the treatment and the control is the absence of control.

The obvious ambiguity in the previous sentence can easily lead to misunderstandings. For example, in an experiment testing the effect on riverside vegetation of controlling (reducing, i.e. second meaning) hippopotami, they were shot (controlled) periodically in one stretch of river. The vegetation along the bank was compared with that of another stretch of river where the animals were protected (the control stretch, i.e. the third meaning). However, a change of hunting staff led inevitably to the control (protected) stretch being controlled (hunted) one sunny Sunday morning. We have seen similar mistakes (discovered at the last minute) in the testing of rabbit control methods. There is no sure remedy, but the chance of a disaster can be reduced somewhat by always linking "experimental" to "control" when discussing experimental design.

Table 20.1 Differences between donkey populations on two 225 km² blocks in the Northern Territory of Australia, 3–4 years after one population was reduced by 80%.

Measurement	High-density block	Low-density block
Initial densities 1982 (donkeys/km²)	> 10	> 10
Treatment (1983)	None	80% shot
Density (1986)	3.3	1.5
Density (1987)	3.2	1.8
Trend	Non-significant decrease	Significant increase (20%)
Sexual maturity (% male, 2.5 years)	43%	100%
Female fecundity (2.5 years)	30%	50%
Juvenile mortality (0.5 years)	62%	21%

From Choquenot (1991).

20.3 Effects of control

If the density of a population is lowered by control measures, the standing crop of renewable resources (e.g. grass needed by a herbivore) will increase because of the lowered use. Non-renewable resources such as nesting holes will be easier for an individual to find. Hence control, like harvesting, increases the resources available to the survivors of the operation. Their fecundity, and their survival in the face of other mortality agents, is thereby enhanced. For example, an increase in survival of juvenile rabbits compensated for an experimentally reduced reproduction of females (Williams and Twigg 1996; Twigg and Williams 1999; Twigg *et al.* 2000). The reduced density, therefore, generates a potential increase that will become manifest if the control or harvesting is terminated. Table 20.1 shows just such an effect generated by control operations against feral donkeys in Australia.

The enhanced demographic vigor following reduction in density is a desirable outcome of a harvesting operation, and in fact the success of the harvesting is determined by such an effect, but it acts against the success of a control operation. The further density is reduced the more the population seeks to increase. Thus control, in the sense of enforcing a permanently lowered density, is simply a sustained-yield operation that seldom utilizes the harvest. It is an attempt to drive a negative feedback loop in the opposite direction. In other words, density-dependent effects compensate for the imposed mortality of the control operation.

20.4 Objectives of control

More than the other two areas of wildlife management (conservation and sustained-yield harvesting), control is often flawed by a lack of appropriate and clearly stated objectives.

Control, in contrast to conservation and sustained-yield harvesting, is not itself an objective. It is simply a management action. Its use must be legitimized by a technical objective such as increasing the density of a food plant of a particular species of bird, say, from one per hectare to three per hectare. The control operations would be aimed at a herbivore for which that plant was a preferred food. The success of the operation would be measured by the density of plants, not by the density of the herbivore or by the number of herbivores killed.

Control campaigns in many countries share a common characteristic. Very often the original reason for the management action is forgotten and the control itself (lowering density) becomes the objective. The means become the end.

A good example is provided by the history of deer control in New Zealand. It is one of the largest and longest running control operations against vertebrates in any

Table 20.2 Published official justification for government control operations against deer in New Zealand.

Years	Official objectives of deer control
1920–29	Increase the size of antlers
1930–31	Reduce competition with sheep
1932–66	Prevent accelerated erosion generally
1967–80	Prevent accelerated erosion in the heads of rivers that may flood cities
1981–92	No verifiable reason offered

From Annual Reports of Department of Internal Affairs and New Zealand Forest Service.

country. Table 20.2 lists the sequence of official justifications for government-funded control of deer from 1920 onwards (Caughley 1983).

Whereas the stated justification for the control operations changed with time, those changes had virtually no effect upon the management action. There were certainly changes in control techniques but, with the exception of the change of 1967, these were evolutionary adjustments in the management action itself. They were not driven by changes in policy. The means themselves were the end.

Up until 1980 the reasons given for the control operations were that deer and other species caused erosion of the higher slopes and silting of lower rivers (Table 20.2). However, in 1978, new meteorological, hydrological, geomorphological, and stratigraphic research showed that deer, chamois (*Rupicapra rupicapra*), and tahr (*Hemitragus jemlahicus*) had little or no effect on the rate at which river beds silted up or on the frequency and size of floods. Despite these data, deer control continued after 1980 for no verifiable reason. All that changed were the stated objectives, which were variously for "aesthetics," for "proper land use," to "ensure the continuing health of the forest," to "protect intrinsic natural values," and to "maintain the distinctive New Zealand character of our landscapes." These are not open to scientific testing.

Many similar examples could be cited from other countries. Control operations must have clear objectives framed in terms of damage mitigation. Their success must be measured by how closely those objectives are met, not by the number of animals killed. The operations must be costed carefully to ensure that their benefit exceeds their cost. And their success or failure must be capable of independent verification. Table 1.1 gives a matrix of possible objectives and actions. It can be filled in to ensure that the management action is appropriate to the chosen objective.

20.5 Determining whether control is appropriate

There are three circumstances in which control may be an inappropriate management action: (i) where the cost exceeds the benefit; (ii) where the "pest" is not in fact the cause of the perceived problem; and (iii) where the control has an unacceptable effect upon non-target species. These are best investigated experimentally before a control program is instituted. We give two examples.

Cats were introduced to the subantarctic Marion Island in 1947 to deal with house mice marooned by shipwrecks. They increased rapidly to 3000 by 1977 and fed mostly on ground-nesting petrels. The breeding success of the petrels, particularly the great-winged petrel (*Pterodroma macroptera*), seemed to be declining and cats were suspected to be the cause. The neighboring island of Prince Edward was conveniently free of cats and became the experimental control. The objective of reducing cats was to increase the breeding success of the petrels. Hence the success must be defined in terms of the birds' breeding success, not in terms of reduced numbers of cats. An introduced disease, shooting, and trapping reduced the cats. The petrel breeding success increased from 0–23% (1979–84) to 100%; chick mortality decreased from 60% in 1979–84 to 0% in 1990. Comparisons with breeding on Prince Edward Island and within a cat-free enclosure on Marion Island identified the cats as the cause of the initial high mortality and the reduction in cat numbers as the reason for the increase in recruitment (Cooper and Fourie 1991).

The next example deals with non-target species. The insecticide fenitrothion is a well-known organophosphorus pesticide, but its effects on song birds and other non-target animals are little known. The Forestry Commission in Scotland wanted to use fenitrothion to control the pine beauty moth (*Panolis flammea*), and was required to undertake environmental assessment of the effects of spraying on non-target species. For 3 years Spray *et al.* (1987) monitored the effect on forest birds.

Their design comprised two pairs of plots, each plot measuring about 70 ha. The elements of each pair were matched by soil type, age of planting, and tree composition. One element of each pair was sprayed, all plots being monitored before and after spraying to detect annual variation of the density of breeding birds, short-term changes in abundance within 5 days of spraying, and breeding performance of the coal tit (*Parus ater*). They detected no significant difference in these variables between the insecticide-treated plots and the experimental control plots.

20.6 Methods of control

Animal welfare is an important consideration in any control operation. An animal has the right to be treated in a humane manner whether it is to be protected or controlled. Unfortunately the notion of humane treatment is often the first casualty of turning a species into a pest. That is particularly noticeable when the species is an exotic. The wildlife manager's paramount responsibility in any control operation is to ethical conduct rather than to operational efficiency.

Control methods can be divided into those aimed at directly increasing mortality, those aimed at directly reducing fertility, and those that act indirectly to manipulate mortality, fertility, or both. The success of an operation is not gauged by the reduction in the density of the target species but by the reduction in the deleterious effects of the target species. In all cases the prime responsibility of the wildlife manager is to determine whether the control adequately reduces deleterious effects and whether its benefit exceeds its cost.

20.6.1 Control by manipulating mortality

Control by increasing mortality may be direct, as in poisoning, trapping, or shooting, or it may be indirect as in biological control through pathogens.

Direct killing
Five simple principles guide the control of a target population living in an environment that remains reasonably constant from year to year. These are largely independent of a population's pattern of growth and they emphasize the conceptual similarities between control and sustained-yield harvesting.

1 When a constant number of animals is removed from the population each year the size of the population will be stabilized by the control operations unless the annual offtake exceeds the population's maximum sustained yield (MSY).

2 The level at which the population is stabilized by the removal of a constant number each year is equal to or greater than the density from which the MSY is harvested.

3 Density is stabilized by the removal each year of a constant proportion of the population, provided that proportion is lower than the intrinsic rate of increase r_m.

4 The level at which a population is stabilized by removing a constant proportion of the population each year can be at any density above the threshold of extinction.

5 If animals are removed at an annual rate greater than r_m the population will decline to extinction.

These simple rules can be sharpened for those populations whose pattern of population growth is approximated by a logistic curve (Caughley 1977b). In general they will serve for populations of large mammals (i.e. low r_m) feeding on vegetation that recovers rapidly from grazing.

1 When the constant number C removed each year is less than the MSY of $r_mK/4$, the population is stabilized at a size of

$$N = [r_m + \sqrt{(r_m^2 - 4Cr_m/K)}]/(2r_m/K)$$

where K is the ecological carrying-capacity density, corresponding to the asymptote of the logistic curve.

2 When a constant proportion of the population is removed each year at a rate less than r_m, the population is stabilized at a size of:

$$N = K - (KH/r_m)$$

where H is the instantaneous rate of removal.

3 If a constant number of animals greater than $r_mK/4$ is removed each year, the population will eventually become extinct.

Examples of eradication

Even with large mammals the proportion of the population that has to be culled each year to eradicate a population is substantial: 90% of the feral goats in Egmont National Park, New Zealand, had to be culled annually to achieve eradication in 12 years; if only 50% had been culled eradication would have taken over 50 years, if at all (Forsyth et al. 2003).

By far the most common examples of the eradication of pest species are found in the Pacific islands, New Zealand, and Australia, because these places have been subject to the invasion of exotic vertebrates. An important review of these is provided by Veitch and Clout (2002). Many islands were deliberately seeded with pigs, goats, and rabbits by sailors in the 1700s to provide a food source in case of shipwreck. These populations increased rapidly, changed the vegetation, and indirectly caused the extinction of many birds. Possums were introduced to New Zealand for commercial harvesting in the 1800s. Shipwrecks and ordinary landings resulted in the inadvertent introduction of rats and mice to most islands, and snakes to Guam and Mauritius. Control of rats was the motive for introducing cats (e.g. on Marion Island) and mongooses (e.g. Mauritius, some Hawaiian and Caribbean islands). Control of rabbits in New Zealand was the reason for introducing stoats (*Mustella erminea*) and ferrets (*M. furo*). All these predators increased rapidly and exterminated much of the native fauna.

In recent decades there has been much effort to eradicate these exotics and repair the ecosystems (see papers in Veitch and Clout 2002). Cats were successfully removed from islands off Mexico, the British West Indies, Marion Island off South

Africa (Bloomer and Bester 1992), and Lord Howe Island (Hutton 1998). Foxes have been removed from 39 islands of Alaska (Ebbert and Byrd 2002). Pigs were successfully removed on Lord Howe as well as Santa Catalina Island, California, and some of the Mariana Islands. Goats have also been removed from Mauritius and some Pacific islands. However, despite their large size, both pigs and goats are difficult to eradicate because they live in difficult terrain; intensive efforts to eradicate goats on Lord Howe left a core of six living on precipitous cliffs (Parkes *et al.* 2002). In general, animals that are hunted will change their behavior, becoming more shy and using refuge habitats so that a disproportionate effort is required to kill the last remaining animals (Choquenot *et al.* 1999; Forsyth *et al.* 2003): it took 1000 hunter-days to kill the last four goats on Raoul Island, north of New Zealand (Parkes 1984). Other species are difficult to remove because of their particular habitat uses and adaptations. Thus, the Indian musk shrew (*Suncus murinus*) is an insectivore expanding rapidly across Asia, Africa, and many islands, where it competes with endemic skinks and geckos. Difficulties in removing this species include their ability to withstand anticoagulant poisons and the need to use live bait for traps (Varnham *et al.* 2002). Brown tree snakes (*Boiga irregularis*) have been particularly difficult to remove on Guam, as were wolf snakes (*Lycodon aulicus*) on Mauritius, because of their ability to hide in small holes (Rodda *et al.* 2002).

Rats and mice have been removed from several small islands. In general, small species (rodents) with high rates of increase were removed successfully from islands of less than 1000 ha, an exception being Campbell Island, south of New Zealand, which is 11,000 ha. Larger species can be removed from larger areas.

Biological control

Biological control, so effective against insects, has a poor record against pest wildlife. One of the few successes is the use of *Myxoma* against rabbits. It holds the density of rabbits in Australia to about 20% of their uncontrolled density despite a decline in virulence of the virus and of susceptibility of the rabbit, both a product of massive natural selection.

The chances of finding a biological agent to control vertebrates are always low, largely because the pathogen must be highly host specific and highly contagious.

20.6.2 *Control by manipulating fertility*

Population control by manipulating fecundity has several advantages over simply killing animals, but it also has problems of its own. It was first suggested as a control method by E.F. Knipling in 1938 (Marsh 1988) but was not applied for another 20 years. Its first use was against the screwworm fly (*Cochliomyia hominivorax*), a serious pest of livestock in the southeast of the USA. Subsequently it has been used against a number of insect pests in various parts of the world.

The use of contraceptive techniques for population control has been reviewed by Marsh (1988) with respect to rodents and lagomorphs, by Turner and Kirkpatrick (1991) with respect to horses, and by Bomford (1990) for vertebrates in general. Bomford showed that although contraception has often been advocated as a useful control method against vertebrates, and tried from time to time, there is no clear and well-documented example of unqualified success. "Many tests of fertility control have not been robust enough to allow clear conclusions. Experiments have often failed to include treatment replicates, or have relied on small samples. These results cannot be analyzed statistically to estimate the probability of a treatment effect" (Bomford 1990).

Table 20.3
Characteristics of an
ideal chemosterilant
for rodents.

---

1 Orally effective, preferably in a single feeding.
2 Effective in very low doses (not exceeding 10 mg/kg).
3 Permanent or long-lasting sterility (preferably) lasting 6 months of longer, or at least through the major breeding period of the pest species.
4 Effective for both sexes or preferably for females if only one sex.
5 Rodent-specific or genus-specific.
6 Relatively inexpensive.
7 There should be a wide margin between the chemosterilant effects and lethal doses. (If high in specificity, this may be unimportant or the narrow margin be of value.)
8 Well accepted (i.e. highly palatable) in baits at effective concentrations.
9 Biodegradable after a few days in the environment.
10 If not highly specific, rapid elimination from the body of the primary target to avoid secondary effects.
11 No acquired tolerance or genetic or behavioral resistance.
12 Free of behavioral modification (such as altering libido, aggression, or territoriality).
13 Free from producing discomfort or ill feelings that could suppress consumption (i.e. bait shyness) on repeat or subsequent feedings.
14 Humane (i.e. produces no stressful symptoms).
15 Easy to formulate into various kinds of baits.
16 Sufficiently stable when prepared in baits (i.e. adequate shelf life).
17 Not translocated into plants (or at a very low level), thus permitting use on crops.

From Marsh (1988).

The usual method of use against insects – flooding the population with sterile males – is dependent on the females mating only once. That is common behavior amongst insects that live for only 1 year, but is rare amongst vertebrates.

Most attempts at control by contraception or sterilization have utilized chemicals such as bromocriptine, quinestrol, mestranol, and cyprosterone. Table 20.3 gives Marsh's (1988) criteria for an ideal rodent chemosterilant.

The effect of a contraceptive or sterilizing agent upon the population's dynamics depends on the breeding system of the species and particularly upon the form of dominance. In general, a vertebrate population will seldom be controlled adequately by a contraceptive or sterilant specific to males (Bomford 1990) and so the target should be either the female segment of the population or both sexes.

Caughley *et al.* (1992) explored the theoretical effect on productivity of three forms of behavioral dominance, two effects of sterilization on dominance, and four modes of transmission. Seventeen of the 24 combinations are feasible but lead to only four possible outcomes. Three of these result in lowered productivity. The fourth, where the breeding of a dominant female suppresses breeding in the subordinate females of her social group, leads to a perverse outcome. Productivity *increases* with sterilization unless the proportion of females sterilized exceeds $(n - 2)/(n - 1)$, where n is the average number of females in the social group (Fig. 20.1). Hence, a knowledge of social structure and mating system is desirable before population control by suppressing female fertility is attempted. Experimental tests of this dominant female model using artificially sterilized female red foxes has shown that dominance is not an important social suppressor of subordinate female reproduction. Thus, greater female sterility led to lower juvenile recruitment (Saunders and McIlroy 2001).

The theoretically derived examples of reducing litter production exclude the effect of increased fertility consequent upon lowered density. It cannot be modeled from first principles in the same way as the expected reduction in litter production

Fig. 20.1 Mean number of litters produced during a season of births by a group of females of size *n* subject to varying rates of sterilization. One female is dominant in each group and the other females subordinate to her. Only the dominant female breeds. She relinquishes dominance if sterilized, and the subordinates are then free to breed. (After Caughley *et al.* 1992.)

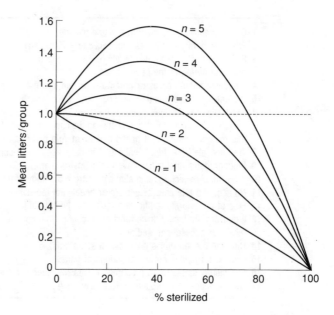

because its effect is specific to particular species and habitats. It must be examined by way of carefully designed field experiments. Such experiments should follow the effect of the treatment on the population's dynamics rather than simply upon reproduction, because compensatory changes to survival through density-dependent effects may also occur and should always be expected. As mentioned above, two rabbit populations in Australia were artificially sterilized to varying degrees. The reduced recruitment of newborn was compensated by a density-dependent survival of the remaining juveniles. Female sterility had to reach 80% before a decline in population was observed (Twigg and Williams 1999; Twigg *et al.* 2000). Similar sterility experiments on brush-tailed possums (*Trichosurus vulpecula*) in New Zealand showed population declines with 50% and 80% sterile females provided immigration was prevented (Ramsey 2000).

Two additional methods of fertility control have been suggested: immunocontraception and genetic engineering. These must await further research to demonstrate their general applicability.

Immunocontraception

Antibodies can be raised in an individual against some protein or peptide involved in reproduction, the antibodies hindering the reproductive process. Immunocontraception has been used to reduce reproductive rates, and hence mean densities, of wild species that have become too numerous, such as elephants, horses, white-tailed deer, fallow deer, and seals (Kirkpatrick and Rutberg 2001). One approach uses porcine zona pellucida (PZP) protein antigens that raise antibodies to block sperm-binding sites on the surface of the ovum in mammals (Kirkpatrick *et al.* 1997; McShea *et al.* 1997; Rudolph *et al.* 2000). This prevents fertilization. The protein must usually be administered by injection or implant because most are broken down by digestion, and the primary inoculation must be boosted 1–2 times in the following few weeks. Generally the duration of these vaccines is less than 1 year, so animals must be reinoculated on an annual basis. This means that the technique is limited to animals that are easy to capture and from small populations. This problem has been overcome partly through the use of PZP proteins encapsulated in liposomes allowing the proteins to be released gradually over a long time span. Thus, 10 years of contraception was achieved in gray seals (*Halichoerus grypus*) with a single dose of encapsulated PZP (Brown *et al.* 1996, 1997), while Fraker *et al.* (2002) report 3 years of contraception in fallow deer (*Dama dama*). Some of the problems are being overcome, but this technique remains limited to small populations where a significant proportion of the females can be captured for inoculation.

Genetic engineering

Tyndale-Biscoe (1994) suggested using a pathogen of low virulence to vector a foreign gene that would disrupt reproduction. He suggested particularly that the *Myxoma* virus could be used to carry an inserted gene that would reduce the birth rate of the European rabbit (*Oryctolagus cuniculus*) in Australia. Similar approaches are being used for red foxes and house mice (Shellam 1994; Pech 2000).

20.6.3 *Control by indirect methods*

Exclusion

The most obvious way of reducing the deleterious effect of wildlife is to exclude animals from the area. That can be achieved by fencing, by chemical repellents, or by deterrents of one sort or another. Often exclusion is necessary for only part of the time. For example, damage by deer to regenerating pines may be limited to only the first few years after establishment.

Exclosures can be as small as a hectare or of mind-boggling proportions. The first of the latter was the Great Wall of China erected at the instigation of Shih Huang Ti, the first Emperor of the Ch'in Dynasty, between 228 BC and 210 BC. It protected his northern and western frontiers, the direction from which he was most frequently attacked. The wall traverses about 2400 km of rough country.

Another big one is the Australian barrier fence built to keep dingoes out of sheep country. It runs from the south coast, divides South Australia into two, skirts around the top inland corner of New South Wales, and then loops to enclose all of central Queensland. It was started in 1914 and built in sections, often as an upgrading of previous state border fences and rabbit fences. At its greatest extent it spanned 8614 km, 3.5 times longer than the Great Wall of China. In 1980 the loop up through inland Queensland was fenced off about half-way up and the upper fence abandoned. The present exclosure has a perimeter of 5614 km. In contrast to the numerous rabbit fences that have been built in Australia, the dingo barrier fence has been relatively successful in reducing the spread of a pest species.

The difficulty in reducing populations of introduced predators in New Zealand, Australia, and other islands has promoted the use of fences in those countries with some success. Although only relatively small areas can be protected, fence maintenance is high, and some predators get through occasionally, this method has worked better than any other so far (Long and Robley 2004).

Sonic deterrents

The modern forms of the scarecrow comprising sonic devices (bangers, clangers, alarm calls, ultrasonics) have been reviewed by Bomford and O'Brien (1990), who suggested that, at best, these achieve only short-term reduction in damage. They were particularly critical of the claims made for commercially produced ultrasonic devices and of the standard of experimental testing in this field.

Habitat and food manipulation

This is certainly the most elegant of control techniques because it does not have to counteract density-dependent compensation within the pest population. The key habitat elements are water and shelter. Red squirrels (*Tamiasciurus hudsonicus*) in British Columbia lodgepole pine forests can be dissuaded from feeding on the stems of very young trees by aerial spreading of sunflower seeds. This alternative food source is preferred over the pines (Sullivan and Klenner 1993). This diversionary feeding has possibly reduced predation by raptors on grouse chicks in Britain (Redpath *et al.* 2001), and by small carnivores and corvids on artificial grassland bird nests in Texas (Vander Lee *et al.* 1999). Supplemental food reduced predation by striped skunks (*Mephitis mephitis*) on duck nests, but other carnivores increased their predation so the results were equivocal (Greenwood *et al.* 1998). Where food is provided in local concentrations at feeders (instead of distributed widely), the high density of carnivores may increase predation of ground-nesting birds (Cooper and Ginnett 2000). Thus, the way food is presented is important to the outcome.

20.7 Summary

The control of a pest species, in the sense of holding its density at a reduced level, is essentially a sustained-yield operation where the yield is not used. Reduction in density is not an end in itself: the success of the operation is measured not by the number of animals removed but by whether the objective was attained, be it the increase in density of an endangered species, an increase in grass biomass, or the reduction of damage to fences. The logic of experimental design must be utilized to determine whether benefits exceed costs, whether the treatment has a deleterious effect on non-target species, and whether the targeted "pest" is really the cause of the perceived problem.

21 Ecosystem management and conservation

21.1 Introduction

Most of the previous chapters have focused management and conservation on individual species and their direct interactions with the next trophic level. Management and conservation threats are usually stated in terms of overharvesting (Chapter 19), overpredation (Chapters 10 and 17), disease (Chapter 11), pests (Chapter 20), or habitat loss (Chapter 17; Griffith *et al.* 1989). Although these issues are correct as far as they go, they hide a common feature: population declines occur through a combination of factors, derived from complex interactions between environment and biota, which together overwhelm the ability of a species to withstand them. These are the effects of ecosystem dynamics and they often dictate the fate of individual species. Natural ecosystem complexity arises from factors such as non-linear biotic interactions, evolutionary history and assembly history of species, and often-unpredictable environmental disturbance.

Human interference, from both exploitation and stewardship, can alter ecosystem dynamics with results that can be opposite to what is intended. For example, fences were put up in central Botswana and Kruger National Park, South Africa intended to protect the cattle industry in the former and wildlife in the latter, but they both altered migrations and caused a collapse of wildlife populations; placement of artificial water holes in Hwangi Park, Zimbabwe, Kruger Park (du Toit *et al.* 2003), and the Sahel of northern Africa resulted in local overuse of resources and unwanted ecosystem effects (Sinclair and Fryxell 1985); and crude attempts at biological control through the introduction of exotic predators to control exotic prey in Hawaii, Australia, New Zealand, and many other areas have resulted in the extinction, or near extinction, of endemic species and the complete failure of the control of pests (Serena 1994). Thus, what at first seems an obvious conservation solution may not be with closer examination.

The fundamental problem is that by ignoring some aspects of ecosystem dynamics – historical legacy, community interactions, and disturbance at different scales – inappropriate conservation efforts can result, curing the symptoms rather than the cause of a threat to a species. Ecosystem dynamics have now been well described (e.g. Boyce and Haney 1997). Here we put the various aspects that we have discussed in the rest of the book into the context of the ecosystem to show how these are pertinent to management and conservation. We start by providing some definitions.

21.2 Definitions

Communities are complexes of interacting populations. They involve both direct and indirect effects of competition, predation, and parasitism. They contain major players, called **dominant** and **keystone** species, and combinations of many other minor species. Although species combinations are fluid, there are usually identifiable characteristic groups that define that community. Communities exist within the abiotic physical and chemical environment, and these together form the **ecosystem**. All ecosystems are maintained by inputs of energy and some nutrients from outside the system, but some receive most of their nutrients from outside (**allochthonous** supply), such as leaf debris from riverbanks for stream communities, or the rain of detritus to the abyssal zone of the ocean; while other ecosystems are more self-contained, depending on nutrient supplies within them (**autochthonous** supply). Ecosystems can be small or large, ranging from a stream complex or watershed to the 25,000 km^2 Serengeti, or even segments of the vast boreal forest of Canada or Russia. Although ecosystems are open, they are identified by some form of boundary across which the combination of abiotic and biotic factors changes. The study of the large-scale ecology of ecosystems is called **landscape ecology**.

21.3 Gradients of communities

The early classic studies of plant communities suggested that communities of plants existed as discrete units, something like a superorganism. These units had sharp boundaries. It was generally agreed that sharp divisions in the environment – different geology, soils, or other environmental factors – caused discrete boundaries between communities. However, it was also suggested that such boundaries could be caused by groups of coevolved species occurring together. Another school of thought suggested that plants generally existed independently of others so that a gradual change of species took place along a gradient in the abiotic environment – for example, gradients of moisture, or of altitude and temperature on mountains, or of exposure and salinity on the seashore (Whittaker 1967). Present understanding of gradients (Austin 1985) suggests that species do form a continuum along gradients, but not uniformly. Groups of species appear and disappear together for two reasons. First, some species depend on each other. Where plants go so do their associated animals, and therefore these groups are found in similar places. Second, the abiotic gradient is not usually uniform: there is often a break or rapid change in geology, soil substrate, or exposure, and at these points there are rapid changes in species complexes.

21.4 Niches

Where a species is found is determined by its tolerance and adaptation to the abiotic environment. This is the **fundamental niche**, which is constrained by the biotic processes of competition and predation to form the **realized niche**. Sections 9.4–9.6 give a more detailed explanation of niches, and how species may divide up the niche space along gradients.

21.5 Food webs and intertrophic interactions

A community can be divided into **trophic levels**, defined as the location of energy or nutrients within a food chain or food web (Fig. 21.1). In terrestrial systems we describe these as plants that capture energy from the sun (**autotrophs**), herbivores that feed on plants, carnivores that feed on herbivores or other carnivores (the last two levels being called **heterotrophs**), and detritivores that receive dead products from all other trophic levels. Energy and nutrients enter a system from outside, and flow through the system. Energy is progressively lost through the levels but much of the nutrients cycle back via the detritivores into the plants. A fraction is lost by water leaching out of soils, runoff into rivers, burning, and wind transport, etc. In a stable system, inputs and outputs at all levels must balance.

Indirect effects are those where one trophic level affects components below or above it. There are two forms of this: trophic cascades where linear predation influences

Fig. 21.1 Hypothetical flow diagram showing the passage of nutrients through the trophic levels (boxes). The numbers are in units/ha/year. Inputs and outputs for each box and for the whole system must balance.

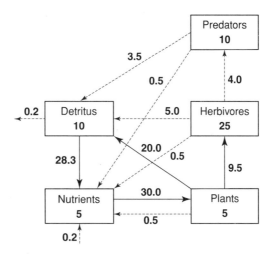

the next-but-one level, and non-linear effects where competitors within the predator or prey levels respond to changes in one of those levels (Wootton 1994a,b). Two examples are illustrated here.

21.5.1 Hyperpredation and apparent competition

Top predators can affect the diversity of their prey through changes in the abundance of primary prey. This is particularly important where both exotic predators and their primary prey result in increased predation of endemic species. The introduction of feral cats and European rabbits on islands has resulted in catastrophic declines of many seabird populations (many examples are given in Courchamp *et al.* 2000b). In New Zealand, introduced ferrets (*Mustela furo*), cats, and rabbits have resulted in increased predation on two endemic skink species. Predation on skinks was inversely density dependent, as one expects for secondary prey (see Section 10.7.2) (Norbury 2001). We saw in Section 17.8.1 what happened when feral pigs (*Sus scrofa*) were released onto the California Channel Islands in the early twentieth century (Roemer *et al.* 2002). As they became more abundant, mainland golden eagles (*Aquila chrysaetos*) were able to colonize the islands in the 1970s, building to high numbers by the 1990s (see Fig. 17.9). These raptors caused a rapid decline in the endemic island fox (*Urocyon littoralis*). In turn, this caused a change in the competitive balance of the fox with the endemic spotted skunk (*Spilogale gracilis amphiala*) that was able to increase. Thus, hyperpredation, where top predators consume lower predators, has resulted in a change in competition between them.

21.5.2 Meso-predator release

The opposite process to that of hyperpredation occurs when top predators are removed and a linear trophic cascade results in the increase of lower predators and a decrease in some of their prey. Consequently, there is a change in species composition. Again on islands where both top predators, such as cats, and their prey, such as rats, are present, removal of cats can increase the rats that then prey on seabird nests (Courchamp *et al.* 1999). In southwest Spain, the Iberian lynx (*Felis pardalis*) depredates both Egyptian mongooses (*Herpestes ichneumon*) and rabbits. In areas where lynx were absent, mongoose predation resulted in rabbit densities being 2–4 times lower than in areas where lynx occurred (Palomares *et al.* 1995). In southern California the disappearance of coyotes (*Canis latrans*) has allowed an increase in several smaller predators (e.g. striped skunk, gray fox, feral cats) that in turn have reduced the diversity of scrub-breeding birds (Crooks and Soule 1999).

A more complex situation is found on Marion Island off South Africa (Huyser *et al.* 2000). There feral cats fed on exotic house mice (*Mus musculus*) and burrowing petrels, reducing petrel numbers considerably. Mice consumed terrestrial macro-invertebrates. The endemic lesser sheathbill (*Chionis minor*), an aberrant shorebird of the Chionididae related to plovers, also depends on these macro-invertebrates in winter. Macro-invertebrates are maintained by the feces from burrowing petrels, and the loss of petrels resulted in the loss of invertebrates. Removal of cats caused an increase in house mice (but not petrels because their habitat for burrows had changed), and a further decrease in macro-invertebrates. The sheathbill population collapsed on that island in comparison to that on neighboring Prince Edward Island where mice were at much lower density. Thus, non-linear indirect effects involved an increase in competition between mice and sheathbills for invertebrates through meso-predator release.

In all of these examples ecosystem disturbances through introduction of a new species resulted in unpredicted outcomes due to indirect food web interactions, with unexpected consequences for conservation.

21.6 Community features and management consequences

21.6.1 *Dominant and keystone species*

Within groups of co-occurring species some of the species play a larger role in defining the group. In terrestrial environments, it is the plants that dictate the structure and function of that community, and these plants are determined by the abiotic environment. Thus, in arctic and alpine areas we get tundra with herbs, grasses, and shrubs determining a low-lying structure. In contrast, in very wet temperate and tropical areas we find forests with tall, woody plants creating a complex three-dimensional structure. Thus, we say that the most common plants (most numerous, highest biomass), such as white spruce in boreal forest, are **dominant** species that determine the structure and function of the ecosystem.

Some species can have a considerable influence on the community even though they are relatively rare. Often these are predators and they can determine not just the species composition of prey but, indirectly, many other components of the ecosystem as well. Such species have been called **keystone** (Paine 1969; Power *et al.* 1996), and are defined as those that have a greater role in maintaining ecosystem structure and function than one would predict based on their abundance or biomass. Top predators are often presented as keystone species: for example, the presence or absence of sea otters (*Enhydra lutris*) as top predators of inshore marine communities determines the abundance and species composition of other members (Estes and Duggins 1995). Herbivores, however, can also act as keystones: for example, rodents can structure desert plant communities (Brown and Heske 1990), snowshoe hares (*Lepus americanus*) structure boreal forest vertebrate communities (Krebs *et al.* 2001b), and wildebeest (*Connochaetes taurinus*) structure the Serengeti ecosystem (Sinclair 2003).

There are two major problems with the keystone concept. First, there are operational problems with identifying keystone species. We need to define which parameter we measure – abundance, biomass, species composition, or something else. We need to specify what is the degree of change in the community expected from losing a keystone species. Communities are open ended and we must state how far into the food web we should trace the impacts. Thus, the impacts of top predators may be traced only as far as the herbivores and plants, or through other indirect links to more distant herbivores, detritivores, protozoans, even microbes (Mills *et al.* 1993; Power *et al.* 1996).

Despite these problems, we recognize that some species define the community composition, and removal of these can produce changes in state, whereas the loss of other species has little effect on the rest of the community. The conservation consequences from the loss of wildebeest as a keystone species in Serengeti would be disproportionately greater than those due to the loss of black rhino (*Diceros bicornis*) and wild dog (*Lycaon pictus*), both of which have occurred with little impact on the system.

Keystone species can also have counterintuitive effects. Bell miners (*Manorina melanophrys*) are dominant territorial insectivorous birds feeding on psyllids (plant-sucking Homoptera) in Australian eucalypt forests. Where bell miners occur trees appear unhealthy, the foliage infested with these insects. When bell miners were removed, 11 other insectivore bird species moved in, fed on the psyllids, and within 4 months eradicated the infestation (Loyn *et al.* 1983). Interspecific territoriality by the miners maintained their food supply but reduced the diversity of competing predators.

The management and conservation implications for keystones such as wildebeest or bell miners are very different from those for other species in their ecosystems. The task of managers is to identify such species.

21.6.2 *Overpredation*

Top predators can increase the diversity of prey species through intermediate disturbance effects (Connell 1978). However, predators can also have the opposite effect and reduce diversity of prey. Such effects arise because rare species are secondary prey, essentially **by-catch** for predators that depend on more common prey (Chapter 10). It is, of course, the rare species that attract the attention of managers.

Exotic predators are of special management concern because they can threaten rare species. We see this where exotic predators such as red fox (*Vulpes vulpes*) in Australia, stoats (*Mustela erminea*) in New Zealand, and many other species on islands in the Pacific and Indian oceans, supported by exotic prey or carrion, have caused the extinction of numerous marsupial mammals, birds, and invertebrates (Serena 1994; Atkinson 2001).

The concept of by-catch stems from harvesting, particularly fisheries harvests. Many fish species are caught as by-catch in fisheries that focus on more abundant species. The latter can be maintained while the by-catch species decline, either because they are easier to catch or because they have a lower intrinsic rate of increase (Hilborn *et al.* 2003). In the Atlantic, both common skates (*Raja batis*) and barndoor skates (*Raja laevis*) almost declined to extinction because of by-catch (Casey and Myers 1998), and shark numbers have declined by over 50% since 1990 due to by-catch in the swordfish and tuna fisheries (Baum *et al.* 2003).

A prey species' role can change depending on the presence of other species. In British Columbia, moose (*Alces alces*) are primary prey for wolves, which in turn are driving mountain caribou (*Rangifer tarandus*) to extinction (Seip 1992; Wittmer *et al.* 2005); but in the nearby Banff ecosystem moose are secondary prey that are being exterminated by wolves dependent primarily on elk (*Cervus elaphus*).

In general, the dynamics of predation, whether by natural predators or by humans, affect prey species differently depending on their role in the system, their abundance, and their intrinsic adaptations (Courchamp *et al.* 1999, 2000b). Therefore, the conservation threat to a species, particularly due to predation, depends on the ecosystem in which it is found.

21.6.3 Coevolved associations

A species may not be a keystone yet can be essential to the survival of other species if it has closely evolved associations with those species. Many tropical flowers have special adaptations to attract specific insect pollinators. In Hawaii, a whole group of plants, *Hibiscadelphus*, have become extinct, or nearly extinct, after the extinction of their honeycreeper pollinators. On Mauritius, the tree *Calvaria major* almost went extinct following the extinction of the large flightless pigeon, the dodo (*Raphus cucullatus*), that ate the seeds and promoted their germination, probably by cleaning the fruit (Temple 1977; Traveset 1998).

Such coevolved associations have important flow-on effects on ecosystem dynamics. New Zealand is relatively depauperate in animal pollinators, allowing the evolution of some unusual coevolved associations (Webb and Kelly 1993). One such involves the rare dioecious obligate root parasite of forest trees, *Dactylanthus taylorii*, which produces strongly scented brownish flowers on the forest floor. This species has adaptations for pollination by short-tailed bats (*Mystacina* species), which in the total absence of terrestrial mammals forage on the forest floor like rodents. Multiple threats such as limited pollination of *Dactylanthus* due to declining bat populations (Ecroyd 1996), and severe herbivory by the exotic brush-tailed possum (*Trichosurus vulpecula*), feral pig (*Sus scrofa*), and rats (*Rattus* species), have put the plant in serious decline.

Even common species can be threatened if there is a coevolved association with a vulnerable species. For example, pollination limitation is becoming increasingly evident in mainland New Zealand flora compared with that on offshore islands (Montgomery *et al.* 2001; Anderson 2003) and the decline in avian pollinators appears to be the main cause. Thus, the mistletoe, *Peraxilla tetrapetala*, is showing declining pollination rates because its main pollinator, the bellbird (*Anthornis melanura*), an endemic honeyeater, is declining in number (Murphy and Kelly 2001). In general, these plant species were not themselves under threat until the species that they depended upon declined. Thus, the conservation needs of one species must take into account the requirements of other species. Prior knowledge of such obligatory associations would allow conservationists to predict threats to survival of species in a wider ecosystem context.

21.7 Multiple states

Operationally, multiple states can be identified when an external perturbation changes a system from one state to another and the system does not return to the original state once the perturbation has ceased. This definition excludes situations where different states occur under different environmental conditions (Holling 1973; Sinclair 1989; Knowlton 1992; Beisner *et al.* 2003). Changes in state are characterized by non-linear dynamics between trophic levels, exhibiting initial slow change followed by fast, catastrophic change (May 1977; Scheffer *et al.* 2001). Predation is one process that can produce such multiple states. Under special circumstances of Type III functional responses, predators can theoretically hold prey populations at two levels under the same environmental conditions (see Section 10.7.1 and Fig. 10.7). Multiple states can also arise through switches in competitive ability between species; environmental disturbances such as storms or fire that change soil conditions can result in permanent changes in state.

Evidence for multiple states in nature is extremely sparse. Forest insects may be held at low density by warblers but can erupt to high density where warblers do not regulate them (Ludwig *et al.* 1978; Crawford and Jennings 1989). There are a few examples where mammals act as the predator. White-tailed deer maintain different plant communities by feeding on young trees. Two tree densities can be found depending on whether young trees can escape this herbivory or not (Augustine *et al.* 1998). Similarly, elephants in Serengeti can maintain two different densities of *Acacia* trees. When fire prevents regeneration and mature trees die of old age a grassland is produced that elephants can maintain by feeding on and regulating juvenile trees. When overhunting by humans removed elephants (in both the 1880s and 1980s), trees escaped herbivory and formed a mature savanna. After both periods of removal elephant numbers increased; they fed on the mature trees but did not return the woodland to grassland (Sinclair and Krebs 2002). Examples of multiple states where mammals are prey are also rare (see Section 10.7.1). Outbreaks of house mice and European rabbits in Australia may be interpreted as changes from a predator-regulated to a food-regulated state (Pech *et al.* 1992). The collapse of the "forty-mile" caribou herd of Yukon may be evidence of multiple states. Similarly, managers culled wildebeest in the Kruger National Park, South Africa to reduce their numbers. When culling ceased wildebeest numbers continued to decline through lion predation (Smuts 1978). In general, these examples illustrate that more than one state can occur under a given set of climatic conditions.

In some cases the change in state has undesirable consequences for management and conservation. For example, in the semi-arid regions of the Negev–Sinai in Israel and Egypt, and in the Sahel of Africa, a shrub and herb layer acts as a blanket on the soil, retaining moisture and heat overnight. During the day thermal upcurrents carry moisture from both the soil and transpiring plants to upper levels where it condenses as rain. This supplies the plants and soil, completing a positive feedback self-sustaining system when in an undisturbed state. In contrast, overgrazing by livestock leaves a denuded soil surface, higher surface albedo, and a cooling at night. There are fewer thermals, and these carry less moisture. Thus, overgrazed areas have much lower precipitation, and this is also a positive feedback (Otterman 1974; Sinclair and Fryxell 1985). The vegetated state switches to the denuded one through the disturbance of overgrazing, that is, there is a threshold level of disturbance (grazing) where one state switches to another. A similar positive feedback switch in vegetation state occurs in Niger where overgrazing has altered vegetation structure leading to reduced water retention, increased soil loss, and further vegetation loss. The system is now locked into this reduced state (Wu *et al.* 2000).

However, good examples of multiple states are rare, a few known from lakes, rivers, coral reefs, grasslands, and forests (Knowlton 1992; Augustine *et al.* 1998; Dent *et al.* 2002). The relevance to management is that multiple states are an emerging property of ecosystems that will rarely be predicted from the study of single species. Some states arise from excessive disturbance (see below). Thus conservation needs to plan for more than one natural state whilst avoiding unnatural states due to excessive human disturbance.

21.8 Regulation of top-down and bottom-up processes

Figure 21.1 illustrates the pathways for energy and nutrients. However, it does not indicate where the regulation occurs in the system. It can come via the food supply (bottom-up), from predation (top-down), or both. In the absence of predators (or parasites), bottom-up processes must regulate all populations. Regulation through resources must be the basic rule, and it clearly applies to all top predators. There are several factors that affect resource production and biomass in the world (Polis 1999). However, there are four main conditions that provide a refuge from predation and so allow bottom-up regulation.

21.8.1 Body size

Small prey species are vulnerable to predation whereas very large species, especially in mammals, have outgrown all present-day predators, and so are regulated by food supply. Thus, a suite of predators account for virtually all mortality of adult snow-shoe hares in northern Canada (Hodges *et al.* 2001). In contrast, the wood bison (*Bison bison athabascae*) population in the Mackenzie Bison Sanctuary of Canada appears to be regulated by food supply, despite wolf predation of juveniles (Larter *et al.* 2000).

In Africa, we see a similar effect of body size on causes of regulation. Elephants, rhinos, and hippos are too large for predators. Although predators kill a few new-born animals they have no effect on the population (Sinclair 1977). Even animals the size of African buffalo and giraffe are large enough that predators have difficulty killing them, so that predation accounts for a small proportion of adult mortality, and undernutrition is the predominant cause of mortality (Sinclair 1977, 1979b).

21.8.2 Migration

Migration is an adaptation that overcomes the constraints imposed by body size (see Section 10.8.1). Predators cannot follow migrating herbivores because they are confined to territories to raise and protect their young. This general rule is evident in all mammal migration systems such as the wildebeest and gazelles in Serengeti and Botswana, white-eared kob (*Kobus kob*) in Sudan, caribou (*Rangifer tarandus*) in northern Canada, and most probably the original plains bison of the North American prairies (Fryxell and Sinclair 1988a). It might also apply to the migration of marine mammals. Migrating species, therefore, escape from predator regulation even when they are relatively small in size, as in the gazelles. In addition, migration is an adaptation to access ephemeral, high-quality food resources not available to non-migrants. These two features of migration systems allow populations to become an order of magnitude greater in number compared with resident populations.

21.8.3 Low-diversity ecosystems

In higher latitudes there are often predator–prey systems with only one major predator and one or a few mammal prey species. We see such systems in temperate woodlands and tundra, and even in mammals of tropical forest (though not in other groups). In these ecosystems we normally see bottom-up regulation of the prey. Nevertheless, there are a few cases of top-down regulation of prey. Wolves might regulate moose in some parts of Canada and Alaska (Gasaway *et al.* 1992; Messier and Joly 2000). In contrast, on Isle Royale in Lake Superior, wolf numbers appeared to track moose numbers and did not regulate that population (Peterson and Vucetich 2003). Thus, we have evidence of regulation of herbivores by both predators and food supply. Particular features of the ecosystem and the species involved determine the direction of regulation. In addition, multiple states (see Section 21.6) may occur where regulation can switch in the same system from resource limitation to predation or vice versa. Alternatively, regulation may be determined by the presence or absence of alternative prey for the predator (see Section 10.7.1).

21.8.4 *High-diversity communities*

In some systems there is a high diversity of herbivores and carnivores. Nearly all are associated with tropical ecosystems. Whether a herbivore species is limited by predators is determined by its place in the hierarchy of herbivores. In African savanna there are as many as 10 coexisting canid or felid carnivores feeding on ungulates, lagomorphs, and rodents. They vary in size from the 200 kg lion (*Panthera leo*) to the 10 kg wild cat (*Felis sylvestris*). The larger the carnivore the greater is its range of prey sizes. Thus, the lions' diet ranges from buffalo (450 kg) to dikdik (*Madoqua kirkii*), a small antelope (10 kg), whereas that of the 16 kg caracal (*Felis caracal*) ranges from duiker (15 kg) to 100 g rodents. The consequence of this is that smaller ungulates have many more predators than larger ungulates. Thus, smaller ungulates experience more predation and top-down regulation (Sinclair *et al.* 2003).

21.9 Ecosystem consequences of bottom-up processes

Regulation of herbivore populations through their food supply has profound consequences on the ecosystems where they occur. Mammals may not be numerous compared with other animal groups but their impact is considerable. Perhaps more than any other group they can determine the physical structure of the habitats, alter the rates of ecosystem processes such as nutrient flow, growth rate, or decomposition, and dictate species diversity. These large-scale effects – at the level of ecosystems, watersheds, and biomes – can be thought of as **ecological landscaping** (Sinclair 2003).

21.9.1 *Vegetation structuring*

Plants determine the physical structure of habitats, the particular type being a function of the abiotic conditions. Some periodic environmental effects, such as fire, hurricanes, and floods, can interrupt the normal succession of plant species towards a climax. In savanna, fire typically impedes the succession of trees to produce a "fire disclimax" of grassland and fire-tolerant herbs, shrubs, and trees. Herbivorous mammals can have analogous effects to fire in savanna systems (Hobbs 1996) and so produce a "mammal disclimax." Plant succession is held in a different state as a result of the restructuring imposed by mammals. Such impacts are evident in most terrestrial biomes where mammals are abundant. However, mammals have their greatest impacts in the tropical savannas, particularly through feeding by megaherbivores (Owen-Smith 1988); and in grasslands throughout temperate and tropical regions due to grazing and browsing by ungulates (Sinclair 2003).

In recent times, mammal herbivores have had little structuring effect in the high-latitude tundra biomes. However, the Pleistocene tundra supported a substantial biomass of mammoths, woolly rhinos, and bison that fed upon the shrubs and sedges. Herbivorous mammals also do not substantially alter tropical forest, although mammals do influence the dispersal of tree seedlings (Janzen 1970). In both arctic tundra and tropical forest the low impact of herbivores may be due to the top-down effects of mammal carnivores that limit herbivore densities in these systems (Terborgh 1988; Oksanen 1990).

21.9.2 *Ecosystem rates*

Mammals and birds influence the rates of nutrient cycling in addition to altering physical structure. High densities of mammals and birds can influence the soil processes through their deposition of feces and urine. Before the arrival of Pacific rats (*Rattus exulans*) in New Zealand some 700 years ago, billions of shearwaters lived on the forest floor and provided a considerable nutrient input. The extinction of shearwaters from mainland New Zealand due to the rats has altered the nutrient dynamics of that country (Worthy and Holdaway 2002). The volcanic Serengeti plains have very high nutrient turnover rates due to the large herds of ungulates grazing them in the wet season and the plethora of dung beetle species that act to bury the dung. This process leads to high protein and mineral content of the grasses eaten by the grazing herds. In essence, ungulates fertilize their own food, and so create a positive feedback increasing their own density (Botkin *et al.* 1981; McNaughton *et al.* 1997). In low-nutrient granitic soils of southern Africa, the vegetation is also low in nutrients and only the very large mammals such as elephants can feed in those woodlands. Nutrient recycling is slow (Bell 1982).

On the Canadian arctic shorelines, high densities of snow geese (*Chen caerulescens*) influence the growth of their food plants. Moderate grazing promotes growth through fecal nutrient cycling. Recent population increases of geese have resulted in overgrazing that has overwhelmed the positive effects (Jefferies *et al.* 2004). In boreal forests moose decrease nitrogen mineralization of the soil by decreasing the return of high-quality litter: their browsing on deciduous trees reduces their leaf fall while promoting low-quality white spruce inputs (Pastor *et al.* 1993). In contrast, soil nitrogen cycling in Yellowstone and other prairie areas of the USA is increased by large mammal grazers (Hobbs 1996; Frank and Evans 1997).

21.9.3 *Plant species composition*

Herbivory alters not only structure but also the type of plants that can withstand such impacts. On the North American prairies, rodents such as black-tailed prairie dogs (*Cynomys ludovicianus*) live in large colonies. These species graze grasses to a low level (a few centimeters) around their colonies. Grazing changes the grass species composition to low-growing forms, and many dicot species survive due to reduced competition from grass. American plains bison preferentially graze these short grasses, and pronghorn antelope (*Antilocapra americana*) feed on the dicots (Huntly and Inouye 1988; Miller *et al.* 1994).

Rabbits maintain short grasslands with many dicots on the South Downs of Sussex, England. When the epizootic myxomatosis removed rabbits in 1953, plant species composition changed to one of tall tussock grasses with few dicots, and there were subsequent changes in ants and lizards dependent on these plant forms (Ross 1982a,b). A whole range of plant species evolved in New Zealand with special structural defenses against moa browsing not seen elsewhere (Bond *et al.* 2004).

21.10 **Ecosystem disturbance and heterogeneity**

21.10.1 *Degrees of disturbance*

Ecosystems, left undisturbed, will change through succession to a plant community dominated by good competitors with their associated fauna. A few wet tropical forests may exhibit this situation. However, it is rare that ecosystems experience such constancy of environment. Disturbances disrupt this succession and the community reflects this history of disturbance (Pickett and White 1985). If disturbances are severe and frequent then a few hardy plants that can tolerate these stressful environments characterize the community. The boundary between the alpine tussock grasslands and nival (snow) zone of the southern alps of New Zealand is characterized by a few plants adapted to steep loose scree (tallus) that moves frequently through heavy rain, earthquakes, and trampling or feeding by exotic mountain ungulates, the tahr (*Hemitragus jemlahicus*) and chamois (*Rupicapra rupicapra*).

More commonly disturbances are both less frequent and less extreme, allowing a combination of good dispersing plants and good competitors (see Section 21.12). Forests experience tree-falls that create canopy gaps, letting in light and opportunities for light-seeking species. This disturbance is particularly important in tropical forests where fruiting plants are fed upon by birds (Levey 1990). Temperate conifer forests experience fire at frequencies of 50–200 years, maintaining a mosaic of stands of different age and species composition, and a diversity of habitats for birds and mammals (Bunnell 1995).

In tropical savannas, frequent fire dominates the system. It impedes plant succession, maintaining an open tree canopy (< 30% cover) and a grass understory. This fire regime provides the optimum habitat for the high diversity of ungulates in East Africa (Frost 1985). Fire is also required to maintain specific habitats: in North America the endangered Kirtland's warbler (*Dendroica kirtlandii*) requires fire to create its jack pine (*Pinus banksiana*) habitat (Probst 1986).

Other forms of disturbance are hurricanes and floods. Hurricanes are important at the 10–20° latitudes along coastlines. Their periodicity is usually measured in decades and they have physical restructuring effects by opening up forests, and altering shores, estuaries and riverbanks, and sedimentation rates. Flooding of rivers and estuaries is more frequent and universal. Some river flooding is necessary to maintain nesting habitat on sandbars for least terns (*Sterna albifrons*) and piping plovers (*Charadrius melodius*) in the USA, and on braided rivers of New Zealand for endangered black stilts (*Himantopus novaezealandiae*) (Boyce and Payne 1997).

One of the important consequences of disturbance is that it creates heterogeneity, or patchiness, in habitats, particularly because it reverses succession. A mosaic of patches of different age from the time of the disturbance leads to different combinations of habitats and species, an aspect that is important for ecosystem management and conservation. In particular, heterogeneity creates the mosaic of sources and sinks (Pulliam 1988; see Section 7.7.4). Sources are good habitats where species are self-supporting and surplus animals emigrate from these to sinks. The latter are poorer habitats that are not self-supporting. However, sinks provide a vital role in allowing non-breeding individuals to survive while they wait for opportunities to obtain territories in source habitats. Sinks provide the compensation in a population for unpredictable disturbances that reduce breeding populations. Protected areas, therefore, should contain both source and sink habitats.

21.10.2 *Disturbance and ecosystem management*

Disturbances that are too frequent or too extreme can radically alter an ecosystem, changing it to a different state. Persistent overgrazing can result in denudation, as mentioned earlier for semi-arid areas (Wu *et al.* 2000), or in a change from grassland to woodland, as in savanna areas (Walker *et al.* 1981). Other forms of human overdisturbance are often the underlying cause for invasions of exotic species that can take over and maintain new states (Vitousek *et al.* 1996).

More rarely, single extreme natural disturbances can change ecosystem states: the 'Wahine' storm of 1968 destroyed the beds of aquatic macrophytes in Lake Ellesmere in the South Island of New Zealand, and the physical change in the lake sediments has prevented their return. The resident population of black swans (*Cygnus atratus*) numbered 40,000–80,000 in the 1950s and 1960s. They used the weed beds for food and to raise young. Mortality from the storm itself, starvation, and reduced breeding because of a lack of suitable habitat rapidly reduced the population to less than 10,000 in subsequent years, and they have never returned to their original numbers (Williams 1979; A. Byrom, pers. comm.).

Earthquakes and their resulting tsunamis are another form of disturbance that can cause sudden effects in an ecosystem with long-term consequences. Botswana in southern Africa is so flat that even minor tectonic shifts can change the direction of river flows (Cooke 1980; Shaw 1985). The Savuti River is a side channel from the Kwando/Linyanti river system that flows into the Zambezi. The Savuti was dry from the 1860s to the 1950s. In 1956 the channel started flowing due to a tectonic shift and continued to flow, with floods in 1979, until 1983 when just as suddenly it stopped. The channel has been dry since then, and now both the Kwando and Linyanti are drying, exacerbated by climate change and human use. The Savuti was used extensively by large numbers of ungulate species in the dry season. When the river dried, it altered the ecology of a large region, with ungulate migration patterns changing to other areas on the Linyanti (M. Vanderwalle, pers. comm.). In New Zealand, earthquake disturbances are relatively frequent and their ecosystem effects can be of a lower intensity, but widespread. Landslides resulting from earthquakes cause sudden catastrophic mortality of forest trees that is scale dependent and has important effects on forest dynamics: multiple small patches have high mortality resulting in a mosaic of different-aged stands in the forest (Allen *et al.* 1999).

Climatic and temperature changes in physical oceanic conditions, as occurred suddenly in the North Pacific in the late 1970s, appear to have long-term ecosystem consequences. In the North Pacific the complex of fish species changed after the regime shift and it appears that the new fish community cannot provide sufficient quality food for the Steller's sea lion (*Eurometopias jubatus*). In the mid-1970s this sea lion population had been increasing and was around 250,000. It dropped rapidly to 100,000 by 1990 and 50,000 by 2000 (Trites and Donnelly 2003). It is possible, though not yet established, that killer whales (*Orcinus orca*) are exacerbating the decline through predation on sea lions at these low numbers (Springer *et al.* 2003).

Conservation of ecosystems, therefore, has to be sufficiently flexible to accommodate major natural disturbances such as earthquakes, fires, floods, and storms, and allow recovery from human disturbances such as overgrazing and overharvesting. Such approaches will require a better understanding not only of the impacts of disturbances, but also of the temporal and spatial scales at which those disturbances operate.

21.11 Ecosystem management at multiple scales

21.11.1 Management at large spatial scales

Ecosystems function at multiple scales, small scales affecting large scales and vice versa. The Serengeti wildebeest migration covers the entire ecosystem of 25,000 km². Wildebeest move several hundred kilometers to the short grass plains in the wet season because these plains support the most nutritious grasses in the system (Fryxell 1995). Dung beetles, of which there are some 80 species, rapidly bury feces (within a few minutes) and hence expedite nutrient cycling on these plains. They promote the high-quality nutrition of the grasses, producing a positive feedback. Dung beetles can function only when the soil is damp, so they have a negligible effect on returning nutrients to the soil in the dry season when wildebeest are in woodland areas. Thus, the very local-scale functions of the beetles influence the large-scale movements of the ungulates.

The recent collapse of the Canadian arctic grazing ecosystem has occurred through subsidies to snow geese (*Chen caerulescens*) on winter feeding grounds as far away as the southern USA resulting in overpopulation, overgrazing, and a new ecosystem state in the Arctic (Jefferies *et al.* 2004).

Although population declines can often be attributed to immediate proximate causes such as predation and habitat loss, ultimately large-scale, remote causes may underlie these events. Such fundamental causes become apparent only when the large-scale ecosystem is considered. Thus, in North America the brown-headed cowbird (*Molothrus ater*) is a nest parasite of many small passerines. Its population is increasing and it has caused the decline of at least two species, Least Bell's vireo (*Vireo bellii*) and the black-capped vireo (*V. atricapillus*) (Smith *et al.* 2000). The ultimate cause is due to events that date back to the 1800s (Rothstein 1994). This cowbird's original range lies in forests of the American northeast where it prefers open areas for foraging. Expansion of open land through agriculture across North America has allowed this species to spread into new areas and parasitize species that have few adaptive traits to counter it. Both large-scale and long-term events underlie the spread of this nest parasite.

21.11.2 *Management at long time scales*

Large-scale temporal patterns are also important in ecosystem dynamics. The classic snowshoe hare (*Lepus americanus*) cycle of North America is synchronized spatially by decadal weather events (Stenseth *et al.* 2002). This cycle of numbers then influences the rest of the ecosystem (Krebs *et al.* 2001b). Synchrony is enhanced by environmental correlation across sites and reduced by dispersal between sites (Kendall *et al.* 2000). There is increasing evidence that complex ecosystem processes are influenced by multi-year fluctuations in climate such as the North Atlantic Oscillation and the southern and northern Pacific oscillations (Post and Stenseth 1998; Coulson *et al.* 2001a). In the context of long-term conservation, spatial and temporal synchronies are of particular concern when a species is rare because of its increased probability of extinction.

Human-induced changes in climate can be considered very long-term, persistent disturbance. Global climate change is now having measurable effects on ecosystems, altering community composition by shifting species ranges differentially towards the poles, higher in altitude (especially in the tropics; Pounds *et al.* 1999), and away from the tropics (Schneider and Root 2002; Parmesan and Yohe 2003). In Britain, birds and other groups are breeding earlier. Changes in community structure mean that species will experience changes in food supply and predation rates (Crick *et al.* 1997).

In general, ecosystem processes at different spatial and temporal scales, including disturbance and long-term trends, must be considered as part of any conservation strategy for the ecosystem in which a threatened species exists. Such processes operate at larger scales than we have traditionally planned for. Moreover, conservation is no longer a short-term exercise: we have to consider time scales of 100 years or more.

21.12 Biodiversity

21.12.1 *Determinants of diversity*

Biodiversity is defined as the complement of living organisms in an ecosystem. In terrestrial systems there is a general tendency for animal groups to be more species diverse in tropical latitudes (MacArthur 1972). Figure 21.2 shows such a distribution for birds. There are local anomalies in this pattern, for example for reptiles in Australia (Schall and Pianka 1978).

There have been many hypotheses offered to explain this latitudinal gradient in biodiversity from the tropics to the poles. Some of these are:

1 *Structural heterogeneity.* Tropical vegetation is more complex structurally, providing more niches for animals compared with arctic tundra or even boreal forest.

2 *Community age.* Ice caps covered large areas of northern North America and Eurasia, and these melted only 10,000 years ago. There has been insufficient time to reinvade these higher latitudes and evolve new species. This process of eradication

Fig. 21.2 The number of terrestrial bird species in North America declines from the tropics to the Arctic. (Data from MacArthur 1972.)

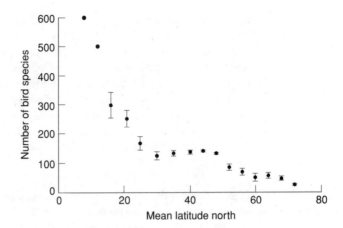

and replacement has been repeated many times in the Pliocene and Pleistocene and probably earlier, whilst the tropics have been extant throughout.

3 *Productivity.* The warmer temperatures and higher rainfall of tropical regions allows species to fit into narrow niches over small areas while still maintaining a large enough population to avoid extinction. In contrast, low productivity at high latitudes means that fewer species must maintain broad niches over wide areas to maintain the same population size.

4 *Predation.* The higher biodiversity of the tropics includes a higher number of predators. These impose a top-down regulation on prey species (see Section 21.8.4), which allows a greater number of prey species to coexist because they are not competing with each other. This process is called **predator-mediated coexistence.**

5 *Environmental stability.* High latitudes fluctuate considerably between summer and winter in environmental parameters such as temperature. Large populations are required to withstand such fluctuations and this means few species can live there. Stable environments in the tropics allow smaller populations to survive. These small populations can have smaller niches and so more species can fit into the system.

6 *Intermediate disturbance.* This is a variation of (5). While agreeing that higher latitudes experience frequent major disturbances, and so support fewer species, it differs in considering that the tropics have some smaller disturbances (hence intermediate between few and many) that allow both early succession and highly competitive species to coexist (Connell 1978).

In summary, no single hypothesis explains all of the observed distributions. It is likely that different hypotheses apply in different locations, and also that more than one process occurs at one location.

21.12.2 *Local and regional diversity*

The total diversity within a large area, a region, is called the **gamma** diversity. This is determined by two components: (i) **alpha** diversity, which is the number of species in a local area or habitat; and (ii) **beta** diversity, which is the reciprocal of the mean number of habitats or localities occupied by a species. Thus:

gamma diversity = average diversity per habitat (alpha) × 1/mean number of
                  habitats occupied by a species (beta) × total number of
                  habitats

This allows us to differentiate between the separate aspects of species (habitat breadth through beta diversity, and the regional heterogeneity through the number of available habitats). These values allow a mechanism for comparing different communities (Schluter and Ricklefs 1993).

If a species lives in a local area or habitat independent of any other species present, then one would expect that the more species there are in the surrounding region the more there will be in any particular local area. Thus, there should be a linear relationship between species richness locally and that on the larger regional scale. Such an assumption, at face value, may seem unlikely since we know that there are many interactions between species. However, a linear relationship could occur where a young community that is still evolving may not yet have achieved its full complement of species, so that local area richness reflects that in the wider region – biotic interactions may not have come fully into play. Alternatively, a local patch that receives a large number of immigrants (such as one with many sink populations) relative to the competitive abilities of the residents will also reflect the richness of the region that produces the dispersers – dispersing species overwhelm the residents. This linear relationship has been termed an **unsaturated** pattern.

In contrast, if there are strong biotic interactions between species locally such that many species are excluded from the community, then one would expect that after an initial colonization period a limit to the number of species would occur locally irrespective of the number available in the region. A plot of local versus regional species richness would produce a curve with an upper limit. This has been called the **saturated** pattern because no more species can be added to the local area (Srivastava 1999; Hillebrand and Blenckner 2002).

Processes such as dispersal and interspecific competition for space underlie these patterns. Initially, these patterns were used to infer the mechanism. However, several different processes can produce either of the above patterns (Chave *et al.* 2002; Shurin and Srivastava 2005). For example, facilitation through a keystone predator can increase local richness, the amount dependent on the regional pool of species. This process could override the saturating pattern of strong interspecific competition.

Another problem arises when one tries to identify local and regional scales. Shurin and Srivastava (2005) showed that the pattern changes from saturated to unsaturated as the ratio of local to region increases. Thus, the pattern depends on the scale of the study. Many saturated communities have been overlooked because the local scale was too large. In general, such patterns of biodiversity provide only weak evidence for the underlying mechanisms structuring that diversity.

21.13 Island biogeography and dynamic processes of diversity

MacArthur and Wilson (1967) and MacArthur (1972) proposed that the diversity of species in discrete ecosystems can be considered as a dynamic equilibrium, a balance between the rate of immigrating species and the rate of local extinction of species. They chose islands to illustrate this principle (Fig. 21.3). Starting with a newly formed island, the rate of immigration of new species declines as the full complement of species from the source (the mainland) is achieved. However, as species accumulate on the island some of them are going extinct, and the probability of extinction increases as more species arrive. Thus, where the two rates are equal we achieve the equilibrium of species number, S, for that island.

The immigration rate must be determined by the **distance** the island lies from the mainland, so a distant island should receive a lower immigration rate than a near

Fig. 21.3 Hypothetical relationship between the rates of immigration or extinction of species and the number of species in a habitat patch or island. Where the immigration and extinction curves cross we obtain an equilibrium number of species, S. For a given extinction rate, islands closer to the source of immigrants have a higher number (S_{near}) than islands further away (S_{far}). For a given immigration rate, larger islands have a higher number of species (S_{large}) than smaller islands (S_{small}).

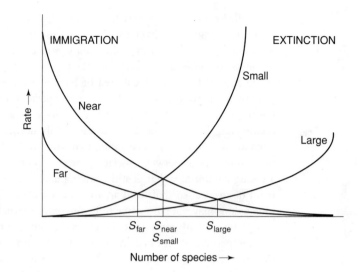

island. Thus, for islands of equal size the expected number of species for distant islands, S_{far}, should be lower than that for near islands, S_{near}.

The extinction rate is determined by several factors such as the size of the population and the number of competing and predatory species. All of these are related to the area of the ecosystem such that islands of larger **size** should have lower extinction rates. Thus, the expected number of species for a large island, S_{large}, should be larger than that for a small island, S_{small}. This leads to the classic species–area equation that we explored in Section 18.5.3.

Some classic studies provided circumstantial evidence for this **theory of island biogeography**. Diamond (1969) showed that bird species on the California Channel Islands maintained similar numbers of species over a 50-year period. There was a turnover of species because some went extinct in the interval but others arrived to replace them. Simberloff and Wilson (1970) showed that small mangrove islands, artificially denuded of their invertebrates, regained a similar number of species as occurred before the removal but the types of species differed to some extent. These results concur with the theory.

We should recognize that this is an idealized concept and there are several factors that could distort it. The complexity of the habitat is one such factor (Lack 1971), but several of the other factors determining biodiversity that we discussed in Section 21.12.1 also apply. The important point is to understand that communities are dynamic, with species coming and going, and not to place too much emphasis on looking for an equilibrium.

One important prediction from this theory concerns what happens when a piece of a large mainland ecosystem is suddenly isolated. Isolation means that the immigration rate should be reduced and hence the species number should decline to a lower value. This process, called **faunal relaxation** (Diamond, 1972), was documented for islands that were isolated from the mainland by the 100 m rise in seawater level after the ice age, 10,000 years ago. Examples of such islands, called **landbridge islands**, are those in the Malay archipelago (the Sunda Islands), amongst others (MacArthur 1972).

The process of faunal relaxation has important consequences for conservation. Protected areas (national parks and other reserves) are often small islands carved out of much larger ecosystems. A barrier is created by altered habitat surrounding them that many species find difficult or impossible to cross – particularly large mammals and sedentary birds, reptiles, and amphibians. Thus, a park should lose species over time. The rate of loss should be higher and the eventual number should be lower for smaller parks. There is some evidence that this loss is already occurring (Newmark 1987).

21.14 Ecosystem function

Ecosystem function is a general term that covers three different components: processes, products or "goods," and services. Box 21.1 lists some examples of each. Processes involve rates of flow of nutrients, growth of biomass, etc. Services are processes that are of benefit to human society, for example pollution control, and goods are the end result of these, such as clean water (Schulze and Mooney 1993; Mooney and Ehrlich 1997; Kinzig *et al.* 2001; Loreau *et al.* 2002; Srivastava and Vellend 2005).

Ecosystem function is often linked to biodiversity (Schulze and Mooney 1993; du Toit and Cumming 1999). The argument is that the more species there are in a system, the greater the ability of the system to withstand shocks to it (Walker 1992, 1995; Naeem 1998; Tilman 1999; Loreau 2000). This can be illustrated in two ways (Fig. 21.4). First, ecosystem function would be related to the number of species linearly if all species had equal contribution to the function (e.g. productivity) and acted independently (Fig. 21.4a). However, we know this equality of function is unlikely

Box 21.1 The three categories of ecosystem function.

Ecosystem processes
1 Hydrology
2 Biological productivity
3 Biogeographical cycling and storage
4 Decomposition
5 Resilience
6 Robustness–fragility

Ecosystem services
1 Maintaining hydrological cycles
2 Regulating climate
3 Purifying water and air
4 Pollinating crops and natural systems
5 Generating and maintaining soils
6 Cycling nutrients
7 Detoxifying pollutants
8 Providing aesthetic quality
9 Providing baseline research

Ecosystem goods
1 Food and clean water
2 Construction materials
3 Medicinal plants
4 Wild genes
5 Replacement species
6 Biological control agents
7 Tourism

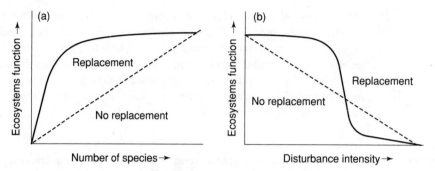

Fig. 21.4 (a) Hypothetical relationship between ecosystem function and species number. If species can replace the function of lost species then system function follows the solid line. If there is no replacement then function follows the broken line. (b) Ecosystem function relative to the degree of disturbance that can deplete species diversity. Low disturbance (left on *x*-axis) is tolerated because species can replace each other. Higher disturbance can cause a rapid decline in function when redundancy of species is used up.

to occur because of dominance and keystone effects. Thus, if we lose species from a system, initially we are unlikely to see much reduction in function for two reasons. One is that the species lost could be minor players in the system, and a drop in function is not observed until a keystone or dominant is lost. Thus, a drop in function occurs anywhere in the sequence (even at the beginning) depending on when the important species are lost. A second reason, however, is that in a diverse community the first few species to be lost can be replaced by other species that take over their function. This involves the idea that there are redundant species in the system (Walker 1992, 1995). An indication of this effect is seen in the progressive loss of fish species due to human harvesting of the oceans. Initially, productivity of the fisheries was maintained through the replacement of lost species by others that increased. However, after a delay of several decades, so many species were lost that the remaining ones were unable to replace them and fish harvests collapsed (Jackson *et al.* 2001).

The second way of looking at ecosystem function is to consider how it changes with the degree of disturbance (Fig. 21.4b). With minor disturbances some species can compensate for lost ones and maintain the function. However, at some point the disturbance is so large that there are insufficient species to replace the lost ones. So far most of the evidence for these concepts has come from plants and the stability of productivity and nutrient cycles (Vitousek and Hooper 1993; Naeem *et al.* 1995; Naeem and Li 1997; Naeem 1998, 2002).

The distortion of ecosystem processes can lead to unwanted ecosystem effects, conservation threats to individual species, and expensive ecosystem management. For example, Australia before European settlement was largely covered in eucalypt woodland. In the past century agriculture has removed nearly all of this woodland, especially in Western Australia. Originally eucalypts kept groundwater levels down through transpiration processes. Once the trees were removed, groundwater levels rose, water evaporated at the soil surface, and saline deposits made the soil unsuitable not only for crops but also for native biota. Now large areas of Australia have a major problem with salinization of soil and groundwater upwelling, with a resultant decline in agricultural productivity. In response to this ecological (and economic) problem, Australia has had to adopt the expensive policy of revegetation (McFarlane *et al.* 1993; Nulsen 1993).

Disturbed systems provide evidence for the function of biota in creating resilience to disturbance. Thus, intact eucalypt forest in Australia supports many species of coexisting endemic honeyeaters. In contrast, the noisy miner (*Manorina melanocephala*), the geographical replacement to the bell miner mentioned in Section 21.6.1, dominates the bird community in fragmented forest in which an open canopy has developed through a combination of logging, persistent livestock grazing, and agriculture. The noisy miner aggressively excludes most of the smaller honeyeaters. Consequently, exposed trees suffer chronic infestations of psyllids, dieback of their main limbs, and death. In essence, exposed trees are no longer viable due to the dearth of insectivorous birds (Landsberg 1988; Grey *et al.* 1997, 1998).

We still need to evaluate whether it is the biodiversity itself (richness) or the composition of the community that is important, that is, whether who is present is as important as how many species are present (Srivastava and Vellend 2005). Much still needs to be done to determine whether biodiversity alters the processes within ecosystems and if so what the mechanisms are. Nevertheless, evidence is accumulating to suggest that where ecosystem processes have been ignored or distorted they can lead to perverse effects for both ecosystems and individual species, and expensive remedial action for conservation (Schwartz *et al.* 2000; Hector *et al.* 2001).

21.15 **Summary**

Management of populations has to take into account that they are embedded within a matrix of other competitors, predators, and prey. These form the community and their environments. Thus, we need to consider the management of individuals, populations, and species in the context of the ecosystem. The main points are:

1 Ecosystems involve long-term events related to the environment. Management needs to account for infrequent and unpredictable events, such as earthquakes, hurricanes, floods, fire, and droughts.

2 These events provide insight into mechanisms of ecosystem regulation and stability. The term "long term" is a function of the slowest variable in the system and is not related to the life history of the organisms concerned. Management should maintain these regimes either naturally or by mimicking their effects. Planning should consider natural periodicities from the very long term (200 years for earthquake and fire cycles) to the short term (of a few years for the effects of the El Niño Southern Oscillation and the North Atlantic Oscillation). Thus, the periods of both unpredictable, sudden events and slow change dictate the time scale for conservation planning. In most cases planning should be for 30–50-year periods or longer.

3 Ecosystem management needs to consider that there are slow trends due to environmental change, plant succession, and animal population fluctuation. These slow trends show an interaction between abiotic and biotic processes, each affecting the other.

4 Ecosystems should be managed at an appropriate spatial scale. Small patches of forest are insufficient to support viable populations of predators such as northern spotted owls (*Strix occidentalis*). Large areas are required for migrating ungulates moving between summer and winter ranges. Sufficient area is required to produce the mosaic of burns of different age. This mosaic from disturbances creates habitat heterogeneity that is used as sources and sinks for animals. Sinks are required as holding areas for non-breeding animals waiting to obtain territories.

5 Management of target populations, such as pest species, can result in indirect interactions through hyperpredation, apparent competition, and meso-predator release. These can produce unexpected consequences.

6 Slow change can become an irreversible rapid shift into a new state. Thus, management needs to consider that the ecosystem can occur in multiple states. Some of these can be natural but others can be artifacts of human disturbance.

7 Ecosystems are not static and, therefore, management cannot aim to maintain the status quo, but rather should allow natural change to take place. It is likely that these changes are oscillatory in that they return to previous conditions after a time.

8 Within protected areas, management should distinguish between natural change and direct human-induced change. Protected areas can act as ecological baselines where human-induced change is kept to a minimum, and the system can then be compared with areas outside, influenced by human activity.

9 Long-term baseline data are fundamental to conservation management, because they provide the background to interpret causes of change and hence determine the course of management.

Appendices

Appendix 1 Cumulative F distribution. The body of the table contains critical values of F for both "one-sided" and "two-sided" significance probabilities.

d.f. in denominator	One-sided tests	1	2	3	4	5	6	7	8	9	10	12	15	20	30	60	∞	Two-sided tests
1	0.10	39.9	49.5	53.6	55.8	57.2	58.2	58.9	59.4	59.9	60.2	60.7	61.2	61.7	62.3	62.8	63.3	0.20
	0.05	161	200	216	225	230	234	237	239	241	242	244	246	248	250	252	254	0.10
	0.01	4050	5000	5400	5620	5760	5860	5930	5980	6020	6060	6110	6160	6210	6260	6310	6370	0.02
2	0.10	8.53	9.00	9.16	9.24	9.29	9.33	9.35	9.37	9.38	9.39	9.41	9.42	9.44	9.46	9.47	9.49	0.20
	0.05	18.5	19.0	19.2	19.2	19.3	19.3	19.4	19.4	19.4	19.4	19.4	19.4	19.5	19.5	19.5	19.5	0.10
	0.01	98.5	99.0	99.2	99.2	99.3	99.3	99.4	99.4	99.4	99.4	99.4	99.4	99.4	99.5	99.5	99.5	0.02
3	0.10	5.54	5.46	5.39	5.34	5.31	5.28	5.27	5.25	5.24	5.23	5.22	5.20	5.18	5.17	5.15	5.13	0.20
	0.05	10.1	9.55	9.28	9.12	9.01	8.94	8.89	8.85	8.81	8.79	8.74	8.70	8.66	8.62	8.57	8.53	0.10
	0.01	34.1	30.8	29.5	28.7	28.2	27.9	27.7	27.5	27.3	27.2	27.1	26.9	26.7	26.5	26.3	26.1	0.02
4	0.10	4.54	4.32	4.19	4.11	4.05	4.01	3.98	3.95	3.94	3.92	3.90	3.87	3.84	3.82	3.79	3.76	0.20
	0.05	7.71	6.94	6.59	6.39	6.26	6.16	6.09	6.04	6.00	5.96	5.91	5.86	5.80	5.75	5.69	5.63	0.10
	0.01	21.2	18.0	16.7	16.0	15.5	15.2	15.0	14.8	14.7	14.5	14.4	14.2	14.0	13.8	13.7	13.5	0.02
5	0.10	4.06	3.78	3.62	3.52	3.45	3.40	3.37	3.34	3.32	3.30	3.27	3.24	3.21	3.17	3.14	3.10	0.20
	0.05	6.61	5.79	5.41	5.19	5.05	4.95	4.88	4.82	4.77	4.74	4.68	4.62	4.56	4.50	4.43	4.37	0.10
	0.01	16.3	13.3	12.1	11.4	11.0	10.7	10.5	10.3	10.2	10.1	9.89	9.72	9.55	9.38	9.20	9.02	0.02
6	0.10	3.78	3.46	3.29	3.18	3.11	3.05	3.01	2.98	2.96	2.94	2.90	2.87	2.84	2.80	2.76	2.72	0.20
	0.05	5.99	5.14	4.76	4.53	4.39	4.28	4.21	4.15	4.10	4.06	4.00	3.94	3.87	3.81	3.74	3.67	0.10
	0.01	13.7	10.9	9.78	9.15	8.75	8.47	8.26	8.10	7.98	7.87	7.72	7.56	7.40	7.23	7.06	6.88	0.02
7	0.10	3.59	3.26	3.07	2.96	2.88	2.83	2.78	2.75	2.72	2.70	2.67	2.63	2.59	2.56	2.51	2.47	0.20
	0.05	5.59	4.74	4.35	4.12	3.97	3.87	3.79	3.73	3.68	3.64	3.57	3.51	3.44	3.38	3.30	3.23	0.10
	0.01	12.2	9.55	8.45	7.85	7.46	7.19	6.99	6.84	6.72	6.62	6.47	6.31	6.16	5.99	5.82	5.65	0.02
8	0.10	3.46	3.11	2.92	2.81	2.73	2.67	2.62	2.59	2.56	2.54	2.50	2.46	2.42	2.38	2.34	2.29	0.20
	0.05	5.32	4.46	4.07	3.84	3.69	3.58	3.50	3.44	3.39	3.35	3.28	3.22	3.15	3.08	3.01	2.93	0.10
	0.01	11.3	8.65	7.59	7.01	6.63	6.37	6.18	6.03	5.91	5.81	5.67	5.52	5.36	5.20	5.03	4.86	0.02

d.f. in numerator

df	p																	p
9	0.10	3.36	3.01	2.81	2.69	2.61	2.55	2.51	2.47	2.44	2.42	2.38	2.34	2.30	2.25	2.21	2.16	0.20
	0.05	5.12	4.26	3.86	3.63	3.48	3.37	3.29	3.23	3.18	3.14	3.07	3.01	2.94	2.86	2.79	2.71	0.10
	0.01	10.6	8.02	6.99	6.42	6.06	5.80	5.61	5.47	5.35	5.26	5.11	4.96	4.81	4.65	4.48	4.31	0.02
10	0.10	3.28	2.92	2.73	2.61	2.52	2.46	2.41	2.38	2.35	2.32	2.28	2.24	2.20	2.16	2.11	2.06	0.20
	0.05	4.96	4.10	3.71	3.48	3.33	3.22	3.14	3.07	3.02	2.98	2.91	2.84	2.77	2.70	2.62	2.54	0.10
	0.01	10.0	7.56	6.55	5.99	5.64	5.39	5.20	5.06	4.94	4.85	4.71	4.56	4.41	4.25	4.08	3.91	0.02
12	0.10	3.18	2.81	2.61	2.48	2.39	2.33	2.28	2.24	2.21	2.19	2.15	2.10	2.06	2.01	1.96	1.90	0.20
	0.05	4.75	3.89	3.49	3.26	3.11	3.00	2.91	2.85	2.80	2.75	2.69	2.62	2.54	2.47	2.38	2.30	0.10
	0.01	9.33	6.93	5.95	5.41	5.06	4.82	4.64	4.50	4.39	4.30	4.16	4.01	3.86	3.70	3.54	3.36	0.02
15	0.10	3.07	2.70	2.49	2.36	2.27	2.21	2.16	2.12	2.09	2.06	2.02	1.97	1.92	1.87	1.82	1.76	0.20
	0.05	4.54	3.68	3.29	3.06	2.90	2.79	2.71	2.64	2.59	2.54	2.48	2.40	2.33	2.25	2.16	2.07	0.10
	0.01	8.68	6.36	5.42	4.89	4.56	4.32	4.14	4.00	3.89	3.80	3.67	3.52	3.37	3.21	3.05	2.87	0.02
20	0.10	2.97	2.59	2.38	2.25	2.16	2.09	2.04	2.00	1.96	1.94	1.89	1.84	1.79	1.74	1.68	1.61	0.20
	0.05	4.35	3.49	3.10	2.87	2.71	2.60	2.51	2.45	2.39	2.35	2.28	2.20	2.12	2.04	1.95	1.84	0.10
	0.01	8.10	5.85	4.94	4.43	4.10	3.87	3.70	3.56	3.46	3.37	3.23	3.09	2.94	2.78	2.61	2.42	0.02
30	0.10	2.88	2.49	2.28	2.14	2.05	1.98	1.93	1.88	1.85	1.82	1.77	1.72	1.67	1.61	1.54	1.46	0.20
	0.05	4.17	3.32	2.92	2.69	2.53	2.42	2.33	2.27	2.21	2.16	2.09	2.01	1.93	1.84	1.74	1.62	0.10
	0.01	7.56	5.39	4.51	4.02	3.70	3.47	3.30	3.17	3.07	2.98	2.84	2.70	2.55	2.39	2.21	2.01	0.02
60	0.10	2.79	2.39	2.18	2.04	1.95	1.87	1.82	1.77	1.74	1.71	1.66	1.60	1.54	1.48	1.40	1.29	0.20
	0.05	4.00	3.15	2.76	2.53	2.37	2.25	2.17	2.10	2.04	1.99	1.92	1.84	1.75	1.65	1.53	1.39	0.10
	0.01	7.08	4.98	4.13	3.65	3.34	3.12	2.95	2.82	2.72	2.63	2.50	2.35	2.20	2.03	1.84	1.60	0.02
120	0.10	2.75	2.35	2.13	1.99	1.90	1.82	1.77	1.72	1.68	1.65	1.60	1.54	1.48	1.41	1.32	1.19	0.20
	0.05	3.92	3.07	2.68	2.45	2.29	2.18	2.09	2.02	1.96	1.91	1.83	1.75	1.66	1.55	1.43	1.25	0.10
	0.01	6.85	4.79	3.95	3.48	3.17	2.96	2.79	2.66	2.56	2.47	2.34	2.19	2.03	1.86	1.66	1.38	0.02
∞	0.10	2.71	2.30	2.08	1.94	1.85	1.77	1.72	1.67	1.63	1.60	1.55	1.49	1.42	1.34	1.24	1.00	0.20
	0.05	3.84	3.00	2.60	2.37	2.21	2.10	2.01	1.94	1.88	1.83	1.75	1.67	1.57	1.46	1.32	1.00	0.10
	0.01	6.63	4.61	3.78	3.32	3.02	2.80	2.64	2.51	2.41	2.32	2.18	2.04	1.88	1.70	1.47	1.00	0.02

Appendix 2 Critical 5% values for Cochran's test for homogeneity of variance. $C = (\text{largest } s^2)/(\sum s_j^2)$.

d.f. for s_j^2	$k =$ number of variances										
	2	3	4	5	6	7	8	9	10	15	20
1	0.9985	0.9669	0.9065	0.8412	0.7808	0.7271	0.6798	0.6385	0.6020	0.4709	0.3894
2	0.9750	0.8709	0.7679	0.6838	0.6161	0.5612	0.5157	0.4775	0.4450	0.3346	0.2705
3	0.9392	0.7977	0.6841	0.5981	0.5321	0.4800	0.4377	0.4027	0.3733	0.2758	0.2205
4	0.9057	0.7457	0.6287	0.5441	0.4803	0.4307	0.3910	0.3584	0.3311	0.2419	0.1921
5	0.8772	0.7071	0.5895	0.5065	0.4447	0.3974	0.3595	0.3286	0.3029	0.2195	0.1735
6	0.8534	0.6771	0.5598	0.4783	0.4184	0.3726	0.3362	0.3067	0.2823	0.2034	0.1602
7	0.8332	0.6530	0.5365	0.4564	0.3980	0.3535	0.3185	0.2901	0.2666	0.1911	0.1501
8	0.8159	0.6333	0.5175	0.4387	0.3817	0.3384	0.3043	0.2768	0.2541	0.1815	0.1422
9	0.8010	0.6167	0.5017	0.4241	0.3682	0.3259	0.2926	0.2659	0.2439	0.1736	0.1357
10	0.7880	0.6025	0.4884	0.4118	0.3568	0.3154	0.2829	0.2568	0.2353	0.1671	0.1303

Glossary

Adapted from Watt *et al.* (1995), Ricklefs & Miller (2000), and Krebs (2001).

Abiotic factors Characterized by the absence of life; include temperature, humidity, pH, and other physical and chemical influences. (Cf. *biotic factors.*)

Adaptation A genetically determined characteristic that enhances the ability of an individual to cope with its environment; an evolutionary process by which organisms become better suited to their environments.

Adaptive radiation The evolution of ecological and phenotypic diversity within a rapidly multiplying lineage. It is the differentiation of a single ancestor into an array of species that inhabit a variety of environments and that differ in the morphological, physiological, and behavioral traits used to exploit those environments.

Age class The individuals in a population of a particular age.

Age structure The relative proportions of a population in different age classes.

Aggregation Organisms show an aggregated spatial distribution when they co-occur significantly more than would be expected from a (completely random) Poisson distribution. This clumping is reflected in a variance mean ratio significantly greater than unity. *Macroparasites* are usually aggregated in their host population, the majority of hosts harboring a few or no parasites and a few hosts harboring large parasite burdens. Aggregated distributions are often well described empirically by the negative binomial distribution; the degree of aggregation is inversely proportional to the negative binomial parameter, k.

Allele One of a pair of characters that are alternative to each other in inheritance, being governed by genes situated at the same locus on homologous chromosomes.

Allochthonous Originating outside a system, such as minerals and organic matter transported from marine to terrestrial habitats or from land into streams and lakes. (Cf. *autochthonous.*)

Allopatric Occurring in different places; usually referring to geographic separation of populations. (Cf. *sympatric.*)

Alpha diversity The mean variety of organisms occurring in a particular place or habitat; often called local diversity.

Ambient Referring to conditions of the *abiotic* environment surrounding the organism.

Antibody A protein produced in the blood of vertebrates in response to an *antigen*. The antibody produced is able to bind specifically to that antigen, and plays a role in its inactivation or removal by the immune system.

Antigen A substance, generally foreign, capable of inducing *antibody* formation.

Apparent competition A situation in which two or more species negatively affect one another indirectly through their interaction with a common predator.

Assimilation efficiency A percentage expressing the proportion of ingested energy that is absorbed into the bloodstream.

Autecology Study of the individual in relation to environmental conditions.

Autochthonous Originating within a system, such as organic matter produced, and minerals cycled, within streams and lakes. (Cf. *allochthonous*.)

Autotroph Organism that obtains energy from the sun and materials from inorganic sources. Most plants are autotrophs. (Cf. *heterotroph*.)

Basal metabolic rate (BMR) The energy expenditure of an organism that is at rest, fasting, and in a thermally neutral environment.

Beta diversity The reciprocal of the mean number of habitats or localities occupied by a species. Also defined as the proportional difference in species composition between habitats, that is, the turnover of species among habitats.

Biodiversity The variety of types of organisms, habitats, and ecosystems on earth or in a particular place.

Biomass Weight of living material, usually expressed as a dry weight, in all or part of an organism, population, or community. Commonly presented as weight per unit of area, or biomass density.

Biome A major category of ecological communities (e.g. tundra biome).

Biosphere The whole-earth ecosystem, also called the *ecosphere*. Divided into *biomes*.

Biota All species living in a defined area.

Biotic factors Environmental influences caused by plants or animals. (Cf. *abiotic factors*.)

Birth rate (b_x) The average number of offspring produced per individual per unit time, often expressed as a function of age (x).

Browsers Organisms that consume parts of woody plants. (Cf. *grazers*.)

By-catch The incidental capture of prey by predators or humans whose efforts are dependent on, or focus on, other more abundant prey.

Carnivore Flesh eater; organism that eats other animals. (Cf. *herbivore*.)

Carrying capacity The number of individuals in a population that the resources of a habitat can support; the asymptote, or plateau, of the logistic and other sigmoid equations for population growth.

Chromosomes Rod-like structures in eukaryotic cells on which genes reside.

Climax The endpoint of a successional sequence, or sere; a community that has reached a steady state under a particular set of environmental conditions.

Cline A gradual change in population characteristics or adaptations over a geographic area.

Coexistence Occurrence of two or more species in the same habitat; usually applied to potentially competing species.

Cohort A group of individuals of the same age recruited into a population at the same time.

Cohort life table See *dynamic life table*.

Community An association of interacting populations, usually defined by the nature of their interaction or by the place in which they live.

Competition Occurs when a number of organisms of the same or different species utilize common resources that are in short supply (exploitation competition); if the resources are not in short supply, competition occurs when the organisms seeking that resource harm one another in the process (interference competition).

Competitive exclusion principle The hypothesis that two or more species cannot coexist on a single resource that is scarce relative to the demand for it. Also called *Gause's principle.*

Consumer–resource interactions Interactions in which individuals of one species consume individuals of another. Consumer–resource interactions affect the consumer positively and the resource negatively.

Contact rate The average frequency per unit time with which infected individuals contact, or otherwise put themselves in a position to transmit an *infection* to, *susceptible* individuals.

Contagious disease See *infectious disease.*

Death rate (d_x) The percentage of newborns dying during a specified interval, often expressed as a function of age (x). (Cf. *mortality.*)

Deme A local population within which mating occurs among individuals more or less at random.

Demographic stochasticity Random variation in birth and death rates in a population.

Demography The study of population structure and growth.

Density Number of individuals in relation to the space, volume, or other resources that they need.

Density dependent Having an influence on individuals in a population that varies with the density of that population. Often applied to birth and death rates.

Density independent Having an influence on individuals in a population that does not vary with the density of that population.

Depensatory See *inverse density dependent.*

Deterministic Having an outcome that is not subject to stochastic (random) variation.

Deterministic model Mathematical model in which all the relationships are fixed and the concept of probability does not enter; a given input produces an exact prediction as output. (Cf. *stochastic model.*)

Detritus Freshly dead or partially decomposed organic matter.

Direct interaction An interaction between organisms that involves direct physical contact between the interactors (e.g. predation and herbivory). (Cf. *indirect interaction.*)

Disease The debilitating effects of *infection* by a *parasite*; sometimes incorrectly used to refer to the disease-causing parasite. It is possible for a host to be infected by a parasite but to show no symptoms of disease.

Dispersal Movement of organisms away from their place of birth or from centers of population density.

Dispersion The spatial pattern of distribution of individuals within populations.

Distribution The geographic extent of a population or other ecological unit.

Diversity The number of species in a local area (*alpha diversity*) or region (*gamma diversity*). Also, a measure of the variety of species in a community that takes into account the relative abundance of each species.

Dominance condition Communities or vegetational strata in which one or more species, by means of their number, coverage, or size, have considerable influence upon or control of the conditions of existence of associated species.

Dominants The few species that attain high abundances in a community.

Dynamic life table The age-specific survival and fecundity of a cohort followed from birth to the death of the last individual. Also called a *cohort life table*.

Ecological efficiency The percentage of energy in the biomass produced by one trophic level that is incorporated into the biomass produced by the next higher trophic level.

Ecological longevity Average length of life of individuals of a population under stated conditions.

Ecosphere See *biosphere*.

Ecosystem Biotic community and its abiotic environment. Can be at various scales, but at the larger scale is synonymous with a *landscape*.

Ecosystem diversity The variety of different ecosystems.

Ecosystems ecology The study of ecosystems, particularly the interactions of the whole biota and their environments.

Ecotone Transition zone between two diverse communities (e.g. the tundra–boreal forest ecotone).

Effective population size (N_e) The average size of a population expressed in terms of the number of individuals assumed to contribute genes equally to the next generation; generally smaller than the actual size of the population, depending on the variation in reproductive success among individuals.

Emergent property A feature of a system not deducible from lower-order processes.

Endemic (i) In biodiversity, a species whose range is confined to a defined area. (Cf. *exotic*, *indigenous*). (ii) In epidemiology, a *parasite* whose *prevalence* does not exhibit wide fluctuations through time in a defined host, host species, or host population.

Environment All the biotic and abiotic factors that actually affect an individual organism at any point in its life cycle.

Environmental stochasticity Random variation in the abiotic environment.

Epidemic A sudden, rapid spread or increase of a disease-causing *parasite* through a human population. An epidemic is often the result of a change in circumstances that favor pathogen transmission such as a rapid increase of host population density, or the introduction of a new parasite (or genetic strain of a parasite) to a previously unexposed host population.

Epizootic The sudden spread of a disease-causing *parasite* through a non-human population; equivalent to an *epidemic* in human populations.

Equilibrium isocline A line on a population graph designating combinations of competing populations, or predator and prey populations, for which the growth rate of one of the populations is zero.

Eruption A sudden increase in a species population in a defined area. (Cf. *irruption*.)

Eutrophic Rich in the mineral nutrients required by green plants; usually applied to an aquatic habitat with high productivity. (Cf. *oligotrophic*.)

Evapotranspiration Sum total of water lost from the land by evaporation and plant transpiration.

Exotic A species found outside its normal habitats. (Cf. *endemic*, *indigenous*.)

Experiment A test of a hypothesis, either observational or manipulative. The experimental method is the scientific method.

Exponential rate of increase (r) The natural log of the *finite rate of increase*. Also called the *instantaneous rate of increase*.

Extinction The disappearance of a species or other taxon from a region or biota.

Facilitation Enhancement of conditions for a population of one species by the activities of another, particularly during early succession.

Fecundity The potential reproductive capacity of an organism, measured by the number of gametes produced.

Fertility An ecological concept of the actual number of viable offspring produced by an organism, equivalent to realized fecundity.

Finite rate of increase (λ) The ratio of the population density in one year to that in the previous year (N_t/N_{t-1}). (Cf. *exponential rate of increase, intrinsic rate of increase*.)

Fitness The genetic contribution by an individual's descendants to future generations of a population relative to that of other individuals.

Food chain The sequence of energy or nutrient transfer through the trophic levels, beginning with plants and ending with the largest carnivores.

Food web A representation of the various paths of energy flow through populations in a community; the complex of *food chains*.

Fragility Referring to a habitat with narrow tolerance to disturbance. (Cf. *resilience, robust*.)

Functional response A change in the rate of exploitation of prey by an individual predator as a result of a change in prey density. (Cf. *numerical response*.)

Fundamental niche See *niche*.

Gamma diversity The average diversity per habitat (*alpha*) × 1/the mean number of habitats occupied by a species (*beta*) × the total number of habitats.

Gause's principle See *competitive exclusion principle*.

Gene A unit of genetic inheritance. (i) In biochemistry, the part of the DNA molecule that encodes a single enzyme or structural protein. (ii) In evolutionary ecology, that which segregates independently.

Generation time The average age at which a female gives birth to her first offspring, or the average time for a population to increase by a factor equal to the net reproductive rate.

Genotype Entire genetic constitution of an organism. (Cf. *phenotype*.)

Global stability Ability to withstand perturbations of a large magnitude and not be affected. (Cf. *local stability*.)

Grazers (i) Organisms that eat grasses or non-woody herbs. (ii) Organisms that feed on many other individuals but usually do not kill them. (Cf. *browsers*.)

Gross production Production before respiration losses are subtracted; photosynthetic production for plants and metabolizable production for animals.

Habitat The place where an animal or plant normally lives, often characterized by a dominant plant form or physical characteristic (e.g. soil habitat, forest habitat).

Harvesting Removing animals or plants from a population, usually by humans.

Helminths Members of the five classes of parasitic worms: Monogenea, Digenea, Cestodes, Nematodes, and Acanthocephalans.

Herbivore Organism that eats plants. (Cf. *carnivore*.)

Herbivory A consumer–resource interaction involving the consumption of plants or plant parts.

Heterosis A situation in which the heterozygous genotype is more fit than either homozygote.

Heterotroph Organism that obtains energy and materials by eating other organisms. (Cf. *autotroph*.)

Heterozygous Having two different alleles of a gene, one derived from each parent. (Cf. *homozygous*.)

Homeothermic Pertaining to warm-blooded animals that regulate their body temperature. (Cf. *poikilothermic*.)

Homozygous Having two identical alleles of a gene. (Cf. *heterozygous*.)

Hypothesis Universal proposition that suggests an explanation for some observed ecological situation.

Hysteresis In ecosystems, the time lag in the effect of a disturbance differs between the increase in disturbance and the decrease in disturbance.

Immunity (Sometimes confusingly termed "resistance," or more correctly, **specific resistance**.) The ability to combat *infection* or *disease* due to the presence of *antibodies* or activated cells. Essentially it can be divided into three types: (i) **acquired immunity** is conferred on an individual after recovery from a disease; (ii) **natural or innate immunity** is inherited from parents, or in some cases antibodies may be passed across the placenta and therefore are present in the blood at birth; (iii) **artificial immunity** may be induced by the injection of either a vaccine, denatured *antigens* of a *parasite* (which induces production of antibodies and thus gives active artificial immunity), or antiserum which contains antibodies and thus may be used when the host is already infected. As well as strengthening the host's resistance, this also confers passive artificial immunity against any subsequent infection.

Inbreeding Mating among related individuals.

Inbreeding depression Reduction in fitness caused by inbreeding.

Inclusive fitness The sum of fitness of an individual and the fitnesses of its relatives, the latter weighted according to the coefficient of relationship.

Indigenous A species that occurs in its normal habitats but is not necessarily confined to that area. (Cf. *endemic*, *exotic*.)

Indirect interaction An interaction between two individuals that involves one or more intermediary species. (Cf. *direct interaction*.)

Infection The presence of a *parasite* within a host, where it may not cause *disease*.

Infectious disease *Disease* caused by *infection* with a *parasite* that can be transmitted from one individual to another, either directly (e.g. measles) or indirectly, by a *vector* (e.g. malaria). *Contagious disease* is a specific subset of infectious disease and pertains exclusively to those diseases transmitted directly between hosts through close bodily contact (this includes aerosols).

Infectious period Period during which the infected individual is able to transmit an *infection* to a *susceptible* host or *vector*. The infectious period may or may not coincide with the disease.

Instantaneous rate of increase (r) The rate of increase of a population undergoing exponential growth under a given set of ecological conditions; can be positive, negative, or zero, and if birth and death rates are constant for sufficient time will produce a *stable age distribution*. In the *logistic equation*, $r = r_{max}(1 - N/K)$. (Cf. *exponential rate of increase*, *intrinsic rate of increase*.)

Intensity Either (i) the mean number of *parasites* within infected members of the host population, or (ii) the mean parasite burden of the entire population. It is important to distinguish between these two usages, because unless *prevalence* is 100%, the latter will be smaller than the former.

Interaction See *direct interaction, indirect interaction.*

Intermediate host A host organism which acquires an infectious agent and in which an obligatory period of development or multiplication/replication occurs before the agent becomes infectious to another host. For example, tabanid flies are vectors (only) of surra (*Trypanosoma evansi*). Their habit of interrupted feeding allows the direct mechanical transmission of the trypanosome from one host to another on the flies' mouthparts. Tabanids are both vectors and intermediate hosts of the nematode (*Pelecitus roemeri*) that act in a similar fashion to mosquitos playing the role of intermediate hosts and vectors of malaria. Snails are the intermediate hosts of schistosomes but they are not vectors; the infectious stage of the parasite escapes from the snail and finds and penetrates the definitive host.

Interspecific competition Competition between members of different species.

Intrinsic rate of increase (r_{max}) The rate of increase of a population undergoing exponential growth under optimum ecological conditions; the maximum instantaneous rate that a species is capable of. It is a characteristic of a species. (Cf. *exponential rate of increase, instantaneous rate of increase.*)

Inverse density dependent A rate (births or deaths) that decreases proportionately as population size or density increases. Also called *depensatory.*

Irruption A sudden expansion of the range of a species that may or may not be accompanied by an increase in population. (Cf. *eruption.*)

Isocline On a population graph, a line designating combinations of competing populations, or predator and prey populations, for which the growth rate of one of the populations is zero.

Keystone species A species whose functional role in a community is disproportionately greater than that predicted by its abundance; usually whose removal has strong effects on community diversity and composition. Keystone species are often top predators.

Landscape A large-scale *ecosystem.*

Leslie matrix A matrix of values of age-specific fecundity and survivorship used to project the size and age structure of a population through time; a population matrix.

Life history The set of adaptations of an organism that influence the life-table values of age-specific survival and fecundity, such as reproductive rate or age at first reproduction.

Life table A summary by age of the survivorship and fecundity of individuals in a population.

Limitation A process that determines the size of the equilibrium population.

Limiting factor Any factor that causes population *limitation.*

Limiting resource A resource that is scarce relative to the demand for it.

Local stability Ability to withstand perturbations of a small magnitude and not be affected. (Cf. *fragility, global stability, robust.*)

Logistic equation Model of *population growth rate* (dN/dt) described by the two constants r_{max} and K, the *carrying capacity.* Thus, $dN/dt = r_{max}N(1 - N/K)$. It produces a symmetrical S-shaped curve with K the upper asymptote.

Log-normal distribution Frequency distribution of species abundances in which the *x*-axis is expressed as a logarithmic scale; *x*-axis is (log) number of individuals represented in the sample, *y*-axis is number of species.

Macroparasites *Parasites* which in general do not multiply within their definitive hosts but instead produce transmission stages (eggs and larvae) that pass into the external environment or to *vectors* (i.e. the parasitic *helminths* and arthropods). The immune response elicited by these metazoans generally depends on the number of parasites present in a given host, and tends to be of a relatively transient nature.

Maximum sustained yield (MSY) The greatest sustainable rate at which individuals may be harvested from a population without causing a decline to extinction; that is, the harvest at which recruitment equals harvesting.

Mesic Referring to habitats with plentiful rainfall and well-drained soils. (Cf. *xeric*.)

Metapopulation A set of local populations linked together through dispersal.

Microparasites *Parasites* that undergo direct multiplication within their definitive hosts (e.g. viruses, rickettsia, bacteria, fungi, and protozoa). Microparasites are characterized by small size, short generation times, and a tendency to induce *immunity* to reinfection in those hosts that survive the primary *infection*. Duration of infection is usually short in relation to the expected life span of the host (there are, however, important exceptions, e.g. the slow viruses).

Mineral nutrients Elements such as nitrogen, phosphorus, sulfur, potassium, calcium, and others that are necessary for the growth and development of plants.

Minimum viable population (MVP) The smallest population that can persist for some arbitrarily long time, usually 1000 years.

Mortality (m_x) Ratio of the number of deaths to individuals at risk, often described as a function of age (x). (Cf. *death rate*.)

Multiple states Communities can exist in different combinations of species abundances under the same environmental conditions. These multiple states within an ecosystem are detected when a perturbation radically alters the abundance of many species in the community, which then do not return to their original abundance when the perturbation is removed.

Mutation Any change in the genotype of an organism occurring at the gene, chromosome, or genome level; usually applied to changes in genes to new allelic forms.

Mutualism A relationship between two species that benefits both.

Natal dispersal Dispersal of young animals from their place of birth.

Natural selection Change in the frequency of genetic traits in a population through differential survival and reproduction of individuals bearing those traits.

Net production Production after respiration losses are subtracted.

Net reproductive rate (R_e) The expected number of offspring produced by a female during her lifetime.

Niche The set of resources and environmental conditions that allow a single species to persist in a particular region, often conceived as a multidimensional space. Also called *fundamental niche*. (Cf. *realized niche*.)

Niche breadth The variety of resources utilized and range of conditions tolerated by an individual, population, or species. (Cf. *niche width*.)

Niche complementarity A situation in which species that overlap extensively in their use of one resource differ substantially in their use of another resource.

Niche overlap The sharing of niche space by two or more species; similarity of resource requirements and tolerance of ecological conditions.

Niche width The standard deviation of the distribution of resource use. (Cf. *niche breadth*.)

Numerical response A change in the population size of a predator species as a result of a change in the density of its prey. (Cf. *functional response*.)

Nutrient Any substance required by organisms for normal growth and maintenance.

Nutrient cycle The path of an element through the ecosystem, including its assimilation by organisms and its regeneration in a reusable inorganic form.

Oligotrophic Poor in the mineral nutrients required by green plants; applied to an aquatic habitat with low productivity. (Cf. *eutrophic*.)

Omnivore An organism whose diet is broad, including both plant and animal foods; specifically, an organism that feeds at more than one trophic level.

Pandemic A widely distributed *epidemic*.

Panmixia Random mating within a population, a panmictic population.

Panzootic A widely distributed epizootic, often affecting more than one host species.

Parameter In statistics, an unknown true characteristic of a statistical population. It is usually impossible to know the value of a parameter. A statistic estimates a parameter.

Parasite An organism exhibiting a varying but obligatory dependence on another organism, its host, which is detrimental to the survival and/or *fecundity* of that host. (See also *macroparasites, microparasites*.)

Parasitoid Any of a number of insect species whose larvae live within and consume their host, usually another insect.

Perennial Referring to an organism that lives for more than one year; lasting throughout the year.

Phenology Study of the periodic (seasonal) phenomena of animal and plant life and their relations to the weather and climate (e.g. the time of flowering in plants).

Phenotype Expression of the characteristics of an organism as determined by the interaction of its genetic constitution and the environment. (Cf. *genotype*.)

Phylogeny The pattern of evolutionary relationships among species or other taxonomic groups.

Pleiotropy The influence of one gene on the expression of more than one trait in the phenotype.

Poikilothermic Referring to cold-blooded animals, organisms that have no rapidly operating heat-regulatory mechanism. (Cf. *homeothermic*.)

Polygenic Determined by the expression of more than one gene.

Polymorphism The occurrence together in the same habitat of two or more discontinuous forms of a species in such proportions that the rarest of them cannot merely be maintained by recurrent mutation or immigration.

Population Group of individuals of a single species in a defined area.

Population growth rate (dN/dt) The rate of growth of a population over a short period of time, defined by the product of population size, N, and the *instantaneous rate of increase*, r. (See *logistic equation*.)

Population viability analysis (PVA) The strategic analysis of the ecological, economic, and political issues and challenges related to the conservation of an endangered species, community, or ecosystem.

Prevalence The proportion of the host population with *infection* or *disease*, often expressed as a percentage. A measure of how widespread is the infection or disease.

Primary production Production by green plants. (Cf. *secondary production*.)

Production Accumulation of energy or biomass.

Productivity The rate at which energy is accumulated.

Realized niche The set of resources and environmental conditions constrained by competition or predation that allow a single species to persist in a particular region. A subset of the *fundamental niche*.

Recruitment Increment to a natural population, usually from young animals or plants entering the adult population.

Regulation Occurs when a population experiences density-dependent mortality or birth rates.

Rescue effect Prevention of the extinction of a local population by immigration of individuals from elsewhere, often from a more productive habitat.

Resilience (i) The rate at which a population returns to equilibrium after a disturbance. (ii) The ability to withstand disturbance. (Cf. *fragility*, *robust*.)

Resource A substance or object required by an organism for normal maintenance, growth, and reproduction. (See also *limiting resource*.)

Respiration Complex series of chemical reactions in all organisms by which energy is made available for use; carbon dioxide, water, and energy are the end products.

Robust Referring to a habitat with relatively wide tolerance to disturbance. (See *fragility*, *resilience*.)

Secondary plant compounds Chemical products of plant metabolism specifically for the purpose of defense against herbivores and disease organisms.

Secondary production Production by herbivores, carnivores, or detritus feeders. (Cf. *primary production*.)

Senescence Process of aging.

Sere A series of stages of community change in a particular area leading towards a stable state, or climax.

Sex ratio Ratio of the number of individuals of one sex to that of the other sex in a population.

Shared predation A type of *apparent competition* in which two species fall victim to a single predator and may compete for enemy-free space in which to avoid the predator.

Sink An ecosystem, habitat, population, or community that receives input of materials or individual organisms. The population is not self-sustaining. (Cf. *source*.)

Source An ecosystem, habitat, population, or community from which materials or organisms move. The population is self-sustaining. (Cf. *sink*.)

Source–sink metapopulation A metapopulation in which some local populations (*sources*) have a positive growth rate at low densities and others (*sinks*) have a negative growth rate in the absence of immigration.

Stability Absence of fluctuations in populations; ability to withstand perturbations without large changes in composition. (See also *resilience*.)

Stable age distribution The proportions of individuals in various age classes in a population that has a constant instantaneous rate of growth, r.

Stable equilibrium The state to which a system returns if displaced by an outside force.

Stage-classified population A population containing individuals of different developmental states (e.g. adults and larvae) in the same or different habitats.

Static life table The age-specific survival and fecundity of individuals of different ages within a population at a given time; also called a *time-specific life table*. It is a cross-section of all cohorts in the population at a given time.

Stochastic model Mathematical model based on probabilities; the prediction of the model is not a single fixed number but a range of possible numbers. (Cf. *deterministic model*.)

Succession Replacement of one kind of community by another kind; the progressive changes in vegetation and animal life that may culminate in the climax community.

Survival (l_x) Proportion of newborn individuals alive at age x; also called survivorship.

Survivorship curve Curve showing the number of individuals surviving to age x (log scale) plotted against age.

Susceptible Accessible to or liable to infection by a particular *parasite*.

Susceptible individual Either naive (previously uninfected) or having lost *immunity*.

Switching A change in diet to favor food items of increasing suitability or abundance.

Sympatric Occurring in the same place; usually refers to areas of overlap in species distributions. (Cf. *allopatric*.)

Synecology Study of groups of organisms in relation to their environment; includes population, community, and ecosystem ecology.

Time-specific life table See *static life table*.

Total response of predator The product of the functional and numerical responses plotted as per capita mortality of prey against prey density.

Transmission The process by which a *parasite* passes from a source of *infection* to a new host. There are two major types: **horizontal** and **vertical** transmission. The majority of transmission processes operate horizontally, for example by direct contact between infected and *susceptible* individuals or between disease *vectors* and susceptibles. There are six main methods of horizontal transmission: (i) ingestion of contaminated food or drink; (ii) inhalation of contaminated air droplets; (iii) direct contact; (iv) injection into a tissue via an animal's saliva or bite; (v) invasion via open wounds; and (vi) penetration of the host by active parasite transmission stages (e.g. schistosome miracidia or cercariae). Vertical transmission occurs when a parent conveys an infection to its unborn offspring, as occurs in HIV in humans or in many arboviruses.

Transmission threshold Level of transmission below which an *infection* is unable to maintain itself within the host population (or populations, in the case of indirectly transmitted infections).

Trophic level Position in the food chain, determined by the number of energy transfer steps to that level. The first trophic level includes green plants; the second tropic level includes herbivores, and so on.

Type 1 error The rejection of a true null hypothesis.

Type 2 error The acceptance of a false null hypothesis.

Unstable equilibrium The state of a system at which forces are precisely balanced, but away from which the system moves when displaced.

Variable A characteristic or measure of the natural world that may take on any of a number of different values.

Vector An organism that acquires, transports, and delivers an infectious agent to a host. Sometimes there is development or multiplication/replication of the infectious agent in a vector but this is not obligatory. (See also *intermediate host*).

Watershed (i) In North America, the drainage basin of a stream or river. (ii) In Europe and elsewhere, the line demarcating different drainage basins.

Wildlife Wild animals, usually terrestrial vertebrates whose populations are monitored and managed for exploitation or conservation.

Xeric Referring to habitats where water is in short supply to plants. (Cf. *mesic*.)

Zoonosis A *parasite* occurring naturally in animals and naturally transmitted between animals and humans.

References

Abaturov, B.D. & Magomedov, M.-R.D. 1988. Food value and dynamics of food resources as a factor characterizing the state of populations of herbivorous mammals. *Zoologicheskii Zhurnal* 76:223–234.

Abrams, P.A. 1993. Why predation rate should not be proportional to predator density. *Ecology* 74:726–733.

Abramsky, Z. 1981. Habitat relationships and competition in two Mediterranean *Apodemus* spp. *Oikos* 36:219–225.

Abramsky, Z. & Sellah, C. 1982. Competition and the role of habitat selection in *Gerbillus allenbyi* and *Meriones tristrami*: a removal experiment. *Ecology* 63:1242–1247.

Abramsky, Z., Bowers, M.A. & Rosenzweig, M.L. 1986. Detecting interspecific competition in the field: testing the regression method. *Oikos* 47:199–204.

Abramsky, Z., Dyer, M.I. & Harrison, P.D. 1979. Competition among small mammals in experimentally perturbed areas of the shortgrass prairie. *Ecology* 60:530–536.

Akaike, H. 1973. Information theory as an extension of the maximum likelihood principle. In: Petrov, B.N. & Csaki, F., eds. *Second International Symposium on Information Theory*, pp. 267–281. Akademiai Kiado, Budapest.

Akçakaya, H.R. & Atwood, J.L. 1997. A habitat-based metapopulation model of the California gnatcatcher. *Conservation Biology* 11:422–434.

Alexander, R.D. 1974. The evolution of social behavior. *Annual Review of Ecology and Systematics* 5:325–383.

Allee, W.C. 1938. *The Social Life of Animals*. W.W. Norton & Co. Inc., New York.

Allen, R.B., Bellingham, P.J. & Wiser, S.K. 1999. Immediate damage by an earthquake to a temperate montane forest. *Ecology* 80:708–714.

Anderson, A.E., Bowden, D.C. & Medin, D.E. 1990. Indexing the annual fat cycle in a mule deer population. *Journal of Wildlife Management* 54:550–556.

Anderson, A.E., Medin, D.E. & Bowden, D.C. 1972. Indices of carcass fat in a Colorado mule deer population. *Journal of Wildlife Management* 36:579–594.

Anderson, A.E., Medin, D.E. & Ochs, D.P. 1969. Relationships of carcass fat indices in 18 wintering mule deer. *Proceedings of the Western Association of State Game and Fish Commissioners* 49:329–340.

Anderson, D.R. 1975. Optimal exploitation strategies for an animal population in a Markovian environment: a theory and an example. *Ecology* 56:1281–1297.

Anderson, D.R., Burnham, K.P. & Thompson, W.L. 2000. Null hypothesis testing: problems, prevalence, and an alternative. *Journal of Wildlife Management* 64:912–923.

Anderson, D.R., Burnham K.P., White, G.C. & Otis, D.L. 1983. Density estimation of small-mammal populations using a trapping web and distance sampling methods. *Ecology* 64:674–680.

Anderson, E.W. & Sherzinger, R.J. 1975. Improving quality of winter forage for elk by cattle grazing. *Journal of Range Management* 28:120–125.

Anderson, R.C. 1972. The ecological relationships of meningeal worm and native cervids in North America. *Journal of Wildlife Diseases* 8:304–310.

Anderson, R.M. 1991. Populations and infectious diseases: ecology or epidemiology? *Journal of Animal Ecology* 60:1–50.

Anderson, R.M. & May, R.M. 1979. Population biology of infectious diseases: Part I. *Nature* 280:361–367.

Anderson, R.M. & May, R.M. 1986. The invasion, persistence and spread of infectious diseases within animal and plant communities. *Philosophical Transactions of the Royal Society of London B, Biological Sciences* 314:533–570.

Anderson, S.H. 2003. The relative importance of birds and insects as pollinators of the New Zealand flora. *New Zealand Journal of Ecology* 27:83–94.

Andow, D.A., Karieva, P.M., Levin, S.A. & Okubo, A. 1990. Spread of invading organisms. *Landscape Ecology* 4:177–188.

Andren, H. 1992. Corvid density and nest predation in relation to forest fragmentation: a landscape perspective. *Ecology* 73:794–804.

Arcese, P. & Sinclair, A.R.E. 1997. The role of protected areas as ecological baselines. *Journal of Wildlife Management* 61:587–602.

Arcese, P., Smith, J.N.M., Hochachka, W.M., Rogers, C.M. & Ludwig, D. 1992. Stability, regulation, and the determination of abundance in an insular song sparrow population. *Ecology* 73:805–822.

Arditi, R. & Dacorogna, B. 1988. Optimal foraging on arbitrary food distributions and the definition of habitat patches. *American Naturalist* 131:837–846.

Arlettaz, R., Perrin, N. & Hausser, J. 1997. Trophic resource partitioning and competition between the two sibling bat species *Myotis myotis* and *Myotis blythii*. *Journal of Animal Ecology* 66:897–911.

Arneberg, P., Folstad, I. & Karter, A.J. 1996. Gastrointestinal nematodes depress food intake in naturally infected reindeer. *Parasitology* 112:213–219.

Arnold, G.W., Steven, D.E. & Weeldenburg, J.R. 1993. Influences of remnant size, spacing pattern and connectivity on population boundaries and demography in euros (*Macropus robustus*) living in a fragmented landscape. *Biological Conservation* 64:219–230.

Arnold, T.W. 1994. A roadside transect for censusing breeding coots and grebes. *Wildlife Society Bulletin* 22:437–443.

Arthur, S.M., Paragi, T.F. & Krohn, W.B. 1993. Dispersal of juvenile fishers in Maine. *Journal of Wildlife Management* 57:868–874.

Atkinson, I.A.E. 2001. Introduced mammals and models for restoration. *Biological Conservation* 99:81–86.

Augustine, D.J., Frelich, L.E. & Jordan, P.A. 1998. Evidence for two alternate stable states in an ungulate grazing system. *Ecological Applications* 8:1260–1269.

Austin, M.P. 1985. Continuum concept, ordination methods, and niche theory. *Annual Review of Ecology and Systematics* 16:39–61.

Bailey, J.A. 1968. A weight–length relationship for evaluating physical condition of cottontails. *Journal of Wildlife Management* 32:835–841.

Bailey, N.T.J. 1951. On estimating the size of mobile populations from recapture data. *Biometrika* 38:293–306.

Bailey, N.T.J. 1952. Improvements in the interpretation of recapture data. *Journal of Animal Ecology* 21:120–127.

Baker, C.S., Perry, A., Bannister, J.L., *et al.* 1993. Abundant mitochondrial variation and world-wide population structure in humpback whales. *Proceedings of the National Academy of Sciences of the USA* 90:8239–8243.

Baker, J.R. 1938. The evolution of breeding seasons. In: de Beer, G.R., ed. *Evolution: Essays on Aspects of Evolutionary Biology Presented to Professor E.S. Goodrich on his Seventieth Birthday*, pp. 161–177. Clarendon Press, Oxford.

Ballard, W.B., Whitman, J.S. & Gardner, C.L. 1987. Ecology of an exploited wolf population in south-central Alaska. *Wildlife Monographs* 98:1–54.

Bamford, J. 1970. Estimating fat reserves in the brush-tailed possum, *Trichosurus vulpecula* Kerr (Marsupialia: Phalangeridae). *Australian Journal of Zoology* 18:415–425.

Barker, S.C., Singleton, G.R. & Spratt, D.M. 1991. Can the nematode *Capillaria hepatica* regulate abundance of wild house mice? Results of enclosure experiments in southeastern Australia. *Parasitology* 103:439–449.

Barr, W. 1991. *Back from the Brink: The Road to Muskox Conservation in the Northwest Territories.* Komatik Series no. 3. The Arctic Institute of North America, University of Calgary.

Basson, M., Beddington, J.R. & May, R.M. 1991. An assessment of the maximum sustainable yield of ivory from African elephant populations. *Mathematical Biosciences* 104:73–95.

Baum, J.K., Myers, R.A., Kehler, D.G., Worm, B., Harley, S.J. & Doherty, P.A. 2003. Collapse and conservation of shark populations in the northwest Atlantic. *Science* 299:389–392.

Bayliss, M., Mellor, P.S. & Meiswinkel, R. 1999. Horse sickness and ENSO in South Africa. *Nature* 397:574.

Bayliss, P. 1987. Kangaroo dynamics. In: Caughley, G., Shepherd, N. & Short, J., eds. *Kangaroos: Their Ecology and Management in the Sheep Rangelands of Australia*, pp. 119–134. Cambridge University Press, Cambridge.

Bazely, D.R. & Jefferies, R.L. 1989. Lesser snow geese and the nitrogen economy of a grazed salt marsh. *Journal of Ecology* 77:24–34.

Beasom, S.L. & Roberson, S.F., eds. 1985. *Game Harvest Management.* Caesar Kleberg Wildlife Research Institute, Kingsville, TX.

Beddington, J.R. 1975. Mutual interference between parasites or predators and its effect on searching efficiency. *Journal of Animal Ecology* 44:331–340.

Beier, P. 1996. Metapopulation models, tenacious tracking, cougar conservation. In: McCullough, D., ed. *Metapopulations and Wildlife Conservation*, pp. 293–323. Island Press, Washington, DC.

Beisner, B.E., Haydon, D.T. & Cuddington, K. 2003. Alternative stable states in ecology. *Frontiers in Ecology and Environment* 1:376–382.

Bell, D.J., Oliver, W.L.R. & Goose, R.K. 1990. The hispid hare *Caprolagus hispidus*. In: Chapman, J.A. & Flux, J.E.C., eds. *Rabbits, Hares and Pikas: Status Survey and Conservation Action Plan*, pp. 128–136. IUCN, Gland, Switzerland.

Bell, R.H.V. 1970. The use of the herb layer by grazing ungulates in the Serengeti. In: Watson, A., ed. *Animal Populations in Relation to their Food Resources*, pp. 111–124. 10th Symposium of the British Ecological Society. Blackwell, Oxford.

Bell, R.H.V. 1971. A grazing ecosystem in the Serengeti. *Scientific American* 225:86–93.

Bell, R.H.V. 1981. An outline of a management plan for Kasungu National Park, Malawi. In: Jewell, P.A., Holt, S. & Hart, D., eds. *Problems in Management of Locally Abundant Wild Mammals*, pp. 69–89. Academic Press, New York.

Bell, R.H.V. 1982. The effect of soil nutrient availability on community structure in African ecosystems. In: Huntly, B.J. & Walker, B.H., eds. *Ecology of Tropical Savannas*, pp. 193–216. Springer-Verlag, New York.

Belovsky, G.E. 1978. Diet optimization in a generalist herbivore: the moose. *Theoretical Population Biology* 14:105–134.

Belovsky, G.E. 1986. Generalist herbivore foraging and its role in competitive interactions. *American Zoologist* 26:51–69.

Belovsky, G.E. 1988. Food plant selection by a generalist herbivore: the moose. *Ecology* 62:1020–1030.

Belovsky, G.E., Ritchie, M.E. & Moorehead, J. 1989. Foraging in complex environments: when prey availability varies over time and space. *Theoretical Population Biology* 36:144–160.

Belsky, A.J. 1987. The effects of grazing: confounding of ecosystem, community and organism scales. *American Naturalist* 129:777–783.

Bender, E.A., Case, T.J. & Gilpin, M.E. 1984. Perturbation experiments in community ecology: theory and practice. *Ecology* 65:1–13.

Bennett, A.F., Henein, K. & Merriam, G. 1994. Corridor use and the elements of corridor quality: chipmunks and fencecrows in a farmland mosaic. *Biological Conservation* 68:155–165.

Bent, A.C. 1932. *Life Histories of North American Gallinaceous Birds*. US National Museum Bulletin no. 162. Smithsonian Institution, Washington, DC.

Berger, L., Speare, R., Daszak, P., *et al.* 1998. Chytridiomycosis causes amphibian mortality associated with population declines in the rain forests of Australia and Central America. *Proceedings of the National Academy of Sciences of the USA* 95:9031–9036.

Bergerud, A.T. & Page, R.E. 1987. Displacement and dispersion of parturient caribou at calving as antipredator tactics. *Canadian Journal of Zoology* 65:1597–1606.

Bergman, C.M., Fryxell, J.M., Gates, C.C. & Fortin, D. 2001. Ungulate foraging strategies: energy maximizing or time minimizing? *Journal of Animal Ecology* 70:289–300.

Biggins, D.E., Vargas, A., Godbey, J.L. & Anderson, S.H. 1999. Influence of prerelease experience on reintroduced black-footed ferrets (*Mustela nigripes*). *Biological Conservation* 89:121–129.

Birch, L.C. 1957. The meanings of competition. *American Naturalist* 91:5–18.

Björnhag, G. 1987. Comparative aspects of digestion in the hindgut of mammals. The colonic separation mechanism (CSM) (a review). *Deutsche Tieraerztliche Wochenschrift* 94:33–36.

Bjørnstad, O.N. & Grenfell, B.T. 2001. Noisy clockwork: time series analysis of population fluctuations in animals. *Science* 293:638–643.

Blank, T.H., Southwood, T.R.E. & Cross, D.J. 1967. The ecology of the partridge: I. Outline of the population processes with particular reference to chick mortality and nest density. *Journal of Animal Ecology* 36:549–556.

Bloomer, J.P. & Bester, M.N. 1992. Control of feral cats on sub-Antarctic Marion Island, Indian Ocean. *Biological Conservation* 60:211–219.

Blower, J.G., Cook, L.M. & Bishop, J.A. 1981. *Estimating the Size of Animal Populations*. George Allen and Unwin, London.

Bobek, B. 1977. Summer food as the factor limiting roe deer population size. *Nature* 268:47–49.

Bock, C.E., Smith, H.M. & Bock, J.H. 1990. The effect of livestock grazing upon abundance of the lizard, *Sceloporus scalaris*, in southeastern Arizona. *Journal of Herpetology* 24:445–446.

Boertje, R.D., Valkenburg, P. & MacNay, M.E. 1996. Increases in moose, caribou, and wolves following wolf control in Alaska. *Journal of Wildlife Management* 60:474–489.

Bomford, M. 1988. Effect of wild ducks on rice production. In: Norton, G.A. & Pech, R.P., eds. *Vertebrate Pest Management in Australia: A Decision Analysis/Systems Analysis Approach*, pp. 53–57. Project Report no. 5. CSIRO, Australia.

Bomford, M. 1990. *A Role for Fertility Control in Wildlife Management*. Bulletin no. 7. Bureau of Rural Resources, Canberra.

Bomford, M. & O'Brien, P.H. 1990. Sonic deterrents in animal damage control: a review of device tests and effectiveness. *Wildlife Society Bulletin* 18:411–422.

Bond, W.J., Lee, W.G. & Craine, J.M. 2004. Plant structural defences against browsing birds: a legacy of New Zealand's extinct moas. *Oikos* 104:500–508.

Boonstra, R., Krebs, C.J. & Beacham, T.D. 1980. Impact of botfly parasitism on *Microtus townsendii* populations. *Canadian Journal of Zoology* 58:1683–1692.

Boscovik, R., Kovacs, K.M., Hammill, M.O. & White, B.N. 1996. Geographic distribution of mitochondrial DNA haplotypes in grey seals (*Halichoereus grypus*). *Canadian Journal of Zoology* 74:1787–1796.

Bothma, J. du P., ed. 1989. *Game Ranch Management*. J.L. van Schaik, Pretoria.

Botkin, D.B., Mellilo, J.M. & Wu, L.S.-Y. 1981. How ecosystem processes are linked to large mammal population dynamics. In: Fowler, C.W. & Smith, T.D., eds. *Dynamics of Large Mammal Populations*, pp. 373–387. John Wiley & Sons, New York.

Boutin, S. 1992. Predation and moose population dynamics: a critique. *Journal of Wildlife Management* 56:116–127.

Boutin, S., Krebs, C.J., Sinclair, A.R.E. & Smith, J.N.M. 1986. Proximate causes of losses in a snowshoe hare population. *Canadian Journal of Zoology* 64:606–610.

Bowden, D.C. & Kufeld, R.C. 1995. Generalized mark–sight population size estimation applied to Colorado moose. *Journal of Wildlife Management* 59:840–851.

Bowen, B.H. & Avise, J.C. 1996. Conservation genetics of marine turtles. In: Avise, J.C. & Hamrick, J.L., eds. *Conservation Genetics: Case Histories from Nature*, pp. 190–237. Chapman & Hall, New York.

Boyce, M. 1992. Population viability analysis. *Annual Reviews of Ecology and Systematics* 23:481–506.

Boyce, M. 1998. Ecological-process management and ungulates: Yellowstone's conservation paradigm. *Wildlife Society Bulletin* 26:381–398.

Boyce, M.S. 1989. *The Jackson Elk Herd: Intensive Wildlife Management in North America.* Cambridge University Press, Cambridge.

Boyce, M.S. & Haney, A. 1997. *Ecosystem Management.* Yale University Press, New Haven, CT.

Boyce, M.S. & McDonald, L.L. 1999. Relating populations to habitats using resource selection functions. *Trends in Ecology and Evolution* 14:268–272.

Boyce, M.S. & Payne, N.F. 1997. Applied disequilibriums: Riparian habitat management for wildlife. In: Boyce, M.S. & Haney, A., eds. *Ecosystem Management*, pp. 133–146. Yale University Press, New Haven, CT.

Boyd, D.K. & Pletscher, D.H. 1999. Characteristics of dispersal in a colonizing wolf population in the central Rocky Mountains. *Journal of Wildlife Management* 63:1094–1108.

Bredon, R.M., Harker, K.W. & Marshall, B. 1963. The nutritive value of grasses grown in Uganda when fed to Zebu cattle: I. The relation between the percentage of crude protein and other nutrients. *Journal of Agricultural Science (Cambridge)* 61:101–104.

Briggs, S.V. 1991. Effects of breeding and environment on body condition of maned ducks *Chenonetta jubata. Wildlife Research* 18:577–588.

Brightwell, L.R. 1951. Some experiments with the common hermit crab (*Eupagurus bernhardus*) Linn., and transparent univalve shells. *Proceedings of the Zoological Society of London* 121:279–283.

Bromley, R.G., Heard, D.C. & Croft, B. 1995. Visibility bias in aerial surveys relating to nest success of Arctic geese. *Journal of Wildlife Management* 59:364–371.

Brook, B.W., O'Grady, J.J., Chapman, A.P., Burgman, M.A., Akçakaya, H.R. & Frankham, R. 2000. Predictive accuracy of population viability analysis in conservation biology. *Nature* 404:385–387.

Brooks, T.M., Mittermeier, R.A., Mittermeier, C.G., *et al.* 2002. Habitat loss and extinction in the hotspots of biodiversity. *Conservation Biology* 16:900–923.

Brower, L.P. 1984. Chemical defense in butterflies. In: Vane-Wright, R.I. & Ackery, P.R., eds. *The Biology of Butterflies*, pp. 109–134. Academic Press, London.

Brown, J.H. 1995. *Macroecology.* The University of Chicago Press, Chicago, IL.

Brown, J.H. & Heske, E.J. 1990. Control of a desert–grassland transition by a keystone rodent guild. *Science* 250:1705–1707.

Brown, J.L. 1964. The evolution of diversity in avian territorial systems. *Wilson Bulletin* 76:160–169.

Brown, J.L. 1969. The buffer effect and productivity in tit populations. *American Naturalist* 103:347–354.

Brown, J.S. 1988. Patch use as an indicator of habitat preference, predation risk, and competition. *Behavioral Ecology and Sociobiology* 22:37–47.

Brown, J.S., Kotler, B.P. & Mitchell, W.A. 1994. Foraging theory, patch use, and the structure of a Negev desert granivore community. *Ecology* 75:2286–2300.

Brown, K.S., Jr. 1987. Conclusions, synthesis, and alternative hypotheses. In: Whitmore, T.C. & Prance, G.T., eds. *Biogeography and Quaternary History in Tropical America*, pp. 175–191. Oxford University Press, New York.

Brown, R.D., Hellgren, E.C., Abbott, M., Ruthven, D.C. & Bingham, R.L. 1995. Effects of dietary energy and protein restriction on nutritional indices of female white-tailed deer. *Journal of Wildlife Management* 59:595–609.

Brown, R.G., Bowen, W.D., Eddington, J.D., *et al.* 1997. Evidence for a long-lasting single administration contraceptive vaccine in wild grey seals. *Journal of Reproductive Immunology* 35:43–51.

Brown, R.G., Kimmins, W.C., Mezei, M., Parsons, J.L. & Pohajdak, B. 1996. Birth control for grey seals. *Nature* 379:30–31.

Bryant, D.M. 1989. Determination of respiration rates of free-living animals by the double-labelling technique. In: Grubb, P.J. & Whittaker, J.B., eds. *Toward a More Exact Ecology*, pp. 95–109. 30th Symposium of the British Ecological Society. Blackwell Scientific Publications, Oxford.

Bryant, J.P. 1981. Phytochemical deterrence of snowshoe hare browsing by adventitious shoots of four Alaskan trees. *Science* 213:889–890.

Bryant, J.P. & Kuropat, P.J. 1980. Selection of winter forage by subarctic browsing vertebrates: the role of plant chemistry. *Annual Review of Ecology and Systematics* 11:261–285.

Buckland, S.T., Anderson, D.R., Burnham, K.P. & Laake, J.L. 1993. *Distance Sampling: Estimating Abundance of Biological Populations*. Chapman & Hall, London.

Buckland, S.T., Anderson, D.R., Burnham, K.P., Laake, J.L., Borchers, D.L. & Thomas, L. 2001. *Introduction to Distance Sampling: Estimating Abundance of Biological Populations*. Oxford University Press, New York.

Buckner, C.H. & Turnock, W.J. 1965. Avian predation on the larch sawfly, *Pristiphora erichsonii* (Htg.) (Hymenoptera: Tenthredinidae). *Ecology* 46:223–236.

Bunnell, F.L. 1995. Forest-dwelling vertebrate faunas and natural fire regimes in British Columbia: patterns and implications for conservation. *Conservation Biology* 9:636–644.

Burnham, K.P. & Anderson, D.R. 1998. *Model Selection and Inference*. Springer-Verlag, New York.

Burnham, K.P. & Overton, W.S. 1978. Estimation of the size of a closed population when capture probabilities vary among animals. *Biometrika* 65:625–633.

Burnham, K.P., Anderson, D.R. & Laake, J.L. 1980. Estimation of density from line transect sampling of biological populations. *Wildlife Monographs* 72:1–202.

Byrom, A.E. 2002. Dispersal and survival of juvenile feral ferrets *Mustela furo* in New Zealand. *Journal of Applied Ecology* 39:67–78.

Byrom, A.E. & Krebs, C.J. 1999. Natal dispersal of juvenile arctic ground squirrels in the boreal forest. *Canadian Journal of Zoology* 77:1048–1059.

Cade, T.J. & Jones, C.G. 1993. Progress in restoration of the Mauritius kestrel. *Conservation Biology* 7:169–175.

Cade, T.J., Lincer, J.L. & White, C.M. 1971. DDE residues and eggshell changes in Alaskan falcons and hawks. *Science* 172:955–957.

Calaby, J.H. & Grigg, C.C. 1989. Changes in macropodoid communities and populations in the past 200 years, and the future. In: Grigg, G., Jarman, P. & Hume, I., eds. *Kangaroos, Wallabies and Rat-kangaroos*, pp. 813–820. Surrey Beatty & Sons, Chipping Norton, New South Wales.

Cann, R.L. & Douglas, L.J. 1999. Parasites and conservation of Hawaiian birds. In: Landweber, L.F. & Dobson, A.P., eds. *Genetics and the Extinction of Species*, pp. 121–161. Princeton University Press, Princeton, NJ.

Carpenter, J.W., Clark, G.G. & Watts, D.M. 1989. The impact of eastern equine encephalitis virus on efforts to recover the endangered whooping crane. In: Cooper, J.E., ed. *Disease and Threatened Birds*, pp. 115–120. Technical Publication no. 10. International Council for Bird Preservation, Cambridge.

Carr, S.M., Ballinger, S.W., Derr, J.N., Blankenship, L.H. & Bickham, J.W. 1986. Mitochondrial DNA analysis of hybridization between sympatric white-tailed deer and mule deer in west Texas, USA. *Proceedings of the National Academy of Sciences of the USA* 83:9576–9580.

Carson, R. 1962. *Silent Spring*. Hamish Hamilton, London.

Case, T.J. 2000. *An Illustrated Guide to Theoretical Ecology*. Oxford University Press, New York.

Casey, J.M. & Myers, R.A. 1998. Near extinction of a large widely distributed fish. *Science* 281:690–692.

Caswell, H. 1978. A general formula for the sensitivity of population growth rate to changes in life history parameters. *Theoretical Population Biology* 14:215–230.

Caswell, H. 2001. *Matrix Population Models*. Sinauer Associates, Sunderland, MA.

Catchpole, E.A., Morgan, B.J.T., Coulson, T.N., Freeman, S.N. & Albon, S.D. 2000. Factors influencing Soay sheep survival. *Applied Statistics* 49:453–472.

Caughley, G. 1970. Population statistics of chamois. *Mammalia* 34:194–199.

Caughley, G. 1974. Bias in aerial survey. *Journal of Wildlife Management* 38:921–933.

Caughley, G. 1976. Wildlife management and the dynamics of ungulate populations. In: Coaker, T.H., ed. *Applied Biology*, Vol. 1, pp. 183–246. Academic Press, New York.

Caughley, G. 1977a. Sampling in aerial survey. *Journal of Wildlife Management* 41:605–615.

Caughley, G. 1977b. *Analysis of Vertebrate Populations*. John Wiley & Sons, London.

Caughley, G. 1981. Overpopulation. In: Jewell, P.A., Holt, S. & Hart, D., eds. *Problems in Management of Locally Abundant Wild Mammals*, pp. 7–19. Academic Press, New York.

Caughley, G. 1983. *The Deer Wars: The Story of Deer in New Zealand*. Heinemann, Auckland.

Caughley, G. 1985. Harvesting of wildlife: past, present and future. In: Beasom, S.L. & Roberson, S.F., eds. *Game Harvest Management*, pp. 3–14. Caesar Kleberg Wildlife Research Institute, Kingsville, TX.

Caughley, G. 1987. Ecological relationships. In: Caughley, G., Shepherd, N. & Short, J., eds. *Kangaroos: Their Ecology and Management in the Sheep Rangelands of Australia*, pp. 159–187. Cambridge University Press, Cambridge.

Caughley, G. 1994. Directions in conservation biology. *Journal of Animal Ecology* 63:215–244.

Caughley, G. & Goddard, J. 1972. Improving the estimates from inaccurate censuses. *Journal of Wildlife Management* 36:135–140.

Caughley, G. & Grice, D. 1982. A correction factor for counting emus from the air, and its application to counts in Western Australia. *Australian Wildlife Research* 9:253–259.

Caughley, G. & Krebs, C.J. 1983. Are big mammals simply little mammals writ large. *Oecologia* 59:7–17.

Caughley, G., Grice, D., Barker, R. & Brown, R. 1988. The edge of the range. *Journal of Animal Ecology* 57:771–785.

Caughley, G., Pech, R. & Grice, D. 1992. Effect of fertility control on a population's productivity. *Wildlife Research* 19:623–627.

Channell, R. & Lomolino, M.V. 2000. Trajectories to extinction: spatial dynamics of the contraction of geographical ranges. *Journal of Biogeography* 27:169–179.

Chapman, D.G. 1951. Some properties of the hypergeometric distribution with application to zoological sample censuses. *University of California Publications in Statistics* 1:131–160.

Charnov, E.L. 1976a. Optimal foraging: attack strategy of a mantid. *American Naturalist* 110:141–151.

Charnov, E.L. 1976b. Optimal foraging: the marginal value theorem. *Theoretical Population Biology* 9:129–136.

Chase, A. 1987. *Playing God in Yellowstone*. Harcourt Brace Jovanovich, San Diego, CA.

Chave, J., Muller-Landau, H.C. & Levin, S.A. 2002. Comparing classical community models: theoretical consequences for patterns of diversity. *American Naturalist* 159:1–23.

Cheatum, E.L. 1949. Bone marrow as an index of malnutrition in deer. *New York State Conservationist* 3(5):19–22.

Chepko-Sade, B.D., Shields, W.M., Berger, J., *et al.* 1987. The effects of dispersal and social structure on effective population size. In: Chepko-Sade, B.D. & Halpin, Z.T., eds. *Mammalian Dispersal Patterns: The Effects of Social Structure on Population Genetics*, pp. 287–321. The University of Chicago Press, Chicago, IL.

Cherfas, J. 1988. *The Hunting of the Whale*. The Bodley Head, London.

Chilcott, M.J. & Hume, I.D. 1985. Coprophagy and selective retention of fluid digesta: their role in the nutrition of the common ringtail possum, *Pseudocheirus peregrinus*. *Australian Journal of Zoology* 33:1–15.

Child, G. & von Richter, W. 1969. Observations on ecology and behaviour of lechwe, puku, and waterbuck along the Chobe River, Botswana. *Zeitschrift für Säugetierkunde* 34:275–295.

Chitty, D. 1967. The natural selection of self-regulatory behavior in animal populations. *Proceedings of the Ecological Society of Australia* 2:51–78.

Chivers, D.J. & Langer, P. 1994. *The Digestive System in Mammals: Food, Form and Function*. Cambridge University Press, Cambridge.

Choquenot, D. 1991. Density-dependent growth, body condition, and demography in feral donkeys: testing the food hypothesis. *Ecology* 72:805–813.

Choquenot, D., Hone, J. & Saunders, G. 1999. Using aspects of predator–prey theory to evaluate helicopter shooting for feral pig control. *Wildlife Research* 26:251–261.

Clark, C.W. 1976. *Mathematical Bioeconomics: The Optimal Management of Renewable Resources*. John Wiley & Sons, New York.

Clark, C.W. 1990. *Mathematical Bioeconomics: The Optimal Management of Renewable Resources*, 2nd edn. Wiley-Interscience, New York.

Clark, C.W. & Mangel, M. 2000. *Dynamic State Variable Models in Ecology*. Oxford University Press, Oxford.

Clark, W.C., Jones, D.D. & Holling, C.S. 1979. Lessons for ecological policy design: a case study of ecosystem management. *Ecological Modelling* 7:1–53.

Clayton, L., Keeling, M. & Milner-Gulland, E.J. 1997. Bringing home the bacon: a spatial model of wild pig harvesting in Sulawesi, Indonesia. *Ecological Applications* 7:642–652.

Clifton-Hadley, R.S., Sauter-Louis, C.M., Lugton, I.W., Jackson, R., Durr, P.A. & Wilesmith, J.W. 2001. Mycobacterial diseases. In: Williams, E.S. & Barker, I.K., eds. *Infectious Diseases of Wild Mammals*, pp. 340–371. Manson Publishing/The Veterinary Press, London.

Clutton-Brock, T.H. & Harvey, P.H. 1983. The functional significance of variation in body size among mammals. In: Eisenberg, J.F. & Kleiman, D.G., eds. *Advances in the Study of Mammalian Behaviour*, pp. 632–663. Special Publication no. 7. The American Society of Mammalogists, Shippensburg State College, Shippensburg, PA.

Clutton-Brock, T.H., Coulson, T.N., Milner-Gulland, E.J., Thomson, D. & Armstrong, H.M. 2002. Sex differences in emigration and mortality affect optimal management of deer populations. *Nature* 415:633–637.

Clutton-Brock, T.H., Guinness, F.E. & Albon, S.D. 1982. *Red Deer: Behaviour and Ecology of Two Sexes*. The University of Chicago Press, Chicago, IL.

Clutton-Brock, T.H., Illius, A.W., Wilson, K., Grenfell, B.T., MacColl, A.D.C. & Albon, S.D. 1997. Stability and instability in ungulate populations: an empirical analysis. *American Naturalist* 149:195–219.

Clutton-Brock, T.H., Major, M. & Guinness, F.E. 1985. Population regulation in male and female red deer. *Journal of Animal Ecology* 54:831–846.

Clutton-Brock, T.H., Price, O.F., Albon, S.D. & Jewell, P.A. 1991. Persistent instability and population regulation in Soay sheep. *Journal of Animal Ecology* 60:593–608.

Cochran, W.G. 1954. The combination of estimates from different experiments. *Biometrics* 10:101–129.

Cochran, W.G. 1977. *Sampling Techniques*, 3rd edn. John Wiley & Sons, New York.

Cockburn, A., Scott, M.P. & Scotts, D.J. 1985. Inbreeding avoidance and male-biased natal dispersal in *Antechinus* sp. (*Marsupialia: Dasyuridae*). *Animal Behaviour* 33:908–915.

Cohn, J.P. 1986. Surprising cheetah genetics. *BioScience* 36:358–362.

Cohn, J.P. 1991. Ferrets return from near extinction. *BioScience* 41:132–135.

Collar, N.J., Gonsaga, L.P., Krabbe, N., *et al.* 1992. *Threatened Birds of the North Americas.* International Council for Bird Preservation, Cambridge.

Connell, J.H. 1978. Diversity in tropical rainforests and coral reefs. *Science* 199:1302–1310.

Conner, M.L. & Leopold, B.D. 2001. A Euclidean distance metric to index dispersion from radiotelemetry data. *Wildlife Society Bulletin* 29:783–786.

Cooke, A.S. 1973. Shell thinning in avian eggs by environmental pollutants. *Environmental Pollution* 4:85–152.

Cooke, F. 1988. Genetic studies of birds – the goose with blue genes. In: Quellet, H., ed. *Acta 19th Congressus Internationalis Ornithologicus,* pp. 189–214. University of Ottawa Press, Ottawa.

Cooke, H.J. 1980. Landform evolution in the context of climatic change and neo-tectonism in the Middle Kalahari of north-central Botswana. *Transactions of the Institute of British Geography, New Series* 5:80–99.

Cooper, A.B., Hilborn, R. & Unsworth, J.W. 2003. An approach for population assessment in the absence of abundance indices. *Ecological Applications* 13:814–828.

Cooper, J. & Fourie, A. 1991. Improved breeding success of Great-winged Petrels *Pterodroma macroptera* following control of feral cats *Felis catus* at subantarctic Marion Island. *Bird Conservation International* 1:171–175.

Cooper, S.M. & Ginnett, T.F. 2000. Potential effects of supplemental feeding of deer on nest predation. *Wildlife Society Bulletin* 28:660–666.

Cooper, S.M., Owen-Smith, N. & Bryant, J.P. 1988. Foliage acceptability to browsing ruminants in relation to seasonal changes in the leaf chemistry of woody plants in a South African savanna. *Oecologia* 75:336–342.

Coppock, D.L., Ellis, J.E., Detling, J.K. & Dyer, M.I. 1983. Plant–herbivore interactions in a North American mixed-grass prairie: II. Responses of bison to modification of vegetation by prairie dogs. *Oecologia* 56:10–15.

Corfield, T.F. 1973. Elephant mortality in Tsavo National Park, Kenya. *East African Wildlife Journal* 11:339–368.

Cork, S.J. & Warner, A.C.I. 1983. The passage of digesta markers through the gut of a folivorous marsupial, the koala *Phascolarctos cinereus. Journal of Comparative Physiology B* 152:43–51.

Coughenour, M.B. & Singer, F.J. 1996. Elk population processes in Yellowstone National Park under the policy of natural regulation. *Ecological Applications* 6:573–593.

Coulson, T., Catchpole, E.A., Albon, S.D., *et al.* 2001a. Age, sex, density, winter weather, and population crashes in Soay sheep. *Science* 292:1528–1531.

Coulson, T., Mace, G.M., Hudson, E. & Possingham, H. 2001b. The use and abuse of population viability analysis. *Trends in Ecology and Evolution* 16:219–221.

Coulson, T., Guinness, F., Pemberton, J. & Clutton-Brock, T. 2003. The demographic consequences of releasing a population of red deer from culling. *Ecology* 85:411–422.

Courchamp, F., Grenfell, B.T. & Clutton-Brock, T.H. 2000a. Inverse density dependence and the Allee effect. *Trends in Ecology and Evolution* 14:405–410.

Courchamp, F., Langlais, M. & Sugihara, G. 2000b. Rabbits killing birds: modelling the hyperpredation process. *Journal of Animal Ecology* 69:154–164.

Courchamp, F., Langlais, M. & Sugihara, G. 1999. Cats protecting birds: modelling the mesopredator release effect. *Journal of Animal Ecology* 68:282–292.

Courchamp, F., Woodruffe, R. & Roemer, G. 2003. Removing protected populations to save endangered species. *Science* 302:1532.

Cowie, R.J. 1977. Optimal foraging in great tits (*Parus major*). *Nature* 268:137–139.

Cox, D.R. 1969. Some sampling problems in technology. In: Johnson, N.L. & Smith, H., eds. *New Developments in Survey Sampling,* pp. 506–527. John Wiley & Sons, New York.

Cramp, S. 1972. One hundred and fifty years of Mute Swans on the Thames. *Wildfowl* 23:119–124.

Crawford, H.S. & Jennings, D.T. 1989. Predation by birds on spruce budworm *Choristoneura fumiferana*: functional, numerical and total responses. *American Naturalist* 70:152–163.

Crawley, M.J. 1987. Benevolent herbivores? *Trends in Ecology and Evolution* 2:167–168.

Crête, M. & Bédard, J. 1975. Daily browse consumption by moose in the Gaspé Peninsula, Quebec. *Journal of Wildlife Management* 39:368–373.

Crick, Q.P., Dudley, C., Glue, D.E. & Thomson, D.L. 1997. UK birds are laying eggs earlier. *Nature* 388:526.

Cronin, M.A., Vyse, E.R. & Cameron, D.G. 1988. Genetic relationships between mule deer and white-tailed deer in Montana. *Journal of Wildlife Management* 52:320–328.

Crooks, K.R. & Soule, M.E. 1999. Mesopredator release and avifaunal extinctions in a fragmented system. *Nature* 400:563–566.

Crouse, D., Crowder, L. & Caswell, H. 1987. A stage-based population model for loggerhead sea turtles and implications for conservation. *Ecology* 68:1412–1423.

Crowell, K.L. & Pimm, S.L. 1976. Competition and niche shifts of mice introduced onto small islands. *Oikos* 27:251–258.

Dailey, T.V., Thompson Hobbs, N. & Woodard, T.N. 1984. Experimental comparisons of diet selection by mountain goats and mountain sheep in Colorado. *Journal of Wildlife Management* 48:799–806.

Dale, B.W., Adams, L.G. & Bowyer, R.T. 1994. Functional response of wolves preying on barren-ground caribou in a multiple-prey ecosystem. *Journal of Animal Ecology* 63:644–652.

Dark, J., Forger, N.G. & Zucker, I. 1986. Regulation and function of lipid mass during the annual cycle of the golden-mantled ground squirrel. In: Heller, H.C., Musacchia, X.J. & Wang, L.C.H., eds. *Living in the Cold: Physiological and Biochemical Adaptations*, pp. 445–451. Elsevier Scientific, New York.

Dasmann, R.F. 1956. Fluctuations in a deer population in California chaparral. *Transactions of the North American Wildlife Conference* 21:487–499.

Daszak, P., Cunningham, A.A. & Hyatt, A. 2000. Emerging infectious diseases of wildlife: threats to biodiversity and human health. *Science* 287:443–449.

Davidson, N.C. & Evans, P.R. 1988. Prebreeding accumulation of fat and muscle protein by arctic-breeding shorebirds. *Proceedings of the International Ornithological Congress* 19:342–363.

Davies, S.J.J.F. 1976. Environmental variables and the biology of Australian arid zone birds. In: Frith, H.J. & Calaby, J.H., eds. *Proceedings of the 16th International Ornithological Congress, Canberra*, pp. 481–488. Australian Academy of Science, Canberra.

Davis, M.B. 1986. Climatic instability, time lags, and community disequilibrium. In: Diamond, J. & Case, T.J., eds. *Community Ecology*, pp. 269–284. Harper & Row, New York.

Dawson, T.J. & Ellis, B.A. 1979. Comparison of the diets of yellow-footed rock-wallabies and sympatric herbivores in western New South Wales. *Australian Wildlife Research* 6:245–254.

Dawson, T.J. & Hulbert, A.J. 1970. Standard metabolism, body temperature, and surface areas of Australian marsupials. *American Journal of Physiology* 218:1233–1238.

DeAngelis, D.L. & Waterhouse, J.C. 1987. Equilibrium and nonequilibrium concepts in ecological models. *Ecological Monographs* 57:1–21.

DeAngelis, D.L., Goldstein, R.A. & O'Neill, R.V. 1975. A model for trophic interactions. *Ecology* 56:881–892.

Debinski, D.M. & Holt, R.D. 2000. A survey and overview of habitat fragmentation experiments. *Conservation Biology* 14:342–355.

De Kroon, H., Plaisier, A., van Groenendael, J. & Caswell, H. 1986. Elasticity: the relative contribution of demographic parameters to population growth. *Ecology* 67:1427–1431.

DelGiudice, G.D. & Seal, U.S. 1988. Classifying winter undernutrition in deer via serum and urinary urea nitrogen. *Wildlife Society Bulletin* 16:27–32.

DelGiudice, G.D., Asleson, M.A., Varner, L., Hellgren, E.C. & Riggs, M.R. 1996. Creatinine ratios in random sampled and 24-hour urines of white-tailed deer. *Journal of Wildlife Management* 60:381–387.

DelGiudice, G.D., Mech, L.D. & Seal, U.S. 1990. Effects of winter undernutrition on body composition and physiological profiles of white-tailed deer. *Journal of Wildlife Management* 54:539–550.

Dent, C.L., Cumming, G.S. & Carpenter, S.R. 2002. Multiple states in river and lake ecosystems. *Philosophical Transactions of the Royal Society B* 357:635–645.

Depperschmidt, J.D., Torbit, S.C., Alldredge, A.W. & Deblinger, R.D. 1987. Body condition indices for starved pronghorns. *Journal of Wildlife Management* 51:675–678.

de Roos, A., McCauley, E. & Wilson, W. 1991. Mobility versus density limited predator–prey dynamics on different spatial scales. *Proceedings of the Royal Society of London B* 246:117–122.

Derr, J.N. 1991. Genetic interactions between white-tailed and mule deer in the southwestern United States. *Journal of Wildlife Management* 55:228–237.

Detling, J.K. 1998. Mammalian herbivores: ecosystem-level effects in two grassland national parks. *Wildlife Society Bulletin* 26:438–448.

de Vries, M.F.W., Laca, E.A. & Demment, M.W. 1999. The importance of scale of patchiness for selectivity in grazing herbivores. *Oecologia* 121:355–363.

Diamond, J. & Case, T.J. 1986. Overview: introductions, extinctions, exterminations and invasions. In: Diamond, J. & Case, T.J., eds. *Community Ecology*, pp. 65–79. Harper & Row, New York.

Diamond, J.M. 1969. Avifaunal equilibria and species turnover rates on the Channel Islands of California. *Proceedings of the National Academy of Sciences of the USA* 64:57–63.

Diamond, J.M. 1972. Biogeographic kinetics: estimation of relaxation times for avifaunas of southwest Pacific islands. *Proceedings of the National Academy of Sciences of the USA* 69:3199–3202.

Diamond, J.M. 1975. Assembly of species communities. In: Cody, M.L. & Diamond, J.M., eds. *Ecology and Evolution of Communities*, pp. 342–444. Belknap, Cambridge, MA.

Diamond, J.M. 1983. Taxonomy by nucleotides. *Nature* 305:17–18.

Dice, L.R. 1938. Some census methods for mammals. *Journal of Wildlife Management* 2:119–130.

Dill, L.M. 1978. An energy-based model of optimal feeding-territory size. *Theoretical Population Biology* 14:396–429.

Dinius, D.A. & Baumgardt, B.R. 1970. Regulation of food intake in ruminants: 6. Influence of caloric density of pelleted rations. *Journal of Dairy Science* 53:311–316.

Distel, R.A., Laca, E.A., Griggs, T.C. & Demment, M.W. 1995. Patch selection by cattle: maximization of intake rate in horizontally heterogeneous pastures. *Applied Behaviour Science* 45:11–21.

Dixon, A.M., Mace, G.M., Newby, J.E. & Olney, P.J.S. 1991. Planning for the re-introduction of scimitar-horned oryx (*Oryx dammah*) and addax (*Addax nasomaculatus*) into Niger. *Symposium of the Zoological Society of London* 62:201–216.

Doak, D.F. 1989. Spotted owls and old growth logging in the Pacific Northwest. *Conservation Biology* 3:389–396.

Doak, D.F. & Mills, L.S. 1994. A useful role for theory in conservation. *Ecology* 75:615–626.

Dobson, A. & Hudson, P. 1994. The interaction between the parasites and predators of red grouse *Lagopus lagopus scoticus*. *Ibis* 137:S87–S96.

Dobson, A. & Meagher, M. 1996. The population dynamics of brucellosis in the Yellowstone National Park. *Ecology* 77:1026–1036.

Dobson, A.P. & Hudson, P.J. 1986. Parasites, disease and the structure of ecological communities. *Trends in Ecology and Evolution* 1:11–15.

Dobson, A.P. & Hudson, P.J. 1992. Regulation and stability of a free-living host–parasite system: *Trichostrongylus tenuis* in red grouse. II. Population models. *Journal of Animal Ecology* 61:487–498.

Dobson, A.P. & May, R.M. 1986a. Disease and conservation. In: Soulé, M.E., ed. *Conservation Biology: The Science of Scarcity and Diversity*, pp. 345–365. Sinauer Associates, Sunderland, MA.

Dobson, A.P. & May, R.M. 1986b. Patterns of invasions by pathogens and parasites. In: Mooney, H.A. & Drake, J.A., eds. *Ecology of Biological Invasions of North America and Hawaii*, pp. 58–76. Ecological Studies no. 58. Springer-Verlag, New York.

Dobson, A.P., Rodriguez, J.P., Roberts, W.M. & Wilcove, D.S. 1997. Geographic distribution of endangered species in the United States. *Science* 275:550–553.

Dobson, F.S. 1982. Competition for mates and predominant juvenile male dispersal in mammals. *Animal Behaviour* 30:1183–1192.

Dobson, F.S. & Jones, W.T. 1985. Multiple causes of dispersal. *American Naturalist* 126:855–858.

Dodd, M.G. & Murphy, T.M. 1995. Accuracy and precision of techniques for counting great blue heron nests. *Journal of Wildlife Management* 59:667–673.

Donovan, T.M., Thompson, F.R., III, Faaborg, J. & Probst, J.R. 1995. Reproductive success of migratory birds in habitat sources and sinks. *Conservation Biology* 9:1380–1395.

Droege, S. & Sauer, J.R. 1989. North American Breeding Bird Survey annual summary 1988. *United States Fish and Wildlife Service Biological Report* 89:13.

Dublin, H.T., Sinclair, A.R.E. & McGlade, J. 1990. Elephants and fire as causes of multiple stable states in the Serengeti–Mara woodlands. *Journal of Animal Ecology* 59:1147–1164.

DuBowy, P.J. 1988. Waterfowl communities and seasonal environments: temporal variability in interspecific competition. *Ecology* 69:1439–1453.

du Toit, J.T. & Cumming, D.H.M. 1999. Functional significance of ungulate diversity in African savanna and the ecological implications of the spread of pastoralism. *Biodiversity and Conservation* 8:1643–1661.

du Toit, J.T., Rogers, K.H. & Biggs, H.C., eds. 2003. *The Kruger Experience*. Island Press, Washington, DC.

Eason, C.T., Murphy, E.C., Wright, G.R.G. & Spurr, E.B. 2002. Assessment of risks of brodifacoum to non-target birds and mammals in New Zealand. *Ecotoxicology* 11:35–48.

Ebbert, S.E. & Byrd, G.V. 2002. Eradications of invasive species to restore natural biological diversity on Alaska Maritime National Wildlife Refuge. In: Veitch, C.R. & Clout, M.N., eds. *Turning the Tide: The Eradication of Invasive Species*, pp. 102–109. Occasional Paper of the IUCN Species Survival Commission no. 27. IUCN, Gland, Switzerland.

Eberhardt, L.L. 1969. Population estimates from recapture frequencies. *Journal of Wildlife Management* 33:28–39.

Eberhardt, L.L. 1978. Transect methods for population studies. *Journal of Wildlife Management* 42:1–31.

Eberhardt, L.L. 1982. Calibrating an index by using removal data. *Journal of Wildlife Management* 46:734–740.

Eberhardt, L.L. & Simmons, M.A. 1987. Calibrating population indices by double sampling. *Journal of Wildlife Management* 51:665–675.

Eberhardt, L.L. & Thomas, J.M. 1991. Designing environmental field studies. *Ecological Monographs* 6:53–73.

Eberhardt, L.L., Knight, R.R. & Blanchard, B. 1986. Monitoring grizzly bear population trends. *Journal of Wildlife Management* 50:613–618.

Ecroyd, C.E. 1996. The ecology of *Dactylanthus taylorii* and threats to its survival. *New Zealand Journal of Ecology* 20:81–100.

Edwards, W.R. & Eberhardt, L. 1967. Estimating cottontail abundance from livetrapping data. *Journal of Wildlife Management* 31:87–96.

Elliott, G.P. 1996. Mohua and stoats: a population viability analysis. *New Zealand Journal of Zoology* 23:239–247.

Elton, C. 1927. *Animal Ecology*. Macmillan, New York.

Elton, C.S. 1924. Periodic fluctuations in the numbers of animals: their causes and effects. *British Journal of Experimental Biology* 2:119–163.

Elton, C.S. 1958. *The Ecology of Invasions by Animals and Plants*. Methuen, London.

Emlen, J.T., Dejong, M.J., Jaeger, M.J., Moermond, T.C., Rusterholz, K.A. & White, R.P. 1986. Density trends and range boundary constraints of forest birds along a latitudinal gradient. *Auk* 103:791–803.

Erlinge, S. 1975. Feeding habits of the weasel *Mustela nivalis* in relation to prey abundance. *Oikos* 26:378–384.

Erlinge, S., Goransson, G., Hansson, L., *et al.* 1983. Predation as a regulating factor on small rodent populations in southern Sweden. *Oikos* 40:36–52.

Estes, J.A. & Duggins, D.O. 1995. Sea otters and kelp forests in Alaska: generality and variation in a community ecology paradigm. *Ecological Monographs* 65:75–100.

Estes, R.D. 1976. The significance of breeding synchrony in the wildebeest. *East African Wildlife Journal* 14:135–152.

Ewins, P.J. & Miller, M.J.R. 1995. Measurement error in aerial surveys of osprey productivity. *Journal of Wildlife Management* 59:333–338.

Fa, J.E. & Peres, C.A. 2001. Game vertebrate extraction in African and Neotropical forests: an intercontinental comparison. In: Reynolds, J.D., Mace, G.M., Redford, K.H. & Robinson, J.G., eds. *Conservation of Exploited Species*, pp. 203–241. Cambridge University Press, Cambridge.

Fancy, S.G. & White, R.G. 1985. Energy expenditures by caribou while cratering in snow. *Journal of Wildlife Management* 49:987–993.

Farentinos, R.C., Capretta, P.J., Kepner, R.E. & Littlefield, R.E. 1981. Selective herbivory in tassle-eared squirrels: role of monoterpenes in ponderosa pines chosen as feeding trees. *Science* 213:1273–1275.

Farley, S.D. & Robbins, C.T. 1994. Development of two methods to predict body composition of bears. *Canadian Journal of Zoology* 72:220–226.

Fedak, M.A. & Seeherman, H.J. 1979. Reappraisal of energetics of locomotion shows identical cost in bipeds and quadrupeds including ostrich and horse. *Nature* 282:713–716.

Feeny, P.P. & Bostock, H. 1968. Seasonal changes in the tannin content of oak leaves. *Phytochemistry* 7:871–880.

Fenner, F. 1983. The Florey lecture, 1983: Biological control, as exemplified by smallpox eradication and myxomatosis. *Proceedings of the Royal Society of London B, Biological Sciences* 218:259–286.

Fenner, F. & Fantini, B. 1999. *Biological Control of Vertebrate Pests: The History of Myxomatosis, an Experiment in Evolution*. CABI Publishing, Wallingford.

Ferrar, A.A. & Walker, B.H. 1974. An analysis of herbivore/habitat relationships in Kyle National Park, Rhodesia. *Journal of the Southern African Wildlife Management Association* 4:137–147.

Fieberg, J. & Ellner, S.P. 2000. When is it meaningful to estimate an extinction probability? *Ecology* 81:2040–2047.

Fielder, P.C. 1986. Implications of selenium levels in Washington Mountain goats, mule deer, and Rocky Mountain goats. *Northwest Science* 60:15–20.

Finch, V.A. 1972. Energy exchanges with the environment of two East African antelopes, the eland and the hartebeest. In: Maloiy, G.M.O., ed. *Comparative Physiology of Desert Animals*, pp. 315–326. Symposia of the Zoological Society of London no. 31. Academic Press, London.

Findlay, C.S. & Cooke, F. 1982. Synchrony in the lesser snow goose (*Anser caerulescens*): II. The adaptive value of reproductive synchrony. *Evolution* 36:786–799.

Finger, S.E., Brisbin, I.L., Jr., Smith, M.H. & Urbston, D.F. 1981. Kidney fat as a predictor of body condition in white-tailed deer. *Journal of Wildlife Management* 45:964–968.

Fisher, R.A. 1930. *The Genetical Theory of Natural Selection*. Clarendon Press, Oxford.

Flueck, W.T. 1994. Effect of trace elements on population dynamics: selenium deficiency in free-ranging black-tailed deer. *Ecology* 75:807–812.

Focardi, S., Marcellini, P. & Montanaro, P. 1996. Do ungulates exhibit a food density threshold? A field study of optimal foraging and movement patterns. *Journal of Animal Ecology* 65:606–620.

Fogden, M.P.L. & Fogden, P.M. 1979. The role of fat and protein reserves in the annual cycle of the grey-backed camaroptera in Uganda (Aves: Sylvidae). *Journal of Zoology (London)* 189:233–258.

Foley, W.J. & Cork, S.J. 1992. Use of fibrous diets by small herbivores: how far can the rules be 'bent'? *Trends in Ecology and Evolution* 7:159–172.

Foley, W.J. & Hume, I.D. 1987. Passage of digesta markers in two species of arboreal folivorous marsupials – the great glider (*Petauroides volans*) and the brushtail possum (*Trichosurus vulpecula*). *Physiological Zoology* 60:103–113.

Foltz, D.W. & Schwagmeyer, P.L. 1989. Sperm competition in the thirteen-lined ground squirrel: differential fertilization success under field conditions. *American Naturalist* 133:257–265.

Ford, E.B. 1940. Polymorphism and taxonomy. In: Huxley, J., ed. *The New Systematics*, pp. 493–513. Clarendon Press, Oxford.

Forsyth, D.M. & Hickling, G.J. 1998. Increasing Himalayan tahr and decreasing chamois densities in the eastern Southern Alps, New Zealand: evidence for interspecific competition. *Oecologia* 113:377–382.

Forsyth, D.M., Hone, J., Parkes, J.P., Reid, G.H. & Stronge, D. 2003. Feral goat control in Egmont National Park, New Zealand. *Wildlife Research* 30:437–450.

Fortin, D. & Arnold, G.W. 1997. The influence of road verges on the use of nearby small shrubland remnants by birds in the central wheatbelt of Western Australia. *Wildlife Research* 24:679–689.

Forys, E.A. & Humphrey, S.R. 1997. Comparison of two methods to estimate density of an endangered lagomorph. *Journal of Wildlife Management* 61:86–92.

Fowler, C.W. 1981. Density dependence as related to life history strategy. *Ecology* 62:602–610.

Fowler, C.W. 1987. A review of density dependence in populations of large mammals. In: Genoways, H.H., ed. *Current Mammalogy*, pp. 401–441. Plenum Press, New York.

Fox, G.A. & Gurevitch, J. 2000. Population numbers count: tools for near-term demographic analysis. *American Naturalist* 156:242–256.

Fraker, M.A., Brown, R.G., Gaunt, G.E., Kerr, J.A. & Pohajdak, B. 2002. Long-lasting, single-dose immunocontraception of feral fallow deer in British Columbia. *Journal of Wildlife Management* 66:1141–1147.

Frank, D.A. 1998. Ungulate regulation of ecosystem processes in Yellowstone National Park: direct and feedback effects. *Wildlife Society Bulletin* 26:410–418.

Frank, D.A. & Evans, R.D. 1997. Effects of native grazers on grassland N cycling in Yellowstone National Park. *Ecology* 78:2238–2248.

Frankel, O.H. & Soulé, M.E. 1981. *Conservation and Evolution.* Cambridge University Press, Cambridge.

Franklin, I.R. 1980. Evolutionary change in small populations. In: Soulé, M.E. & Wilcox, B.A., eds. *Conservation Biology: An Evolutionary-Ecological Perspective*, pp. 135–149. Sinauer Associates, Sunderland, MA.

Frantz, A.C., Schaul, M., Pope, L.C., *et al.* 2004. Estimating population size by genotyping remotely plucked hair: the Eurasian badger. *Journal of Applied Ecology* 41:985–995.

Franzmann, A.W. & LeResche, R.E. 1978. Alaskan moose blood studies with emphasis on condition evaluation. *Journal of Wildlife Management* 42:334–351.

Fraser, D. & Reardon, E. 1980. Attraction of wild ungulates to mineral-rich springs in central Canada. *Holarctic Ecology* 3:36–39.

Fretwell, S.D. 1972. *Populations in a Seasonal Environment.* Princeton University Press, Princeton, NJ.

Fretwell, S.D. & Lucas, H.L. 1970. On territorial behavior and other factors influencing habitat distribution in birds. I. Theoretical development. *Acta Biotheoretica* 19:16–36.

Friesen, L.E., Eagles, P.F. & Mackay, R.J. 1995. Effects of residential development on forest-dwelling neotropical migrant songbirds. *Conservation Biology* 9:1408–1414.

Frost, P.G.H. 1985. The responses of savanna organisms to fire. In: Tothill, J.C. & Mott, J.J., eds. *Ecology and Management of the World's Savannas*, pp. 232–237. Australian Academy of Science, Canberra.

Fryxell, J.M. 1995. Aggregation and migration by grazing ungulates in relation to resources and predators. In: Sinclair, A.R.E. & Arcese, P., eds. *Serengeti II: Dynamics, Management and Conservation of an Ecosystem*, pp. 257–273. The University of Chicago Press, Chicago, IL.

Fryxell, J.M. 1999. Functional responses to resource complexity: an experimental analysis of foraging by beavers. In: Olff, H., Brown, V.K. & Drent, R.H., eds. *Herbivores: Between Plants and Predators*, pp. 371–396. Blackwell Science, Oxford.

Fryxell, J.M. 2001. Habitat suitability and source–sink dynamics of beavers. *Journal of Animal Ecology* 70:310–316.

Fryxell, J.M. & Doucet, C.M. 1993. Diet choice and the functional response of beavers. *Ecology* 74:1297–1306.

Fryxell, J.M. & Lundberg, P. 1993. Optimal patch use and metapopulation dynamics. *Evolutionary Ecology* 7:379–393.

Fryxell, J.M. & Lundberg, P. 1994. Optimal diet choice and predator–prey dynamics. *Evolutionary Ecology* 8:407–421.

Fryxell, J.M. & Lundberg, P. 1997. *Individual Behavior and Community Dynamics*. Chapman & Hall, New York.

Fryxell, J.M. & Sinclair, A.R.E. 1988a. Causes and consequences of migration by large herbivores. *Trends in Ecology and Evolution* 3:237–241.

Fryxell, J.M. & Sinclair, A.R.E. 1988b. Seasonal migration by white-eared kob in relation to resources. *African Journal of Ecology* 26:17–31.

Fryxell, J.M., Falls, J.B., Falls, E.A. & Brooks, R.J. 1998. Long-term dynamics of small-mammal populations in Ontario. *Ecology* 79:213–225.

Fryxell, J.M., Falls, J.B., Falls, E.A., Brooks, R.J., Dix, L. & Strickland, M.A. 1999. Density-dependence, prey-dependence and population dynamics of martens in Ontario. *Ecology* 80:1311–1321.

Fryxell, J.M., Falls, J.B., Falls, E.A., Brooks, R.J., Dix, L. & Strickland, M. 2001. Harvest dynamics of mustelid carnivores in Ontario, Canada. *Wildlife Biology* 7:151–159.

Fryxell, J.M., Greever, J. & Sinclair, A.R.E. 1988a. Why are migratory ungulates so abundant? *American Naturalist* 131:781–798.

Fryxell, J.M., Mercer, W.E. & Gellately, R.B. 1988b. Population dynamics of Newfoundland moose using cohort analysis. *Journal of Wildlife Management* 52:14–21.

Fryxell, J.M., Hussell, D.J.T., Lambert, A.B. & Smith, P.C. 1991. Time lags and population fluctuations in white-tailed deer. *Journal of Wildlife Management* 55:377–385.

Fryxell, J.M., Wilmshurst, J.F. & Sinclair, A.R.E. 2004. Predictive models of movement by Serengeti grazers. *Ecology* 85:2429–2435.

Fryxell, J.M., Wilmshurst, J.F., Sinclair, A.R.E., Haydon, D.T., Holt, R.D. & Abrams, P.A. 2005. Landscape scale, heterogeneity, and the viability of Serengeti grazers. *Ecology Letters* 8:328–335.

Fuller, T.K. 1989. Population dynamics of wolves in north-central Minnesota. *Wildlife Monographs* 105:1–41.

Fuller, T.K. & Keith, L.B. 1980. Wolf population dynamics and prey relationships in north-eastern Alberta. *Journal of Wildlife Management* 44:583–602.

Futuyma, D.J. 1986. *Evolutionary Biology*, 2nd edn. Sinauer Associates, Sunderland, MA.

Gaillard, J.M., Festa-Bianchet, M., Yoccoz, N.G. & Toigo, C. 2000. Temporal variation in fitness components and population dynamics of large herbivores. *Annual Review of Ecology and Systematics* 31:367–393.

Galbreath, R. 1989. The lighthouse keeper's cat. In: *Walter Buller, The Reluctant Conservationist*, pp. 207–215. G.P. Books, Wellington.

Gales, R., Renouf, D. & Worthy, G.A.J. 1994. Use of bioelectrical impedance analysis to assess body composition of seals. *Marine Mammal Science* 10:1–12.

Gasaway, W.C., Boertje, R.D., Grangaard, D.V., Kelleyhouse, D.G., Stephenson, R.O. & Larsen, D.G. 1992. The role of predation in limiting moose at low densities in Alaska and Yukon and implications for conservation. *Wildlife Monographs* 120:1–59.

Gascoigne, J.C. & Lipcius, R.N. 2004. Allee effects driven by predation. *Journal of Applied Ecology* 41:801–810.

Gass, C.L., Angehr, G. & Centa, J. 1976. Regulation of food supply by feeding territoriality in the rufous hummingbird. *Canadian Journal of Zoology* 54:2046–2054.

Gaston, K.J. & Blackburn, T.M. 2000. *Pattern and Process in Macroecology*. Blackwell Science, Oxford.

Gates, C.C. & Hudson, R.J. 1981. Weight dynamics of wapiti in the boreal forest. *Acta Theriologica* 26:407–418.

Gause, G.F. 1934. *The Struggle for Existence*. Williams and Wilkins, Baltimore, MD. (Reprinted 1964 by Hafner, New York.)

Gavin, T.A. & May, B. 1988. Taxonomic status and genetic purity of Columbian white-tailed deer. *Journal of Wildlife Management* 52:1–10.

Geist, V. & McTaggert-Cowan, I., eds. 1995. *Wildlife Conservation Policy*. Deselig Press, Calgary.

Gentilli, J. 1992. Numerical clines and escarpments in the geographical occurrence of avian species; and a search for relevant environmental factors. *Emu* 92:129–140.

Gerhart, K.L., White, R.G., Cameron, R.D. & Russell, D.E. 1996. Estimating fat content of caribou from body condition scores. *Journal of Wildlife Management* 60:713–718.

Gibson, L.A., Wilson, B.A., Cahill, D.M. & Hill, J. 2004. Spatial prediction of rufous bristlebird habitat in a coastal heathland: a GIS-based approach. *Journal of Applied Ecology* 41:213–223.

Gill, R.B., Carpenter, L.H. & Bowden, D.C. 1983. Monitoring large animal populations: the Colorado experience. In: *Transactions of the 48th North American Wildlife and Natural Resources Conference*, pp. 330–441. Wildlife Management Institute, Washington, DC.

Gillingham, M.P., Parker, K.L. & Hanley, T.A. 1997. Forage intake by large herbivores in a natural environment: bout dynamics. *Canadian Journal of Zoology* 75:1118–1128.

Ginsberg, J.R. & Milner-Gulland, E.J. 1993. Sex-biased harvesting and population dynamics: implications for conservation and sustainable use. *Conservation Biology* 8:157–166.

Ginsberg, J.R., Mace, G.M. & Albon, S. 1995. Local extinction in a small and declining population: wild dogs in the Serengeti. *Proceedings of the Royal Society of London B* 262:221–228.

Gleeson, S.K. & Wilson, D.S. 1986. Equilibrium diet: optimal foraging and prey coexistence. *Oikos* 46:139–144.

Gochfeld, M. 1980. Mechanisms and adaptive value of reproductive synchrony in colonial seabirds. In: Burger, J., Olla, B.L. & Winn, H.E., eds. *Behavior of Marine Animals: Current Perspectives in Research, Vol. 4, Marine Birds*, pp. 207–270. Plenum Press, New York.

Godin, J.-G. & Keenleyside, M.H.A. 1984. Foraging on patchily distributed prey by a cichlid fish (Teleostei: Cichlidae): a test of the ideal free distribution theory. *Animal Behaviour* 32:120–131.

Golley, F.B. 1961. Energy values of ecological materials. *Ecology* 42:581–584.

Goodman, S.J., Barton, N.J., Swanson, G., Abernethy, K. & Pemberton, J.M. 1999. Introgression through rare hybridization: a genetic study of a hybrid zone between red and sika deer (genus *Cervus*) in Argyll, Scotland. *Genetics* 152:355–371.

Gordon, I.J. 1988. Facilitation of red deer grazing by cattle and its impact on red deer performance. *Journal of Applied Ecology* 25:1–10.

Gordon, I.J. 1991. Ungulate re-introductions: the case of the scimitar-horned oryx. *Symposium of the Zoological Society of London* 62:217–240.

Goss-Custard, J.D. & Durrell, S.E.A. le V. di. 1987. Age-related effects in oystercatchers, *Haematopus ostralegus*, feeding on mussels, *Mytilus edulis*. I. Foraging efficiency and interference. *Journal of Animal Ecology* 56:521–536.

Goss-Custard, J.D. & Sutherland, W.J. 1997. Individual behaviour, populations and conservation. In: Krebs, J.R. & Davies, N.B., eds. *Behavioural Ecology*, pp. 375–395. Blackwell Science, Oxford.

Grant, P. & Grant, R. 2002. Unpredictable evolution in a 30-year study of Darwin's finches. *Science* 296:707–711.

Grant, P.R. 1986. *Ecology and Evolution of Darwin's Finches*. Princeton University Press, Princeton, NJ.

Green, A.J. 1998. Comparative feeding behaviour and niche organization in a Mediterranean duck community. *Canadian Journal of Zoology* 76:500–507.

Green, B. 1978. Estimation of food consumption in the dingo, *Canis familiaris dingo*, by means of ^{22}Na turnover. *Ecology* 59:207–210.

Green, B., Anderson, J. & Whateley, T. 1984. Water and sodium turnover and estimated food consumption in free-living lions (*Panthera leo*) and spotted hyaenas (*Crocuta crocuta*). *Journal of Mammalogy* 65:593–599.

Green, G. & Brothers, N. 1989. Water and sodium turnover and estimated food consumption rates in free-living fairy prions (*Pachyptila turtur*) and common diving petrels (*Pelecanoides urinatrix*). *Physiological Zoology* 62:702–715.

Green, R.H., Newton, I., Shutlz, S., *et al.* 2004. Diclofenac poisoning as a cause of vulture population declines across the Indian subcontinent. *Journal of Applied Ecology* 41:793–800.

Greenwood, P.J. 1980. Mating systems, philopatry and dispersal in birds and mammals. *Animal Behaviour* 28:1140–1162.

Greenwood, P.J. 1983. Mating systems and the evolutionary consequences of dispersal. In: Swingland, I.R. & Greenwood, P.J., eds. *The Ecology of Animal Movement*, pp. 116–131. Clarendon Press, Oxford.

Greenwood, P.J. & Harvey, P.H. 1982. The natal and breeding dispersal of birds. *Annual Review of Ecology and Systematics* 13:1–21.

Greenwood, R.J., Pietruszewski, D.G. & Crawford, R.D. 1998. Effects of food supplementation on depredation of duck nests in upland habitat. *Wildlife Society Bulletin* 26:219–226.

Grenfell, B.T., Price, O.F., Albon, S.D. & Clutton-Brock, T.H. 1992. Overcompensation and population cycles in ungulates. *Nature* 355:823–826.

Grenfell, B.T., Wilson, K., Finkenstädt, B.F., *et al.* 1998. Noise and determinism in synchronized sheep dynamics. *Nature* 394:674–677.

Grey, M.J., Clarke, M.F. & Loyn, R.H. 1997. Initial changes in the avian community of remnant eucalypt woodlands following a reduction in the abundance of Noisy Miners, *Manorina melanocephala*. *Wildlife Research* 24:631–648.

Grey, M.J., Clarke, M.F. & Loyn, R.H. 1998. Influence of the Noisy Miner, *Manorina melanocephala*, on avian diversity and abundance in remnant Grey Box woodland. *Pacific Conservation Biology* 4:55–69.

Griffith, B., Scott, J.M., Carpenter, J.W. & Reed, C. 1989. Translocation as a species conservation tool: status and strategy. *Science* 245:477–480.

Grigg, G.C., Taplin, L.E., Green, B. & Harlow, P. 1986. Sodium and water fluxes in free-living *Crocodylus porosus* in marine and brackish conditions. *Physiological Zoology* 59:240–253.

Gulland, F.M.D. 1992. The role of nematode parasites in Soay sheep (*Ovis aries* L.) mortality during a population crash. *Parasitology* 105:493–503.

Gullion, G.W. 1965. A critique concerning foreign game bird introductions. *Wilson Bulletin* 77:409–414.

Gullison, R.E. 1998. Bigleaf mahogany. In: Milner-Gulland, E.J. & Mace, R., eds. *Conservation of Biological Resources*, pp. 193–205. Blackwell Science, Oxford.

Gunn, A., Shank, C. & McLean, B. 1991. The history, status and management of muskoxen on Banks Island. *Arctic* 44:188–195.

Guthery, F.S., Lusk, J.L. & Peterson, M.L. 2001. The fall of the null hypothesis: liabilities and opportunities. *Journal of Wildlife Management* 65:379–384.

Guthrie, R.D., ed. 1990. *Frozen Fauna of the Mammoth Steppe: The Story of Blue Babe*. The University of Chicago Press, Chicago, IL.

Gyllenberg, M., Hastings, A. & Hanski, I.A. 1997. Structured metapopulation models. In: Hanski, I.A. & Gilpin, M.E., eds. *Metapopulation Biology*, pp. 93–122. Academic Press, San Diego, CA.

Haila, Y. 1986. North European land birds in forest fragments: evidence for area effects? In: Verner, J., Morrison, M.L. & Ralph, J.C., eds. *Wildlife 2000: Modeling Habitat Relationships of Terrestrial Vertebrates*, pp. 315–319. University of Wisconsin Press, Madison, WI.

Hallett, J.G. 1982. Habitat selection and the community matrix of a desert small-mammal fauna. *Ecology* 63:1400–1410.

Hamilton, W.D. 1971. Geometry of the selfish herd. *Journal of Theoretical Biology* 31:295–311.

Hanks, J. 1981. Characterization of population condition. In: Fowler, C.W. & Smith, T.D., eds. *Dynamics of Large Mammal Populations*, pp. 47–73. John Wiley & Sons, New York.

Hansen, R.M., Mugambi, M.M. & Bauni, S.M. 1985. Diets and trophic ranking of ungulates of the northern Serengeti. *Journal of Wildlife Management* 49:823–829.

Hanski, I. 1994. A practical model of metapopulation dynamics. *Journal of Animal Ecology* 63:151–162.

Hanski, I. 1998. Metapopulation dynamics. *Nature* 396:41–49.

Hanski, I., Kuussaari, M. & Nieminen, M. 1994. Metapopulation structure and migration in the butterfly *Melitaea cinxia*. *Ecology* 75:747–762.

Hanski, I.A. & Gilpin, M.E. 1997. *Metapopulation Biology*. Academic Press, San Diego, CA.

Hanson, W.R. 1967. Estimating the density of an animal population. *Journal of Research on the Lepidoptera* 6:203–247.

Hansson, L. 1987. An interpretation of rodent cycles as due to trophic interactions. *Oikos* 50:308–318.

Hansson, L. & Henttonen, H. 1985. Gradients in density variations of rodents: the importance of latitude and snow cover. *Oecologia* 67:394–402.

Harcourt, A.H. 1995. Population viability estimates: theory and practice for a wild gorilla population. *Conservation Biology* 9:134–142.

Hardin, G. 1960. The competitive exclusion principle. *Science* 131:1292–1297.

Haroldson, M. 1999. Bear monitoring and population trend: unduplicated females. In: Schwartz, C.C. & Haroldson, M.A., eds. *Yellowstone Grizzly Bear Investigations: Annual Report of the Interagency Grizzly Bear Study Team, 1999*, pp. 3–17. US Geological Survey, Bozeman, MT. (http://www.nrmsc.usgs.gov/research/igbstpub.htm)

Harper, D.G.C. 1982. Competitive foraging in mallards: "ideal free" ducks. *Animal Behaviour* 30:575–584.

Harrington, R.N., Owen-Smith, N., Viljoen, P., Biggs, H. & Mason, D. 1999. Establishing the causes of the roan antelope decline in the Kruger National Park, South Africa. *Biological Conservation* 90:69–78.

Harrison, S., Murphy, D.D. & Ehrlich, P.R. 1988. Distribution of the bay checkerspot butterfly, *Euphydryas editha bayensis*: evidence for a metapopulation model. *American Naturalist* 132:360–382.

Hastings, A. & Powell, T. 1991. Chaos in a three-species food chain. *Ecology* 72:896–903.

Haydon, D.T. & Fryxell, J.M. 2004. Using knowledge of recruitment to manage harvesting. *Ecology* 85:78–85.

Hayes, R.D. & Harestad, A.S. 2000. Wolf functional response and regulation of moose in the Yukon. *Canadian Journal of Zoology* 78:60–66.

Hayes, R.D., Baer, A.M., Wotschikowsky, U. & Harestad, A.S. 2000. Kill rate by wolves on moose in the Yukon. *Canadian Journal of Zoology* 78:49–59.

Hayne, D.W. 1949. An examination of the strip census method for estimating animal populations. *Journal of Wildlife Management* 13:145–157.

Hayward, B.J., Heske, E.J. & Painter, C.W. 1997. Effects of livestock grazing on small mammals at a desert cienaga. *Journal of Wildlife Management* 61:123–129.

Heard, D.C. 1992. The effect of wolf predation and snow cover on musk-ox group size. *American Naturalist* 139:190–204.

Heard, D.C. & Williams, W.D. 1991. Wolf den distribution on migratory Barren Ground caribou ranges in the N.W.T. In: Butler, C.E. & Mahoney, S.P., eds. *Proceedings of the 4th North American Caribou Workshop, St. John's, Newfoundland*, pp. 249–250. Newfoundland Labrador Wildlife Division, St. John's.

Hebblewhite, M., Pletscher, D.H. & Paquet, P.C. 2002. Elk population dynamics in Banff National Park, Alberta. *Canadian Journal of Zoology* 80:789–799.

Hector, A., Joshi, J., Lawler, E., Spehn, E. & Wilby, A. 2001. Conservation implications of the link between biodiversity and ecosystem functioning. *Oecologia* 129:624–628.

Henny, C.J., Byrd, M.A., Jacobs, J.A., McLain, P.D., Todd, M.R. & Halla, B.F. 1977. Mid-Atlantic coast osprey population: present numbers, productivity, pollutant contamination, and status. *Journal of Wildlife Management* 41:254–265.

Hersteinsson, P. & Macdonald, D.W. 1992. Interspecific competition and the geographical distribution of red and arctic foxes *Vulpes vulpes* and *Alopex lagopus*. *Oikos* 64:505–515.

Heske, E.J. & Campbell, M. 1991. Effects of an 11-year livestock exclosure on rodent and ant numbers in the Chihuahuan Desert, southeastern Arizona. *Southwestern Naturalist* 36:89–93.

Hickey, J.J. & Anderson, D.W. 1968. Chlorinated hydrocarbons and eggshell changes in raptorial and fish-eating birds. *Science* 162:271–273.

Higuchi, R., Bowman, B., Freilberger, M., Ryder, O.A. & Wilson, A.C. 1984. DNA sequences from the quagga, an extinct member of the horse family. *Nature* 312:282–284.

Hik, D.S. 1995. Does risk of predation influence population dynamics? Evidence from the cyclic decline of snowshoe hares. *Wildlife Research* 22:115–129.

Hik, D.S. & Jefferies, R.L. 1990. Increases in the net above-ground primary production of a salt-marsh forage grass: a test of the predictions of the herbivore-optimization model. *Journal of Ecology* 78:180–195.

Hilborn, R. & Mangel, M. 1997. *The Ecological Detective*. Princeton University Press, Princeton, NJ.

Hilborn, R., Branch, T.A., Ernst, B., et al. 2003. State of the world's fisheries. *Annual Review of Environmental Resources* 28:15.1–15.40.

Hill, C.J. 1995. Linear strips of rain forest vegetation as potential dispersal corridors for rain forest insects. *Conservation Biology* 9:1559–1566.

Hillebrand, H. & Blenckner, T. 2002. Regional and local impact on species diversity – from pattern to process. *Oecologia* 132:479–491.

Hilton-Taylor, C. 2000. *The IUCN Red List of Threatened Species*. IUCN, Gland, Switzerland.

Hjeljord, O., Saether, B.-E. & Anderson, R. 1994. Estimating energy intake of free-ranging moose cows and calves through collection of feces. *Canadian Journal of Zoology* 72:1409–1415.

Hobbs, N.T. 1996. Modification of ecosystems by ungulates. *Journal of Wildlife Management* 60:695–713.

Hobbs, N.T., Baker, D.L., Bear, G.D. & Bowden, D.C. 1996. Ungulate grazing in sagebrush grassland: mechanisms of resource competition. *Ecological Applications* 6:200–217.

Hobbs, N.T., Gross, J.E., Shipley, L.A., Spalinger, D.E. & Wunder, B.A. 2003. Herbivore functional response in heterogeneous environments: a contest among models. *Ecology* 84:666–681.

Hobbs, R.J. 2001. Synergisms among habitat fragmentation, livestock grazing, and biotic invasions in Southwestern Australia. *Conservation Biology* 15:1522–1528.

Hochachka, W.M. & Dhondt, A.A. 2000. Density-dependent decline of host abundance resulting from a new infectious disease. *Proceedings of the National Academy of Sciences of the USA* 97:5303–5306.

Hodges, K.E. & Sinclair, A.R.E. 2003. Does predation risk cause snowshoe hares to modify their diets? *Canadian Journal of Zoology* 81:1973–1985.

Hodges, K.E., Krebs, C.J., Hik, D.S., Stefan, C.I., Gillis, E.A. & Doyle, C.E. 2001. Snowshoe hare demography. In: Krebs, C.J., Boutin, S. & Boonstra R., eds. *Ecosystem Dynamics of the Boreal Forest*, pp. 142–178. Oxford University Press, Oxford.

Hofmann, R.R. 1973. *The Ruminant Stomach: Stomach Structure and Feeding Habits of East African Game Ruminants*. East African Monographs in Biology, Vol. 2. East African Literature Bureau, Nairobi.

Holekamp, K.E. 1986. Proximal causes of natal dispersal in Belding's ground squirrels (*Spermophilus beldingi*). *Ecological Monographs* 56:365–391.

Holling, C.S. 1959. The components of predation as revealed by a study of small-mammal predation of the European pine sawfly. *Canadian Entomologist* 91:293–320.

Holling, C.S. 1973. Resilience and stability of ecological systems. *Annual Review of Ecology and Systematics* 4:1–23.

Holmgren, N. 1995. The ideal free distribution of unequal competitors: predictions from a behaviour-based functional response. *Journal of Animal Ecology* 64:197–212.

Holt, R.D. 1977. Predation, apparent competition and the structure of prey communities. *Theoretical Population Biology* 12:197–229.

Holt, R.D. 1984. Spatial heterogeneity, indirect interactions, and the coexistence of prey species. *American Naturalist* 124:377–406.

Hone, J. 1994. *Analysis of Vertebrate Pest Control*. Cambridge University Press, Cambridge.

Hone, J. 2004. *Wildlife Damage Control: Principles for the Management of Damage by Vertebrate Pests*. CSIRO Publishing, Melbourne.

Hone, J., Pech, R. & Yip, P. 1992. Estimation of the dynamics and rate of transmission of classical swine fever (hog cholera) in wild pigs. *Epidemiological Infection* 108:377–386.

Hoogland, J.L. 1982. Prairie dogs avoid extreme inbreeding. *Science* 215:1639–1641.

Hooper, P.T., Lunt, R.A., Gould, A.R., *et al.* 1999. Epidemic blindness in kangaroos – evidence of a viral aetiology. *Australian Veterinary Journal* 77:529–536.

Hope, J.H. 1972. Mammals of the Bass Strait islands. *Proceedings of the Royal Society of Victoria* 85:163–195.

Hornicke, H. & Björnhag, G. 1980. Coprophagy and related strategies for digesta utilization. In: Ruckebusch, Y. & Thivend, P., eds. *Digestive Physiology and Metabolism in Ruminants*, pp. 707–730. Proceedings of the 5th International Symposium on Ruminant Physiology, Clermont Ferrand, France.

Hoskinson, R.L. & Mech, L.D. 1976. White-tailed deer migration and its role in wolf predation. *Journal of Wildlife Management* 40:429–441.

Houston, D.B. 1982. *The Northern Yellowstone Elk: Ecology and Management*. Macmillan, New York.

Houston & McNamara 1990.

Howery, L.D. & Pfister, J.A. 1990. Dietary and fecal concentrations of nitrogen and phosphorus in penned white-tailed deer does. *Journal of Wildlife Management* 54:383–389.

Hudson, P. & Greenman, J. 1998. Competition mediated by parasites: biological and theoretical progress. *Trends in Ecology and Evolution* 13:387–390.

Hudson, P.J. 1986. The effect of a parasitic nematode on the breeding production of red grouse. *Journal of Animal Ecology* 55:85–92.

Hudson, P.J. & Dobson, A.P. 1988. The ecology and control of parasites in gamebird populations. In: Hudson, P.J. & Rands, M.R.W., eds. *Ecology and Management of Gamebirds*, pp. 98–133. BSP Professional Books, Oxford.

Hudson, P.J. & Dobson, A.P. 1991. Control of parasites in natural populations: nematode and virus infections of red grouse. In: Perrins, C.M., Lebreton, J.D. & Hirons, G.J.M., eds. *Bird Population Studies*, pp. 413–432. Oxford University Press, Oxford.

Hudson, P.J., Dobson, A.P. & Newborn, D. 1992. Do parasites make prey vulnerable to predation? Red grouse and parasites. *Journal of Animal Ecology* 61:681–692.

Hudson, P.J., Dobson, A.P. & Newborn, D. 1998. Prevention of population cycles by parasite removal. *Science* 282:2256–2258.

Hudson, P.J., Newborn, D. & Dobson, A.P. 1992. Regulation and stability of a free-living host–parasite system: *Trichostrongylus tenuis* in red grouse. I. Monitoring and parasite reduction experiments. *Journal of Animal Ecology* 61:477–486.

Hudson, P.J., Norman, R., Laurenson, M.K., *et al.* 1995. Persistence and transmission of tick-borne viruses: *Ixodes ricinus* and louping-ill virus in red grouse populations. *Parasitology* 111:S49–S58.

Hudson, R.J., Drew, K.R. & Baskin, L.M., eds. 1989. *Wildlife Production Systems*. Cambridge University Press, Cambridge.

Hume, I.D. & Warner, C.I. 1980. Evolution of microbial digestion in mammals. In: Ruckebusch, Y. & Thivend, P., eds. *Digestive Physiology and Metabolism in Ruminants*, pp. 669–684. Proceedings of the 5th International Symposium on Ruminant Physiology, Clermont Ferrand, France.

Huntly, N. & Inouye, R. 1988. Pocket gophers in ecosystems: patterns and mechanisms. *BioScience* 38:786–793.

Hurlbert, S.H. 1984. Pseudoreplication and the design of ecological field experiments. *Ecological Monographs* 54:187–211.

Hutchinson, G.E. 1957. Concluding remarks. *Cold Spring Harbor Symposia on Quantitative Biology* 22:415–427.

Hutton, I. 1991. *Birds of Lord Howe Island Past and Present*. Ian Hutton, PO Box 6367, Coffs Harbour Plaza, Australia.

Hutton, I. 1998. *The Australian Geographic Book of Lord Howe Island*. Australian Geographic Pty Ltd, PO Box 321, Terrey Hills, Australia.

Huyser, O., Ryan, P.G. & Cooper, J. 2000. Changes in population size, habitat use and breeding biology of lesser sheathbills (*Chionis minor*) at Marion island: impacts of cats, mice and climate change? *Biological Conservation* 92:299–310.

Imbeau, L. & Desroches, A. 2002. Area sensitivity and edge avoidance: the case of the Three-toed Woodpecker (*Picoides tridactylus*) in a managed forest. *Forest Ecology and Management* 164:249–256.

Ims, R.A. 1990. On the adaptive value of reproductive synchrony as a predator-swamping strategy. *American Naturalist* 136:485–498.

Ims, R.A., Bondrup-Nielsen, S. & Stenseth, N.C. 1988. Temporal patterns of breeding events in small rodent populations. *Oikos* 53:229–234.

Inchausti, P. & Ginzburg, L.R. 1998. Small mammal cycles in northern Europe: patterns and evidence for a maternal effects hypothesis. *Journal of Animal Ecology* 67:180–194.

Inman, A.J. 1990. Group foraging in starlings – distribution of unequal competitors. *Animal Behaviour* 40:801–810.

Innes, J., Hay, R., Flux, I., Bradfield, P., Speed, H. & Jansen, P. 1999. Successful recovery of North Island kokako (*Callaeas cinerea wilsoni*) populations, by adaptive management. *Biological Conservation* 87:201–214.

Irwin, D. 2000. Song variation in an avian ring species. *Evolution* 54:998–1010.

Irwin, D.E., Bensch, S., Irwin, J.H. & Price, T.D. 2005. Speciation by distance in a ring species. *Science* 307:414–416.

Irwin, D.E., Bensch, S. & Price, T.D. 2001a. Speciation in a ring. *Nature* 409:333–337.

Irwin, D.E., Irwin, J.H. & Price, T.D. 2001b. Ring species as bridges between microevolution and speciation. *Genetica* 112/113:223–243.

Isenberg, A.C. 2000. *The Destruction of the Bison*. Cambridge University Press, Cambridge.

Jackman, T.R. 1998. Molecular and historical evidence for the introduction of clouded salamanders (genus *Aneides*) to Vancouver Island, British Columbia, Canada, from California. *Canadian Journal of Zoology* 76:1570–1580.

Jackson, J.B.C., Kirby, M.X., Berger, W.H., *et al.* 2001. Historical overfishing and the recent collapse of coastal ecosystems. *Science* 293:629–638.

Janczewski, D., Yuhki, N., Gilbert, D.A., Jefferson, G.T. & O'Brien, S.J. 1992. Molecular phylogenetic inference from saber-toothed cat fossils of Rancho La Brea. *Proceedings of the National Academy of Sciences of the USA* 89:9769–9773.

Janzen, D.H. 1970. Herbivores and the number of tree species in tropical forests. *American Naturalist* 104:501–528.

Janzen, F.J. 1994. Climate change and temperature-dependent sex determination in reptiles. *Proceedings of the National Academy of Sciences of the USA* 91:7487–7490.

Jarman, P.J. 1973. The free water intake of impala in relation to the water content of their food. *East African Agricultural and Forestry Journal* 38:343–351.

Jarman, P.J. 1974. The social organisation of antelope in relation to their ecology. *Behaviour* 48:215–266.

Jarman, P.J. & Sinclair, A.R.E. 1979. Feeding strategy and the pattern of resource partitioning in ungulates. In: Sinclair, A.R.E. & Norton-Griffiths, M., eds. *Serengeti: Dynamics of an Ecosystem*, pp. 130–163. The University of Chicago Press, Chicago, IL.

Jefferies, R.L. & Gottlieb, L.D. 1983. Genetic variation within and between populations of the asexual plant *Puccinellia* × *phryganodes*. *Canadian Journal of Botany* 61:774–779.

Jefferies, R.L., Rockwell, R.F. & Abraham, K.F. 2004. Agricultural food subsidies, migratory connectivity and large-scale disturbance in Arctic coastal systems: a case study. *Integrative and Comparative Biology* 44:130–139.

Jenkins, S.H. 1980. A size–distance relation in food selection by beavers. *Ecology* 61:740–746.

Jenks, J.A., Soper, R.B., Lochmiller, R.L. & Leslie, D.M., Jr. 1990. Effect of exposure on nitrogen and fiber characteristics of white-tailed deer feces. *Journal of Wildlife Management* 54:389–391.

Johnson, C.N. 1994. Nutritional ecology of a mycophagous marsupial in relation to production of hypogeous fungi. *Ecology* 75:2015–2021.

Johnson, D.H. 1999. The insignificance of statistical significance testing. *Journal of Wildlife Management* 63:763–772.

Johnson, D.H. 2002. The importance of replication in wildlife research. *Journal of Wildlife Management* 66:919–932.

Johnson, D.H., Krapu, G.L., Reinecke, K.J. & Jorde, D.G. 1985. An evaluation of condition indices for birds. *Journal of Wildlife Management* 49:569–575.

Johnson, J.B. & Omland, K.S. 2004. Model selection in ecology and evolution. *Trends in Ecology and Evolution* 19:101–108.

Johnson, M.L. & Gaines, M.S. 1990. Evolution of dispersal: theoretical models and empirical tests using birds and mammals. *Annual Review of Ecology and Systematics* 21:449–480.

Jolly, G.M. 1969. Sampling methods for aerial censuses of wildlife populations. *East African Agricultural Forestry Journal* 33:46–49.

Joly, D.O. & Messier, F. 2004. Testing hypotheses of bison population decline (1970–1999) in Wood Buffalo National Park: synergism between exotic disease and predation. *Canadian Journal of Zoology* 82:1165–1176.

Jones, C. 2003. Safety in numbers for secondary prey populations: an experimental test using egg predation by small mammals in New Zealand. *Oikos* 102:57–66.

Jones, D.M. 1982. Conservation in relation to animal diseases in Africa and Asia. In: Edwards, M.A. & McDonnell, U., eds. *Animal Disease in Relation to Animal Conservation*, pp. 271–285. Symposium of the Zoological Society of London no. 50. Academic Press, London.

Jones, M.M. 1980. Nocturnal loss of muscle protein from house sparrows (*Passer domesticus*). *Journal of Zoology (London)* 192:33–39.

Jones, W.T. 1987. Dispersal patterns in kangaroo rats (*Dipodomys spectabilis*). In: Chepko-Sade, B.D. & Halpin, Z.T., eds. *Mammalian Dispersal Patterns: The Effects of Social Structure on Population Genetics*, pp. 119–127. The University of Chicago Press, Chicago, IL.

Jorgenson, J.T., Festa-Bianchet, M., Gaillard, J.-M. & Wishart, W.D. 1997. Effects of age, sex, disease, and density on survival of bighorn sheep. *Ecology* 78:1019–1032.

Joubert, S.C.J. 1983. A monitoring programme for an extensive national park. In: Owen-Smith, R.N., ed. *Management of Large Mammals in African Conservation Areas*, pp. 201–212. HAUM Educational Publishers, Pretoria.

Källén, A., Arcuri, P. & Murray, J.D. 1985. A simple model for the spatial spread and control of rabies. *Journal of Theoretical Biology* 116:377–393.

Kat, P.W., Alexander, K.A., Smith, J.S. & Munson, L. 1995. Rabies and African wild dogs in Kenya. *Proceedings of the Royal Society of London B* 262:229–233.

Keane, B., Creel, S.R. & Waser, P.M. 1996. No evidence of inbreeding avoidance or inbreeding depression in a social carnivore. *Behavioral Ecology* 7:480–489.

Keen, C.L. & Graham, T.W. 1989. Trace elements. In: Kaneko, J.J., ed. *Clinical Biochemistry of Domestic Animals*, 4th edn. Academic Press, New York.

Keith, L.B., Cary, J.R., Rongstad, O.J. & Brittingham, M.C. 1984. Demography and ecology of a declining snowshoe hare population. *Wildlife Monographs* 90:1–43.

Kelker, G.H. 1940. Estimating deer populations by a differential hunting loss in the sexes. *Proceedings of the Utah Academy of Sciences, Arts and Letters* 17:65–69.

Kelker, G.H. 1944. Sex ratio equations and formulas for determining wildlife populations. *Proceedings of the Utah Academy of Sciences, Arts and Letters* 20:189–198.

Keller, L.F., Arcese, P., Smith, J.N.M., Hochachka, W.M. & Stearns, S.C. 1994. Selection against inbred song sparrows during a natural population bottleneck. *Nature* 372:356–357.

Kelsall, J.P. & Prescott, W. 1971. *Moose and Deer Behaviour in Snow in Fundy National Park, New Brunswick*. Canadian Wildlife Service Report Series no. 15. Information Canada, Ottawa.

Kenagy, G.J. 1989. Daily and seasonal uses of energy stores in torpor and hibernation. In: Malan, A. & Canguilhem, B., eds. *Second International Symposium on Living in the Cold*, pp. 17–24. John Libbey Eurotext, London.

Kenagy, G.J., Sharbaugh, S.M. & Nagy, K.A. 1989. Annual cycle of energy and time expenditure in a golden-mantled ground squirrel population. *Oecologia* 78:269–282.

Kendall, B.E., Bjørnstad, O.N., Bascompte, J., Keitt, T.H. & Fagan, W.F. 2000. Dispersal, environmental correlation, and spatial synchrony in population dynamics. *American Naturalist* 155:628–636.

Kennedy, S. 1990. A review of the 1988 European morbillivirus epizootic. *The Veterinary Record* 127:563–567.

Keymer, A.E. & Dobson, A.P. 1987. The ecology of helminthes in populations of small mammals. *Mammal Review* 17:105–116.

Kiff, L.F. 1989. DDE and the California condor *Gymnogyps californianus*: the end of a story? In: Meyburg, B.U. & Chancellor, R.D., eds. *Raptors in the Modern World*, pp. 477–480. World Working Group on Birds of Prey and Owls, Berlin.

King, A.A. & Schaffer, W.M. 2001. The geometry of a population cycle: a mechanistic model of snowshoe hare demography. *Ecology* 82:814–830.

King, C.M. 1983. The relationships between beech (*Nothofagus* sp.) seedfall and populations of mice (*Mus musculus*) and the demographic and dietary responses of stoats (*Mustela erminea*), in three New Zealand forests. *Journal of Animal Ecology* 52:141–166.

Kingsland, S.E. 1985. *Modeling Nature*. The University of Chicago Press, Chicago, IL.

Kinnear, J.E., Onus, M.L. & Sumner, N.R. 1998. Fox control and rock-wallaby population dynamics – II. An update. *Wildlife Research* 25:81–88.

Kinzig, A.P., Pacala, S.W. & Tilman, D., eds. 2001. *The Functional Consequences of Biodiversity*. Princeton University Press, Princeton, NJ.

Kirkpatrick, J.F. & Rutberg, A.T. 2001. Fertility control in animals. In: Salem, D.S. & Rowan, A.N., eds. *State of the Animals 2001*, pp. 183–198. Humane Society Press, Washington, DC.

Kirkpatrick, J.F., Turner, J.W., Jr., Liu, I.K.M., Fayrer-Hosken, R.A. & Rutberg, A. 1997. Case studies in wildlife immunocontraception: Wild and feral equids and white-tailed deer. *Reproduction, Fertility, and Development* 9:105–110.

Kirkpatrick, R.L. 1980. Physiological indices in wildlife management. In: Schemnitz, S.D., ed. *Wildlife Management Techniques Manual*, 4th edn, pp. 99–112. The Wildlife Society, Washington, DC.

Kistner, T.P., Trainer, C.E. & Hartmann, N.A. 1980. A field technique for evaluating physical condition of deer. *Wildlife Society Bulletin* 8:11–17.

Kleiber, M. 1947. Body size and metabolic rate. *Physiological Reviews* 27:511–541.

Klein, D.R. & Olson, S.T. 1960. Natural mortality patterns of deer in southeast Alaska. *Journal of Wildlife Management* 24:80–88.

Knopf, F.L., Sedgwick, J.A. & Cannon, R.W. 1988. Guild structure of a riparian avifauna relative to seasonal cattle grazing. *Journal of Wildlife Management* 52:280–290.

Knowlton, N. 1992. Thresholds and multiple stable states in coral reef community dynamics. *American Zoologist* 32:674–682.

Kodric-Brown, A. & Brown, J.H. 1978. Influence of economics, interspecific competition, and sexual dimorphism on territoriality of migrant rufous hummingbirds. *Ecology* 59:285–296.

Koford, C.B. 1953. *The California Condor*. Research Report no. 4. National Audubon Society, New York.

Komdeur, J. 1992. Importance of habitat saturation and territory quality for evolution of cooperative breeding in the Seychelles warbler. *Nature* 358:493–495.

Komdeur, J. 1993. Fitness-related dispersal. *Nature* 366:23–24.

Koopman, M.E., Cypher, B.L.K. & Scrivner, J.H. 2000. Dispersal patterns of San Joaquin Kit foxes (*Vulpes macrotis mutica*). *Journal of Mammalogy* 81:213–222.

Korpimäki, E. & Norrdahl, K. 1991. Numerical and functional responses of kestrels, short-eared owls, and long-eared owls to vole densities. *Ecology* 72:814–826.

Kovaliski, J. 1998. Monitoring the spread of rabbit hemorrhagic disease virus as a new biological agent for control of wild European rabbits in Australia. *Journal of Wildlife Diseases* 34:421–428.

Kraft, K.M., Johnson, D.H., Samuelson, J.M. & Allen, S.H. 1995. Using known populations of pronghorn to evaluate sampling plans and estimators. *Journal of Wildlife Management* 59:129–137.

Krause, R.M. 1992. The origin of plagues: old and new. *Science* 257:1073–1078.

Krebs, C.J. 1999. *Ecological Methodology*, 2nd edn. Benjamin/Cummings, Menlo Park, CA.

Krebs, C.J. 2001. *Ecology: The Experimental Analysis of Distribution and Abundance*, 5th edn. Addison-Wesley, San Francisco, CA.

Krebs, C.J. & Myers, J.H. 1974. Population cycles in small mammals. *Advances in Ecological Research* 8:267–399.

Krebs, C.J., Boutin, S., Boonstra, R., *et al.* 1995. Impact of food and predation on the snowshoe hare cycle. *Science* 269:1112–1115.

Krebs, C.J., Dale, M.R.T., Vilis, O., Sinclair, A.R.E. & O'Donoghue, M. 2001a. Shrubs. In: Krebs, C.J., Boutin, S. & Boonstra R., eds. *Ecosystem Dynamics of the Boreal Forest: The Kluane Project*, pp. 92–115. Oxford University Press, New York.

Krebs, C.J., Boutin, S. & Boonstra, R., eds. 2001b. *Ecosystem Dynamics of the Boreal Forest: The Kluane Project*. Oxford University Press, New York.

Krebs, C.J., Gilbert, B.S., Boutin, S., Sinclair, A.R.E. & Smith, J.N.M. 1986. Population biology of snowshoe hares: I. Demography of food-supplemented populations in the southern Yukon, 1976–84. *Journal of Animal Ecology* 55:963–982.

Krebs, J.R. 1971. Territory and breeding density in the great tit, *Parus major* L. *Ecology* 52:2–22.

Krebs, J.R., Erichsen, J.T. & Webber, M.I. 1977. Optimal prey selection in the great tit (*Parus major*). *Animal Behaviour* 25:30–38.

Krivan, V. 1996. Optimal foraging and predator–prey dynamics. *Theoretical Population Biology* 49:265–290.

Krivan, V. 1997. Dynamic ideal-free distribution: effects of optimal patch choice on predator–prey dynamics. *American Naturalist* 149:164–178.

Kruuk, H., Conroy, J.W.H. & Moorhouse, A. 1987. Seasonal reproduction, mortality and food of otters (*Lutra lutra* L.) in Shetland. *Symposium of the Zoological Society of London* 58:263–278.

Laake, J.L., Buckland, S.T., Anderson, D.R. & Burnham, K.P. 1993. DISTANCE *User's Guide*. Colorado Cooperative Fish and Wildlife Research Unit, Colorado State University, Fort Collins, CO.

Laake, J.L., Calambokdis, J., Osmek, S.D. & Rugh, D.J. 1997. Probability of detecting harbor porpoise from aerial surveys: estimating g(0). *Journal of Wildlife Management* 61:63–75.

Laca, E.A., Distel, R.A., Griggs, T.C., Deo, G. & Demment, M.W. 1993. Field test of optimal foraging with cattle: the marginal value theorem successfully predicts patch selection and utilisation. In: *Proceedings of the 17th International Grassland Congress, New Zealand*, pp. 709–710.

Lack, D. 1971. *Ecological Isolation in Birds*. Blackwell, Oxford.

Lafferty, K.D. & Morris, K. 1996. Altered behaviour of parasitized killifish increases susceptibility to predation by bird final hosts. *Ecology* 77:1390–1397.

Laikre, L. & Ryman, N. 1991. Inbreeding depression in a captive wolf (*Canis lupus*) population. *Conservation Biology* 5:33–40.

Lamberson, R.H., McKelvey, R., Noon, B.R. & Voss, C. 1992. A dynamic analysis of northern spotted owl viability in a fragmented forest landscape. *Conservation Biology* 6:505–512.

Lamberson, R.H., Noon, B.R., Voss, C. & McKelvey, K.S. 1994. Reserve design for territorial species: the effects of patch size and spacing on the viability of the northern spotted owl. *Conservation Biology* 8:185–195.

Lamprey, H.F. 1963. Ecological separation of the large mammal species in the Tarangire game reserve, Tanganyika. *East African Wildlife Journal* 1:63–92.

Lande, R. 1987. Extinction thresholds in demographic models of territorial populations. *American Naturalist* 130:624–635.

Lande, R. 1988. Demographic models of the northern spotted owl (*Strix occidentalis caurina*). *Oecologia* 75:601–607.

Lande, R. 1993. Risks of population extinction from demographic and environmental stochasticity and random catastrophes. *American Naturalist* 142:911–927.

Lande, R., Engen, S. & Sæther, B. 1994. Optimal harvesting, economic discounting, and extinction risk in fluctuating populations. *Nature* 373:88–90.

Lande, R., Engen, S. & Sæther, B. 1997. Threshold harvesting for sustainability of fluctuating resources. *Ecology* 78:1341–1350.

Lande, R., Sæther, B., Filli, F., Matthysen, E. & Weimerskirch, H. 2002. Estimating density dependence from population time series using demographic theory and life-history data. *American Naturalist* 159:321–337.

Landsberg, J. 1988. Dieback of rural eucalypts: tree phenology and damage caused by leaf-feeding insects. *Australian Journal of Ecology* 13:251–267.

Landweber, L.F. 1999. Something old for something new: the future of ancient DNA in conservation. In: Landweber, L.F. & Dobson, A.P., eds. *Genetics and the Extinction of Species*, pp. 163–187. Princeton University Press, Princeton, NJ.

Langvatn, R. & Hanley, T.A. 1993. Feeding-patch choice by red deer in relation to foraging efficiency. *Oecologia* 95:164–170.

La Polla, V.N. & Barrett, B.W. 1993. Effects of corridor width and presence on the population dynamics of the meadow vole (*Microtus pennsylvanicus*). *Landscape Ecology* 8:25–37.

Larter, N.C., Sinclair, A.R.E., Ellsworth, T., Nishi, J. & Gates, C.C. 2000. Dynamics of reintroduction in an indigenous large ungulate: the wood bison of northern Canada. *Animal Conservation* 3:299–309.

Laurance, W.F. & Cochrane, M.A. 2001. Synergistic effects in fragmented landscapes. *Conservation Biology* 15:1488–1489.

Laurance, W.F., Lovejoy, T.E., Vasconcelos, H.L., *et al.* 2002. Ecosystem decay of Amazonian forest fragments: a 22-year investigation. *Conservation Biology* 16:605–618.

Lavigne, D.M., Brooks, R.J., Rosen, D.A. & Galbraith, D.A. 1989. Cold, energetics and populations. In: Wange, L.C.H., ed. *Advances in Comparative and Environmental Physiology: Vol. 4, Animal Adaptation to Cold*, pp. 403–432. Springer-Verlag, Berlin.

Laws, R.M. 1969. The Tsavo Research Project. *Journal of Reproduction and Fertility* (Suppl.) 6:495–531.

Leader-Williams, N. 1988. *Reindeer on South Georgia*. Cambridge University Press, Cambridge.

Lebreton, J.-D., Burnham, K.P., Clobert, J. & Anderson, D.R. 1992. Modeling survival and testing biological hypotheses using marked animals: a unified approach with case studies. *Ecological Monographs* 62:67–118.

Lefkovitch, L.P. 1965. The study of population growth in organisms shaped by stages. *Biometrics* 21:1–18.

Lehtonen, A. 1998. Managing moose, *Alces alces*, populations in Finland: hunting virtual animals. *Annales Zoologici Fennici* 35:173–179.

Leibold, M.A. 1995. The niche concept revisited: mechanistic models and community context. *Ecology* 76:1371–1382.

Leirs, H., Stenseth, N.C., Nichols, J.D., Hines, J.E., Verhagen, R. & Verheyen, W. 1997. Stochastic seasonality and nonlinear density-dependent factors regulate population size in an African rodent. *Nature* 389:176–180.

Lens, L. & Dhondt, A.A. 1994. Effects of habitat fragmentation on the timing of crested tit *Parus cristatus* natal dispersal. *Ibis* 136:147–152.

Lesica, P. & Alendorf, R.W. 1995. When are peripheral populations valuable for conservation? *Conservation Biology* 9:753–760.

Leslie, D.M., Jr. & Starkey, E.E. 1985. Fecal indices to dietary quality of cervids in old-growth forests. *Journal of Wildlife Management* 49:142–146.

Leslie, D.M., Jr., Starkey, E.E. & Vavra, M. 1984. Elk and deer diets in old-growth forests in western Washington. *Journal of Wildlife Management* 48:762–775.

Leslie, P.H. 1945. On the use of matrices in certain population mathematics. *Biometrika* 33:183–212.

Leslie, P.H. 1948. Some further notes on the use of matrices in population mathematics. *Biometrika* 35:213–245.

Letcher, A.J. & Harvey, P.H. 1994. Variation in geographical range size among mammals of the palearctic. *American Naturalist* 144:30–42.

Levey, D.J. 1990. Habitat-dependent fruiting behavior of an understory tree, *Miconia centrodesma*, and tropical treefall gaps as keystone habitats for frugivores in Costa Rica. *Journal of Tropical Ecology* 6:409–420.

Levins, R. 1969. Some demographic and genetic consequences of environmental heterogeneity for biological control. *Bulletin of the Entomological Society of America* 15:237–240.

Lewis, M.A. & Murray, J.D. 1993. Modelling territoriality and wolf–deer interactions. *Nature* 366:738–740.

Lima, S.L. & Dill, L.M. 1990. Behavioral decisions made under the risk of predation: a review and prospectus. *Canadian Journal of Zoology* 68:619–640.

Lindenmayer, D.B. 1994. Wildlife corridors and the mitigation of logging impacts on fauna in wood-production in south-eastern Australia: a review. *Wildlife Research* 21:323–340.

Lindenmayer, D.B. & Possingham, H.P. 1996. Ranking conservation and timber management options for Leadbetter's possum in southeastern Australia using population viability analysis. *Conservation Biology* 10:235–251.

Lindenmayer, D.B., Possingham, H.P., Lacy, R.C., McCarthy, M.A. & Pope, M.L. 2003. How accurate are population models? Lessons from landscape-scale tests in a fragmented system. *Ecology Letters* 6:41–47.

Lindstrom, E.R., Andren, H., Angelstam, P., *et al.* 1994. Disease reveals the predator: sarcoptic mange, red fox predation, and prey populations. *Ecology* 75:1042–1049.

Lockyer, C. 1987. Evaluation of the role of fat reserves in relation to the ecology of North Atlantic fin and sei whales. In: Huntley, A.C., Costa, D.P., Worthy, G.A.G. & Castellini, M.A., eds. *Approaches to Marine Mammal Energetics*, pp. 183–203. Special Publication no. 1. Society for Marine Mammalogy, Lawrence, KS.

Lomolino, M.V. & Channell, R. 1995. Splendid isolation: patterns of geographic range collapse in endangered mammals. *Journal of Mammalogy* 76:335–347.

Long, K. & Robley, A. 2004. *Effective Feral Animal Exclusion Fencing for Areas of High Conservation Value in Australia*. A Report for the Commonwealth of Australia, the Department of Environment and Heritage, Environment Australia, Canberra.

Loreau, M. 2000. Biodiversity and ecosystem functioning: recent theoretical advances. *Oikos* 91:3–17.

Loreau, M., Naeem, S. & Inchausti, P., eds. 2002. *Biodiversity and Ecosystem Functioning*. Oxford University Press, Oxford.

Lotka, A.J. 1925. *Elements of Physical Biology*. Williams and Wilkins, Baltimore, MD.

Loye, J. & Carroll, S. 1995. Birds, bugs and blood: avian parasitism and conservation. *Trends in Ecology and Evolution* 10:232–235.

Loyn, R.H., Runnalls, R.G., Forward, G.Y. & Tyers, J. 1983. Territorial bell miners and other birds affecting populations of insect prey. *Science* 221:1411–1412.

Lubina, J.A. & Levin, S.A. 1988. The spread of a reinvading species: range expansion in the California sea otter. *American Naturalist* 131:526–543.

Ludwig, D. 1999. Is it meaningful to estimate a probability of extinction? *Ecology* 80:298–310.

Ludwig, D., Jones, D.D. & Holling, C.S. 1978. Quantitative analysis of insect outbreak systems: the spruce budworm and forests. *Journal of Animal Ecology* 47:315–332.

Lundberg, P. 1988. Functional response of a small mammalian herbivore: the disc equation revisited. *Journal of Animal Ecology* 57:999–1006.

Lyster, S. 1985. *International Wildlife Law*. Groteius Publications Ltd, Cambridge.

MacArthur, R.H. 1957. On the relative abundance of bird species. *Proceedings of the National Academy of Sciences of the USA* 43:293–295.

MacArthur, R.H. 1958. Population ecology of some warblers of northeastern coniferous forests. *Ecology* 39:599–619.

MacArthur, R.H. 1960. On the relative abundance of species. *American Naturalist* 94:25–36.

MacArthur, R.H. 1972. *Geographical Ecology: Patterns in the Distribution of Species*. Harper & Row, New York.

MacArthur, R.H. & Levins, R. 1967. The limiting similarity, convergence and divergence of coexisting species. *American Naturalist* 101:377–385.

MacArthur, R.H. & Pianka, E.R. 1966. On optimal use of a patchy environment. *American Naturalist* 100:603–609.

MacArthur, R.H. & Wilson, E.O. 1967. *The Theory of Island Biogeography*. Princeton University Press, Princeton, NJ.

McCallum, H. & Dobson, A. 1995. Detecting disease and parasite threats to endangered species and ecosystems. *Trends in Ecology and Evolution* 10:190–194.

McCann, K. & Yodzis, P. 1994. Biological conditions for chaos in a three-species food chain. *Ecology* 75:561–564.

McCauley, E., Wilson, W.G. & de Roos, A.M. 1993. Dynamics of age-structured and spatially structured predator–prey interactions: individual-based models and population-level interactions. *American Naturalist* 142:412–442.

McCleery, R.H. & Perrins, C.M. 1985. Territory size, reproductive success and population dynamics in the great tit, *Parus major*. In: Sibly, R.M. & Smith, R.H., eds. *Behavioural Ecology: Ecological Consequences of Adaptive Behaviour*, pp. 353–373. 25th Symposium of the British Ecological Society. Blackwell Scientific Publications, Oxford.

McCullough, D.R., Pel, K.C.J. & Wang, Y. 2000. Home range, activity patterns, and habitat relations of Reeves muntjacs in Taiwan. *Journal of Wildlife Management* 64:430–441.

McCutchen, A.A. 1938. Preliminary results of wildlife census based on actual counts compared to previous estimates on national forests. *Transactions of the North American Wildlife Conference* 3:407–414.

McDonald, D.B. & Potts, W.K. 1994. Cooperative display and relatedness among males in a lek-mating bird. *Science* 266:1030–1032.

McFarlane, D.J., George, R.J. & Farrington, P. 1993. Changes in the hydrologic cycle. In: Hobbs, R.J. & Saunders, D.A., eds. *Reintegrating Fragmented Landscapes*, pp. 147–186. Springer-Verlag, New York.

McGinley, M.A. & Whitham, T.G. 1985. Central place foraging by beavers (*Castor canadensis*): a test of foraging predictions and the impact of selective feeding on the growth form of cottonwoods (*Populus fremontii*). *Oecologia* 66:558–562.

McKilligan, N.G. 1984. The food and feeding ecology of the cattle egret, *Ardeola ibis*, when nesting in south-east Queensland. *Australian Wildlife Research* 11:133–144.

McLaren, B.E. & Peterson, R.O. 1994. Wolves, moose, and tree rings on Isle Royale. *Science* 266:1555–1558.

McLellan, C.H., Dobson, A.P., Wilcove, D.S. & Lynch, J.F. 1986. Effects of forest fragmentation on New- and Old-World bird communities: empirical observations and theoretical implications. In: Verner, J., Morrison, M.L. & Ralph, J.C., eds. *Wildlife 2000: Modeling Habitat Relationships of Terrestrial Vertebrates*, pp. 305–313. University of Wisconsin Press, Madison, WI.

McLeod, M.N. 1974. Plant tannins – their role in forage quality. *Nutrition Abstracts and Reviews* 44:803–815.

McLoughlin, C.A. & Owen-Smith, N. 2003. Viability of a diminishing roan antelope population: predation is the threat. *Animal Conservation* 6:231–236.

MacLulich, D.A. 1937. *Fluctuations in the Numbers of the Varying Hare (Lepus americanus)*. The University of Toronto Press, Toronto.

McNab, B.K. 1988. Complications inherent in scaling the basal rate of metabolism in mammals. *Quarterly Review of Biology* 63:25–54.

McNamara, J.M. & Houston, A.I. 1987. Starvation and predation as factors limiting population size. *Ecology* 68:1515–1519.

McNaughton, S.J. 1976. Serengeti migratory wildebeest: facilitation of energy flow by grazing. *Science* 191:92–94.

McNaughton, S.J. 1986. On plants and herbivores. *American Naturalist* 128:765–770.

McNaughton, S.J., Banykwa, F.F. & McNaughton, M.M. 1997. Promotion of the cycling of diet-enhancing nutrients by African grazers. *Science* 278:1798–1800.

MacPhee, R.D.E., ed. 1999. *Extinctions in Near Time: Causes, Contexts, and Consequences.* Kluwer Academic/Plenum Publishers, New York.

McShea, W., Underwood, H.B. & Rappole, J.H., eds. 1997. *The Science of Overabundance.* Smithsonian Institution Press, Washington, DC.

McWilliams, S.R., Afik, D. & Secor, S. 1997. Patterns and processes in the vertebrate digestive system. *Trends in Ecology and Evolution* 12:420–422.

Mace, G.M. 1979. *The evolutionary ecology of small mammals.* PhD thesis, University of Sussex, Brighton.

Macnab, J. 1985. Carrying capacity and related slippery shibboleths. *Wildlife Society Bulletin* 13:403–410.

Magnusson, W.E., Caughley, G.J. & Grigg, G.C. 1978. A double-survey estimate of population size from incomplete counts. *Journal of Wildlife Management* 42:174–176.

Maloiy, G.M.O. 1973. The water metabolism of a small East African antelope: the dikdik. *Proceedings of the Royal Society of London B, Biological Sciences* 184:167–178.

Malthus, T.R. 1798. *An essay on the principle of population as it affects the future improvement of society: with remarks on the speculation of Mr Godwin, M. Condoroet, and other writers.* Printed for J. Johnson, 1798. London.

Manel, S., Berthier, P. & Luikart, G. 2002. Detecting wildlife poaching: identifying the origin of individuals with Bayesian assignment tests and multilocus genotypes. *Conservation Biology* 16:650–659.

Mangel, M. & Clark, C.W. 1986. Towards a unified foraging theory. *Ecology* 67:1127–1138.

Manly, B.F., McDonald, L.L. & Thomas, D.L. 1993. *Resource Selection by Animals*. Chapman & Hall, London.

Manseau, M. 1996. *Relationships between the rivière George caribou herd and its summer range*. PhD thesis, University of Laval.

Marcogliese, D.J. 2001. Implications of climate change for parasitism of animals in the aquatic environment. *Canadian Journal of Zoology* 79:1331–1352.

Margules, C., Higgs, A.J. & Rafe, R.W. 1982. Modern biogeographical theory: are there any lessons for nature reserve design? *Biological Conservation* 24:115–128.

Marsh, R.E. 1988. Chemosterilants for rodent control. In: Prakash, I., ed. *Rodent Pest Management*, pp. 353–367. CRC Press, Boca Raton, FL.

Martin, P.S. & Klein, R.G., eds. 1984. *Quaternary Extinctions: A Prehistoric Revolution*. The University of Arizona Press, Tucson, AZ.

Martin, P.S. & Steadman, D.W. 1999. Prehistoric extinctions on islands and continents. In: MacPhee, R.D.W., ed. *Extinctions in Near Time: Causes, Contexts, and Consequences*, pp. 17–55. Kluwer Academic/Plenum Publishers, New York.

Martinez del Rio, C. 1990. Dietary, phylogenetic, and ecological correlates of intestinal sucrose and maltase activity in birds. *Physiological Zoology* 63:987–1011.

May, R.M. 1972. Limit cycles in predator–prey communities. *Science* 177:900–902.

May, R.M. 1973. *Stability and Complexity in Model Systems*. Princeton University Press, Princeton, NJ.

May, R.M. 1976. Simple mathematical models with very complicated dynamics. *Nature* 261:459–467.

May, R.M. 1977. Thresholds and breakpoints in ecosystems with a multiplicity of stable states. *Nature* 269:471–477.

May, R.M. & Oster, G. 1976. Bifurcations and dynamic complexity in simple ecological models. *American Naturalist* 110:573–599.

Maynard-Smith, J. 1982. *Evolution and the Theory of Games*. Cambridge University Press, Cambridge.

Mduma, S.A.R., Sinclair, A.R.E. & Hilborn, R. 1999. Food regulates the Serengeti wildebeest population: a 40-year record. *Journal of Animal Ecology* 68:1101–1122.

Meagher, M. & Meyer, M.E. 1994. On the origin of brucellosis in bison of Yellowstone National Park: A review. *Conservation Biology* 8:645–653.

Mech, L.D. 1994. Buffer zones of territories of gray wolves as regions of intraspecific strife. *Journal of Mammalogy* 75:199–202.

Mech, L.D. & DelGiudice, G.D. 1985. Limitations of the marrow-fat technique as an indicator of body condition. *Wildlife Society Bulletin* 13:204–206.

Mendel, G. 1959. Experiments in plant-hybridization. In: Peters, J.A., ed. *Classic Papers in Genetics*, pp. 1–19. Prentice-Hall, Englewood Cliffs, NJ.

Messier, F. 1994. Ungulate population models with predation: a case study with the North American moose. *Ecology* 75:478–488.

Messier, F. & Crete, M. 1985. Moose–wolf dynamics and the natural regulation of moose populations. *Oecologia* 65:503–512.

Messier, F. & Joly, D.O. 2000. Comment: Regulation of moose populations by wolf predation. *Canadian Journal of Zoology* 78:506–510.

Messier, F., Huot, J., LeHenaff, D. & Luttich, S. 1988. Demography of the George River caribou herd: evidence of population regulation by forage exploitation and range expansion. *Arctic* 41:279–287.

Meyer, M.E. & Meagher, M. 1995. Brucellosis in free-ranging bison (*Bison bison*) in Yellowstone, Grand Teton, and Wood Buffalo National Parks: a review. *Journal of Wildlife Diseases* 31:579–598.

Milinski, M. 1979. An evolutionarily stable feeding strategy in sticklebacks. *Zeitschrift für Tierpsychologie* 51:36–40.

Miller, B., Ceballos, G. & Reading, R. 1994. The prairie dog and biotic diversity. *Conservation Biology* 8:677–681.

Miller, F.L., Edmonds, E.J. & Gunn, A. 1982. Foraging behaviour of Peary caribou in response to springtime snow and ice conditions. *Canadian Wildlife Service Occasional Paper* 48:1–39.

Miller, W.M. 2001. Pasteurellosis. In: Williams, E.S. & Barker, I.K., eds. *Infectious Diseases of Wild Mammals*, pp. 330–339. Manson Publishing/The Veterinary Press, London.

Mills, L.S., Citta, J.J., Lair, K.P., Schwarz, M.K. & Tallmon, D.A. 2000. Estimating animal abundance using noninvasive DNA sampling: promise and pitfalls. *Ecological Applications* 10:283–294.

Mills, L.S., Soule, M.E. & Doak, D.F. 1993 The keystone-species concept in ecology and conservation. *BioScience* 43:219–224.

Milner-Gulland, E.J. 1993. An econometric analysis of consumer demand for ivory and rhino horn. *Environmental and Resource Economics* 3:73–95.

Milner-Gulland, E.J. 1997. A stochastic dynamic programming model for the management of the saiga antelope. *Ecological Applications* 7:130–142.

Milner-Gulland, E.J. & Beddington, J.R. 1993. The exploitation of the elephant for the ivory trade – an historical perspective. *Proceedings of the Royal Society of London B* 252:29–37.

Milner-Gulland, E.J. & Mace, R. 1998. *Conservation of Biological Resources*. Blackwell Science, Oxford.

Milner-Gulland, E.J., Bukreeva, O.M., Coulson, T., *et al.* 2003. Reproductive collapse in saiga antelope harems. *Nature* 422:135.

Milton, K. 1979. Factors influencing leaf choice by howler monkeys: a test of some hypotheses of food selection by generalist herbivores. *American Naturalist* 114:362–378.

Minchella, D.J. & Scott, M.E. 1991. Parasitism: a cryptic determinant of animal community structure. *Trends in Ecology and Evolution* 6:250–254.

Mladenoff, D.J. & Sickley, T.A. 1998. Assessing potential gray wolf restoration in the northeastern United States: a spatial prediction of favorable habitat and potential population levels. *Journal of Wildlife Management* 62:1–10.

Mladenoff, D.J., Sickley, T.A., Haight, R.G. & Wydeven, A.P. 1995. A regional landscape analysis and prediction of favorable gray wolf habitat in the northern Great Lakes region. *Conservation Biology* 9:279–294.

Monello, R.J., Murray, D.L. & Cassirer, E.F. 2001. Ecological correlates of pneumonia epizootics in bighorn sheep herds. *Canadian Journal of Zoology* 79:1423–1432.

Montgomery, B.R., Kelly, D. & Ladley, J.J. 2001. Pollinator limitation of seed set in *Fuchsia perscandens* (Onagraceae) on Banks Peninsula, South Island, New Zealand. *New Zealand Journal of Botany* 29:559–565.

Mooney, H.A. & Ehrlich, P.R. 1997. Ecosystem services: a fragmentary history. In: Daily, G.C., ed. *Nature's Services: Societal Dependence on Natural Ecosystems*, pp. 11–19. Island Press, Washington, DC.

Morell, V. 1999. Are pathogens felling frogs? *Science* 284:728.

Moritz, C. & Lavery, S. 1996. Molecular ecology: contributions from molecular genetics to population ecology. In: Floyd, R.B., Sheppard, A.W. & De Barro, P.J., eds. *Frontiers of Population Ecology*, pp. 433–450. CSIRO Publishing, Melbourne.

Morris, W.F. & Doak, D.F. 2002. *Quantitative Conservation Biology*. Sinauer Associates, Sunderland, MA.

Morrison, M.L., Marcot, B.G. & Mannan, R.W. 1992. *Wildlife–Habitat Relationships: Concepts and Applications*. University of Wisconsin Press, Madison, WI.

Moss, R. 1974. Winter diets, gut lengths, and interspecific competition in Alaskan ptarmigan. *Auk* 91:737–746.

Mould, E.D. & Robbins, C.T. 1981. Nitrogen metabolism in elk. *Journal of Wildlife Management* 45:323–334.

Mould, E.D. & Robbins, C.T. 1982. Digestive capabilities in elk compared to white-tailed deer. *Journal of Wildlife Management* 46:22–29.

Moulton, M.P. & Sanderson, J. 1999. *Wildlife Issues in a Changing World*, 2nd edn. CRC Press, Boca Raton, FL.

Mountainspring, S. & Scott, J.M. 1985. Interspecific competition among Hawaiian forest birds. *Ecological Monographs* 55:219–239.

Mundy, P.J. & Ledger, J.A. 1976. Griffon vultures, carnivores and bones. *South African Journal of Science* 72:106–110.

Munger, J.C. & Brown, J.H. 1981. Competition in desert rodents: an experiment with semi-permeable exclosures. *Science* 211:510–512.

Murphy, B.C. & Dowding, J.E. 1995. Ecology of the stoat in *Nothofagus* forest: home range, habitat use and diet at different stages of the beech mast cycle. *New Zealand Journal of Ecology* 19:97–109.

Murphy, D.D. & Noon, B.R. 1992. Integrating scientific methods with habitat conservation planning: reserve design for northern spotted owls. *Ecological Applications* 2:3–17.

Murphy, D.J. & Kelly, D. 2001. Scarce or distracted? Bellbird (*Anthornis melanura*) foraging and diet in an area of inadequate mistletoe pollination. *New Zealand Journal of Ecology* 25:69–81.

Murray, D.L. 2002. Differential body condition and vulnerability to predation in snowshoe hares. *Journal of Animal Ecology* 71:614–625.

Murray, D.L., Cary, J.R. & Keith, L.B. 1997. Interactive effects of sublethal nematodes and nutritional status on snowshoe hare vulnerability to predation. *Journal of Animal Ecology* 66:250–264.

Murray, D.L., Kapke, C.A., Evermann, J.F. & Fuller, T.K. 1999. Infectious disease and the conservation of free-ranging carnivores. *Animal Conservation* 2:241–254.

Murton, R.K., Isaacson, A.J. & Westwood, N.J. 1966. The relationships between wood-pigeons and their clover food supply and the mechanism of population control. *Journal of Applied Ecology* 3:55–93.

Mutze, G., Cooke, B. & Alexander, P. 1998. The initial impact of rabbit hemorrhagic disease on European rabbit population in South Australia. *Journal of Wildlife Diseases* 34:221–227.

Myers, J.P., Connors, P.G. & Pitelka, F.A. 1979. Territory size in wintering sanderlings: the effects of prey abundance and intruder density. *Auk* 96:551–561.

Myers, K. & Bults, H.G. 1977. Observations on changes in the quality of food eaten by the wild rabbit. *Australian Journal of Ecology* 2:215–229.

Naeem, S. 1998. Species redundancy and ecosystem reliability. *Conservation Biology* 12:39–45.

Naeem, S. 2002. Ecosystem consequences of biodiversity loss: the evolution of a paradigm. *Ecology* 83:1537–1552.

Naeem, S. & Li, S. 1997. Biodiversity enhances ecosystem reliability. *Nature* 390:507–509.

Naeem, S., Thompson, L.J., Lawler, S.P., Lawton, J.H. & Woodfin, R.M. 1995. Empirical evidence that declining species diversity may alter the performance of terrestrial ecosystems. *Proceedings of the Royal Society of London B* 347:249–262.

Nagy, K.A. 1980. CO_2 production in animals: analysis of potential errors in the doubly labeled water method. *American Journal of Physiology* 238:R466–R473.

Nagy, K.A. 1983. *The Doubly Labeled Water Method: A Guide to its Use*. UCLA Publication no. 12-1417. University of California, Los Angeles, CA.

Nagy, K.A. 1989. Field bioenergetics: accuracy of models and methods. *Physiological Zoology* 62:237–252.

Nagy, K.A. & Peterson, C.C. 1988. *Scaling of Water Flux Rate in Animals*. University of California Publications in Zoology, Vol. 120. University of California, Los Angeles, CA.

Nelson, M.E. & Mech, L.D. 1981. Deer social organization and wolf predation in northeastern Minnesota. *Wildlife Monographs* 77:1–53.

Nevo, E., Beiles, A. & Ben-Shlomo, R. 1984. The evolutionary significance of genetic diversity: ecological, demographic and life history correlates. *Lecture Notes in Biomathematics* 53:13–213.

Newmark, W.D. 1987. A land-bridge island perspective on mammalian extinctions in western North American parks. *Nature* 325:430–432.

Newton, I. 1972. *Finches.* Collins, London.

Nichols, J.D. 1992. Capture–recapture models: using marked animals to study population dynamics. *BioScience* 42:94–102.

Nichols, J.D., Johnson, F.A. & Willimas, B.K. 1995. Managing North American waterfowl in the face of uncertainty. *Annual Review of Ecology and Systematics* 26:177–199.

Noon, B.R. & McKelvey, K.S. 1996. A common framework for conservation planning: linking individual and metapopulation models. In: McCullough, D.R., ed. *Metapopulations and Wildlife Conservation*, pp. 139–166. Island Press, Washington, DC.

Norbury, G. 2001. Conserving dryland lizards by reducing predator-mediated apparent competition and direct competition with introduced rabbits. *Journal of Applied Ecology* 38:1350–1361.

Norbury, G.L., Norbury, D.C. & Heyward, R.P. 1998. Behavioral responses of two predator species to sudden declines in primary prey. *Journal of Wildlife Management* 62:45–58.

Norbury, G.L., Norbury, D.C. & Oliver, A.J. 1994. Facultative behaviour in unpredictable environments: mobility of red kangaroos in arid Western Australia. *Journal of Animal Ecology* 63:410–418.

Norton, G.A. 1988. Philosophy, concepts and techniques. In: Norton, G.A. & Pech, R.P., eds. *Vertebrate Pest Management in Australia: A Decision Analysis/Systems Analysis Approach*, pp. 1–17. Project Report no. 5. CSIRO, Australia.

Norton, I.O. & Sclater, J.G. 1979. A model for the evolution of the Indian Ocean and the breakup of Gondwanaland. *Journal of Geophysical Research* 84:6803–6830.

Norton-Griffiths, M. 1973. Counting the Serengeti migratory wildebeest using two-stage sampling. *East African Wildlife Journal* 11:135–149.

Norton-Griffiths, M. 1978. Counting animals. In: Grimsdell, J.J.R., ed. *Counting Animals*, 2nd edn. Handbook no. 1. African Wildlife Leadership Foundation, Nairobi.

Noss, R.F. 1987. Corridors in real landscapes: a reply to Simberloff and Cox. *Conservation Biology* 1:159–164.

Nowak, R.M. 1992. The red wolf is not a hybrid. *Conservation Biology* 6:593–595.

Nudds, T.D. 1990. Retroductive logic in retrospect: the ecological effects of meningeal worms. *Journal of Wildlife Management* 54:396–402.

Nudds, T.D., Sjoberg, K. & Lundberg, P. 1994. Ecomorphological relationships among Palearctic dabbling ducks on Baltic coastal wetlands and a comparison with the Nearctic. *Oikos* 69:295–303.

Nulsen, R.A. 1993. Changes in soil properties. In: Hobbs, R.J. & Saunders, D.A., eds. *Reintegrating Fragmented Landscapes*, pp. 107–145. Springer-Verlag, New York.

O'Brien, S.J., Nash, W.G., Wildt, D.E., Bush, M.E. & Benveniste, R.E. 1985a. A molecular solution to the riddle of the giant panda's phylogeny. *Nature* 317:140–144.

O'Brien, S.J., Roelke, M.E., Marker, L., *et al.* 1985b. Genetic basis for species vulnerability in the cheetah. *Science* 227:1428–1434.

O'Brien, S.J., Wildt, D.E. & Bush, M. 1986. The cheetah in genetic peril. *Scientific American* 254:68–76.

O'Brien, S.J., Wildt, D.E., Bush, M., *et al.* 1987. East African cheetahs: evidence for two population bottlenecks? *Proceedings of the National Academy of Sciences of the USA* 84:508–511.

O'Brien, S.J., Wildt, D.E., Goldman, D., Merril, C.R. & Bush, M. 1983. The cheetah is depauperate in genetic variation. *Science* 221:459–462.

O'Donoghue, M. 1991. *Reproduction, juvenile survival and movements of snowshoe hares at a cyclic population peak*. MSc thesis, University of British Columbia.

Ojasti, J. 1983. Ungulates and large rodents of South America. In: Bourlière, F., ed. *Ecosystems of the World: 13. Tropical Savannas*, pp. 427–439. Elsevier, Amsterdam.

O'Kelly, J.C. 1973. Seasonal variations in the plasma lipids of genetically different types of cattle: steers on different diets. *Comparative Biochemistry and Physiology* 44:303–312.

Oksanen, L. 1990. Predation, herbivory, and plant strategies along gradients of primary productivity. In: Tilman, D. & Grace, J, eds. *Perspectives on Plant Competition*, pp. 445–474. Academic Press, New York.

Orians, G.H. & Willson, M.F. 1964. Interspecific territories of birds. *Ecology* 45:736–745.

Orians, G.H. & Pearson, N.E. 1979. On the theory of central place foraging. In: Horn, D.J., Stairs, G.R. & Mitchell R.D, eds. *Analysis of Ecological Systems*, pp. 155–179. Ohio University Press, Columbus, OH.

O'Ryan, C., Flamand, J.R.B. & Harley, E.H. 1994. Mitochondrial DNA variation in black rhinoceros (*Diceros bicornis*): conservation management implications. *Conservation Biology* 8:495–500.

Osborne, P. 1985. Some effects of Dutch elm disease on the birds of a Dorset dairy farm. *Journal of Applied Ecology* 22:681–691.

Otis, D.L., Burnham, K.P., White, G.C. & Anderson, D.R. 1978. Statistical interference from capture data on closed animal populations. *Wildlife Monographs* 62:1–135.

Otterman, J. 1974. Baring high-albedo soils by overgrazing: a hypothesized desertification mechanism. *Science* 186:531–533.

Owen, R.B., Jr. & Reinecke, K.J. 1979. Bioenergetics of breeding dabbling ducks. In: Bookhout, T.A., ed. *Waterfowl and Wetlands: An Integrated Review*, pp. 71–93. Proceedings of a Symposium held at the 39th Midwestern Fish and Wildlife Conference, Madison, Wisconsin. The Wildlife Society, Washington, DC.

Owen-Smith, N. 1988. *Megaherbivores: The Influence of Very Large Body Size on Ecology*. Cambridge University Press, Cambridge.

Owen-Smith, N. 1990. Demography of a large herbivore, the greater kudu *Tragelaphus strepsiceros*, in relation to rainfall. *Journal of Animal Ecology* 59:893–913.

Owen-Smith, N. 1999. The interaction of humans, megaherbivores, and habitats in the late Pleistocene extinction event. In: MacPhee, R.D.W., ed. *Extinctions in Near Time: Causes Contexts, and Consequences*, pp. 17–55. Kluwer Academic/Plenum Publishers, New York.

Owen-Smith, N. & Cooper, S.M. 1987. Palatability of woody plants to browsing ruminants in a South African savanna. *Ecology* 68:319–331.

Owen-Smith, N. & Cooper, S.M. 1989. Nutritional ecology of a browsing ruminant, the kudu (*Tragelaphus strepsiceros*), through the seasonal cycle. *Journal of Zoology (London)* 219:29–43.

Packer, C., Hilborn, R., Mosser, A., *et al.* 2005. Ecological change, group territoriality, and population dynamics in Serengeti lions. *Science* 307:390–393.

Pagel, M., May, R.M. & Collie, A.R. 1991. Ecological aspects of the geographic distribution and diversity of mammal species. *American Naturalist* 137:791–815.

Paine, R.T. 1969. A note on trophic complexity and species diversity. *American Naturalist* 103:91–93.

Palomares, F., Gaona, P., Ferreras, P. & Delibes, M. 1995. Positive effects on game species of top predators by controlling smaller predator populations: an example with lynx, mongooses, and rabbits. *Conservation Biology* 9:295–305.

Parer, I. 1987. Factors influencing the distribution and abundance of rabbits (*Oryctolagus cuniculus*) in Queensland. *Proceedings of the Royal Society of Queensland* 98:73–82.

Parer, I., Conolly, D. & Sobey, W.R. 1985. Myxomatosis: the effects of annual introductions of an immunizing strain and a highly virulent strain of myxoma virus into rabbit populations at Urana, N.S.W. *Australian Wildlife Research* 12:407–423.

Parker, P.G., Snow, A.A., Schug, M.D., Booton, G.F. & Fuerst, P.A. 1998. What molecules can tell us about populations: choosing and using a molecular marker. *Ecology* 79:361–382.

Parkes, J.P. 1984. Feral goats on Raoul Island. I. Effect of control methods on their density, distribution, and productivity. *New Zealand Journal of Ecology* 7:85–94.

Parkes, J.P. & Tustin, K.G. 1985. A reappraisal of the distribution and dispersal of female Himalayan tahr in New Zealand. *New Zealand Journal of Ecology* 8:5–10.

Parkes, J.P., Macdonald, N. & Leaman, G. 2002. An attempt to eradicate feral goats from Lord Howe Island. In: Veitch, C.R. & Clout, M.N., eds. *Turning the Tide: The Eradication of Invasive Species*, pp. 233–239. Occasional Paper of the IUCN Species Survival Commission no. 27. IUCN, Gland, Switzerland.

Parmesan, C. & Yohe, G. 2003. A globally coherent fingerprint of climate change impacts across natural systems. *Nature* 421:37–42.

Pastor, J., Dewey, B., Naiman, R.J., McInnes, P.F. & Cohen, Y. 1993. Moose browsing and soil fertility in the boreal forests of Isle Royale National Park. *Ecology* 74:467–480.

Patterson, I.J. 1965. Timing and spacing of broods in the black-headed gull. *Ibis* 107:433–459.

Paulus, S.L. 1982. Gut morphology of gadwalls in Louisiana in winter. *Journal of Wildlife Management* 46:483–489.

Pease, J.L., Vowles, R.H. & Keith, L.B. 1979. Interaction of snowshoe hares and woody vegetation. *Journal of Wildlife Management* 43:43–60.

Pech, R.P. 2000. Biological control of vertebrate pests. In: Salmon, T.P. & Crabb, A.C., eds. *Proceedings of the 19th Vertebrate Pest Conference*, pp. 206–211. University of California, Davis, CA.

Pech, R.P. & McIlroy, J.C. 1990. A model of the velocity of advance of foot and mouth disease in feral pigs. *Journal of Applied Ecology* 27:635–650.

Pech, R.P., Sinclair, A.R.E., Newsome, A.E. & Catling, P.C. 1992. Limits to predator regulation of rabbits in Australia: evidence from predator-removal experiments. *Oecologia* 89:102–112.

Pehrsson, O. 1984. Relationship of food to spatial and temporal breeding strategies of mallards in Sweden. *Journal of Wildlife Management* 48:322–339.

Pelletier, L. & Krebs, C.J. 1997. Line-transect sampling for estimating ptarmigan (*Lagopus* spp.) density. *Canadian Journal of Zoology* 75:1185–1192.

Penning, S.C., Nadeau, M.T. & Paul, V.J. 1993. Selectivity and growth of the generalist herbivore *Dolabella auricularia* feeding upon complementary resources. *Ecology* 74:879–890.

Percival, S.M., Sutherland, W.J. & Evans, P.R. 1998. Intertidal habitat loss and wildfowl numbers: applications of a spatial depletion model. *Journal of Applied Ecology* 35:57–63.

Perrins, C.M. 1970. The timing of birds' breeding seasons. *Ibis* 112:242–255.

Petersen, C.W. & Levitan, D.R. 2001. The Allee effect: a barrier to recovery by exploited species. In: Reynolds, J.D., Mace, G.M., Redford, K.H. & Robinson, J.G., eds. *Conservation of Exploited Species*, pp. 281–300. Cambridge University Press, Cambridge.

Peterson, M.J., Grant, W.E. & Davis, D.S. 1991. Bison-brucellosis management: simulation of alternative strategies. *Journal of Wildlife Management* 55:205–213.

Peterson, R.O. 1999. Wolf–moose interaction on Isle Royale: the end of natural regulation? *Ecological Applications* 9:10–16.

Peterson, R.O. & Vucetich, J.A. 2003. *Ecological Studies of Wolves on Isle Royale: Annual Report 2002–2003*. Michigan Technological University, Houghton, MI.

Peterson, R.O., Thomas, N.J., Thurber, J.M., Vucetich, J.A. & Waite, T.A. 1998. Population limitation and the wolves of Isle Royale. *Journal of Mammalogy* 79:828–841.

Pianka, E.R., Huey, R.B. & Lawlor, L.P. 1979. Niche segregation in desert lizards. In: Horn, D.J., Stairs, G.R. & Mitchell, R.D., eds. *Analysis of Ecological Systems*, pp. 67–116. Ohio State University Press, Columbus, OH.

Pickett, S.T.R. & White, P.S., eds. 1985. *The Ecology of Natural Disturbance and Patch Dynamics*. Academic Press, Orlando, FL.

Pierce, G.J. & Ollason, J.G. 1987. Eight reasons why optimal foraging is a complete waste of time. *Oikos* 49:111–118.

Piersma, T. & Lindstrom, A. 1997. Rapid reversible changes in organ size as a component of adaptive behaviour. *Trends in Ecology and Evolution* 12:134–138.

Pimm, S.L. & Pimm, J.W. 1982. Resource use, competition and resource availability in Hawaiian honeycreepers. *Ecology* 63:1468–1480.

Plowright, W. 1982. The effects of rinderpest and rinderpest control on wildlife in Africa. *Symposium of the Zoological Society of London* 50:1–28.

Pojar, T.M., Bowden, D.C. & Gill, R.B. 1995. Aerial counting experiments to estimate pronghorn density and herd structure. *Journal of Wildlife Management* 59:117–128.

Polis, G.A. 1999. Why are parts of the world green? Multiple factors control productivity and the distribution of biomass. *Oikos* 86:3–15.

Pollock, K.H. 1974. *The assumption of equal catchability in tag–recapture experiments.* PhD dissertation, Cornell University, Ithaca, NY.

Pollock, K.H. & Kendall, W.L. 1987. Visibility bias in aerial surveys: a review of estimation procedures. *Journal of Wildlife Management* 51:502–510.

Pollock, K.H., Winterstein, S.R., Bunck, C.M. & Curtis, P.D. 1989. Survival analysis in telemetry studies: the staggered entry design. *Journal of Wildlife Management* 53:7–15.

Porter, R.N. 1977. *Wildlife Management Objectives and Practices for the Hluhluwe Game Reserve and the Northern Corridor.* Natal Parks Board (unpublished).

Possingham, H.P. & Davies, I. 1995. ALEX: a model for the viability analysis of spatially structured populations. *Biological Conservation* 73:143–150.

Post, E. & Stenseth, N.C. 1998. Large-scale fluctuation and population dynamics of moose and white-tailed deer. *Journal of Animal Ecology* 67:537–543.

Post, E., Peterson, R.O., Stenseth, N.C. & McLaren, B.E. 1999. Ecosystem consequences of wolf behavioural response to climate. *Nature* 401:905–907.

Poulin, R. 1995. "Adaptive" changes in the behaviour of parasitized animals: a critical review. *International Journal for Parasitology* 25:1371–1383.

Poulin, R. 1999. The functional importance of parasites in animal communities: many roles at many levels? *International Journal for Parasitology* 29:903–914.

Pounds, J.A., Fogden, M.P.L. & Campbell, J.H. 1999. Biological response to climate change on a tropical mountain. *Nature* 398:611–615.

Power, M.E., Tilman, D., Estes, J.A., *et al.* 1996. Challenges in the quest for keystones. *BioScience* 46:609–620.

Pratt, H.D., Bruner, P.L. & Berrett, D.G., eds. 1987. *A Field Guide to the Birds of Hawaii and the Tropical Pacific.* Princeton University Press, Princeton, NJ.

Prenzlow, D.M. & Lovvorn, J.R. 1996. Evaluation of visibility correction factors for waterfowl surveys in Wyoming. *Journal of Wildlife Management* 60:286–297.

Probst, J.R. 1986. A review of factors limiting the Kirtland's warbler on its breeding grounds. *American Midland Naturalist* 116:87–100.

Provenza, F.D., Burritt, E.A., Clausen, T.P., Bryant, J.P., Reichardt, P.B. & Distel, R.A. 1990. Conditioned flavor aversion: a mechanism for goats to avoid condensed tannins in blackbrush. *American Naturalist* 136:810–828.

Prugh, L.R. 2004. *Foraging ecology of coyotes in the Alaska Range.* PhD thesis, University of British Columbia.

Prugh, L.R., Ritland, C.E., Arthur, S.M. & Krebs, C.J. 2005. Monitoring coyote population dynamics by genotyping feces. *Molecular Biology* 14:1585–1596.

Pulliam, H.R. 1974. On the theory of optimal diets. *American Naturalist* 108:59–75.

Pulliam, H.R. 1988. Sources, sinks, and population regulation. *American Naturalist* 132:652–661.

Pulliam, H.R. & Danielson, B.J. 1991. Sources, sinks and habitat selection: a landscape perspective. *American Naturalist* 137:50–66.

Pusey, A.E. 1987. Sex-biased dispersal and inbreeding avoidance in birds and mammals. *Trends in Ecology and Evolution* 2:295–299.

Pusey, A.E. 1992. The primate perspective on dispersal. In: Stenseth, N.C. & Lidicker, W.Z., Jr., eds. *Animal Dispersal*, pp. 243–259. Chapman & Hall, London.

Quinn, T.W., Quinn, J.S., Cooke, F. & White, B.N. 1987. DNA marker analysis detects multiple maternity and paternity in single broods of the lesser snow goose. *Nature* 326:392–394.

Raj, D. & Khamis, S.H. 1958. Some remarks on sampling with replacement. *Annals of Mathematical Statistics* 29:550–557.

Ralls, K., Ballou, J.D. & Templeton, A. 1988. Estimates of lethal equivalents and the cost of inbreeding in mammals. *Conservation Biology* 2:185–193.

Ramakrishnan, U., Coss, R.G. & Pelkey, N.W. 1999. Tiger decline caused by the reduction of large ungulate prey: evidence from a study of leopard diets in southern India. *Biological Conservation* 89:113–120.

Ramsey, D.S.L. 2000. The effect of fertility control on the population dynamics and behavior of brushtail possums (*Trichosurus vulpecula*) in New Zealand. In: Salmon, T.P. & Crabb, A.C.; eds. *Proceedings of the 19th Vertebrate Pest Conference*, pp. 206–211. University of California, Davis, CA.

Ransom, A.B. 1965. Kidney and marrow fat as indicators of white-tailed deer condition. *Journal of Wildlife Management* 29:397–399.

Rapoport, E.H. 1982. *Aereography*. Pergamon Press, Oxford.

Rasmussen, D.I. & Doman, E.R. 1943. Census methods and their application in the management of mule deer. In: *Transactions of the 8th North American Wildlife Conference*, pp. 369–380. American Wildlife Institute, Washington, DC.

Ratcliffe, D.A. 1970. Changes attributable to pesticides in egg breakage frequency and eggshell thickness in some British birds. *Journal of Applied Ecology* 7:67–115.

Redfield, J.A., Krebs, C.J. & Taitt, M.J. 1977. Competition between *Peromyscus maniculatus* and *Microtus townsendii* in grasslands of coastal British Columbia. *Journal of Animal Ecology* 46:607–616.

Redhead, T.D. 1982. *Reproduction, growth and population dynamics of house mice in irrigated and non-irrigated cereal farms in New South Wales*. PhD thesis, Australian National University, Canberra.

Redpath, S.M. & Thirgood, S.J. 1999. Numerical and functional responses in generalist predators: hen harriers and peregrines on Scottish grouse moors. *Journal of Animal Ecology* 68:879–892.

Redpath, S.M., Thirgood, S.J. & Leckie, F.M. 2001. Does supplementary feeding reduce predation of red grouse by hen harriers? *Journal of Applied Ecology* 38:1157–1168.

Reed, J.E., Baker, R.J., Ballard, W.B. & Kelly, B.T. 2004. Differentiating Mexican gray wolf and coyote scats using DNA analysis. *Wildlife Society Bulletin* 32:685–692.

Relyea, R.A., Lawrence, R.K. & Demarias, S. 2000. Home range of desert mule deer: testing the body-size and habitat-productivity hypothesis. *Journal of Wildlife Management* 64:146–153.

Rettie, W.J. & Messier, F. 1998. Dynamics of woodland caribou populations at the southern limit of their range in Saskatchewan. *Canadian Journal of Zoology* 76:251–259.

Reynolds, P.E. 1998. Dynamics and range expansion of a reestablished muskox population. *Journal of Wildlife Management* 62:734–744.

Richards, S.A., Nisbet, R.M., Wilson, W.G. & Possingham, H.P. 2000. Grazers and diggers: exploitation competition and coexistence among foragers with different feeding strategies on a single resource. *American Naturalist* 155:266–279.

Richman, L.K., Montali, R.J., Garber, R.L., *et al.* 1999. Novel endotheliotropic herpesvirus fatal for Asian and African elephants. *Science* 283:1171–1176.

Ricker, W.E. 1954. Stock and recruitment. *Journal of the Fisheries Research Board of Canada* 11:559–623.

Ricklefs, R.E. & Miller, G.L. 2000. *Ecology*, 4th edn. W.H. Freeman & Co., New York.

Ringelman, J.K. & Szymczak, M.R. 1985. A physiological condition index for wintering mallards. *Journal of Wildlife Management* 49:564–568.

Rivest, L.-P. & Crepeau, H. 1990. A two-phase sampling plan for the estimation of the size of a moose population. *Biometrics* 46:163–176.

Robbins, C.T. 1983. *Wildlife Feeding and Nutrition*. Academic Press, New York.

Robbins, C.T., Hanley, T.A., Hagerman, A.E., *et al.* 1987. Role of tannins in defending plants against ruminants: reduction in protein availability. *Ecology* 68:98–107.

Robertshaw, D. & Taylor, C.R. 1969. A comparison of sweat gland activity in eight species of East African bovids. *Journal of Physiology* 203:135–143.

Robertson, G. 1987. Plant dynamics. In: Caughley, G., Shepherd, N. & Short, J., eds. *Kangaroos: Their Ecology and Management in the Sheep Rangelands of Australia*, pp. 50–68. Cambridge University Press, Cambridge.

Robertson, G., Short, J. & Wellard, G. 1987. The environment of the Australian sheep rangelands. In: Caughley, G., Shepherd, N. & Short, J., eds. *Kangaroos: Their Ecology and Management in the Sheep Rangelands of Australia*, pp. 14–34. Cambridge University Press, Cambridge.

Robertson, P.A. & Rosenberg, A.A. 1988. Harvesting gamebirds. In: Hudson, P.J. & Rands, M.R.W., eds. *Ecology and Management of Gamebirds*, pp. 177–201. BSP Professional Books, Oxford.

Robinette, W.L., Jones, D.A. & Loveless, C.M. 1974. Field tests of strip census methods. *Journal of Wildlife Management* 38:81–96.

Robley, A., Short, J. & Bradley, S. 2001. Dietary overlap between the burrowing bettong (*Bettongia lesurur*) and the European rabbit (*Oryctolagus cuniculus*) in semi-arid coastal Western Australia. *Wildlife Research* 28:341–349.

Robson, D.S. & Whitlock, J.H. 1964. Estimation of a truncation point. *Biometrika* 51:33–39.

Rodda, G.H., Fritts, T.H., Campbell, E.W., III, Dean-Bradley, K., Perry, G. & Qualls, C.P. 2002. Practical concerns in the eradication of island snakes. In: Veitch, C.R. & Clout, M.N., eds. *Turning the Tide: The Eradication of Invasive Species*, pp. 260–265. Occasional Paper of the IUCN Species Survival Commission no. 27. IUCN, Gland, Switzerland.

Rodgers, J.A., Linda, S.B. & Nesbitt, S.A. 1995. Comparing aerial estimates with ground counts of nests in wood stork colonies. *Journal of Wildlife Management* 59:656–666.

Roelke-Parker, M.E., Munson, L., Packer, C., *et al.* 1996. A canine distemper virus epidemic in Serengeti lions (*Panthera leo*). *Nature* 379:441–445.

Roemer, G.W., Donlan, C.J. & Courchamp, F. 2002. Golden eagles, feral pigs, and insular carnivores: How exotic species turn native predators into prey. *Proceedings of the National Academy of Sciences of the USA* 99:791–796.

Rogers, L.L., Mech, L.D., Dawson, D.K., Peek, J.M. & Korb, M. 1980. Deer distribution in relation to wolf pack territory edges. *Journal of Wildlife Management* 44:253–258.

Rohner, C., Doyle, F.I. & Smith J.N.M. 2001. Great horned owls. In: Krebs, C.J., Boutin, S. & Boonstra R., eds. *Ecosystem Dynamics of the Boreal Forest: The Kluane Project*, pp. 339–376. Oxford University Press, New York.

Rolley, R.E. & Keith, L.B. 1980. Moose population dynamics and winter habitat use at Rochester, Alberta, 1965–1979. *Canadian Field-Naturalist* 94:9–18.

Rominger, E.M., Whitlaw, H.A., Weybright, D.L., Dunn, W.C. & Ballard, W.B. 2004. The influence of mountain lion predation on bighorn sheep translocations. *Journal of Wildlife Management* 68:993–999.

Root, A. 1972. Fringe-eared oryx digging for tubers in the Tsavo National Park (East). *East African Wildlife Journal* 10:155–157.

Root, A. 1988. *Atlas of Wintering North American Birds: An Analysis of Christmas Bird Count Data*. The University of Chicago Press, Chicago, IL.

Rosenberg, A.A., Fogarty, M.J., Sissenwine, M.P., Beddington, J.R. & Shepherd, J.G. 1993. Achieving sustainable use of renewable resources. *Science* 262:828–829.

Rosenzweig, M.L. 1971. Paradox of enrichment: destabilization of exploitation ecosystems in ecological time. *Science* 171:385–387.

Rosenzweig, M.L. 1981. A theory of habitat selection. *Ecology* 62:327–335.

Rosenzweig, M.L. & MacArthur, R.H. 1963. Graphical representation and stability conditions of predator–prey interaction. *American Naturalist* 97:209–223.

Ross, J. 1982a. Myxomatosis: The natural evolution of the disease. In: Edwards, M.A. & McDonnell, U., eds. *Animal Disease in Relation to Animal Conservation*, pp. 77–95. Academic Press London.

Ross, J. 1982b. *Myxomatosis: The Natural Evolution of the Disease.* Proceedings of the Zoological Society of London no. 50. Academic Press, London.

Rossiter, P. 2001. Rinderpest. In: Williams, E.S. & Barker, I.K., eds. *Infectious Diseases of Wild Mammals*, pp. 39–45. Manson Publishing/The Veterinary Press, London.

Rothstein, S.I. 1994. The cowbird's invasion of the far west: history, causes and consequences experienced by host species. *Studies in Avian Biology* 15:301–315.

Routledge, R.D. 1981. The unreliability of population estimates from repeated, incomplete aerial surveys. *Journal of Wildlife Management* 45:997–1000.

Routledge, R.D. 1982. The method of bounded counts: when does it work? *Journal of Wildlife Management* 46:757–761.

Rubsamen, K., Heller, R., Lawrenz, H. & Engelhardt, W.V. 1979. Water and energy metabolism in the rock hyrax (*Procavia habessinica*). *Journal of Comparative Physiology B* 131:303–309.

Rudolph, B.A., Porter, W.F. & Underwood, H.B. 2000. Evaluating immunocontraception for managing suburban white-tailed deer in Irondequoit, New York. *Journal of Wildlife Management* 64:463–473.

Rupprecht, C.E., Smith, J.S., Fekadu, M. & Childs, J.E. 1995. The ascension of wildlife rabies: a cause for public health concern or intervention? *Emerging Infectious Diseases* 4:107–113.

Ruxton, G.D., Gurney, W.S.C. & de Roos, A.M. 1992. Interference and generation cycles. *Theoretical Population Biology* 42:235–253.

Ryder, O.A. 1993. Prsewalski's Horse: Prospects for reintroduction into the wild. *Conservation Biology* 7:13–15.

Saccheri, I., Kuussaari, M., Kankare, M., Vikman, P., Fortelius, W. & Hanski, I. 1998. Inbreeding and extinction in a butterfly population. *Nature* 392:491–494.

Sæther, B.E., Engen, S. & Solberg, J. 2001. Optimal harvest of age-structured populations of moose *Alces alces* in a fluctuating environment. *Wildlife Biology* 7:171–179.

Sæther, B.E., Ringsby, T.H., Bakke, Ø. & Solberg, E.J. 1999. Spatial and temporal variation in demography of a house sparrow metapopulation. *Journal of Animal Ecology* 68:628–637.

Sæther, B.E., Tufto, J., Engen, S., Jerstad, K., Røstad, O.W. & Skåtan, J.E. 2000. Population dynamical consequences of climate change for a small temperate songbird. *Science* 287:854–856.

Samuel, W.M., Pybus, M.J., Welch, D.A. & Wilke, C.J. 1992. Elk as a potential host for meningeal worm: implications for translocation. *Journal of Wildlife Management* 56:629–639.

Sauer, J.R., Dolton, D.D. & Droege, S. 1994. Mourning dove population trend estimates from call-count and North American breeding bird surveys. *Journal of Wildlife Management* 58:506–515.

Saunders, D.A. & Hobbs, R.J., eds. 1991. *Nature Conservation 2: The Role of Corridors.* Surrey Beatty & Sons, Chipping Norton, New South Wales.

Saunders, D.A., Hobbs, R.J. & Arnold, G.W. 1993. The Kellerberrin project on fragmented landscapes: a review of current information. *Biological Conservation* 64:185–192.

Saunders, G. & McIlroy, J. 2001. Fertility control of foxes using immunocontraception – ecologically feasible? In: *Proceedings of the 12th Australasian Vertebrate Pest Conference*, pp. 169–172. Melbourne.

Savidge, J. 1987. Extinction of an island forest avifauna by an introduced snake. *Ecology* 68:660–668.

Schall, J.J. 1992. Parasite-mediated competition in *Anolis* lizards. *Oecologia* 92:58–64.

Schall, J.J. & Pianka, E.R. 1978. Geographical trends in numbers of species. *Science* 201:679–686.

Schaller, G.B. 1972. *The Serengeti Lion*. The University of Chicago Press, Chicago, IL.

Schaller, G.B., Jinchu, H., Wenski, P. & Jing, Z. 1985. *The Giant Pandas of Wolong*. The University of Chicago Press, Chicago, IL.

Scheffer, M., Carpenter, S.R., Foley, J.A., Folke, C. & Walker, B. 2001. Catastrophic shifts in ecosystems. *Nature* 413:591–596.

Schluter, D. 1981. Does the theory of optimal diets apply in complex environments? *American Naturalist* 118:139–147.

Schluter, D. 1988. The evolution of finch communities on islands and continents: Kenya vs. Galapagos. *Ecological Monographs* 58:229–249.

Schluter, D. 2000. *The Ecology of Adaptive Radiation*. Oxford University Press, Oxford.

Schluter, D. & Ricklefs, R.E. 1993. Species diversity: an introduction to the problem. In: Ricklefs, R.E. & Schluter, D., eds. *Species Diversity in Ecological Communities*, pp. 1–10. The University of Chicago Press, Chicago, IL.

Schmitz, O.J. & Nudds, T.D. 1994. Parasite-mediated competition in deer and moose: how strong is the effect of meningeal worm on moose? *Ecological Applications* 4:91–103.

Schmitz, O.J., Beckerman, A.P. & O'Brien, K.M. 1997. Behaviorally mediated trophic cascades: effects of predation risk on food web interactions. *Ecology* 78:1388–1399.

Schneider, S.H. & Root, T.L., eds. 2002. *Wildlife Responses to Climate Change*. Island Press, Washington, DC.

Schoener, T.W. 1971. Theory of feeding strategies. *Annual Review of Ecology and Systematics* 2:369–404.

Schoener, T.W. 1983. Simple models of optimal feeding-territory size: a reconciliation. *American Naturalist* 121:608–629.

Schoener, T.W. 1989. The ecological niche. In: Cherrett, J.M., ed. *Ecological Concepts: The Contribution of Ecology to an Understanding of the Natural World*, pp. 79–113. 29th Symposium of the British Ecological Society. Blackwell Scientific Publications, Oxford.

Schoener, T.W. & Spiller, D.A. 1987. High population persistence in a system with high turnover. *Nature* 330:474–477.

Schrag, S.J. & Wiener, P. 1995. Emerging infectious disease: what are the relative roles of ecology and evolution? *Trends in Ecology and Evolution* 10:319–324.

Schullery, P. 1984. *Mountain Time*. Nick Lyons Books, New York.

Schulze, E.D. & Mooney, H.A., eds. 1993. *Biodiversity and Ecosystem Function*. Springer-Verlag, Berlin.

Schwagmeyer, P.L. & Woontner, S.J. 1985. Mating competition in an asocial ground squirrel, *Spermophilus tridecemlineatus*. *Behavioural Ecology and Sociobiology* 17:291–296.

Schwartz, C.C., Nagy, J.G. & Regelin, W.L. 1980. Juniper oil yield, terpenoid concentration, and antimicrobial effects on deer. *Journal of Wildlife Management* 44:107–113.

Schwartz, M.W., Brigham, C.A., Hoeksema, J.D., Lyons, K.G., Mills, M.H. & van Mantgem, P.J. 2000. Linking biodiversity to ecosystem function: implications for conservation ecology. *Oecologia* 122:297–305.

Scott, J.M., Mountainspring, S., Ramsey, F.L. & Kepler, C.B. 1986. *Forest Bird Communities of the Hawaiian Islands: Their Dynamics, Ecology, and Conservation*. Studies in Avian Biology no. 9. Cooper Ornithological Society, University of California, Los Angeles, CA.

Scott, M.E. 1987. Regulation of mouse colony abundance by *Heligmosomoides polygyrus* (Nematoda). *Parasitology* 95:111–124.

Scott, M.E. & Dobson, A. 1989. The role of parasites in regulating host abundance. *Parasitology Today* 5:176–183.

Scott, M.E. & Lewis, J.W. 1987. Population dynamics of helminth parasites in wild and laboratory rodents. *Mammal Review* 17:95–103.

Seal, U.S., Thorne, E.T., Bogan, M.A. & Anderson, S.H. 1989. *Conservation Biology and the Black-footed Ferret*. Yale University Press, New Haven, CT.

Seber, G.A.F. 1982. *The Estimation of Animal Abundance and Related Parameters*, 2nd edn. Macmillan, New York.

Seip, D.R. 1991. Predation and caribou populations. Proceedings of the Fifth North American Caribou Workshop, Yellowknife, Northwest Territories, Canada. *Rangifer* Special Issue no. 7:46–52.

Seip, D.R. 1992. Factors limiting woodland caribou populations and their inter-relationships with wolves and moose in southeastern British Columbia. *Canadian Journal of Zoology* 70:1494–1503.

Semel, B. & Andersen, D.C. 1988. Vulnerability of acorn weevils (Coleoptera: Curculionidae) and attractiveness of weevils and infested *Quercus alba* acorns to *Peromyscus leucopus* and *Blarina brevicauda*. *American Midland Naturalist* 119:385–393.

Serena, M., ed. 1994. *Reintroduction Biology of Australian and New Zealand Fauna*. Surrey Beatty & Sons, Chipping Norton, New South Wales.

Shaw, P. 1985. Late Quaternary landforms and environmental changes in northwest Botswana: the evidence of Lake Ngami and Mababe Depression. *Transactions of the Institute of British Geography, New Series* 10:333–346.

Shellam, G.R. 1994. The potential of murine cytomegalovirus as a viral vector for immuno-contraception. *Reproduction, Fertility and Development* 6:401–409.

Shepherd, N. & Caughley, G. 1987. Options for management of kangaroos. In: Caughley, G., Shepherd, N. & Short, J., eds. *Kangaroos: Their Ecology and Management in the Sheep Rangelands of Australia*, pp. 188–219. Cambridge University Press, Cambridge.

Short, J. 1987. Factors affecting food intake of rangeland herbivores. In: Caughley, G., Shepherd, N. & Short, J., eds. *Kangaroos: Their Ecology and Management in the Sheep Rangelands of Australia*, pp. 84–99. Cambridge University Press, Cambridge.

Short, J. & Smith, A. 1994. Mammal decline and recovery in Australia. *Journal of Mammalogy* 75:288–297.

Short, J., Bradshaw, S.D., Giles, J., Prince, R.I.T. & Wilson, G.R. 1992. Reintroduction of macropods (Marsupialia: Macropodoidea) in Australia – a review. *Biological Conservation* 62:189–204.

Short, J., Kinnear, J.E. & Robley, A. 2000. Surplus killing by introduced predators in Australia – evidence for ineffective anti-predator adaptations in native prey species? *Biological Conservation* 103:283–301.

Shurin, J.B. & Srivastava, D.S. 2005. New perspectives on local and regional diversity: beyond saturation. In: Holyoak, M., Leibold, M.A. & Hold, R.D., eds. *Metacommunities*. The University of Chicago Press, Chicago, IL (pp. 399–417).

Sibly, R.M. 1981. Strategies of digestion and defecation. In: Townsend, C.R. & Calow, P., eds. *Physiological Ecology: An Evolutionary Approach to Resource Use*, pp. 109–139. Blackwell Scientific Publications, Oxford.

Sih, A. 1980. Optimal behavior: can foragers balance two conflicting demands? *Science* 210:1041–1042.

Sih, A. & Christensen, B. 2001. Optimal diet theory: when does it work and when and why does it fail? *Animal Behaviour* 61:379–390.

Simberloff, D. 1988. The contribution of population and community biology to conservation science. *Annual Review of Ecology and Systematics* 19:473–511.

Simberloff, D. & Cox, J. 1987. Consequences and costs of conservation corridors. *Conservation Biology* 1:63–71.

Simberloff, D.S. & Wilson, E.O. 1970. Experimental zoogeography of islands: a two-year record of colonization. *Ecology* 51:934–937.

Sinclair, A.R.E. 1973. Population increases of buffalo and wildebeest in the Serengeti. *East African Wildlife Journal* 11:93–107.

Sinclair, A.R.E. 1977. *The African Buffalo: A Study of the Resource Limitation of Populations.* The University of Chicago Press, Chicago, IL.

Sinclair, A.R.E. 1978. Factors affecting the food supply and breeding season of resident birds and movements of palaearctic migrants in a tropical African savannah. *Ibis* 120:480–497.

Sinclair, A.R.E. 1979a. Dynamics of the Serengeti ecosystem: process and pattern. In: Sinclair, A.R.E. & Norton-Griffiths, M., eds. *Serengeti: Dynamics of an Ecosystem*, pp. 1–30. The University of Chicago Press, Chicago, IL. (2nd edn 1995.)

Sinclair, A.R.E. 1979b. The eruption of the ruminants. In: Sinclair, A.R.E. & Norton-Griffiths, M., eds. *Serengeti: Dynamics of an Ecosystem*, pp. 82–103. The University of Chicago Press, Chicago, IL. (2nd edn 1995.)

Sinclair, A.R.E. 1983. The adaptations of African ungulates and their effects on community function. In: Bourlière, F., ed. *Ecosystems of the World: 13. Tropical Savannas*, pp. 401–426. Elsevier, Amsterdam.

Sinclair, A.R.E. 1985. Does interspecific competition or predation shape the African ungulate community? *Journal of Animal Ecology* 54:899–918.

Sinclair, A.R.E. 1989. Population regulation in animals. In: Cherrett, J.M., ed. *Ecological Concepts: The Contribution of Ecology to an Understanding of the Natural World*, pp. 197–241. 29th Symposium of the British Ecological Society. Blackwell Scientific Publications, Oxford.

Sinclair, A.R.E. 1992. Do large mammals disperse like small mammals? In: Stenseth, N.C. & Lidicker, W.Z., Jr., eds. *Animal Dispersal*, pp. 229–242. Chapman & Hall, London.

Sinclair, A.R.E. 1995. Serengeti past and present. In: Sinclair, A.R.E. & Arcese, P., eds. *Serengeti II: Dynamics, Management and Conservation of an Ecosystem*, pp. 3–30. The University of Chicago Press, Chicago, IL.

Sinclair, A.R.E. 1996. Mammal populations: fluctuation, regulation, life history theory and their implications for conservation. In: Floyd, R.B. & Sheppard, A.W, eds. *Frontiers and Applications of Population Ecology*, pp. 127–154. CSIRO, Melbourne.

Sinclair, A.R.E. 1998. Natural regulation of ecosystems in protected areas as ecological baselines. *Wildlife Society Bulletin* 26:399–409.

Sinclair, A.R.E. 2003. Mammal population regulation, keystone processes and ecosystem dynamics. *Philosophical Transactions of the Royal Society of London B* 358:1729–1740.

Sinclair, A.R.E. & Arcese, P. 1995. Population consequences of predation-sensitive foraging: the Serengeti wildebeest. *Ecology* 76:882–891.

Sinclair, A.R.E. & Duncan, P. 1972. Indices of condition in tropical ruminants. *East African Wildlife Journal* 10:143–149.

Sinclair, A.R.E. & Fryxell, J.M. 1985. The Sahel of Africa: ecology of a disaster. *Canadian Journal of Zoology* 63:987–994.

Sinclair, A.R.E. & Krebs, C.J. 2002. Complex numerical responses to top-down and bottom-up processes in vertebrate populations. *Philosophical Transactions of the Royal Society of London B* 357:1221–1231.

Sinclair, A.R.E. & Smith, J.N.M. 1984. Do plant secondary compounds determine feeding preferences of snowshoe hares? *Oecologia* 61:403–410.

Sinclair, A.R.E., Dublin, H. & Borner, M. 1985. Population regulation of Serengeti wildebeest: a test of the food hypothesis. *Oecologia* 65:266–268.

Sinclair, A.R.E., Gosline, J.M., Holdsworth, G., *et al.* 1993. Can the solar cycle and climate synchronize the snowshoe hare cycle in Canada? Evidence from tree rings and ice cores. *American Naturalist* 141:173–198.

Sinclair, A.R.E., Krebs, C.J. & Smith, J.N.M. 1982. Diet quality and food limitation in herbivores: the case of the snowshoe hare. *Canadian Journal of Zoology* 60:889–897.

Sinclair, A.R.E., Krebs, C.J., Smith, J.N.M. & Boutin, S. 1988. Population biology of snowshoe hares: III. Nutrition, plant secondary compounds and food limitation. *Journal of Animal Ecology* 57:787–806.

Sinclair, A.R.E., Mduma, S.A.R. & Arcese, P. 2000. What determines the phenology and synchrony of births in Serengeti ungulates? *Ecology* 81:2100–2111.

Sinclair, A.R.E., Mduma, S.A.R. & Brashares, J.S. 2003. Patterns of predation in a diverse predator–prey system. *Nature* 425:288–290.

Sinclair, A.R.E., Olsen, P.D. & Redhead, T.D. 1990. Can predators regulate small mammal populations? Evidence from house mouse outbreaks in Australia. *Oikos* 59:382–392.

Sinclair, A.R.E., Pech, R.P., Dickman, C.R., Hik, D., Mahon, P. & Newsome, A.E. 1998. Predicting effects of predation on conservation of endangered prey. *Conservation Biology* 12:564–575.

Singer, F.J., Swift, D.M., Coughenour, M.B. & Varley, J.D. 1998a. Thunder on the Yellowstone revisited: an assessment of native ungulates by natural regulation, 1968–1993. *Wildlife Society Bulletin* 26:375–390.

Singer, F.J., Zeigenfuss, L.C., Gates, R.G. & Barnett, D.T. 1998b. Elk, multiple factors, and persistence of willows in national parks. *Wildlife Society Bulletin* 26:419–428.

Singleton, G.R. & Spratt, D.M. 1986. The effects of *Capillaria hepatica* (Nematoda) on natality and survival to weaning in BALB/c mice. *Australian Journal of Zoology* 34:677–681.

Sjögren Gulve, P. 1994. Distribution and extinction patterns within a northern metapopulation: case of the pool frog, *Rana lessonae*. *Ecology* 75:1357–1367.

Skellam, J.G. 1951. Random dispersal in theoretical populations. *Biometrika* 38:196–218.

Skogland, T. 1985. The effects of density-dependent resource limitations on the demography of wild reindeer. *Journal of Animal Ecology* 54:359–374.

Sloane, M.A., Sunnucks, P., Alpers, D., Beheregaray, B. & Taylor, A.C. 2000. Highly-reliable genetic identification of individual northern hairy-nosed wombats from single remotely-collected hairs: a feasible censusing method. *Molecular Ecology* 9:1233–1240.

Smith, A.P. & Quin, D.G. 1996. Patterns and causes of extinction and decline in Australian conilurine rodents. *Biological Conservation* 77:243–267.

Smith, A.T. & Gilpin, M.E. 1997. Spatially correlated dynamics in a pika metapopulation. In: Hanski, I.A. & Gilpin, M.E., eds. *Metapopulation Biology*, pp. 407–428. Academic Press, San Diego, CA.

Smith, F.D.M., May, R.M. & Harvey, P.H. 1994. Geographical ranges of Australian mammals. *Journal of Animal Ecology* 63:441–450.

Smith, J.N.M., Cook, T.L., Rothstein, S.I., Robinson, S.K. & Sealy, S.G., eds. 2000. *Ecology and Management of Cowbirds and their Hosts*. University of Texas Press, Austin, TX.

Smith, J.N.M., Krebs, C.J., Sinclair, A.R.E. & Boonstra, R. 1988. Population biology of snowshoe hares: II. Interactions with winter food plants. *Journal of Animal Ecology* 57:269–286.

Smith, N.S. 1970. Appraisal of condition estimation methods for East African ungulates. *East African Wildlife Journal* 8:123–129.

Smuts, G.L. 1978. Interrelations between predators, prey, and their environment. *BioScience* 28:316–320.

Snell, G.P. & Hlavachick, B.D. 1980. Control of prairie dogs – the easy way. *Rangelands* 2:239–240.

Snyder, N.F.R. & Johnson, E.V. 1985. Photographic censusing of the 1982–1983 California condor population. *Condor* 87:1–13.

Snyder, N.F.R., Wiley, J.W. & Kepler, C.B. 1987. *The Parrots of Luqillo: Natural History and Conservation of the Puerto Rican Parrot*. Western Foundation of Vertebrate Zoology, Los Angeles, CA.

Solomon, M.E. 1949. The natural control of animal populations. *Journal of Animal Ecology* 18:1–35.

Southern, H.N. 1951. Change in status of the bridled guillemot after ten years. *Proceedings of the Zoological Society of London* 121:657–671.

Spinage, C.A. 1973. The role of photoperiodism in the seasonal breeding of tropical African ungulates. *Mammal Review* 3:71–84.

Spowart, R.A. & Thompson Hobbs, N. 1985. Effect of fire on diet overlap between mule deer and mountain sheep. *Journal of Wildlife Management* 49:942–946.

Spratt, D.M. 1990. The role of helminths in the biological control of mammals. *International Journal for Parasitology* 20:543–550.

Spratt, D.M. & Presidente, P.J.A. 1981. Prevalence of *Fasciola hepatica* infection in native mammals in southeastern Australia. *Australian Journal of Experimental Biology and Medical Science* 59:713–721.

Spratt, D.M. & Singleton, G.R. 1986. Studies on the life cycle, infectivity and clinical effects of *Capillaria hepatica* (Bancroft) (Nematoda) in mice, *Mus musculus. Australian Journal of Zoology* 34:663–675.

Spray, C.J., Crick, H.Q.P. & Hart, A.D.M. 1987. Effects of aerial applications of fenitrothion on bird populations of a Scots pine plantation. *Journal of Applied Ecology* 24:29–47.

Springer, A.M., Estes, J.A., van Vliet, G.B., *et al.* 2003. Sequential megafaunal collapse in the North Pacific Ocean: an ongoing legacy of industrial whaling? *Proceedings of the National Academy of Sciences of the USA* 100:12223–12228.

Spurr, E.B. 1994. Review of the impacts on non-target species of sodium monofluoroacetate (1080) in baits used for brushtail possum control in New Zealand. In: Seawright, A.A. & Eason, C.T., eds. *Proceedings of the Science Workshop on 1080*, pp. 124–133. Miscellaneous Series no. 28. Royal Society of New Zealand.

Spurr, E.B. 2000. Impacts of possum control on non-target species. In: Montague, T.L., ed. *The Brushtail Possum – Biology, Impact and Management of an Introduced Marsupial*, pp. 175–186. Manaaki Whenua Press, Lincoln, New Zealand.

Srivastava, D.S. 1999. Using local–regional richness plots to test for species saturation: pitfalls and potentials. *Journal of Animal Ecology* 68:1–16.

Srivastava, D.S. & Vellend, M. 2005. Biodiversity–ecosystem function research: is it relevant to conservation? *Annual Reviews of Ecology and Systematics* (in press).

Stanley Price, M.R. 1989. *Animal Re-introductions: The Arabian Oryx in Oman.* Cambridge University Press, Cambridge.

Stenseth, N.C. & Lidicker, W.J., Jr., eds. 1992. Presaturation and saturation dispersal 15 years later: some theoretical considerations. In: Stenseth, N.C. & Lidicker, W.Z., Jr., eds. *Animal Dispersal*, pp. 201–223. Chapman & Hall, London.

Stenseth, N.C., Mysterud, A., Ottersen, G., Hurrell, J.W., Chan, K.-S. & Lima, M. 2002. Ecological effects of climate fluctuations. *Science* 297:1292–1296.

Stephens, D.W. & Dunbar, S.R. 1993. Dimensional analysis in behavioral ecology. *Behavioral Ecology* 4:172–183.

Stephens, D.W. & Krebs, J. 1986. *Foraging Theory.* Princeton University Press, Princeton, NJ.

Stevens, G.C. 1989. The latitudinal gradient in geographical range: how so many species coexist in the tropics. *American Naturalist* 133:240–256.

Stillman, R.A., Goss-Custard, J.D., West, A.D., *et al.* 2000. Predicting mortality in novel environments: tests and sensitivity of a behaviour-based model. *Journal of Applied Ecology* 37:564–588.

Stillman, R.A., Goss-Custard, J.D., West, A.D., *et al.* 2001. Predicting shorebird mortality and population size under different regimes of shellfishery management. *Journal of Applied Ecology* 38:857–868.

Stonehouse, B. 1967. The general biology and thermal balances of penguins. In: Cragg, J.B., ed. *Advances in Ecological Research*, Vol. 4, pp. 131–196. Academic Press, London.

Storfer, A. 1996. Quantitative genetics: a promising approach for the assessment of genetic variation in endangered species. *Trends in Ecology and Evolution* 11:343–348.

Sullivan, T.P. & Klenner, W. 1993. Influence of diversionary food on red squirrel populations and crop tree damage in young lodgepole pine forest. *Ecological Applications* 3:708–718.

Sutherland, W.J. 1996. *From Individual Behaviour to Population Biology.* Oxford University Press, Oxford.

Sutherland, W.J. & Allport, G.A. 1994. A spatial depletion model of the interaction between bean geese and wigeon with the consequences for habitat management. *Journal of Animal Ecology* 63:51–59.

Sutherland, W.J. & Anderson, C.W. 1993. Predicting the distribution of individuals and the consequences of habitat loss: the role of prey depletion. *Journal of Theoretical Biology* 160:223–230.

Swanson, G.A., Adomaitis, V.A., Lee, F.B., Serie, J.R. & Shoesmith, J.A. 1984. Limnological conditions influencing duckling use of saline lakes in south-central north Dakota. *Journal of Wildlife Management* 48:340–349.

Taber, R.D. 1956. Deer nutrition and population dynamics in the north coast range of California. *Transactions of the North American Wildlife Conference* 21:159–172.

Taitt, M.J. & Krebs, C.J. 1981. The effect of extra food on small rodent populations: II. Voles (*Microtus townsendii*). *Journal of Animal Ecology* 50:125–137.

Talbot, L.M. & Stewart, D.R.M. 1964. First wildlife census of the entire Serengeti–Mara region, East Africa. *Journal of Wildlife Management* 28:815–827.

Taper, M.L. & Gogan, P.J.P. 2002. The northern Yellowstone elk: density dependence and climatic conditions. *Journal of Wildlife Management* 66:106–122.

Taylor, C.R. 1968a. The minimum water requirements of some East African bovids. In: Crawford, M.A., ed. *Comparative Nutrition of Wild Animals*, pp. 195–206. Symposia of the Zoological Society of London. Academic Press, London.

Taylor, C.R. 1968b. Hygroscopic food: a source of water for desert antelopes? *Nature* 219:181–182.

Taylor, C.R. 1969. Metabolism, respiratory changes, and water balance of an antelope, the eland. *American Journal of Physiology* 217:317–320.

Taylor, C.R. 1970a. Strategies of temperature regulation: effect on evaporation in East African ungulates. *American Journal of Physiology* 219:1131–1135.

Taylor, C.R. 1970b. Dehydration and heat: effects on temperature regulation of East African ungulates. *American Journal of Physiology* 219:1136–1139.

Taylor, C.R. 1972. The desert gazelle: a paradox resolved. In: Maloiy, G.M.O., ed. *Comparative Physiology of Desert Animals*, pp. 215–227. Symposia of the Zoological Society of London no. 31. Academic Press, London.

Taylor, C.R., Robertshaw, D. & Hofmann, R. 1969b. Thermal panting: a comparison of wildebeest and zebu cattle. *American Journal of Physiology* 217:907–910.

Taylor, C.R., Spinage, C.A. & Lyman, C.P. 1969a. Water relations of the waterbuck, an East African antelope. *American Journal of Physiology* 217:630–634.

Taylor, R.H. 1979. How the Macquarie Island parakeet became extinct. *New Zealand Journal of Ecology* 2:42–45.

Tedman, R. & Green, B. 1987. Water and sodium fluxes and lactational energetics in suckling pups of Weddell seals (*Leptonychotes weddelli*). *Journal of Zoology (London)* 212:29–42.

Telfer, E.S. 1970. Winter habitat selection by moose and white-tailed deer. *Journal of Wildlife Management* 34:553–559.

Temple, S.A. 1977. Plant–animal mutualism: coevolution with dodo leads to near extinction of plant. *Science* 197:885–886.

Temple, S.A. 1986. Predicting impacts of habitat fragmentation on forest birds: a comparison of two models. In: Verner, J., Morrison, M.L. & Ralph, J.C., eds. *Wildlife 2000: Modeling Habitat Relationships of Terrestrial Vertebrates*, pp. 301–304. University of Wisconsin Press, Madison, WI.

Temple, S.A. 1991. The role of dispersal in the maintenance of bird populations in a fragmented landscape. In: *Acta XX Congressus Internationalis Ornithologici*, pp. 2298–2305. New Zealand Ornithological Congress Trust Board, Wellington.

Temple, S.A. & Cary, J.R. 1988. Modeling dynamics of habitat–bird populations in fragmented landscapes. *Conservation Biology* 2:340–427.

Terborgh, J. 1988. The big things that run the world – a sequel to E.O. Wilson. *Conservation Biology* 2:402–403.

Terborgh, J. 1989. *Where Have All the Birds Gone? Essays on the Biology and Conservation of Birds that Migrate to the American Tropics.* Princeton University Press, Princeton, NJ.

Terborgh, J. 1992. Why American songbirds are vanishing. *Scientific American* 266(5):56–62.

Terborgh, J.W. & Janson, C.H. 1986. The socioecology of primate groups. *Annual Review of Ecology and Systematics* 17:111–135.

Thill, R.E. 1984. Deer and cattle diets on Louisiana pine-hardwood sites. *Journal of Wildlife Management* 48:788–798.

Thirgood, S.J., Redpath, S.M., Rothery, P. & Aebischer, N.J. 2000. Raptor predation and population limitation in red grouse. *Journal of Animal Ecology* 69:504–516.

Thomas, V.G. 1988. Body condition, ovarian hierarchies, and their relation to egg formation in Anseriform and Galliform species. *Proceedings of the International Ornithological Congress* 19:353–363.

Thorne, E.T. & Williams, E.S. 1988. Disease and endangered species: the black-footed ferret as a recent example. *Conservation Biology* 2:66–74.

Thorne, E.T., Morton, J.K., Blunt, F.M. & Dawson, H.A. 1978. Brucellosis in elk. II. Clinical effects and means of transmission as determined through artificial infections. *Journal of Wildlife Diseases* 14:280–291.

Thornhill, N.W. 1993. *The Natural History of Inbreeding and Outbreeding: Theoretical and Empirical Perspectives.* The University of Chicago Press, Chicago, IL.

Tilman, D. 1999. The ecological consequences of changes in biodiversity: a search for general principles. *Ecology* 80:1455–1474.

Tompkins, D.M. & Wilson, K. 1998. Wildlife disease ecology: from theory to policy. *Trends in Ecology and Evolution* 13:476–478.

Torbit, S.C., Carpenter, L.H., Bartmann, R.M., Alldredge, A.W. & White, C.G. 1988. Calibration of carcass fat indices in wintering mule deer. *Journal of Wildlife Management* 52:582–588.

Traveset, A. 1998. Effect of seed passage through vertebrate frugivores' guts on germination: a review. *Perspectives in Plant Ecology, Evolution and Systematics* 1:151–190.

Travis, S.E. & Keim, P. 1995. Differentiating individuals and populations of mule deer using DNA. *Journal of Wildlife Management* 59:824–831.

Trites, A.W. & Donnelly, C.P. 2003. The decline of Steller sea lions *Eumetopias jubatus* in Alaska: a review of the nutritional stress hypothesis. *Mammal Review* 33:3–28.

Trostel, K., Sinclair, A.R.E., Walters, C.J. & Krebs, C.J. 1987. Can predation cause the 10-year hare cycle? *Oecologia* 74:185–192.

Turchin, P. 1998. *Quantitative Analysis of Movement.* Sinauer Associates, Sunderland, MA.

Turchin, P. 2003. *Complex Population Dynamics.* Princeton University Press, Princeton, NJ.

Turchin, P. & Batzli, G. 2001. Availability of food and the population dynamics of arvicoline rodents. *Ecology* 82:1521–1534.

Turchin, P. & Ellner, S.P. 2000. Living on the edge of chaos: population dynamics of Fennoscandian voles. *Ecology* 81:3099–3116.

Turchin, P. & Hanski, I. 1997. An empirically-based model for the latitudinal gradient in vole population dynamics. *American Naturalist* 149:842–874.

Turchin, P. & Hanski, I. 2001. Contrasting alternative hypotheses about rodent cycles by translating them into parameterized models. *Ecology Letters* 4:267–276.

Turner, J.W. & Kirkpatrick, J.F. 1991. New developments in feral horse contraception and their potential application to wildlife. *Wildlife Society Bulletin* 19:350–359.

Twigg, L.E. & Williams, C.K. 1999. Fertility control of overabundant species: can it work for rabbits? *Ecology Letters* 2:281–285.

Twigg, L.E., Lowe, T.J., Martin, G.R., *et al.* 2000. Effects of surgically imposed sterility on free-ranging rabbit populations. *Journal of Applied Ecology* 37:16–39.

Tyndale-Biscoe, C.H. 1994. Virus-vectored immunocontraception of feral mammals. *Reproduction, Fertility and Development* 6:281–287.

Urquhart, D. & Farnell, R. 1986. *The Fortymile Herd*. Department of Renewable Resources, Whitehorse, Yukon.

Vander Lee, B.A., Lutz, F.S., Hansen, L.A. & Mathews, N.E. 1999. Effects of supplemental prey, vegetation, and time on success of artificial nests. *Journal of Wildlife Management* 63:1299–1305.

Van der Wal, R., Madan, N., Van Lieshout, S., Dormann, C., Langvatn, R. & Albon, S.D. 2000. Trading forage quality versus quantity? Plant phenology and patch choice by an Arctic ungulate. *Oecologia* 123:108–115.

van Horne, B., Schooley, R.L., Knick, S.T., Olson, G.S. & Burnham, K.P. 1997. Use of burrow entrances to indicate densities of Townsend's ground squirrels. *Journal of Wildlife Management* 61:92–101.

van Riper, C., III & van Riper, S.G. 1986. The epizootiology and ecological significance of malaria in Hawaiian land birds. *Ecological Monographs* 56:327–344.

Varnham, K.J., Roy, S.S., Seymour, A., Mauremootoo, J., Jones, C.G. & Harris, S. 2002. Eradicating Indian musk shrews (*Suncus murinus*, Sloricidae) from Mauritian offshore islands. In: Veitch, C.R. & Clout, M.N., eds. *Turning the Tide: The Eradication of Invasive Species*, pp. 342–349. Occasional Paper of the IUCN Species Survival Commission no. 27. IUCN, Gland, Switzerland.

Vasarhelyi, C. & Thomas, V.G. 2003. Analysis of Canadian and American legislation controlling exotic species in the Great Lakes. *Aquatic Conservation: Marine and Freshwater Ecosystems* 13:417–427.

Veitch, C.R. & Clout, V.N. 2002. *Turning the Tide: The Eradication of Invasive Species*. Occasional Paper of the IUCN Species Survival Commission no. 27. IUCN, Gland, Switzerland.

Veloso, C. & Bozinovic, F. 1993. Dietary and digestive constraints on basal energy metabolism in a small herbivorous rodent. *Ecology* 74:2003–2010.

Verhulst, P.F. 1838. Notice sur la loi que la population suit dans son accroissement. *Correspondence Mathématique et Physique* 10:113–121.

Verme, L.J. & Holland, J.C. 1973. Reagent-dry assay of marrow fat in white-tailed deer. *Journal of Wildlife Management* 37:103–105.

Vesey-Fitzgerald, D.F. 1960. Grazing succession among East African game animals. *Journal of Mammalogy* 41:161–172.

Vincent, J.-P., Hewison, A.J.M., Angibault, J.-M. & Cargnelutti, B. 1996. Testing density estimators on a fallow deer population of known size. *Journal of Wildlife Management* 60:18–28.

Vitousek, P.M. & Hooper, D.U. 1993. Biological diversity and terrestrial ecosystem biogeochemistry. In: Schulze, E.D. & Mooney, H.A., eds. *Biodiversity and Ecosystem Function*, pp. 3–14. Springer-Verlag, Berlin.

Vitousek, P.M., D'Antonio, C.M., Loope, L.L. & Westbrooks, R. 1996. Biological invasions as global environmental change. *American Scientist* 84:468–478.

Vivås, H.J. & Sæther, B.-E. 1987. Interactions between a generalist herbivore, the moose *Alces alces* and its food resources: an experimental study of winter foraging behaviour in relation to browse availability. *Journal of Animal Ecology* 56:509–520.

Volterra, V. 1926a. Variations and fluctuations of the numbers of individuals in animal species living together. In: Chapman, R.N., ed. *Animal Ecology*, pp. 409–448. McGraw-Hill, New York.

Volterra, V. 1926b. Fluctuations in the abundance of a species considered mathematically. *Nature* 118:558–560.

Waits, L.P., Talbot, S.L., Ward, R.H. & Shields, G.F. 1998. Mitochondrial DNA phylogeography of the North American brown bear and implications for conservation. *Conservation Biology* 12:408–417.

Walker, B.H. 1992. Biological diversity and ecological redundancy. *Conservation Biology* 6:18–23.

Walker, B.H. 1995. Conserving biological diversity through ecosystem resilience. *Conservation Biology* 9:747–752.

Walker, B.H., Ludwig, D., Holling, C.S. & Peterman, R.M. 1981. Stability of semi-arid savanna grazing systems. *Journal of Ecology* 69:473–498.

Walsh, J. 1987. War on cattle disease divides the troops. *Science* 237:1289–1291.

Walsh, P.D., Abernathy, K.A., Bermejos, M., *et al.* 2003. Catastrophic ape decline in western equatorial Africa. *Nature* 422:611–614.

Walter, H. 1973. *Vegetation of the Earth in Relation to Climate and the Eco-physiological Conditions.* English Universities Press, London.

Walters, C. 1986. *Adaptive Management of Renewable Resources.* Macmillan, New York.

Walters, C.J. & Holling, C.S. 1990. Large-scale management experiments and learning by doing. *Ecology* 71:2060–2068.

Ward, D. & Salz, D. 1994. Foraging at different spatial scales: dorcas gazelles foraging for lilies in the Negev Desert. *Ecology* 75:48–58.

Ward, P. 1969. The annual cycle of the yellow-vented bulbul *Pycnonotus goiavier* in a humid equatorial environment. *Journal of Zoology (London)* 157:25–45.

Warner, R.E. 1968. The role of introduced diseases in the extinction of the endemic Hawaiian avifauna. *Condor* 70:101–120.

Waser, P. 1996. Patterns and consequences of dispersal in gregarious carnivores. In: Gittleman, J.L., ed. *Carnivore Behavior, Ecology, and Evolution*, pp. 267–295. Cornell University Press, Ithaca, NY.

Watson, A. & Moss, R. 1971. Spacing as affected by territorial behavior, habitat, and nutrition in red grouse (*Lagopus l. scoticus*). In: Esser, A.H., ed. *Behavior and Environment: The Use of Space by Animals and Men*, pp. 92–111. Plenum Press, New York.

Watt, C., Dobson, P. & Grenfell, B.T. 1995. Glossary. In: Grenfell, B.T. & Dobson, A.P., eds. *Ecology of Infectious Diseases in Natural Populations*, pp. 510–521. Cambridge University Press, Cambridge.

Wauters, L.A. & Gurnell, J. 1999. The mechanism of replacement of red squirrels by grey squirrels: a test of the interference competition hypothesis. *Ethology* 105:1053–1071.

Wayne, R.K. 1992. On the use of morphologic and molecular genetic characters to investigate species status. *Conservation Biology* 6:590–592.

Wayne, R.K., Lehman, N., Allard, M.W. & Honeycutt, R.L. 1992. Mitochondrial DNA variability of the gray wolf: genetic consequences of population decline and habitat fragmentation. *Conservation Biology* 6:559–569.

Wayne, R.K., Lehman, N., Girman, D., *et al.* 1991. Conservation genetics of the endangered Isle Royale gray wolf. *Conservation Biology* 5:41–51.

Weatherall, D.J. 1985. *The New Genetics and Clinical Practice*, 2nd edn. Oxford University Press, Oxford.

Webb, C.J. & Kelly, D.K. 1993. The reproductive biology of the New Zealand flora. *Trends in Ecology and Evolution* 8:442–447.

Wegener, A. 1924. *The Origins of Continents and Oceans.* Methuen, London.

Wehausen, J.D. 1995. Fecal measures of diet quality in wild and domestic ruminants. *Journal of Wildlife Management* 59:816–823.

Weir, J.S. 1972. Spatial distribution of elephants in an African National Park in relation to environmental sodium. *Oikos* 23:1–13.

Werner, E.E., Mittelbach, G.G., Hall, D.J. & Gilliam, J.F. 1983. Experimental tests of optimal habitat use in fish: the role of relative habitat profitability. *Ecology* 64:1525–1539.

Western, D. 1975. Water availability and its influence on the structure and dynamics of a savannah large mammal community. *East African Wildlife Journal* 13:265–286.

White, C.A., Olmsted, C.E. & Kay, C.E. 1998. Aspen, elk, and fire in the Rocky Mountain national parks of North America. *Wildlife Society Bulletin* 26:449–462.

White, G.C. & Burnham, K.P. 1999. Program MARK: Survival rate estimation from both live and dead encounters. *Bird Study* 46(Suppl.):S120–S139.

White, G.C., Anderson, D.R., Burnham, K.P. & Otis, D.L. 1982. *Capture–Recapture and Removal Methods for Sampling Closed Populations*. Report LA-8787-NERP. Los Alamos National Laboratory, Los Alamos, NM.

White, R.G., Bunnell, F.L., Gaare, E., Skogland, T. & Hubert, B. 1981. Ungulates on arctic ranges. In: Bliss, L.C., Heal, O.W. & Moore, J.J., eds. *Tundra Ecosystems: A Comparative Analysis*, pp. 397–483. Cambridge University Press, Cambridge.

Whitehead, G.K. 1972. *Deer of the World*. Constable, London.

Whitlock, M.C. & McCauley, D.E. 1999. Indirect measures of gene flow and migration: $F_{ST} \neq 1/(4Nm + 1)$. *Heredity* 82:117–125.

Whitman, K., Starfield, A.M., Quadling, H.S. & Packer, C. 2004. Sustainable trophy hunting of African lions. *Nature* 428:175–178.

Whittaker, R.H. 1967. Gradient analysis of vegetation. *Biological Reviews* 42:207–264.

Whittaker, R.H. 1975. *Communities and Ecosystems*, 2nd edn. Macmillan, New York.

Whyte, R.J. & Bolen, E.G. 1985. Variation in mallard digestive organs during winter. *Journal of Wildlife Management* 49:1037–1040.

Wiemeyer, S.N., Scott, J.M., Anderson, M.P., Bloom, P.H. & Stafford, C.J. 1988. Environmental contaminants in California condors. *Journal of Wildlife Management* 52:238–247.

Wiens, J.A. 1977. On competition and variable environments. *American Scientist* 65:590–597.

Wilbur, S.R., Carrier, W.D. & Borneman, J.C. 1974. Supplemental feeding program for California condors. *Journal of Wildlife Management* 38:343–346.

Wilcove, D.S. 1985. Nest predation in forest tracts and the decline of migratory songbirds. *Ecology* 66:1211–1214.

Williams, B.K., Johnson, F.A. & Wilkins, K. 1996. Uncertainty and the adaptive management of waterfowl harvests. *Journal of Wildlife Management* 60:223–232.

Williams, B.K., Nichols, J.D. & Conroy, M.J. 2002. *Analysis and Management of Animal Populations*. Academic Press, San Diego, CA.

Williams, C.K. & Twigg, L.E. 1996. Responses of wild rabbit populations to imposed sterility. In: Floyd, R.B., Sheppard, A.W. & De Barro, P.J., eds. *Frontiers of Population Ecology*, pp. 547–560. CSIRO Publishing, Australia.

Williams, M. 1979. The status and management of black swans *Cygnus atratus* Lathum at Lake Ellesmere since the 'Wahine' storm, April 1968. *New Zealand Journal of Ecology* 2:34–41.

Wilmshurst, J.F., Fryxell, J.M. & Hudson, R.J. 1995. Forage quality and patch choice by wapiti (*Cervus elaphus*). *Behavioral Ecology* 6:209–217.

Wilson, D.J. & Jeffries, R.L. 1996. Nitrogen mineralization, plant growth and goose herbivory in an arctic coastal system. *Journal of Ecology* 85:841–851.

Windsor, D.A. 1998. Most of the species on Earth are parasites. *International Journal for Parasitology* 28:1939–1941.

Wirtz, P. 1982. Territory holders, satellite males, and bachelor males in a high density population of waterbuck (*Kobus ellipsiprymnus*) and their association with conspecifics. *Zeitschrift für Tierpsychologie* 58:277–300.

Wittmer, H.U., Sinclair, A.R.E. & McLellan, B.N. 2005. The role of predation in the decline and extirpation of woodland caribou. *Oecologia* 144:257–267.

Woodburne, M.O. & Zinsmeister, W.J. 1982. Fossil land mammal from Antarctica. *Science* 218:284–286.

Woolnough, A.P., Foley, J.F., Johnson, C.N. & Evans, M. 1997. Evaluation of techniques for indirect measurement of body composition in a free-ranging large herbivore, the Southern Hairy-nosed Wombat. *Wildlife Research* 24:649–660.

Wootton, J.T. 1994a. Predicting direct and indirect effects: an integrated approach using experiments and path analysis. *Ecology* 75:151–165.

Wootton, J.T. 1994b. The nature and consequences of indirect effects in ecological communities. *Annual Review of Ecology and Systematics* 24:443–466.

Worthy, T.H. & Holdaway, R.N. 2002. *The Lost World of the Moa*. Indiana University Press, Bloomington, IN.

Wright, S.J. & Mulkey, S.S. 1997. Patterns and processes in the vertebrate digestive system. *Trends in Ecology and Evolution* 12:420–422.

Wu, X.B., Thurow, T.L. & Whisenant, S.G. 2000. Fragmentation and changes in hydrologic function of tiger bush landscapes, south-west Niger. *Journal of Animal Ecology* 88:790–800.

Wydeven, A.P. & Dahlgren, R.B. 1985. Ungulate habitat relationships in Wind Cave National Park. *Journal of Wildlife Management* 49:805–813.

Yodzis, P. & Innes, S. 1988. Body size and consumer–resource dynamics. *American Naturalist* 139:1151–1175.

Yuill, T.M. 1987. Diseases as components of mammalian ecosystems: mayhem and subtlety. *Canadian Journal of Zoology* 65:1061–1066.

Zar, J.H. 1996. *Biostatistical Analysis*, 3rd edn. Prentice-Hall, London.

Zeleny, L. 1976. *The Bluebird: How You Can Help its Fight for Survival*. Indiana University Press, Bloomington, IN.

Zhang, Z., Pech, R., Davis, S., Shi, D., Wan, X. & Zhong, W. 2003. Extrinsic and intrinsic factors determine the eruptive dynamics of Brandt's voles *Microtus brandti* in Inner Mongolia, China. *Oikos* 100:299–310.

Index

Page numbers in *italics* refer to figures; those in **bold** refer to tables.

Abies 12
abiotic environment 19, 27
Acacia 15, 17
Acanthis hornimanni 95
Accipiter gentilis 12
accuracy in sampling 221
Acer 13
Acinonyx jubatus 296
Acridotheres tristis 321
adaptation 19
  adaptive radiation 22–3
  applied aspects 33–4
  convergence 21–2
adaptive management 264–7
  active 264
  passive 264
adaptive radiation 22–3
addax 17, 324
Addax nasomaculatus 17, 324
Adélie penguin 220
Adenostema taxiculatum 131
Aepyceros melampus 96
Agathis 13
age distribution 88, 299
  stable 88
  stationary 88
Agelaius phoeniceus 145
age-structured population model 244–7,
  251–2
AIC 85, 258–60, 262–4, 339
AIDS 187
Ailuropoda melanoleuca 4, 23
Aix sponsa 54, 157
Akaike weight 263
Alcanthorhynchus superciliosus 15
Alcelaphus buselaphus 96
Alces alces 25, 95, 236, 369
Alectoris chukar 162

ALEX 310
alkaloids 41
Allactaga elater 17, 43
Allee effect 107
allele 291–2
alligator, American 95
Alligator mississipiensis 95
allochthonous supply 366
allozyme gel electrophoresis 28, 34
Alnus 12
Alopex lagopus 16, 98, 191
Amazona vittata 193, 322
Ameides
  ferreus 34
  vagrans 34
American marten 75
amino acids 37
Ammodorcas clarkei 17
analysis of variance (ANOVA) 264, 269,
  279–87
  one-factor 279–81
  two-factor 281–6
  three-factor 286–8
Anas
  discors 33
  platyrhynchos 33, 54, 141
  strepera 33, 54
Anolis
  gingvivinus 190
  wattsi 190
Anser albifrons 54
Antechinomys 17
Anthornis melanura 370
Antilocapra americana 26, 242, 374
Apodemus
  mystacinus 150
  sylvaticus 150
apparent competition 153–4, 160, 306, 367

Apteryx 29
Aquila chrysaetos 308, 367
Ardea
  herodias 221
  ibis 154
armadillo, nine-banded 27
Artemisia 17
autochthonous supply 366

badger, European 193
Balaenoptera
  acutorostrata 241
  borealis 54
  musculus 81
  physalus 54
banded rail 306
banteng 14
Bartlett's test 284
basal metabolic rate 46–8
bat, short-tailed 370
Batrachochytrium dendrobatidis 187
bear
  brown 25, 30
  grizzly 171, 300–4, *301–4*
beaver 25, 62–3, 106, 108
Bebrornis seychellensis 77
beech 13, 40
behavior 60–77
  contingency model 60–3
  diet selection 60–6
  feeding rates 63–5
  habitat selection 72–4
  interference 74–5, 75
  linear programming 65–6
  optimal patch and habitat use 66–9
  resource selection functions 67–72
  risk-sensitive habitat use 66–70
  social behavior and foraging 72–7
  territoriality 75–7
behavior of prey 176–7
  birth synchrony 177
  herding 176–7
  migration 176
  predator swamping 177
behavioral ecology 60
bellbird 370
bettong
  burrowing 162
  Tasmanian 40
Bettongia
  gainardii 40
  lesurur 162

Betula 12
  glandulosa 42
bias errors 221–2
biodiversity
  and ecosystem function 381–3, **381**
  ecosystems 377–9
bioeconomics 347–53
  costs 348
  discounting 352–3
  open access systems 349
  profit 348
  revenue 348
bioelectrical impedance 57–8
biological control 360
biomes 11–18
  alpine 16
  benthos 18
  boreal forest 12
  communities 11
  continental shelf 18
  deserts 17
  forest 12–14
  grassland 15–17
  habit 11
  marine 17–18
  pelagic 18
  semi-deserts 17
  shrubland 14–15
  temperate forest 13
  temperate grassland 16
biotic environment 19
birch 42
Birgus latro 31
bison 68
  American 15–16, 25, 68, 99, 146, 156, 183, *183*, 193, **325**
  wood 91, 134, 372
Bison 25
  bison 25, 91, 99, 372
blackbird
  red-winged 145, 159, 161
  yellow-headed 145, 159
Blarina 172
blowfly 187
blue jay 42
bluebird 162
bluegill sunfish 69
bluetongue disease 194
body condition 133
body fat 132–3
Boiga irregularis 306, 360
Bombycilla cedrorum 169

bone marrow fat 56–8
boreal forest 12
Borrelia burgdorferi 187
Bos 14
botfly 193
bovine spongiform encephalopathy 188
Brachydura 14
Brachystegia 14
bristlebird, rufous 71
Bromus tectorum 162
brown tree snake 306
Brucella abortus 183–4
brucellosis 180, 183, *183*
BSE 188
Bubalus bubalus 154, 319
Bubo virginianus 112
bubonic plague 191
budgerigar 17
buffalo
    African 34, 43, 45, 53, 86, **87**, 96, 118,
        132, 154, 160, 179, 186, 189, 192, 220,
        373
    water 154, 319
bulbul, yellow-vented 54
bunting, snow 16
bushbuck 52, 147
Buteo 16
    buteo 191, 319
buzzard, common 191, 319

California gnatcatcher 72
Callaeas cinerea 176, 329
Calliphora 187
Calluna vulgaris 131, 190–1
Calvaria major 370
camaroptera, gray-backed 54
Camaroptera brevicaudata 54
camel 17
Camelus 17
Canis
    latrans 34–5, 367
    lupus 16, 25, 34–5, 92
    rufus 34
    simensis 191
Capillaria hepatica 120–1
Capreolus capreolus 76
Caprolagus hispidus 314
capybara 15
caracal 373
Carduelis
    cannabrina 143
    flammea 143

Caretta caretta 34, 248
caribou 15–16, 25, 80, 85, **86**, 164, 167,
        173–7, 194, 318, 322, **323**, **325**, 369,
        371, 372
Carpodacus mexicanus 188
carrying capacity 114–15, 121
    ecological 114
    economic 115
    others 115
Castor canadensis 25, 108
Casuarina 14
cat, domestic 316, 357–9
catastrophe 305
catchability coefficient 343
cattle 68, 192–3, 315, 317, 319
cattle egret 154
Ceanothus 14
cecotrophy 51
cedar waxwing 169
census 219
centripetal systems 205
Centrocercus urophasiaxus 162
Cephalophus 52
Cercomela sordida 17
Certocystis ulmi 191
Cervus 25
    elaphus 30–2, 91, 369
    nippon 30
    timorensis 229
Chalcophaps
    indica 142
    stephani 142
Chamaea fasciata 15
chamise 131
chamois 139, 357, 374
Channel Islands 307
chaos 125–7, 209–11
chaparral 14, 131
Charadrius melodius 375
cheatgrass 162
checkerspot butterfly 108
cheetah 296–7
Chelonia mydas 34
chemosterilant **361**
Chen caerulescens 29, 54, 157, 374,
        376
Chenonetta jubata 54
chickadee 168–9
chinchilla 51
Chionis minor 368
chipmunk 330
Chiroxiphia linearis 28

Chlamydia 187
Chloris chloris 143
Chrysemys picta 35
Chrysopalex 22
chytrid fungus 187–8
Circus cyaneus 171
Clethrionomys 13
   glareolus 167
Coccothraustes coccothraustes 143
Cochliomyia hominivorax 360
coconut crab 31
Coelodonta antiguilatis 12
Colophospermum 14
Columba palumbus 132
community/communities 11
   gradients 366
competition, general
   apparent 367
   disease 134
   exploitative 131
   interference 131
   testing for 131
competition, interspecific
   apparent competition 153–4, 160
   applied aspects 159–62
   assumptions 136
   competitive exclusion 142
   competitive release 141
   and conservation 159–60
   definition 135
   domestic and wildlife 160–1
   exotic species 161–2
   exploitation 36
   facilitation 154–7
   graphical models 136–8
   interference 136
   natural experiments 141–2
   niche 143–5
   perturbation experiments 138–43
   resource partitioning 146–53
   in variable environments 153
competition, intraspecific
   definition 131
   direct measures of food 132–3
   experimental food alteration 131–2
   for food **117**, 131–4
   food, predators and disease 134
   indirect measures of food 133
   problems with measures 133–4
   types of 131
competitive exclusion 142
competitive release 141

condor
   Andean 324
   California 319–20
Connochaetes taurinus 368
conservation in practice 312–34
   CITES 333–4, **334**
   community conservation 332
consumer-resource model 197–8, *199*
   control, statistical 273–5
   dynamics 198, 205–6, 209, 211
   functional response 201
   moose 207
      functional response 208
   numerical response 202
   parameters 197
   plants 207
   rainfall effects 206
   red kangaroo 201–6
   stability of 198, 210
   tri-trophic system 207
   vegetation growth 200
   wolf 207
      functional response 208
      territoriality 210
continental drift 23–5
control, methods 358–64
   biological control 360
   chemosterilant 361
   direct killing 358–9
   eradication 359–60
   exclusion 363
   genetic engineering 363
   habitat and food manipulation 364
   immunocontraception 362–3
   indirect 363–4
   manipulating fertility 360–3, **361**
   manipulating mortality 358–60
   sonic deterrents 363–4
control, wildlife 355–64
   appropriate 357–8
   definitions 355
   effects of 356
   methods 358–64
   objectives of 356–7
convergence 21–2
corridors 329–30, *330*, **331**
   advantages **331**
   disadvantages **331**
   extinction prevention 321–3
   extinction processes 312–21
   international 332–4
   in parks and reserves 324–32

corridors (*cont'd*)
  recovery 323–4
  Red Data Book 332–3
Corvus
  corax 95, 175
  monedula 175
Coryphanta robbinsorum 251
cougar 176
counts 219–34
  change of ratio 235–6
  incomplete 239–41
  index ratio 235
  indices 241–3
  indirect 235–41
  mark-recapture 236–9
  sample 221–34
  total 219–20
cowbird, brown-headed 174, 193, 314,
    377
coyote 34–5, 367
coypu 15, 51
crane
  sandhill 54
  whooping 325
Creatophora cinerea 154
creosote bush 17
crocodile 39
Crocuta crocuta 156
crossbill 40
Culex quinquefasciatus 192
Cupressus 14
Cyanocitta cristata 42
cycles 125–6
cyclic dynamics *198*
Cygnus
  atratus 375
  olor 109
Cynomys
  leucurus 317
  ludovicianus 156

Dactylanthus taylorii 370
Dama dama 67, 243, 363
Damaliscus korrigum 53
Danaus plexippus 42
Dasypus novemcinctus 27
decisions 2–7
  action matrix 4–5
  analysis 4
  feasibility 5–6
  hierarchies 5–7
  payoff 6

deer
  black-tailed 38, 40, 45, 131
  fallow 67, 243, 362–3
  mule 30, 48–9, 51, 55–6, 152, 292
  red 30, 31, 68, 91, 118, *119*, 156, 347,
    357, **357**
  roe 76
  rusa 229
  sika 30
  white-tailed 27, 30, 51, 70, 114, 160–1,
    177, 190, 192, 194, 211, 292, 362
deermouse 139
degrees of freedom 283
deme 30
demographic parameters 82
  age-specific mortality rate q_x 82
  age-specific survival rate p_x 83
  fecundity rate m_x 82
  life table 82–4
  probability of mortality at birth d_x 82
  probability of survivorship at birth l_x 82
demographic stochasticity 123
demography 289
Dendroica petechia 161
density dependence 110–14, *112–13*, *119*,
    133, *211*
  delayed 114, 120, *120*
  density-dependent births 111–14
  density-dependent deaths 111–14
  density-dependent population growth 256
  inverse 114
  overcompensation 125–6
density independence 112, *112–13*
depensatory response 114
desert 17
dibatag 17
Diceros bicornis 34, 369
Didelphis
  marsupialis 24
  virginiana 174
diet
  and bioelectrical impedance 57–8
  and blood 57
  and body condition 53–6
  and fat 53–6, 58
  problems 58
  selection of 52–3
  and urine 57
diet selection 60–6
  optimal feeding rates 63–5
  optimal linear programming 65–6
  optimal selection contingency 60–3

diffusion model 302
digestion
    fermentation 50
    hindgut 50
    morphology 49–51
    ruminants 50
digestive constraint 65
dikdik 33, 52, 96, 373
dingo 363
Dipodomys
    merriami 161
    spectabilis 91
    venustus 15, 17
dispersal 90–2
    and metapopulations 104–8
dispersion 92–3
distemper 188, 193, 317
distribution 93–9
    biotic factors 98
    and day length 97–8
    range collapse 98–9
    reintroduction and invasion 99–104
    and temperature 94–6
    and water or heat 96–7
disturbance, ecosystem 374–6
DNA 292–3
dodo 370
Dolabella auricularia 63
dominant species 365, 368–9
donkey 118, 356, **356**
Douglas fir 13
dove, mourning 242
Dromaeus novaehollandiae 15, 97, 240
Drosophila 298
duck
    eider 54
    gadwall 33, 54
    mallard 33, 54, 74, 141, 265, **265**
    maned 54
    wood 54, 157, 269
dugong 18, 50
Dugong dugon 18, 50
duiker 52, 147
Dutch elm disease 191
d_x 82
dynamic state variable modeling 70
Dyscophus antongilii 325

eagle, golden 308, *308*
eastern equine encephalitis 194
ebola virus 187
ecological baselines **325**

ecosystem management 365–83
    and biodiversity 377–9
    bottom-up processes 373–4
    coevolved associations 370
    and communities 366
    community management 368–70
    definitions 365–6
    disturbance and heterogeneity 374–6
    disturbance and management 375–6
    dominants and keystones 368–9
    ecosystem function 381–3, **381**, *382*
    food webs 366–8
    hyperpredation 367
    island biogeography 379–81, **380**
effective population size 297
eigenvalue 26–48, 251
    dominant 246, 248
eigenvector 246–51
    left 249, 250
    right 246, 249, 250
eland 53, 96
Elanus notatus 116
elasticity 249–51
    economic 351
electrophoresis 292–3, 296
    on cheetahs 296
    methodology 293
elephant
    African 53, 81, 117, 154–5, 173, 188, 220, 351–2, 362, 371
    Asian 188
elk 25, 31, 44, 45, 51, 95, 116–17, 122–5, *123–5*, 131, 152, 157, 183, 332, 369
    Irish 27
elm 13
Elytropappas 14
emerging infectious diseases 187–8
Empidonax traillii 161
emu 15, 97, 240
energy flow 46–7
Enhydra lutris 99, 368
environment
    abiotic 19
    biotic 19
environmental stochasticity 123–4
environmental variability 301
epidemiology 180–4
    infected 180
    models 180–1
    recovered 180
    SIR 180
    susceptibles 180

epidemiology (*cont'd*)
  thresholds of infection 181–3
  transmission coefficient 181–3
epizootic disease 186–7
  emerging infectious diseases 187–8
  myxomatosis 187
  rabbit hemorrhagic disease 187
  rinderpest 186–7
equilibrium/equilibria 113, 114, 199
  economic 349
  constant quota harvest on 337
  fixed proportion harvest on 341
Equus
  asinus 118
  caballus 323–4
  quagga 29
Erethizon dorsatum 35
Eretmocheles imbricata 34
error
  type 1 270–1
  type 2 270–1
escapement 344
estimates of populations 219
Eucalyptus 14
  regnans 13
Euphydryas editha 108
euro 161, 162
Eurometopias jubatus 376
evolutionarily stable strategy 72
experimental design 272–8
  alternative forms of 276
  confounded treatments 273
  one-factor 276
  pseudo-replicated 278
  role of controls 273–5
  role of replication 273–5
  two-factor 277, 278
exponential population growth 300
extinction 312–24
  demographic 312
  driven 312
  and food loss 315
  genetic 312
  habitat change 313–15
  and hunting 317–18
  introduced competitors 318–19
  introduced diseases 320–1
  introduced predators 315–16
  multiple causes 321
  and pest control 316–17
  pesticides and contaminants 319–20
  prevention 312–13

and recovery 323–4
  stochastic 312
extinction risk 289–338
  demographic causes of 289
  effects of harvest 337
  environmental and demographic
    stochasticity 300–5
  genetic causes of 291
  harvesting and 308
  introduced predators 306

F ratio 281–2, 285
facilitation 154–7
  catena 155
  grazing succession 155
factor
  fixed 285
  random 285
Fagus 13, 40
Falco
  peregrinus 171, 319
  punctatus 324
  tinnunculus 171
falcon, peregrine 171, 319
Fasciola hepatica 192
fat
  bone marrow 133
  content 133
  kidney 133
faunal relaxation 380
fecundity 244, 247
  measurement of 115
Felis
  caracal 373
  pardalis 367
  sylvestris 373
ferret 92, 169, 359, 367
  black-footed 157, 193, 316–17, 322
fertility, measurement of 115
Festuca rubra 191
Ficus elastica 13
finch, house 188–9, *189*
fisher 91
fit, best 255, 258
fitness 297
  definition 20
flea, rabbit 187
flycatcher, willow 161
food 36–59
  alkaloids 42
  basal metabolic rate 46–8
  body condition 53–5

body fat 53
bone marrow 56
constituents 36–9
diet selection 52–3
digestion 49–51
doubly labeled water 37
energy 36–7
energy flow 46–7
essential amino acids 37
fecal protein 42–5, 51
fermentation 50–1
mast 40
measurement of 42–6
minerals 38–9
nitrogen 37, 45, 46, 50
passage rate 51
phenols 41
protein 37, 51
requirements 46–9, 51
seasonality 40
secondary compounds 41
supply 40–2
tannins 41–2, 44
terpenes 41
water 37–8
food habit studies 134
food supply 131–3
  direct measures of 132
  experimental alteration of 131
  indirect measures of 133
  problems with measurement 133
food webs 366–8
foraging, predator-sensitive 134
fox
  arctic 16, 98, 191
  Blandford's 191
  island 308, *308*, 367
  kit 92
  red 98, 171–2, 185, 360–1, 363
frog, Madagascan tomato **325**
F_{st} 30–1
functional response 61, 63–5, 69, 165–8,
    178
  of red kangaroos 201
Fundulus parvipinnis 186

Gallicolumba rufigula 142
game cropping 347
game ranching 347
Garrulus glandarius 175
gaur 14
Gause's principle 146–7

Gaussian distribution 302
Gazella 53
  dama 17
  dorcas 68
  granti 96
  thomsonii 96
gazelle
  dama 17
  dorcas 68
  Grant's 96
  Thomson's 42, 68, 96, 155, 310
gecko 360
gemsbok 53
gene pool 291
generalists 60–1
genetic characters of individuals 27–33
  allozyme gel electrophoresis 28
  applied uses 34–5
  divergence 30–1
  within geographic regions 32–3
  island model 31
  isolation-by-distance 31
  methods of study 28–30
  mitochondrial DNA 29–30
  polymerase chain reaction 28–9
  ring species 31–2
genetic diversity 294
genetic drift 294–5
  random 295
genetic variance 293–5
  additive 294–5
genotype 20
genotype frequencies 291
geographical constraints 23–7
  continental drift 23–5
  human expansion 26–7
  ice ages 25–6
  phylogenetic constraints 23
geometric population growth 245
gerbil 17
Gerbillus 17
  allenbyi 141
gerenuk 17, 52
Giraffa camelopardalis 53
giraffe 53, 186
GIS 70
Glanville fritillary butterfly *106*, 310
Glaucomys 22
  sabrinus 157
glider, greater 51
global information system 70–1
gnatcatcher, California 72

goat 360
golden eagle 308
Gondwanaland 24–5
goose
  lesser snow 29, 54, 157
  snow 374, 376
  white-fronted 54
goshawk 12
grasshoppers 69
grasslands 15–17
grazers 68
grazing
  grasses 158
  succession 155
greenfinch 143
grouse
  red 114, *120*, 131–2, 171, 184, **185**,
    189–90, 192
  sage 162
growth 247
Grus
  americana 194, 325
  canadensis 54
Guam 306
guillemot
  bridled 33
  common 33
guinea pig 51
gull
  black-backed 31
  black-headed 95, 177
  herring 31, 95
Gymnogyps californianus 319
Gyps
  bengalensis 319
  coprotheres 38
  indicus 319
  tenuirostris 319

habitat 11
habitat selection 70–3, 149–53
  density-dependent 72–4
  ideal despotic distribution 74
  ideal free distribution 72
    in birds 73
    in fish 73
  marginal value 151
  optimal foraging 151
  shared preference 151
Haematopus ostralagus 74
Haemophysalis spinigera 192
Halichoerus grypus 34, 187, 363

Hardy–Weinberg equilibrium law 292
hare
  arctic 16
  European 187, 191
  hispid 314
  mountain 190
  snowshoe 12, 35, 44, 51, 54, 109–10,
    110, 115, 120, 131–2, 134, 211,
    213–14, 368, 377
harrier, hen 168, 171
hartebeest 96
harvest 346–7
  additive 266, *266*
  commercial 347
  compensatory 266, *266*
  recreational 346
harvest effort, regulating 343
harvest strategy 335–47
  age-biased 347
  constant effort 343–4
  constant quota 335, 340
  fixed escapement 344–5
  fixed proportion 341–2
  sex-biased 347
harvesting 308–9
  effects of longevity on optimal 353
  elephant 351
  game meat 350
  lion 309
  mahogany tree 352
  marten 340, 342–3, 345
  moose 347
  principle of sustainable 335
  red deer 347
  waterfowl 265–7
Hawaii creeper 319
hawfinch 143
heath hen 321
Heligmosomoides polygyrus 184, 189
Helogale parvula 92
Hemitragus jemlahicus 91, 139, 357, 374
heritability 300
heron, great blue 221
Herpestes
  auropunctatus 163, 192
  ichneumon 367
heterogeneity of variance, Bartlett's test of
    284–5
Heterohyrax brucei 21–2
heterozygosity 291–7, *293*
  effects of inbreeding 296
  effects of random genetic drift 295

how much is needed 296
    mammals 292–3, 297
heterozygous 291–2
hibernation 48
Hibiscadelphus 20, 370
hillchat 17
Himantopus novaezealandiae 375
Himatione 152
hippopotamus 220
Hippopotamus amphibius 220
Hippotragus
    equinus 14, 53, 159, 175
    niger 14
Hirundo
    pyrrhonota 184
    rustica 184
holly 14
Homo sapiens 179
homozygous 291
honeycreepers, Hawaii 20, 21, 23, 321
hopping mice 17
horse, Przewalski's 323–4
hummingbird, rufous 76
Hydrochoeris hydrochoeris 15
hyena, spotted 156, 176
Hylocichla mustelina 314
hyperpredation 306–7, *307*, 367
hypothesis 253
    alternative 270
    management policy as 264–7, 268
    null 253, 270–1, 275–6, 281
hyrax 49
    Bruce's 21, 22

ice ages 25–6
ideal despotic distribution 74
ideal free distribution 72–3, *73*
Ilex 14
immunocontraception 362–3
impala 96, 147
inbreeding 296
incidence function 310
information criterion/criteria 258, 339
information theory 264
interaction, statistical 274, 276, 283–4, 287
interference 74
    in birds 74
    population effects 74
island biogeography 379–81
island fox 308
island spotted skunk 308
Isle Royale 210

Isoberlinia 14
Ixodes ricinus 190

jackdaw 175
jackrabbit 156
Jaculus 17
jay, European 174
jerboa 17, 43
judgment
    technical 269
    value 269
Juniperus 14–15

kangaroo
    gray 281, **282**, **288**
    red 17, 97, 200, *202*, *205–6*, 280–1, **280**,
        282, **288**, 336
kangaroo rat 15, 17, 91, 161
kauri 13
kestrel
    European 167
    Mauritius 324
keystone species 365, 368–9
kidney fat 55–6, 58
killifish *186*
Kingia 14
kite, black-shouldered 116
kiwi 29
kleptoparasitism 74
klipspringer 52
knee worm 192
koala 49, 51, 187, 332
kob 53
    white-eared 176, 372
Kobus 53
    defassa 28
    kob 176, 372
    lechee 97
kokako 176, 329
kouprey 14
kudu
    greater 43, 53, 112, 118–19, 147, 186
    lesser 52, 186

λ 79, 246, 251, 255
Lagopus
    lagopus 16, 114, 167–8, 184
    mutus 95
Lagovirus 187
landbridge island 380
landscape ecology 365
Lantana camara 315

Larix 12
  laricina 169
Larrea divaricata 17
Larus
  argentatus 31, 95
  fuscus 31
  ridibundus 95, 177
Lasiorhinus
  krefftii 35
  latifrons 57
lechwe 53, 97–8
Lefkovitch matrix 247
lemming 16, 51
Lemmus 16
Leontopithecus rosalia 330
leopard 169, 315
  snow 16
Lepomis macrochirus 69
Lepus
  americanus 12, 109, 368, 377
  arcticus 16
  californicus 156
  europaeus 187, 191
  timidus 190
Leslie matrix 244, 245–6
life history traits 19
life table 84–8, 115
  direct 84–5
  indirect 85–7
  parameters 87–8
  relationship between parameters 87
life-table estimation 84–7
  African buffalo 86–7
  assumptions 87
  caribou 85–6
  direct methods 84
  indirect methods 85–7
likelihood 256–64
  maximum 258
  model 263
  negative log- 257–62
  profile 258
limitation 111
limiting factor 112
linnet 143
lion 34, 39, 40, 84, 173, 176, 296, 309,
    371, 373
  harvesting of 309
Litocranius walleri 17, 52
liver fluke 192, 195
Lobelia 17
locus 291

loggerhead sea turtle 248, 250–1
logistic model 348
  Gompertz 261
  Ricker 257
  theta- 259–60, 263
logistic population growth model 121–2,
    336
longclaw, yellow-throated 22
louping ill virus 192
Loxia curvirostra 40
Loxodonta africana 81, 188, 220
Loxops 152
Lutra lutra 54
l_x 82
Lycaon pictus 91, 188, 191, 369
Lycodon aulicus 360
Lyme disease 187
lynx
  Canadian 12, 214
  Iberian 367
Lynx canadensis 12

Macaca radiata 192
Macquarie Island 306
Macronyx croceus 22
macroparasites 179
Macropus
  greyi 315
  robustus 161, 330
  rufus 17, 97
Macrozamia 14
Madoqua kirkii 52, 373
magpie, black-billed 175
mahogany tree 352
major histocompatibility complex 29
malaria 190, 192
  avian 321
Malthus, Thomas 88
mammoth 25
Mammut americanum 12
Mammuthus
  columbi 25
  primigenius 25
manakin, long-tailed 28
manatee 18
Manorina
  melanocephala 383
  melanophrys 369
maple 13
Margarops fuscatus 193
marginal value theorem 67
  predictions 67

marine biome 17–18
MARK program 85
mark–recapture 84, 236–9
    density 238–9
    frequency of capture 238
    Peterson–Lincoln 236–8, 237
    program MARK 85
Marmota 16
marrow fat 133
marten, American 337–45, *338–40,*
    342–3
    density dependence 338
    environmental stochasticity 339
    harvesting 340, 342–3, 345
    prey abundance 338
Martes
    americana 337
    pennanti 91
mast 40
mastodon 12
MATHCAD 1
matrix 244, 246
    elasticity 249–50
    payoff 265, **265**
meadowlark, western 22
mean square error 259, 261–2, 283
Mediterranean shrubland 14–15
Megaloceros giganteus 27
Megalonyx 12
Megaptera novaeangliae 29
Melitaea cinxia 105, 310
Melopsittacus undulatus 17
Melospiza melodia 161
Mephitis mephitis 364
Meriones tristrami 141
meso-predator release 367–8
    multiple scales 376–7
    multiple states 370–1
    niche 366
    overpredation 369–70
    top-down and bottom-up 371–3
    trophic cascade 366–7
    trophic levels **367**
meta-analysis 279
metapopulation 30, 104–8, 310
    and dispersal 104–6
    fragmentation *107–8*
    habitat loss 106–7
    source–sink 108
Metrosideros 152
microparasites 179
Microtus 13, 16, 167–8, 184

    agrestis 191
    nesophilus 314
    townsendii 139
migration 372
    of prey 176
miner
    bell 369
    noisy 383
minerals 38–9
minimum viable population (MVP)
    299–300
mistletoe 370
mitochondrial DNA 29–30, 34
moa 25, 374
models 1, 253–64
    alternative 253–4
    evaluation 254, 263–4
    spatially explicit 309
Mohoua ochrocephala 176
mole
    golden 22
    marsupial 22
Molothrus ater 174, 193, 314
monarch butterfly 42
mongoose 359
    dwarf 92
    Egyptian 367
    Indian 163, 192
Monte Carlo model 125, 301, 305
moose 25, 40, 60, 65, *65,* 95, 164, 171,
    173, 190, 192, 207–8, *208–9,* 212, 221,
    236, 347, 369, 372, 374
Morbillivirus 186
mortality
    compensatory 265
    measurement of 115
moth, pine beauty 358
mountain goat 26, 152
mouse, house 109, 120, 176, 357, 359–60,
    363, 368
multiple states 370–1
Mus
    domesticus 109, 120, 176
    musculus 368
muskox 25, 176, 177, 317–18, 322–3,
    323
muskrat 103, *103–4*
muskshrew, Indian 360
Mustela
    erminea 163, 176, 191, 359, 369
    furo 92, 169, 359, 367
    nigripes 157, 193, 316

mutation 294–5
mute swan 109
m_x 82
Mycobacterium bovis 193
Mycoplasma gallisepticum 188, *189*
Mycteria americana 221
myna 321
Myocastor coypus 15
Myotis
  blythii 156
  myotis 156
Mystacina 370
Myxoma 194, 195, 360
myxomatosis 187, 191, 194–5

natural selection
  alternating selection 33
  fitness 20
  frequency-dependent selection 33
  genes 20
  genotype 20
  heterozygote advantages 33
  phenotype 20
  pleiotropy 20
  polygenic effects 20
  theory 19–21
nematode worm 120–1, 184, 192
Nesotragus noschatus 52
nest box 269
net recruitment 336
niche 143–6, 366
  competitive exclusion principle 146
  complementarity 144
  concept 143–6
  fundamental 146
  n-dimensional hypervolume 145
  overlap 147–9
  realized 145
  resource partitioning 146–7
  tautology 146
nitrogen 37
normal distribution 257
North Atlantic Oscillation 252
Nothofagus 13, 40
Notomys 17
Notonecta 69
Notoryctes 22
nullcline 199
  in red kangaroo 205
numerical response 169–78
  in red kangaroo 202
Nyctea scandiaca 16, 95

oak 13, 131, 158
Ochotona 16
Octodon degus 48
Odocoileus
  hemionus 30, 131, 152
  virginianus 26–7
Olea 14
olive 14
Ondatra zebithica 103
opossum 24, 174
opportunity cost 348
optimal diet 62–5
  balanced intake 62
  feeding rates 63–4
  linear programming 65
  model 61
  partial preference 62
  population consequences 65
  predictions 62
  tests 62
optimal foraging theory 60, 151
optimal patch and habitat use 66–9
  by herbivores 67–9
  patch residence 66–7
optimal patch use 66–9
  cumulative energy gain 66
  experimental tests 67–8
  giving-up density 67
  giving-up time 67
  long-term intake 67
Opuntia 17
Orcinus orca 376
Oreamnos 26
  americanus 152
Oreomystis mana 319
Oreotragus oreotragus 52
oribi 52
orthogonal 277
Oryctolagus cuniculus 94, 363
oryx 38
  Arabian 159, 194, 318, 322–3
  beisa 96
Oryx
  gazelle 53
  leucoryx 159, 318
osprey 95, 222
ostrich 40
otter
  European 54
  sea 99, 103, 368
Ourebia ourebia 52
ovenbird 314

overcompensation 125–8
Ovibos moschatus 25, 176, 317
Ovis 25
  canadensis 44, 95, 152, 176, 189
  dalli 95
owl
  great horned 12, 41
  snowy 16, 95
  spotted 313, 383
oystercatcher 74, 310

panda, giant 23, 41, 49, 60
Pandion haliaetus 95, 222
panmixia 31
Panolis flammea 358
Panthera
  leo 34, 373
  pardus 169, 315
  tigris 315
  uncia 16
paradox of enrichment 199
parakeet 306
parameter estimation 258–63
parasites, effects of 179–80
  and biological control 194–5
  and captive breeding 193–4
  and conservation 191–4
  and control of pests 194–5
  and ecosystems 190
  habitat alteration 193
  and host communities 190–1
  interactions 190
  introductions 191–3
  regulation of host 188–9
Parelaphostrongylus tenuis 190, 192
parks, conservation 324–32, **325**
  advantages **325**
  as benchmarks **325**
  and community conservation **325**, 332
  and corridors 329–30
  culling in 332
  disadvantages **325**
  as ecological baselines **325**
  effects of area 327–9, **327–8**
  initial conditions 330–1
  processes 326–7
  purposes 324–6
  size or number 329
parrot, Puerto Rican 193, 322–3
partridge
  chukar 162
  English 118–19, *119*

Parus 168
  ater 358
  cristatus 313
  major 62, 73, 118, *119*, 184
Passer domesticus 54, 162
passion flower 42
Pasteurella 189
patch use 310
pathogens, endemic 184–6
  birth rate 184
  food and predators 184
  with food supply 184–5
  mortality 184
  poultry 188
  with predators 185–6
Pelecitus roemeri 192
penguin 22
Peraxilla tetrapetala 370
Perdix perdix 118
Perognathus 17
  intermedius 149–50
  penicillatus 149–50
Peromyscus 172
  eremicus 149, 150
  maniculatus 139
perturbation 117
Petaurus 22
petrel, great-winged 357–8
Petrogale 315
  penicillata 176
  xanthopus 161
Phascolarctos cinereus 49, 187, 332
Phasianus colchicus 194
pheasant, ring-necked 194
phenols 41
phenotype 20
Phocoena phocoena 232
phylogenetic constraints 23
Phyloscopus trochiloides 31
physiognomy 11
Pica pica 175
Picea
  abies 12
  glauca 12
Picoides tridactylus 313
Picus viridis 95
pig 182, 307–8, *308*, 316, 360, 367, 370
pika 16
pine
  jack 375
  Monterey 13

pine (*cont'd*)
  ponderosa 13
  Scots 12
Pinus
  banksiana 375
  ponderosa 13
  radiata 13
  sylvestris 12
Plasmodium
  azurophilum 190
  relictum 192
Plectrophenax nivalis 16
pleitropy 20
plover, piping 375
pocket mice 17
Podocarpus 13
policy 7–8
  criteria of failure 8
Poliptila californica 72
polygamy 298
polygenic effects 20
polymerase chain reaction (PCR) 28–9
polymorphic loci 292
polymorphism 32–3
  stable 33
  transient 33
population census 254
population cycles 209–14
  deer 211
  lynx 214
  moose 210
  snowshoe hares 212
  stable limit 198, 209
  wolves 209
population growth 78–89
  exponential and geometric 88–9, *89*
  fecundity 82, *83*
  intrinsic rate 80–1, *81–2*
  life table direct 84–5
  life table indirect 85–7
  life-table parameters 87–8
  mortality 82–4
  rate of increase 78–81
population growth model
  exponential 88
  geometric 88, 262
population growth rate, exponential 255
population outbreaks 120
population regulation 109–31
  applications of 120–1
  birth and death rates 115–16
  carrying capacity 114–15
cycles, chaos, stability 125–31
delayed and inverse 114
density dependence 111–12
evidence for 116–20
examples of 118–20
limitation 112
logistic model 121–5
mean density 117–18, **118**
perturbation experiments 116–17
regulation 113–14
stability 109–10
theory of 111–16
population stability 109
population viability analysis (PVA) 124–5,
    300–10
  problems with 304
porcupine 35
porpoise, harbor 232
possum
  brushtail 56, 176, 359, 362, 370
  ringtail 21–2, 51
pouched mouse 17
prairie dog 317, 322
  black-tailed 156
precision in sampling 221
predation
  behavior of prey 176–7
  cannibalism 163
  carnivory 163
  definition 163
  destabilizing 174–6
  herbivory 163
  parasitism 163
  and prey density 164–5
  and regulation 171–3
predator-mediated coexistence 378
predator–prey model 306
predator-sensitive foraging 134
predators
  behavior 165–9
  functional response 165–8, 170,
    174
  interference 169
  numerical response 169–70
  and prey patches 169
  search efficiency 166
  total response 170–6
Presbytis entellus 192
prey
  endemic 306–7
  exotic 306–7
prey preference 307

prickly pear 17
Pristiphora ericksonii 169
Procyon lotor 174, 188
profitability 61
Promerops cafer 15
pronghorn antelope 16, 26, 56, 146, 220, 242, 374
Protea 14
protein 37, 51
Pseudocheirus
  dahlia 21–2
  peregrinus 51
pseudo-replication 278
Pseudotsuga douglasii 13
ptarmigan 16
  rock 95
Pteridium aquilinum 190
Pterodroma macroptera 357
Puccinellia phryganoides 157
Puffinus griseus 174
Puma concolor 176
p_x 83
Pycnonotus goiavier 4
Pygosceles adeliae 220

quagga 29
quail 272
quasi-extinction 300
Quercus 13–15
  robur 158
  wislizenii 131
q_x 82

rabbit
  cottontail 53, 158, 238
  European 44, 94, 161–2, 169, 172, 174, 176, 187, 191, 194–5, 278, *278*, 359–60, 362–3, 367, 374
  marsh 241
rabbit hemorrhagic disease 187, 189
rabies 188
raccoon 174, 188
rainfall in Australia 205
Raja
  batis 369
  laevis 369
Rangifer tarandus 16, 369, 372
Raphiceros campestris 52
Raphus cucullatus 370
rat 359–60, 370
  black 176, 315, 316
  cotton 95

Pacific 373
Polynesian 316
rate of increase 78–82
  allometric relationship 82
  annual (λ) 79
  in caribou 81
  exponential (r) 79
  intrinsic (r_m) 81
Rattus
  exulans 316, 373
  rattus 176, 315
raven 95
red deer 68, 118–19, 309
redpoll 143
  hoary 95
Redunca 52
reedbuck 52, 147
regression 123, 255, 257
  deviates around 257
regulation
  by predators 171–3
  top-down and bottom-up 371–3
regulation of populations 109–31
  applications of 120–1
  birth and death rates 115–16
  carrying capacity 114–15
  cycles, chaos, stability 125–31
  delayed and inverse 114
  density dependence 111–12
  evidence for 116–20
  examples of 118–20
  limitation 112
  logistic model 121–5
  mean density 117–18, **118**
  perturbation experiments 116–17
  regulation 113–14
  stability 109–10
  theory of 111–16
reindeer 68, 118, *119*, 164–5, 347
reintroduced species 99–104
  diffusive spread 99–102
  exponential growth 102–3
  tests of theory 103–4
replicates 273, 275
reproductive value 249–50
residual variation 257, 259–62
resource definition 196
resource partitioning
  applied aspects 159–62
  catena 155
  habitat selection 149–51
  habitat selection theory 151–3

resource partitioning (*cont'd*)
  habitats 146–7
  limiting similarity 147–9
resource selection functions 67–72, 310
  logistic regression 71
  by wolves 71
resource use 196–7
  consumptive 196
  interactive 197
  pre-emptive 196
  reactive 197
response variable 274
rhinoceros
  black 34, 147, 369
  woolly 12
Rhynchotragus kirkii 33 (*see also Madoqua*)
Ricker logistic regression *257–8*
rinderpest 92, 186–8
ring species 31–2
risk 265
  asymmetry of 271
risk-sensitive habitat use 69–70
  by fish 69
  by herbivores 70
  by rodents 70
r_{max} 121, 256, 259–60, 262, 289
roan antelope 14, 53, 159, 175
robin, American 161, 169, 179
rock-wallaby
  black-footed 176
  yellow-footed 161–2
rufous bristlebird 71
Rupicapra rupicapra 139, 357, 374

sabertooth cat 25
sabertooth marsupial 24
sable 14, 53
sagebrush 17
saiga antelope 15–16, 309, **325**
Saiga tatarica 309, **325**
salamander, clouded 34
Salix 12
  glauca 42
sample counting methods 226–34
  comparing estimates 233–4
  equal-sized units 227–9, **230**
  fixed boundaries 227–30
  merging estimates 234
  proportional to size 230, **230**
  stratification 232–3
  unbounded samples 230–2
  unequal-sized units 229, **230**

sample counts 221–6
  bias errors 221–2
  how to sample 226
  precision and accuracy 221, **233**
  random or non-random 225–6
  sampling and replacement 223–4
  sampling frames 222
  sampling strategies 222–3
  transects or quadrats 224–5, **225**
sample size 253, 271
SARS 187
savanna 15
sawfly, larch 169
scalar 246
Sciurus
  carolinensis 154
  vulgaris 154
screwworm fly 360
seal, gray 34, 187, 363
sealions, Steller's 376
Seirus aurocapillus 314
Selasphorus rufus 76
selection 295
  directional 295
  disruptive 295
  natural 295
  stabilizing 295
Senecio 17
sex ratio 298–9
shearwater, sooty 174
sheathbill, lesser 368
sheep 278
  bighorn 44, 152, 176, 189
  Dall 95
  domestic 317
  mountain 95, 177
  Soay 118, 119, 128, *128–9*, **130**, 131,
    185, 251, *251*
    age-structured model 251–2
    effects of age structure 130–1
    effects of weather 128, 130
    model of 130
sheep tick 190
Sialis sialis 162
Sigmodon hispidus 95
significance 253–4
  biological 254
  statistical 253
skate
  barndoor 369
  common 369
skink 360, 367

skunk
    spotted 308, *308*, 367
    striped 364
sloth 14, 49, 51
    giant ground 12
smallpox 191
Smilodon fatalis 25, 29
snake
    brown tree 306, 360
    wolf 360
Somateria molissima 54
Sophora 152
Sorex 13, 172
source–sink 108, 375
sparrow
    house 54, 162
    song 161
    white-crowned 161
specialists 60–1
Spermophilus 17
    parryii 92
    tridecemlineatus 27
Spilogale gracilis 308, 367
Spilopsyllus cuniculus 187
spinebill, western 15
spruce
    Norway 12
    white 12, 40
squirrel
    Belding's ground 91
    flying 22, 157
    gray 154
    ground 17, 27, 92, 242
    red 40, 154, 364
stability 125
stable age distribution 246, 249–50
stage-structured population model 247–8
starling
    European 154, 162, 179
    wattled 154
statistical design 269
statistical inference 253, 264, 268
    weak 279
statistical power 271
statistical test 253, 270
steinbuck 52, 147
Sterna albifrons 375
stickleback 73
stilt, black 375
stoat 163, 176, 191, 359, 369
stochasticity, demographic 289–90
    environmental 291

stork, wood 221
Strix occidentalis 313, 383
Sturnella neglecta 22
Sturnus vulgaris 154, 162, 179
sugarbird, Cape 15
sugarglider 22
Suncus murinus 360
sunfish, bluegill 69
suni antelope 52
sunspot cycle 214
supply and demand 349, *351*
SURGE 85
surplus 336
survival 298
survival rate 128–9, 244, 247
Sus scrofa 182, 307, 367, 370
sustained yield 337–46
    maximum (MSY) 337, 340, 346
    pair 337
swallow
    barn 184
    cliff 184
swan
    black 375
    mute 109, *109*
swine fever 182
Sylvia
    borin 53
    melanocephala 15
Sylvilagus 194
    floridanus 53
    palustris 241
Syncerus caffer 34, 220
synchrony, births 177

θ 259–60
tahr 91, 139, 357, 374
taiga 12
tamarack 169
tamarin, golden lion 330
Tamiasciurus hudsonicus 40, 364
Tatera 17
Taurotragus oryx 53, 96
teal, blue-winged 33
temperate forest 13
tern, least 375
terpenes 41
territoriality 75–7
    costs versus benefits 76
    economic defendability 75
    examples 76
    population effects 77

Thomas Malthus 88
thrasher, pearly eyed 193
thrush, wood 314
Thunnus 18
Thylacosmilus 24
tiger 315
tit
  coal 358
  crested 313
  great 62, 67, 73, 76, 118, *119*, 184
topi 53, 154
total counts 219–20
total response 170–6
  destabilizing 174–6
  and regulation 171–3
Tragelaphus
  imberbis 52
  scriptus 52
  strepsiceros 43, 112
transition matrix 244
Trichechus 18
Trichocereus 14
Tricholimnas sylvestris 315
Trichostrongylus tenuis 184–5, 189
Trichosurus vulpecula 56, 176, 362,
  370
Troglodytes aedon 184
tropical forest 13–14
  savanna 15
  tundra 16
  woodland 14
tsessebe 147, 175
tuberculosis 193
tuna 18
tundra 16
Turdus migratorius 161, 169, 179
turtle
  green 34
  hawksbill 34
  loggerhead 34, 248, **248**, **250**, 251
  painted 35
turtle excluder device (TED) 251
Tympanuchus cupido 321

Ulmus 13, 191
Uria aalge 33
Urocyon littoralis 308, 367
Ursus 16
  arctos 25, 30, 171

value judgments 4
vector 246

Vestaria 152
vireo
  black-capped 377
  least Bell's 377
  red-eyed 314
Vireo
  atricapillus 377
  bellii 377
vital rates 244
vole 13, 16, 131, 139, 140, 167, 184, 191,
  212–13, *213*, 330
  bank 167–8
  Gull Island 314
  Townsend's 131, *132*
Vombatus ursinus 50
Vulpes
  cana 191
  macrotis 92
  vulpes 98, 185
Vultur gryphus 324
vulture
  Cape 38
  long-billed 319
  slender-billed 319
  white-backed 319

warbler
  garden 53
  greenish 31
  Kirtland's 375
  Sardinian 15
  Seychelles brush 77
  yellow 161
warthog 186
water 37–8
waterbuck 28
weasel 337
  least 214
weather 291
  effects on abundance 118
  and elk population growth 123
whale 219
  blue 81
  fin 54
  humpback 29
  killer 376
  minke 241
  sei 54
white-eye, Japanese 319
whooping crane 194
wild cat 373
wild dog 91, 188, 191, 369

wildebeest 44, 53, 56, 96, 117, *117*, 122,
    127, 133–4, 152, 155–6, 159–60, 173,
    176, 186, 190, 226, 229, 254–63, **255**,
    256, *258*, 259, *260–2*, 263, 368–9, 371,
    376
wildlife
  decisions 2–7
  management 2–3
  objective/action matrix 4–5
  policy 7–8
wildlife management
  experimental 268
  traditional 268
willow, gray 42
wolf
  Ethiopian 191
  gray 16, 25, 34, 72, 92, 164, 167, 171–4,
    176, 207, *209*, 210–11, *211–12*, 372
  Mexican 35
  red 34
wombat 50, 57
  hairy-nosed 35
wood duck 269

wood pigeon 132, **133**
woodhen, Lord Howe 315, 322–3
woodpecker
  green 95
  three-toed 313
wren
  house 184
  Stephen Island 316, 321
wrentit 15

Xanthocephalus xanthocephalus 145
Xanthorrhoea 14
Xenicus lyalli 316

yak 14
yellowhead 176

zebra
  Burchell's 53, 147, 155–6, 159
  Grevy's 53
Zenaida macroura 242
Zonotrichia leucophrys 161
Zosterops japonicus 319